Sustainability Sciences in Asia and Africa

Sustainable Agriculture and Food Security

Sustainability as a scientific endeavor requires broad understanding of the interconnections in our global environment. It is therefore positioned at the crossroads of research in natural sciences, technology and human sciences. This book series supports the global efforts by providing timely coverage of the progress, opportunities, and challenges towards achieving sustainable development goals (SDGs) in Asia and Africa. It fosters the research in transdisciplinary academic fields spanning across agriculture and global food security, health and diseases, access to clean water and affordable energy, climate resilience and effects of global climate change on biodiversity, water and land. This series is planned to narrate the stories of scientific development and progression towards SDGs from the regions of Asia and Africa. The overall aim of the series is to provide a comprehensive publication platform and act as a knowledge engine in the growth of sustainability sciences with a special focus on developing nations. The series constitute of four sub-series each focusing on a specific theme closely linked to a set of SDGs. Each sub-series is leaders and experts in agricultural sciences, environmental sciences, biomedicine and medical sciences respectively. Books in this series are useful for environmentalists, biotechnologists, plant scientists, medical practitioners, conservation biologists, policy makers, NGOs and scientific bodies working in the arcas of sustainability sciences in the developing regions.

This book series support the global efforts towards sustainability by providing timely coverage of the progress, opportunities, and challenges of sustainable food production and consumption in Asia and Africa. The series narrates the success stories and research endeavors from the regions of Africa and Asia on issues relating to SDG 2: Zero hunger. It fosters the research in transdisciplinary academic fields spanning across sustainable agriculture systems and practices, post- harvest and food supply chains. It will also focus on breeding programs for resilient crops, efficiency in crop cycle, various factors of food security, as well as improving nutrition and curbing hunger and malnutrition. The focus of the series is to provide a comprehensive publication platform and act as a knowledge engine in the growth of sustainability sciences with a special focus on developing nations. The series will publish mainly edited volumes but some authored volumes. These volumes will have chapters from eminent personalities in their area of research from different parts of the world.

Biswajit Pramanick • Shiv Vendra Singh
Sagar Maitra • Silvia Celletti • Akbar Hossain
Editors

Climate-Smart Agricultural Technologies

Approaches for Field Crops Production Systems

Springer

Editors
Biswajit Pramanick
Department of Agronomy
Dr. Rajendra Prasad Central Agricultural University
Pusa, Bihar, India

Shiv Vendra Singh
College of Agriculture
Rani Lakshmi Bai Central Agricultural University
Jhansi, Uttar Pradesh, India

Sagar Maitra
Department of Agronomy and Agroforestry
Centurion University of Technology and Management
Khurda, Odisha, India

Silvia Celletti
DISAFA Department
University of Turin
Grugliasco, Italy

Akbar Hossain
Bangladesh Wheat and Maize Research Institute
Nashipur, Bangladesh

ISSN 2730-6771 ISSN 2730-678X (electronic)
Sustainability Sciences in Asia and Africa
ISSN 2730-6798 ISSN 2730-6801 (electronic)
Sustainable Agriculture and Food Security
ISBN 978-981-96-7698-9 ISBN 978-981-96-7699-6 (eBook)
https://doi.org/10.1007/978-981-96-7699-6

This Springer imprint is published by the registered company Springer Nature Singapore Pte Ltd.
The registered company address is: 152 Beach Road, #21-01/04 Gateway East, Singapore 189721, Singapore

Series Preface

The global food production systems require expanding their capacity to produce almost twice the current levels to safeguard the food security of the burgeoning population worldwide. More than 800 million people suffering from undernourishment worldwide pose a great risk to the attainment of Sustainable Development Goal (SDG) 2 of the UN, which targets "End hunger, achieve food security and improved nutrition and promote sustainable agriculture" within the next seven years. The challenge is further exacerbated by the rising weather extremities, unpredictability in rainfall patterns, and pest-pathogen dynamics associated with global climate change, which have a profound negative impact on agricultural productivity and farm incomes worldwide. Also, the future targets of food production should be secured in resource-constrained agricultural settings and with the least environment footprint, thus calling for sustainable innovations in agri-farming systems and enhanced participation of women in agriculture. The challenge to reduce hunger is alarming in developing nations, particularly Asia and Africa, which house the largest proportion of people suffering from malnutrition and other nutrition-related issues. Furthermore, the agri-food systems in Asia and Africa are severely constrained by the subsistence nature of farming, declining land and other agricultural resources, and increasing environmental pollution, soil and biodiversity degradation, and climate change. Therefore, this book series, Sustainable Agriculture and Food Security, has been planned to support the global efforts toward sustainability by providing timely coverage of the progress, opportunities, and challenges of sustainable food production and consumption in Asia and Africa. The series narrates the success stories and research endeavors from the regions of Africa and Asia on issues relating to SDG 2: Zero Hunger. It fosters research in transdisciplinary academic fields spanning sustainable agriculture systems and practices, post-harvest, and food supply chains. The focus of the series is to provide a comprehensive publication platform and act as a knowledge engine in the growth of sustainability sciences with a special focus on developing nations.

The growing global population is placing increasing pressure on food production systems. By 2050, the population is projected to reach 9.7 billion people, requiring food production to rise by approximately 70% to meet demands. This challenge has historically been met through scientific and technological advances, often at environmental cost. As climate change intensifies and resources become scarcer, it is critical to redesign agricultural systems to deliver higher yields while maintaining environmental sustainability. Climate-smart agriculture (CSA) addresses this challenge by integrating climate considerations into sustainable agricultural planning and practice. Techniques such as precision agriculture, improved nutrient and water management, regenerative agriculture, and biochar application increase sustainability and crop yields, improve soil health, and contribute to climate change mitigation. As such, CSA has a pivotal role in advancing global food security and environmental stability while helping to counter the impacts of climate change.

Climate change is already impacting agriculture and food security, with more frequent and intense extreme events such as drought, heavy rainfall, flooding, and heat waves expected to worsen in many regions. These events directly affect crop growth and development, exposing plants to biotic and abiotic stresses that significantly reduce productivity. Adopting CSA practices can help mitigate these impacts. Strategies that improve soil and water management and preserve biodiversity enhance carbon sequestration, soil health, and quality while reducing erosion, resulting in more resilient soils and cropping systems capable of supporting food production under climate stress. Precision agriculture, for example, leverages emerging technologies, such as artificial intelligence, sensors, drones, and robotics, to optimize inputs and practices based on the natural variability of the growing region. This improves yields, boosts resource-use efficiency, and supports sustainability. Climate-smart agriculture also aims to reduce agriculture's environmental footprint. While conventional tillage-based agriculture has helped build toward food security, it has also degraded vital natural resources, harmed ecosystems, and contributed up to 20% of global anthropogenic carbon dioxide emissions, according to the EPA. In contrast, practices such as conservation agriculture, carbon farming, biochar application, and agroforestry can reduce energy and nitrogen inputs, lower greenhouse gas emissions, increase carbon sequestration, and stimulate soil biological activity, collectively supporting climate change mitigation.

In view of this, the present book, *Climate Smart Agriculture*, edited by Biswajit Pramanick, Shiv Vendra Singh, Sagar Mitra, Silvia Celletti, and Akbar Hossain, provides a platform to discuss the potential for CSA practices to enhance food production and security, and contribute to climate change mitigation. The book presents 20 chapters written by experts and practitioners, each containing up-to-date and important insights into the topic of CSA. I would like to congratulate the Editors, Biswajit Pramanick, Shiv Vendra Singh, Sagar Mitra, Silvia Celletti, Akbar Hossain and all authors for their valuable contributions. I am sure this book is a great resource for students, researchers, farmers, educators and extension agents, policymakers, and other stakeholders engaged in transforming agriculture around the world.

I wish to extend my sincere thanks and gratitude to the Springer staff, particularly Sanchi Bhimrajka, Associate Editor (Books, Medicine and Life Sciences), and Nick Melchior, Editorial Director (Books, Medicine and Life Sciences, Asia), for their constant support for the accomplishment of this compendium. The cooperation received from my supervisors and research leaders such as Prof. Andrew Deeks

(Vice Chancellor), Prof. Peter Eastwood (Deputy Vice Chancellor- Research), Prof. Peter Davies (Pro Vice Chancellor), and my laboratory colleagues—Rhys G. R. Copeland, Anu Chitikineni, and Vanika Garg from Murdoch University (Australia)—is also gratefully acknowledged. I would like to thank my family members—Monika Varshney, Prakhar Varshney, and Preksha Varshney—for their love and support in discharging my duties as Series Editor.

Perth, WA, Australia Rajeev K. Varshney

Preface

Envision a future where farmers—whether in expansive agricultural lands or small-scale plots in developing regions—are empowered with tools and knowledge to adapt confidently to our changing climate. This is not merely an idealistic vision, nor a distant reality to create; rather, it is a critical mission we must collectively undertake. This purpose drives the concept of Climate-Smart Agriculture. Climate change poses one of the greatest challenges of our time, directly threatening global food security and agricultural systems. With rising temperatures, erratic rainfall, and extreme weather events becoming everyday realities, farmers and communities face profound uncertainties. These shifts affect crop yields, alter growing seasons, and jeopardize millions of livelihoods. Climate-smart agriculture offers a hopeful and innovative blueprint to confront these challenges, embodying resilience through practical solutions and a commitment to sustainability.

The book starts by discussing the direct effects of climate change on food crops, detailing how changing climatic conditions impact plant growth, soil health, and pest dynamics. It introduces climate-smart pulse production and resilient crop technologies, which combine traditional agricultural knowledge with modern advances to create productive and sustainable systems. As a primary focus of this book, precision agriculture presents itself as the transformational approach in the use of data and technological know-how on resources and reduced environmental impact. Another strong pillar of this book is sustainable resource management. This book covers topics such as precision water and nutrient management, conservation agriculture, and organic farming. The importance of such methods extends beyond soil fertility and water availability to greenhouse gas mitigation. Advanced breeding techniques and biotechnological innovations in developing climate-smart crops are another focus area of the book. This time, however, the genetic resilience aspect forms the basis to adapt to changed environmental conditions.

Emerging technologies are featured prominently in this book. For example, nanotechnology is able to provide a revolutionary solution in the improvement of nutrient delivery and pest control. Predictive modeling and decision-making in agriculture are further supported by artificial intelligence and machine learning. It discusses agri-based carbon credit systems, which give an economic opportunity to farmers to sequester carbon while making sustainable practices imperative. Its distinctiveness in taking a holistic approach to emphasize how biodiversity conservation and ecosystem management work together toward achieving sustainable agriculture highlights how building systems may be adapted into a resolute food system withstanding all dimensions of the climatic impacts on agriculture.

This book belongs to everyone involved with the food chain—from those research scientists stuck behind their work benches to hardworking farmers across the agricultural landscape, or policymakers that work on building up regulations to any educator responsible for teaching our new generation of agriculturally motivated people. To many, its solution speaks from and to an agriculturally cultured conscience around the globe. More than just a resource, Climate Smart Agriculture represents the link between science and practice, tradition, and innovation. It's a call to belong to the worldwide community striving toward a common vision of building an agricultural future in which farming doesn't merely adjust to the pace of climate change but assists it, leading toward food security and beyond for all.

Pusa, Bihar, India — Biswajit Pramanick
Jhansi, Uttar Pradesh, India — Shiv Vendra Singh
Khurda, Odisha, India — Sagar Maitra
Grugliasco, Italy — Silvia Celletti
Nashipur, Bangladesh — Akbar Hossain

Contents

Series Editor and Editorial Board Members

About the Series Editor

Rajeev Kumar Varshney FRS is a renowned agricultural scientist with over two decades of experience in genomics and molecular breeding. Currently, he holds multiple prestigious positions at Murdoch University in Australia. He has made a significant impact on global food security by creating genomic resources for major "orphan" tropical crops, using these resources to identify genetic loci and candidate genes for drought and pest tolerance in staple crops in sub-Saharan Africa and India, and spearheading international programs that deliver superior crop varieties to some of the world's most impoverished farmers. He is an accomplished author, with more than 600 publications to his name, and has been recognized as a Clarivate Highly Cited Researcher for ten consecutive years (2014–2023).

Food Futures Institute, Centre for Crop and Food Innovation, State Agricultural Biotechnology Centre, Agriculture and Food Security, Murdoch University, Perth, WA, Australia.

Editorial Board Members

About the Editors

Biswajit Pramanick Post-Doc, University of Nebraska-Lincoln (UNL), USA, is currently serving as an Assistant Professor-cum-Scientist, Agronomy at Dr. Rajendra Prasad Central Agricultural University, Pusa. Previously he served as an Assistant Professor, Agronomy at the harbinger of the green revolution, Govind Ballabh Pant University of Agriculture & Technology, Pantnagar during 2015–2019. His research area is soil fertility management, conservation agriculture, crop modeling, and agricultural systems management. He has published more than 100 research papers, 20 book chapters, and 15 popular articles. His current h-index is 28 with >2200 citations in different international scientific literatures. He is the recipient of the prestigious UNL, USA Fellowship; Excellence in Teaching Practices Award-Pantnagar; Dr. PS Deshmukh Young Agronomist Award of the Indian Society of Agronomy; Dr. Rattan Lal Award from the AEEFWS, Punjab. He is the Associate Editor of the prestigious *Soil Science Society of America Journal*, USA; Editorial Board Member of *BMC Plant Biology*; and *Indian Journal of Agronomy*.

Shiv Vendra Singh earned MSc (Ag.) in Agronomy from N.D. University of Agriculture & Technology, Ayodhya, India, and subsequently, a doctoral degree in Agronomy from G.B. Pant University of Agriculture and Technology, Pantnagar, Uttarakhand, India. Currently, he is working as an Assistant Professor, Agronomy and Agrometeorology at Rani Lakshmi Bai Central Agricultural University, India. He possesses more than 7 years of research experience in the field of crop waste recycling, biochar production, biochar enrichment, soil quality enhancement, carbon sequestration, GHG emission, and carbon footprint mitigation under production system. He has published ~50 scientific publications including full-length articles, one book on biochar, book chapters, conference papers, and popular articles. Besides, he has received reputed awards and fellowships from state and national agencies (e.g., Vice-Chancellor Gold Medal, Acharya Narendra Deva University of Agriculture & Technology (ANDUAT), Ayodhya; INSPIRE Fellowship, Department of Science & Technology (DST), Govt. of India (GOI); Prize in National Essay Writing on Waste to Wealth by Ministry of Agriculture and Farmers Welfare, GOI; Young Scientist Award, Uttarakhand State Council for Science And Technology (UCOST) Uttarakhand State Science & Technology Council, Uttarakhand; Outstanding Doctoral Research Award by Fertilizer Association of India).

Sagar Maitra a doctorate in Agronomy, is a dynamic professional with about 25 years of versatile professional experience in teaching, research, training of farmers and technocrats, and communication and dissemination of proven technologies. Besides, he has experience in developmental aspects including field exposure, project support, and monitoring and transfer of technologies to the farming community. His areas of specialization are cropping husbandry, cropping systems, plant nutrient management, abiotic stress management, precision, and smart agriculture. He is presently serving as Professor and Head, Department of Agronomy and Agroforestry; Coordinator, Research Centre for Smart Agriculture, Centurion University of Technology and Management, Odisha, India. He has published four books, edited three books, and authored 160+ articles and 69 book chapters in publications of national and international repute. He has a G-Scholar citation of more than 4400 with a h-index of 32.

Silvia Celletti Researcher, Agricultural Chemistry, at the Department of Agricultural, Forest and Food Sciences (DISAFA), University of Turin (Italy). In 2016, she received her PhD in Agricultural Chemistry at the University of Tuscia (Italy). She worked as a Research Fellow at the Free University of Bolzano (Italy) from 2017 to 2021. She has been a member of the Study Committee of the Italian Society of Agricultural Chemistry (SICA) and Scientific Responsible of the Joint "BioAgry Lab" at the University of Siena (Italy). She is member of the "BAT Center – Interuniversity Center for Studies on Bioinspired Agro-Environmental Technology, University of Naples Federico II" (Italy). She has experience in different aspects of Plant Mineral Nutrition, such as nutrient interactions, physiological responses to single or combined deficiencies and/or excesses of mineral elements, and mechanisms of nutrient uptake, assimilation, transport, and allocation. Her current research topics concern the study of: (1) the use of biofertilizers based on biochar and wood distillate in crops, and the analysis of the effects of these natural products on the soil-plant system; (2) the effects of bioplastics in agriculture: analysis of the impact on the plant-soil system. In addition, her research interests have focused on: (1) the possible use of hydrothermal carbonization (HTC); (2) the effect of herbicides on non-target crops, particularly on Fe and S homeostasis; (3) the use of biostimulants to mitigate Fe deficiency in plants of agricultural interest; and (4) the problems associated with copper soil contamination. She is a recipient of "Young Minds Awards" at the "III International Symposium on growing media, composting and substrate analysis (Susgro), Milan (Italy)" and winner of a scholarship under the "Erasmus Placement Mobility Program" at Rothamsted Research, West Common Harpenden, London (United Kingdom), within Prof. Malcolm J. Hawkesford's research group (Head of Department of Plant Sciences). She is serving as Review Editor of *Frontiers in Plant Science* and as Topic Editor and Collection Editor of *International Journal of Molecular Sciences* (*IJMS*). She has published 40 scientific articles, has been cited for 633 times, and has an h-index of 17 to date in *Scopus*.

Akbar Hossain is presently working as a principal scientist at Bangladesh Wheat and Maize Research Institute (BWMRI), Bangladesh. He finished his PhD degree in the year 2013 at the Institute of Ecology and Biological Sciences, Astrakhan State University, Astrakhan-Russia. His research interest and expertise are climate change impact assessment on food security, plant physiology, defense mechanisms against stress, crop management, agricultural engineering, weed management and conservation agriculture, crop modelling, nanotechnology, etc. He is also working as a research supervisor of post-graduate students at Department of Agronomy, Soil Science and Department of Agricultural & Industrial Engineering, Hajee Mohammad Danesh Science and Technology University, Dinajpur, and Department of Agronomy, Bangladesh Agricultural University, Mymenisingh, Bangladesh. He is also working on several national international projects related to plant physiology and conservation agriculture funded by International Maize and Wheat Improvement Center (CIMMYT), Australian Centre for International Agricultural Research (ACIAR), International Center for Agricultural Research in the Dry Areas (ICARDA), and Govt. of Bangladesh (GoB). He has authored more than 400 national and international journal publications. Besides these, he is also serving as a voluntary reviewer/ editor of different journals.

Contributors

Adeel Abbas School of Environment and Safety Engineering, Institute of Environment and Ecology, Jiangsu University, Zhenjiang, China

Somdatta Achar Department of Agronomy & Horticulture, University of Nebraska-Lincoln, Lincoln, NE, USA

Sk Asraful Ali Division of Agronomy, ICAR-Indian Agricultural Research Institute, Pusa Campus, New Delhi, India

Maha Awjan Alreshidi Department of Chemistry, University of Ha'il, Ha'il, Saudi Arabia

S. K. Asraful ICAR-Indian Agricultural Research Institute, Pusa Campus, New Delhi, India

Anjana Atapattu Agronomy Division, Coconut Research Institute of Sri Lanka, Lunuwila, Sri Lanka

Shreyas Bagrecha ICAR-National Dairy Research Institute, Karnal, Haryana, India

Anamika Barman ICAR-Indian Agricultural Research Institute, New Delhi, India

Achyuta Basak Department of Genetics and Plant Breeding, Uttar Banga Krishi Viswavidyalaya, Cooch Behar, West Bengal, India

Adima Bharadwaj Department of Agronomy, College of Agriculture, Rani Lakshmi Bai Central Agricultural University, Jhansi, Uttar Pradesh, India

Urjashi Bhattacharya Department of Agronomy, School of Agriculture and Allied Sciences, The Neotia University, Sarisa, West Bengal, India

Malay K. Bhowmick International Rice Research Institute (IRRI)-South Asia Regional Centre, Varanasi, Uttar Pradesh, India

Devilal Birla Department of Agronomy, Anand Agriculture University, Anand, Gujarat, India

Rohit Bapurao Borate Division of Agronomy, ICAR-Indian Agricultural Research Institute, Pusa Campus, New Delhi, India

Marian Brestic Department of Botany and Plant Physiology, Faculty of Agrobiology, Food, and Natural Resources, Czech University of Life Sciences Prague, Prague, Czechia

Shaon Chakraborty Krishi Vigyan Kendra, Deoghar, Jharkhand, India

S. Sai Chandrika Department of Agronomy, Agricultural College, Acharya N.G. Ranga Agricultural University, Guntur, Andhra Pradesh, India

Luciana Raphael Chappa Faculty of Natural Resources and Agricultural Sciences, Swedish University of Agricultural Science, Uppsala, Sweden

Saumi Goswami School of Agriculture and Allied Sciences, The Neotia University, Kolkata, West Bengal, India

Subhendu Bikash Goswami Department of Agronomy, Bidhan Chandra Krishi Viswavidyalaya, Mohanpur, West Bengal, India

P. Gurumurthy Department of Soil Science and Agricultural Chemistry, Agricultural College, Acharya N.G. Ranga Agricultural University, Naira, Andhra Pradesh, India

Rashida Hameed School of Environment and Safety Engineering, Institute of Environment and Ecology, Jiangsu University, Zhenjiang, China

M. Jagadesh Tamil Nadu Agricultural University, Coimbatore, Tamil Nadu, India

Gaikwad Dinkar Jagannath Research Centre for Smart Agriculture, Centurion University of Technology and Management, Paralakhemundi, Odisha, India

Twinkle Jena Division of Agronomy, ICAR-Indian Agricultural Research Institute, Pusa Campus, New Delhi, India

Gaurav Jha Department of Agronomy, Kansas State University, Manhattan, KS, USA

Rashmi Jha Department of Seed Science and Technology, Tamil Nadu Agricultural University, Coimbatore, Tamil Nadu, India

B. Asha Jyothi Geo-spatial Technology Center, Regional Agricultural Research Station, Guntur, Andhra Pradesh, India

Jyotirmay Karforma Regional Research Station (Old Alluvial Zone), Uttar Banga Krishi Viswavidyalaya, Cooch Behar, West Bengal, India

Dhanapati Keerthana College of Agriculture, Rani Lakshmi Bai Central Agricultural University, Jhansi, Uttar Pradesh, India

Jiyung Kim Department of Agronomy, Kansas State University, Manhattan, KS, USA

K. M. Krishna Division of Agronomy, ICAR-Indian Agricultural Research Institute, Pusa Campus, New Delhi, India

Dinesh Kumar Department of Agronomy, Indian Institute of Soil and Water Conservation Research Centre, Datia, Madhya Pradesh, India

Aradhna Kumari College of Agriculture, Jawaharlal Nehru Krishi Vishwa Vidyalaya, Ganj Basoda, Vidisha, MP, India

Megha Kumari Division of Agronomy, ICAR-Indian Agricultural Research Institute, Pusa Campus, New Delhi, India

Vandana Kumari Department of Soil Science, Dr. Rajendra Prasad Central Agricultural University, Samastipur, Bihar, India

Kadagonda Nithin Kumar ICAR-Indian Agricultural Research Institute, Pusa Campus, New Delhi, India

Medida Sunil Kumar Department of Agronomy, Agricultural College, Acharya N.G. Ranga Agricultural University, Guntur, Andhra Pradesh, India

Geo-spatial Technology Center, Regional Agricultural Research Station, Guntur, Andhra Pradesh, India

Nallagatla Vinod Kumar Department of Agronomy, College of Agriculture, College of Agriculture, Professor Jayashankar Telangana State Agricultural University, Rajendranagar, Hyderabad, India

Sanjay Kumar Department of Agronomy, Bihar Agricultural University, Sabour, Bihar, India

Sunil Kumar International Rice Research Institute (IRRI)-South Asia Regional Centre, Varanasi, Uttar Pradesh, India

Ranjan Laik Department of Soil Science, Dr. Rajendra Prasad Central Agricultural University, Samastipur, Bihar, India

Dendukuri Udaya Lakshmi Banaras Hindu University, Varanasi, Uttar Pradesh, India

Jonnada Likhita College of Agriculture, Rani Lakshmi Bai Central Agricultural University, Jhansi, Uttar Pradesh, India

Sagar Maitra Department of Agronomy and Agroforestry, M. S. Swaminathan School of Agriculture, Centurion University of Technology and Management, Paralakhemundi, Odisha, India

Ramyajit Mandal School of Agriculture, Seacom Skills University, Santiniketan, West Bengal, India

Subhadeep Mandal Department of Soil Science and Agricultural Chemistry, Uttar Banga Krishi Viswavidyalaya, Cooch Behar, West Bengal, India

Tanuj Kumar Mandal Department of Agronomy, School of Agriculture and Allied Sciences, The Neotia University, Sarisa, West Bengal, India

Satyabrata Mangaraj Krishi Vigyan Kendra, Sonepur, Odisha University of Agriculture and Technology, Bhubaneswar, Odisha, India

T. H. V. Meghana University of Agricultural Sciences, GKVK, Bangalore, Karnataka, India

Anjali Mehta ICAR-National Dairy Research Institute, Karnal, India

Prasannajit Mishra Odisha University of Agriculture and Technology, Bhubaneswar, Odisha, India

K. A. Mrudula Saline Water Scheme, Acharya NG Ranga Agricultural University, Bapatla, Andhra Pradesh, India

Bishal Mukherjee Department of Agronomy, School of Agriculture and Allied Sciences, The Neotia University, Sarisa, West Bengal, India

B. Sri Sai Siddartha Naik Department of Agronomy, Agriculture College, Acharya N.G. Ranga Agricultural University, Bapatla, Andhra Pradesh, India

Tony Manoj Kumar Nandipamu Department of Agronomy, G.B. Pant University of Agriculture and Technology, Pantnagar, Uttarakhand, India

Dilip Nandwani Department of Agricultural Science and Engineering, College of Agriculture, Tennessee State University, Nashville, TN, USA

Manish Kumar Naskar Department of Agricultural Meteorology and Physics, Bidhan Chandra Krishi Viswavidyalaya, Mohanpur, West Bengal, India

Debabrata Nath International Rice Research Institute (IRRI)-South Asia Regional Centre, Varanasi, Uttar Pradesh, India

Suprava Nath Department of Agronomy and Agroforestry, M. S. Swaminathan School of Agriculture, Centurion University of Technology and Management, Paralakhemundi, Odisha, India

Arati Ningombam ICAR Research Complex for North Eastern Hill Region, Manipur Centre, Lamphelpat, India

Kadagonda Nithinkumar Division of Agronomy, ICAR-Indian Agricultural Research Institute, Pusa Campus, New Delhi, India

Smruti Ranjan Padhan Division of Agronomy, ICAR-Indian Agricultural Research Institute, Pusa Campus, New Delhi, India

Sindhu Palla Department of Agronomy, Kansas State University, Manhattan, KS, USA

Aastika Pandey ICAR-Indian Agricultural Research Institute, New Delhi, India

Bappa Paramanik Subject Matter Specialist, Department of Soil Science, Dakshin Dinajpur KVK, Uttar Banga Krishi Viswavidyalaya, Cooch Behar, West Bengal, India

Partha Sarathi Patra Regional Research Station (Terai Zone), Uttar Banga Krishi Viswavidyalaya, Cooch Behar, West Bengal, India

Biswa Ranjan Pattnayak Krishi Vigyan Kendra, Sonepur, Odisha University of Agriculture and Technology, Bhubaneswar, Odisha, India

Panneerselvam Peramaiyan International Rice Research Institute (IRRI)-South Asia Regional Centre, Varanasi, Uttar Pradesh, India

Masina Sairam Department of Agronomy and Agroforestry, M. S. Swaminathan School of Agriculture, Centurion University of Technology and Management, Paralakhemundi, Odisha, India

Snigdha Samanta Department of Agronomy, School of Agriculture and Allied Sciences, The Neotia University, Sarisa, West Bengal, India

D. T. Santosh Research Centre for Smart Agriculture, Centurion University of Technology and Management, Paralakhemundi, Odisha, India

Guntamukkala Sekhar Centurion University of Technology and Management, Odisha, India

Debadatta Sethi Odisha University of Agriculture and Technology, Bhubaneswar, Odisha, India

Gaurav Sharma College of Horticulture and Agroforestry, Rani Lakshmi Bai Central Agricultural University, Jhansi, Uttar Pradesh, India

Rashmi Sharma Department of Agronomy, G.B. Pant University of Agriculture and Technology, Pantnagar, Uttarakhand, India

S. K. Sharma Indian Council of Agricultural Research, Pusa, New Delhi, India

Shivali Sharma College of Horticulture and Agroforestry, Rani Lakshmi Bai Central Agricultural University, Jhansi, Uttar Pradesh, India

Ashok Kumar Singh Dr. Rajendra Prasad Central Agricultural University, Pusa, Samastipur, Bihar, India

Santosh Kumar Singh Department of Soil Science, Dr. Rajendra Prasad Central Agricultural University, Pusa, Samastipur, Bihar, India

Shashank Shekher Singh Department of Agronomy NIAS, Gopal Narayan Singh University, Sasaram, Rohtas, Bihar, India

Shiv Vendra Singh Department of Agronomy, College of Agriculture, Rani Lakshmi Bai Central Agricultural University, Jhansi, Uttar Pradesh, India

Shubham Singh College of Post Graduate Studies in Agricultural Science, (CAU-Imphal), Umiam, Meghalaya, India

Sudhanshu Singh International Rice Research Institute (IRRI)-South Asia Regional Centre, Varanasi, Uttar Pradesh, India

Yogeshwar Singh Department of Agronomy, College of Agriculture, Rani Lakshmi Bai Central Agricultural University, Jhansi, Uttar Pradesh, India

Milan Skalicky Department of Botany and Plant Physiology, Faculty of Agrobiology, Food, and Natural Resources, Czech University of Life Sciences Prague, Prague, Czechia

Sumit Sow Department of Agronomy, Dr. Rajendra Prasad Central Agricultural University, Pusa, Bihar, India

Ajay Anto Soy ICAR-Indian Agricultural Research Institute, Pusa Campus, New Delhi, India

Ateka Sultana Midnapore City College, Paschim Medinipur, West Bengal, India

Surabhi University of Agricultural Sciences, GKVK, Bangalore, Karnataka, India

M. Sri Surya Department of Agronomy, Agriculture College, Acharya NG Ranga Agricultural University, Bapatla, Andhra Pradesh, India

Sushmita Thokchom Assam Agricultural University, Jorhat, Assam, India

Rupesh Tirunagiri Division of Science and Agricultural Chemistry, Indian Agricultural Research Institute, Pusa, New Delhi, India

Ambikesh Tripathi Department of Agronomy, Institute of Agricultural Sciences, BHU, Varanasi, Uttar Pradesh, India

Nishadini Widanagamage Department of Agronomy, Kansas State University, Manhattan, KS, USA

Chapter 1
Impact of Climate Change on Food Crops Cultivation in the Present Consequence of Global Warming

Aradhna Kumari, Munmun Dash, M. Jagadesh, and Santosh Kumar Singh

Abstract Climate change is having significant impacts on various aspects of the ecosystem, including agriculture. Rising global temperatures are particularly affecting the cultivation of food crops, which are highly sensitive to variations in temperature, rainfall patterns, as well as extreme weather events. The effect of climate change on food crop cultivation is multifaceted but varies depending on the crop, the region, and the specific climate change scenario. This chapter discusses the impacts of climate change on food crop cultivation, the present consequences of global warming, and some potential adaptation strategies that could be employed to mitigate these impacts. The ongoing effects of global warming on food crop cultivation are already being felt in many parts of the world, with reduced crop yields, lower crop quality, increased pest and disease pressure, increased food insecurity, and changes in food availability being some of the most significant consequences. Adaptation strategies such as crop diversification, improved water management, improved crop management, investment in research along with development, and climate information services can be employed to mitigate these impacts. However, concerted global endeavours to decrease greenhouse gas emissions and control the enhancement in global temperatures will also be necessary to prevent the most severe impacts of climate change on food crop cultivation besides ensuring food security for future generations.

Keywords Climate change · Global warming · Crop yield · Crop quality · Adaptation strategies

A. Kumari (✉)
College of Agriculture, Jawaharlal Nehru Krishi Vishwa Vidyalaya,
Ganj Basoda, Vidisha, Madhya Pradesh, India

M. Dash · M. Jagadesh
Tamil Nadu Agricultural University, Coimbatore, Tamil Nadu, India

S. K. Singh
Department of Soil Science, Dr. Rajendra Prasad Central Agricultural University,
Pusa, Samastipur, Bihar, India

B. Pramanick et al. (eds.), *Climate-Smart Agricultural Technologies*, Sustainability Sciences in Asia and Africa, https://doi.org/10.1007/978-981-96-7699-6_1

1.1 Introduction

Climate change is considered to be one of the most burning global challenges in the recent era, with increasing global temperatures which are having significant effects on various aspects of the ecosystem. Agriculture is especially prone to the consequences of climate change, which is sensitive to changes in temperature, rainfall patterns, as well as extreme weather events. The influence of climate change on food crop cultivation is multifaceted and varies depending on the crop, the region, and the specific climate change scenario. Through this chapter, the influences of climate change on food crop cultivation, the recent effects of global warming, and some potential adaptation strategies that could be employed to mitigate these impacts are well furnished. An unusual rainfall pattern, abrupt temperature fluctuations, or catastrophic weather events in a year might negatively affect livestock and agricultural production. Though these detrimental effects on yields can be mitigated by current technologies, we must not ignore the significant effect of recent droughts on cereal production in the world along with its high potential risk. Climate analysis has advanced thanks to advances in machine learning (ML) techniques (Schneider et al. 2017; Reichstein et al. 2019). Then, utilising the newly found climate links, artificial intelligence (AI) can offer improved forewarnings of impending weather phenomena, including extreme events (Huntingford et al. 2019). Worldwide, the efficiency of food production and its quality are already being impacted in a manner similar to that of global warming by climate change (Easterling et al. 2007; FAO 2019; Pramanick et al. 2023). The rise in global food production realised in recent decades has mitigated these disadvantages. According to research, the production of maize and wheat has increased since 1980, but it could have been 5% higher without the impacts of climate change. However, it is challenging to determine how much climate change has affected global agricultural and livestock production, as other factors also play a role. If the average temperature increases by 1–3 °C from 2020, rain-fed agriculture yields in Africa could decrease by almost 50%. Predicting the behaviour of the food and feed production industry is challenging, particularly at the local level. Food price increases are caused by declining global yields and underdeveloped countries. The production of food contributes significantly to greenhouse gas emissions along with environmental deterioration, which means that it can amplify and speed up climate change. About 15% of the world's greenhouse gas emissions are caused by agriculture, which is comparable to transportation. According to more pessimistic estimates, food production can contribute up to 30% of all emissions to the atmosphere. Making food production more climate-resistant and achieving much smaller carbon footprints are thus necessary for effectively limiting the long-term impacts.

1.2 Drivers of Change

1.2.1 Greenhouse Gas Emissions and Rising Temperatures

Without GHGs, the world's surface temperature would be too high, making it impossible for life to exist on the planet. The greenhouse effect is the primary factor that makes the Earth habitable. Global warming, on the other hand, is a catastrophic phenomenon that was caused by an increase in GHGs in the atmosphere (Anderson et al. 2016; Singh et al. 2023). The primary gases that make up the Earth's atmosphere are N_2, O_2, and Ar along with GHGs and other contaminants, also exist in trace amounts. While the percentages of the trace gases (carbon dioxide, methane, nitrous oxides, and ozone) fluctuate daily, seasonally, as well as annually, the percentages of the permanent gases (nitrogen, oxygen, and argon) remain constant (Khandekar et al. 2005). Unlike the other primary components of the atmosphere, GHGs have inherent vibrational modes that allow them to absorb and reradiate infrared light (Oktyabrskiy 2016). Agriculture contributes to more than 25% of the world's greenhouse gas emissions, making it a key driver of global environmental change even though it is susceptible to climate dynamics (Edenhofer et al. 2014; Kar et al. 2021).

1.2.2 Changing Rainfall Patterns

The Earth's climate has passed through noteworthy variations in the current era. One of the most prominent changes is the increase in atmospheric moisture content, resulting from global warming. This increase in moisture content has the potential to create more precipitation. However, the precipitation episodes have become shorter and less frequent due to the time required to replenish the atmosphere with moisture (Stott et al. 2010; Coumou and Rahmstorf 2012). The twentieth century has witnessed a rise in the frequency of floods, relatively because of heavy rainfall that occurs within a less duration. This leads to a situation where the soil's ability to absorb water is inadequate, enhancing the amount of runoff water that leads to floods. The rise in water levels has also resulted in flooding in areas close to water bodies (Parry et al. 2007; Trenberth et al. 2014). Both the southern and northern hemispheres have witnessed drying trends since the 1950s, as well as the size of the world's extremely dry regions has doubled since the 1970s. These trends and extensions cannot be explained by climate models without considering the impact of global warming. The effects of global warmings, such as early snowmelt, enhanced evaporation from vegetation as well as soil, whereas a shift in rainfall trends from average and less intense rains to lesser and more intense rains contributed to the increase in droughts (Dai 2013).

1.2.3 Extreme Weather Events

Mid-latitude storms are fuelled by the temperature variation between the poles and the equator, and the storms' ability to form could be affected if this temperature difference decreases due to global warming. At the poles, where humidity and temperature are lesser, any rise in water vapour, as well as heat, could significantly elevate the temperature. As the temperature of the atmosphere rises, its capacity to carry water vapour also increases. According to Easterling et al. (2000), storms may become more intense due to an increase in moisture absorbed by the atmosphere. However, in areas near the equator where temperature and humidity are already high, the change may not be as significant. As a result, the decrease in temperature fluctuation could lead to a reduction in the number of storms. The development of hurricanes and tropical storms, which are influenced by the temperature and humidity of the air, may also be impacted by global warming. As per the studies conducted by Easterling et al. (2000), the absorption of moisture by the atmosphere could result in an increase in the intensity of storms. Moreover, the rise in the surface temperature of the ocean due to global warming can also lead to stronger winds, thereby increasing the likelihood of hurricanes. According to hurricane strength models, the air humidity has increased by roughly 4% since 1970 (Manabe and Wetherald 1975) while the surface seawater temperature has risen by 0.3 °C since 1980 (Coral Bleaching 2014). These changes may have boosted hurricane wind speeds by 1 knot. However, as hurricane intensity is currently recorded with an accuracy of 5 knots, it is now impossible to verify this (Coral Bleaching 2014).

1.2.4 Heat Waves

According to Easterling et al. (2000), Climate Communication (2011), and Somerville, heat waves are demarcated as extended periods of hot weather that have a significant impact on human health. Exposure to elevated temperatures for an extended period of time can result in serious health problems like heatstroke, fainting, and clams. In the years between 1951 and 1980, just 0.1% of the world was afflicted by extremely hot temperatures; but, since 1980, extreme heat events have become more common. Approximately 10% of the world is already affected by heat waves, and if global warming keeps up, that number might rise to 17% in 10 years (Coumou and Rahmstorf 2012; Wang et al. 2015). According to Figure 8 (Houghton et al. 2001), where the average in the graph represents the mean temperature and the variance indicates the range of the temperature around the mean, as the mean world temperature rises, the climate will change towards a warmer one, increasing the records of extreme weather.

1.3 Consequences of Climate Change on Food Crop Cultivation

The influence of climate change on food crop cultivation is complicated and depends on several aspects, viz., the crop, the region, and the specific climate change scenario. Conversely, there are some common trends that have been observed globally. Growing temperatures, varying rainfall patterns, along with amplified frequency and intensity of extreme weather events are all impacting food crop cultivation in several ways, including:

1.3.1 Changes in Crop Yield

It's been found that crops like corn, soybean, wheat, small grains, rice, cotton, pasture, and fruits may encounter unfavourable conditions due to extreme climate events like droughts, floods, and extreme temperatures (Aryal et al. 2020; Pramanick et al. 2022). A recent study by NASA, published in the Nature Food Journal, reveals that climate change may negatively impact the output of maize (corn) also wheat as early as 2030, particularly under more greenhouse gas emissions scenarios. While wheat crop yields are predicted to increase by roughly 17%, maize crop yields are expected to fall by 24%.

1.3.2 Changes in Crop Quality

Crop quality is shifting due to climate change. While most crops will see a decline in quality due to climate change, certain crops may actually see an improvement. Increased CO_2 concentration lowers the nutritional content of cereals, legumes, and vegetables while raising the essential oil content of anise and *Valeriana* sp. (Kumari et al. 2022; Patni et al. 2022; Dey et al. 2022). Another noteworthy example is the ability of consumers to distinguish between tea harvested from terrace gardens in the sun or shade (Ahmed et al. 2010). Based on the important antioxidant secondary metabolites connected to tea's health-related and flavour attributes, tea from shade-grown agro-forests was also thought to be of greater quality than that from terrace gardens (Ahmed et al. 2014). Additionally, it has been discovered that variations in precipitation, which affect buyer procuring decisions as well as farmer livelihoods, affect the sensory as well as secondary metabolite profiles of tea products (Ahmed et al. 2015).

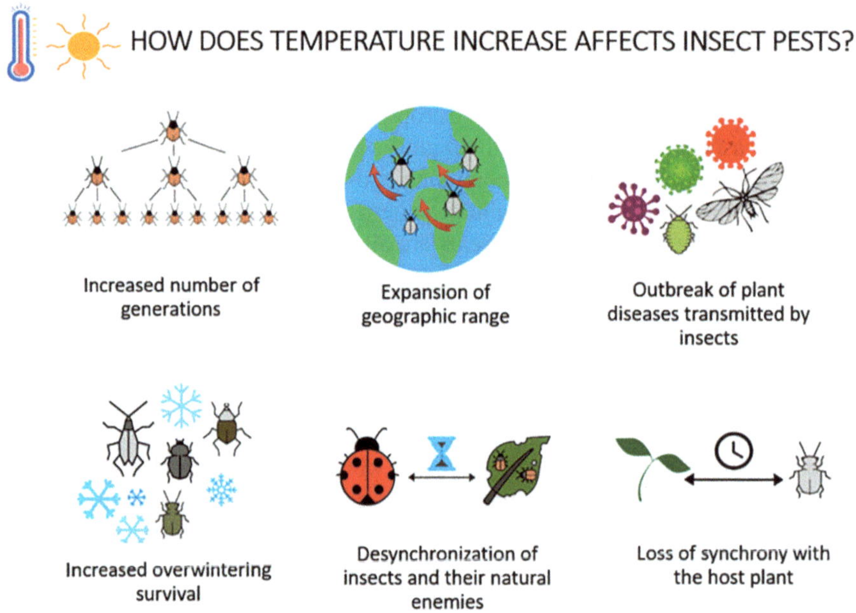

Fig. 1.1 Alteration in agricultural insect pests due to elevated temperature (Skendžić et al. 2021)

1.3.3 Changes in Pests and Diseases

Temperature can have a significant impact on insect behaviour, distribution, development, and reproduction (Kocmánková et al. 2009; Kumari et al. 2022). With the primary drivers of climate change (increased atmospheric CO_2, higher temperatures, and decreased soil moisture), it is expected that insect pest populations and crop losses will be affected (Fand et al. 2012). Climate change may create new biological niches that allow insect pests to grow, proliferate, and move to new areas. The relationship between agricultural crops and insect pests may be profoundly affected by the physiological consequences of increased temperatures and CO_2 (Hare 1992; Caulfield and Bunce 1994; Roth and Lindroth 1995) (Fig. 1.1).

1.3.4 Changes in Water Availability

Due to changes in soil and water quantity and quality, this condition has encouraged unwelcome environmental issues (Thiery et al. 2020; Sloat et al. 2020; Dai et al. 2020). Several researchers have examined how global warming affects agricultural water consumption, including adjustments to net irrigation, water demand, as well as crop water intake. This is crucial because, according to Woznicki et al. (2015), 70% of the water consumed worldwide each year is utilised for irrigation in

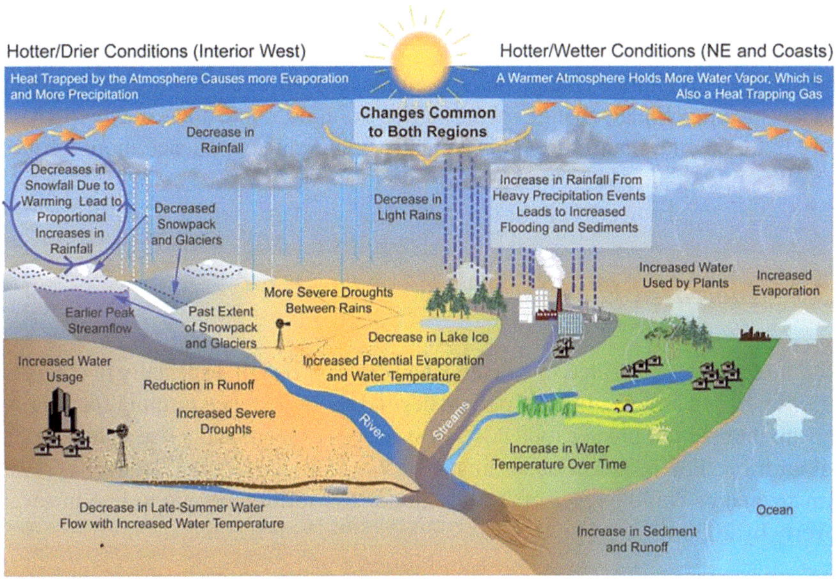

Fig. 1.2 Variation in water cycle

agriculture, which also consumes the freshest water. According to climate estimates, the amount of water needed for irrigation towards the end of this century is predictable to upsurge between 40% and 250%, depending on the crop (Gondim et al. 2012; Bakken et al. 2016). Decreased water availability during the growing seasons, evapotranspiration, as well as fluctuations in crop phenology have all been blamed for the increasing needs (Woznicki et al. 2015; Sloat et al. 2020; Dai et al. 2020; Kumari et al. 2022) (Fig. 1.2).

1.3.5 Impact of Extreme Weather Events

Extreme weather events viz., storms, floods, as well as heat waves can cause significant damage to crop and infrastructure, steering to reduced food availability and increased food prices.

1.3.6 Regional Variations in Impact

The influence of climate change on crop cultivation varies depending on the region and the specific climate change scenario. For example, some areas may experience more frequent and intense droughts, whereas some may experience more frequent as well as intense floods.

1.3.7 Increased Food Insecurity

Food insecurity is projected to worsen when crop yields fall, especially in emerging nations where a large number of people depend on subsistence farming for a living. This might result in greater food costs, less access to food, and a rise in hunger and malnutrition. For instance, according to a World Bank report (Hallegatte 2016), climate change could trigger off an additional 100 million people to live in severe poverty by 2030 (Fig. 1.3).

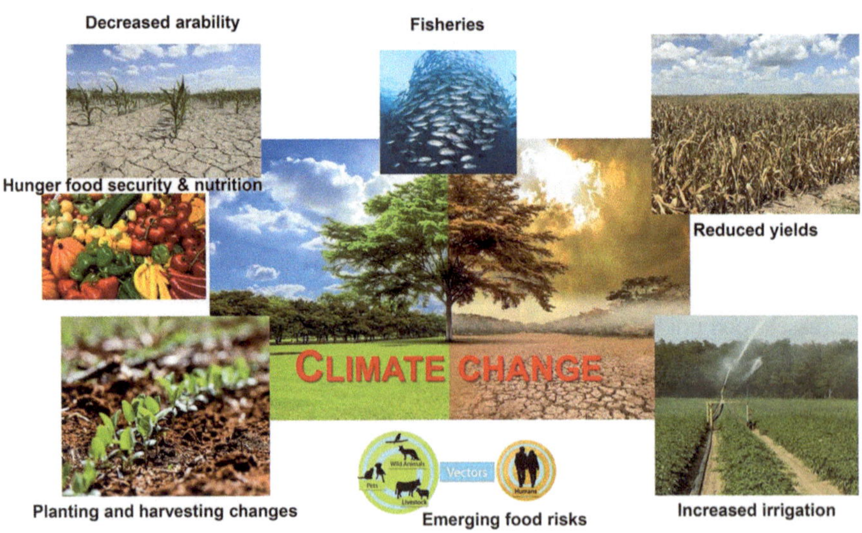

Fig. 1.3 Initial inferences to combat climate change effect on food production (Gomez-Zavaglia et al. 2020)

1.3.8 Changes in Food Availability

In any event, the aforementioned dangers can have an impact on food security, also especially, the accessibility of food for a population of eight billion people around the world at a reasonable price (Hertel 2016). According to a 2011 Foresight report (Government Office for Science 2011), the growth in global food demand anticipated over the following 10 years (United Nations 2017) outweighs the influence of climate change in the short run. The population tends to consume more food (particularly meat) as the level of life rises, which portends a future with unpredictable food costs. According to the Human Development Report (2019), international policy is essential for addressing the concern to rural residents' livelihoods in underdeveloped nations as well as the increases in food prices caused by declines in global yields.

1.3.9 Increased Risk of Crop Failure

Recent research conducted by NASA, which was published in the journal *Nature Food*, suggests that climate change could influence the production of maize (corn) as well as wheat as quickly as 2030, particularly under high greenhouse gas emission scenarios. Agreeing with the study, wheat crop yields may rise by approximately 17%, while maize crop yields could decline by about 24%.

1.4 Potential Adaptation Strategies

To mitigate the impacts of climate change on food crop cultivation, several adaptation strategies could be employed. Some of these include:

1.4.1 Crop Diversification

While crop diversification can address the issue of structural environmental degradation brought on by the overuse of monocultures, results are eventually influenced by the circumstances of the farmers, including age, educational attainment, economic and financial standing, and the size of the farm and the family. Comprehensive long-term research carried out by interdisciplinary experts can be used to support decisions (Schroth and Ruf 2014).

1.4.2 Improved Water Management

Cao et al. (2018) looked at how water stress affected crop productivity in different parts of China between 1996 and 2015. They recommend utilising the water footprint to determine whether irrigation is appropriate in order to put the best crop variety in the world, irrigation, into practice. The greatest strategy to ensure that crops can adjust to environmental changes is this way. Powerful computational resources have made it easier to monitor and put into practice more practical solutions for the rising irrigation demand (Dharminder et al. 2021). Agrohydro-salinity models like SaltMod can currently be used to simulate and predict the impacts of irrigation on the water table, soil salinity, drainage, and groundwater, along with those of mitigating techniques. By using such models, it is now possible to foresee how changes in the climate would affect agriculture (Elliott et al. 2014). For instance, Sun et al. (2019) determined the optimum practices for water management and nitrogen fertilisation for summer cucumber in greenhouses in regions of northern China. They used environmental and economic indices to calibrate and evaluate models built from experimental data (applied to 240 scenarios with varying water uses and diverse fertilisation conditions) in order to establish the optimum management techniques. Mahmoud and Gan (2019) conducted a spatial and temporal analysis in Saudi Arabia between 1950 and 2013 using remote sensing besides a geographic information system (GIS). So as to attain this, scientists gathered data on evapotranspiration as well as simulated crop coefficients as a function of a 16-day time-series of the normalised difference vegetation index from the Moderate Resolution Imaging Spectro-radiometer. With the help of the data they had just collected, they were able to simulate daily evapotranspiration using a model of soil water balance as well as then aggregate it to calculate monthly and annual evapotranspiration. Using computer modelling techniques, Riediger et al. (2014) investigated the requirement for re-adapting agricultural irrigation on sandy soils from Ülzen County (Central Europe). The amount of irrigation water that could be required to avoid crop failures up to the year 2070 was anticipated by combining information on climate changes with evapotranspiration models tailored to certain soil types and crops. Future agricultural practices will not have access to enough groundwater under a scenario of rising temperatures. Therefore, computational tools are crucial for estimating irrigation needs in light of the changing climate.

1.4.3 Improved Crop Management

Both crops and cattle suffer severe economic losses as a result of heat stress. In order to lessen its effects, it is necessary to maintain animal productivity in hot climates through environmental modification, dietary changes, as well as the creation of breeds that are more resistant of heat stress. For greater results, these tactics can be utilised separately or in combination (Collier et al. 2006). Researchers collaborated on the C4 Rice Project to develop high-yielding rice cultivars for smallholder farmers using cutting-edge scientific methods.

1.4.4 Reduce Pest Incidence

To manage grain legume pests, there are effective methods such as intercropping, population monitoring, modifying agricultural systems to rise the action of natural enemies of a particular cultivation, using natural essential oils and biopesticides alone or in combination with traditional pesticides, and cultivating insect varieties with resistance developed through conventional genetic engineering. *Bacillus thuringiensis* crops are genetically modified plants that contain Cry toxins, which are endospore toxins of the bacterium. However, there is a debate regarding the well-being of B. thuringiensis crops for the environment. In 2016, approximately 185 million hectares of GM crops were grown worldwide (Abbas 2018).

1.4.5 Investment in Research and Development

One of the most pressing technological challenges that we face in relation to climate change and its impact on crops is the need to accurately identify, explain, and describe these impacts. By doing so, we can create precise prediction models that will help us to prepare for the future (Iizumi and Ramankutty 2015). However, in order to achieve this, we need to fill in a few knowledge gaps that currently exist. To effectively deal with the influences of climate change, we prerequisite to combine indigenous along with local knowledge with advancements in technology (Painemilla 2010; Cadilhac et al. 2017; Morss et al. 2011). In fact, it is wise to encourage the practice of indigenous knowledge in addressing climate-related issues and to include it in both short- and long-term plans for adapting to anticipated fluctuations (Kangalawe et al. 2011; Cadilhac et al. 2017). By doing so, we can lower uncertainties and create effective strategies for dealing with the challenges ahead.

1.5 Policy Interventions

To address the influences of climate change on the cultivation of food crops, interventions at the policy level can be implemented in addition to the aforementioned techniques. For instance, the creation and implementation of climate-smart agricultural policies can support the development of sustainable farming methods that lower greenhouse gas emissions, boost climate change resistance, and enhance food security. Adopting sustainable land use techniques, such as agroforestry and conservation agriculture, can also promote sustainable development objectives while assisting in reducing the influences of climate change on the cultivation of food crops. Several significant policy changes include:

1.5.1 Climate-Smart Agricultural Policies

Apart from the various methods which are listed above, policy interventions can be utilised to calculate and solve how climate change is influencing the cultivation of food crops. For instance, the development and application of climate-smart agricultural policies can improve the advancement of sustainable farming practices that decrease greenhouse gas emissions, enhance climate change resistance, and secure food security. Adopting sustainable land use practises like conservation agriculture and agroforestry can be utilised to decline in the outcomes of climate change on the production and productivity of food crops while also advancing the goals of sustainable development (Pramanick et al. 2024).

1.5.2 Sustainable Land Use Practices

In addition to supporting the objectives of sustainable development, sustainable land use practices like agroforestry and conservation agriculture can also aid in reducing the effects of climate alteration on the cultivation of food crops. These methods can help to strengthen the health of the soil, boost biodiversity, and lower greenhouse gas emissions.

1.5.3 International Cooperation

In order to address the influence of climate change on the cultivation of food crops, international cooperation is also essential. This covers assistance with global efforts to cut greenhouse gas emissions and funding for adaptation initiatives in developing countries.

1.5.4 Mitigating Greenhouse Gas Emissions

Limiting the effects of climate change on the development of food crops requires the reduction of greenhouse gas emissions. To achieve this, coordinated efforts must be made to cut emissions across the board, especially in agriculture.

1.6 Conclusion

The consequences of climate change on food crop cultivation are varied and significant, with some pros and cons depending on the crop, region, besides specific climate change scenario. The current consequences of global warming on food crop cultivation are already being felt in maximum regions of the globe, including decreased crop yields, augmented food insecurity, and changes in food availability. To mitigate these impacts, adaptation strategies such as crop diversification, improved water management, improved crop management, investment in research as well as development, and climate information services could be employed. However, it will also be necessary to make concerted effort globally to lessen greenhouse gas emissions also restrict the elevation in global temperatures to prevent severe impacts on food crop cultivation and ensure food security for future generations. International cooperation and collaboration are essential to address the global challenge of climate change and its impacts on food crop cultivation. The Paris Agreement, adopted in 2015 by the United Nations Framework Convention on Climate Change (UNFCCC), is a critical international framework for addressing climate change. The agreement targets to restrict global warming to well below 2 °C above pre-industrial levels and to follow efforts to restrict the temperature inclined to 1.5 °C. To obtain these goals, the Paris Agreement calls for the implementation of climate change adaptation and mitigation strategies in all sectors, even in agriculture.

References

Abbas MST (2018) Genetically engineered (modified) crops (Bacillus thuringiensis crops) and the world controversy on their safety. Egypt J Biol Pest Control 28:1–12

Ahmed S, Unachukwu U, Stepp JR, Peters CM, Long C, Kennelly E (2010) Pu-erh tea tasting in Yunnan, China: correlation of drinkers' perceptions to phytochemistry. J Ethnopharmacol 132(1):176–185

Ahmed S, Orians CM, Griffin TS, Buckley S, Unachukwu U, Stratton AE, Stepp JR, Robbat A Jr, Cash S, Kennelly EJ (2014) Effects of water availability and pest pressures on tea (Camellia sinensis) growth and functional quality. AoB Plants 6:plt054

Ahmed S, Stepp JR, Xue D (2015) Cultivating botanicals for sensory quality: from good agricultural practices (GAPs) to taste discernment by smallholder tea farmers. In: Botanicals: methods for quality and authenticity. CRC Press, Taylor & Francis Group, LLC, Boca Raton

Anderson TR, Hawkins E, Jones PD (2016) CO2, the greenhouse effect and global warming: from the pioneering work of Arrhenius and Callendar to today's Earth System Models. Endeavour 40(3):178–187

Aryal JP, Sapkota TB, Khurana R, Khatri-Chhetri A, Rahut DB, Jat ML (2020) Climate change and agriculture in South Asia: adaptation options in smallholder production systems. Environ Dev Sustain 22(6):5045–5075

Bakken TH, Almestad C, Rugelbak JM, Escobar M, Micko S, Alfredsen K (2016) Climate change and increased irrigation demands: what is left for hydropower generation? Results from two semi-arid basins. Energies 9(3):191

Cadilhac L, Torres R, Calles J, Vanacker V, Calderón E (2017) Desafíos para la investigación sobre el cambio climático en Ecuador. Neotrop Biodivers 3(1):168–181
Cao X, Huang X, Huang H, Liu J, Guo X, Wang W, She D (2018) Changes and driving mechanism of water footprint scarcity in crop production: a study of Jiangsu Province, China. Ecol Indic 95:444–454
Caulfield F, Bunce JA (1994) Elevated atmospheric carbon dioxide concentration affects interactions between Spodoptera exigua (Lepidoptera: Noctuidae) larvae and two host plant species outdoors. Environ Entomol 23(4):999–1005
Cutting, H., Chung, S., & Hassol, S. (2011). Current extreme weather and climate change. Climate Communication Report.
Collier RJ, Dahl GE, VanBaale MJ (2006) Major advances associated with environmental effects on dairy cattle. J Dairy Sci 89(4):1244–1253
Coumou D, Rahmstorf S (2012) A decade of weather extremes. Nat Clim Chang 2(7):491–496
Dai A (2013) Increasing drought under global warming in observations and models. Nat Clim Chang 3(1):52–58
Dai C, Qin XS, Lu WT, Huang Y (2020) Assessing adaptation measures on agricultural water productivity under climate change: a case study of Huai River Basin, China. Sci Total Environ 721:137777
Dey P, Mahapatra BS, Pramanick B, Pyne S, Pandit P (2022) Optimization of seed rate and nutrient management levels can reduce lodging damage and improve yield, quality and energetics of sub-tropical flax. Biomass Bioenergy 157:106355. https://doi.org/10.1016/j.biombioe.2022.106355
Dharminder, Singh RK, Kumar V, Pramanick B, Alsanie WF, Gaber A, Hossain A (2021) The use of municipal solid waste compost in combination with proper irrigation scheduling influences the productivity, microbial activity and water use efficiency of direct seeded rice. Agriculture 11:941. https://doi.org/10.3390/agriculture11100941
Easterling DR, Meehl GA, Parmesan C, Changnon SA, Karl TR, Mearns LO (2000) Climate extremes: observations, modeling, and impacts. Science 289(5487):2068–2074
Easterling WE, Aggarwal PK, Batima P, Brander KM, Lin Erda S, Howden M, Kirilenko A et al (2007) Food, fibre and forest products. In: Climate change 2007. Cambridge University Press, Cambridge, pp 273–313
Edenhofer, O., Pichs-Madruga, R., Sokona, Y., Agrawala, S., Bashmakov, I. A., Blanco, G., ... & Zwickel, T. (2014). Summary for policymakers. Germany
Elliott J, Deryng D, Müller C, Frieler K, Konzmann M, Gerten D, Glotter M et al (2014) Constraints and potentials of future irrigation water availability on agricultural production under climate change. Proc Natl Acad Sci 111(9):3239–3244
Fand BB, Kamble AL, Kumar M (2012) Will climate change pose serious threat to crop pest management: a critical review. Int J Sci Res Publ 2(11):1–14
FAO, World Health Organization (2019) The state of food security and nutrition in the world 2019: safeguarding against economic slowdowns and downturns, vol 2019. Food & Agriculture Org
Gomez-Zavaglia A, Mejuto JC, Simal-Gandara J (2020) Mitigation of emerging implications of climate change on food production systems. Food Res Int 134:109256
Gondim RS, de Castro MAH, de HN Maia A, Evangelista SRM, Fuck SC d F (2012) Climate change impacts on irrigation water needs in the Jaguaribe River Basin 1. J Am Water Resour Assoc 48(2):355–365
Government Office for Science (2011) Future of food and farming. Open Government Licence v 3.0. https://www.gov.uk/government/publications/future-of-food-andfarming
Great Barrier Marine Park Authority. Coral bleaching 2014. http://www.gbrmpa.gov.au/managing-the-reef/threats-tothe-reef/climate-change/what-does-this-mean-for-species/corals/what-is-coral-bleaching
Hallegatte S (2016) Shock waves: managing the impacts of climate change on poverty. World Bank Publications, Washington, DC
Hare JD (1992) Herbivore-natural enemy. In: Plant resistance to herbivores and pathogens: ecology, evolution, and genetics. University of Chicago Press, Chicago, p 278

Hertel TW (2016) Food security under climate change. Nat Clim Chang 6(1):10–13

Houghton JT, Albritton DL, Allen MR, Baede APM, Church JA, Cubasch U, Xiaosu D et al (2001) Summary for policymakers. In: Climate change 2001: the scientific basis. Contributions of Working Group I to the third assessment report of the Intergovernmental Panel on Climate Change. Cambridge University Press, Cambridge, pp 1–20

Human Development Report (2019) Climate change and inequalities in the Anthropocene. Chapter 5. In: Human development report, Newyork pp 175–196

Huntingford C, Jeffers ES, Bonsall MB, Christensen HM, Lees T, Yang H (2019) Machine learning and artificial intelligence to aid climate change research and preparedness. Environ Res Lett 14(12):124007

Iizumi T, Ramankutty N (2015) How do weather and climate influence cropping area and intensity? Glob Food Sec 4:46–50

Kangalawe R, Mwakalila S, Masolwa P (2011) Climate change impacts, local knowledge and coping strategies in the great Ruaha river catchment area, Tanzania. Nat Resour 2:212–223

Kar S, Pramanick B, Brahmachari K, Saha G, Mahapatra BS, Saha A, Kumar A (2021) Exploring the best tillage option in rice based diversified cropping systems in alluvial soil of eastern India. Soil Tillage Res 205:104761. https://doi.org/10.1016/j.still.2020.104761

Khandekar ML, Murty TS, Chittibabu P (2005) The global warming debate: a review of the state of science. Pure Appl Geophys 162:1557–1586

Kocmánková E, Trnka M, Juroch J, Dubrovský M, Semerádová D, Možný M, Žalud Z (2009) Impact of climate change on the occurrence and activity of harmful organisms. Plant Prot Sci 45(10):S48–S52

Kumari A, Lakshmi GA, Krishna GK, Patni B, Prakash S, Bhattacharyya M, Singh SK, Verma KK (2022) Climate change and its impact on crops: a comprehensive investigation for sustainable agriculture. Agronomy 12(12):3008

Mahmoud SH, Gan TY (2019) Irrigation water management in arid regions of Middle East: assessing spatio-temporal variation of actual evapotranspiration through remote sensing techniques and meteorological data. Agric Water Manag 212:35–47

Manabe S, Wetherald RT (1975) The effects of doubling the CO2 concentration on the climate of a general circulation model. J Atmos Sci 32(1):3–15

Morss RE, Wilhelmi OV, Meehl GA, Dilling L (2011) Improving societal outcomes of extreme weather in a changing climate: an integrated perspective. Annu Rev Environ Resour 36:1–25

Oktyabrskiy VP (2016) A new opinion of the greenhouse effect. St Petersburg Polytech Univ J: Phys Math 2:124–126

Painemilla KW (ed) (2010) Indigenous peoples and conservation: from rights to resource management. Conservation International, Arlington

Parry ML, Canziani O, Palutikof J, Van der Linden P, Hanson C (eds) (2007) Climate change 2007-impacts, adaptation and vulnerability: working group II contribution to the fourth assessment report of the IPCC, vol 4. Cambridge University Press, Cambridge

Patni B, Bhattacharyya M, Kumari A, Purohit VK (2022) Alarming influence of climate change and compromising quality of medicinal plants. Plant Physiol Rep 7(2):1–5. https://doi.org/10.1007/s40502-021-00616-x

Pramanick B, Kumar M, Naik BM, Kumar M, Singh SK, Maitra S, Naik BSSS, Rajput VD, Minkina T (2022) Long-term conservation tillage and precision nutrient management in maize–wheat cropping system: effect on soil properties, crop production, and economics. Agronomy 12:2766. https://doi.org/10.3390/agronomy12112766

Pramanick B, Mahapatra BS, Datta D, Dey P, Singh SP, Kumar A, Paramanik B, Awasthi N (2023) An innovative approach to improve oil production and quality of mustard (*Brassica juncea* L.) with multi-nutrient-rich polyhalite. Heliyon 9(3):e13997. https://doi.org/10.1016/j.heliyon.2023.e13997

Pramanick B, Kumar M, Naik BM, Singh SK, Kumar M, Singh S (2024) Soil carbon-nutrient cycling, energetics, and carbon footprint in calcareous soils with adoption of long-term conservation tillage practices and cropping systems diversification. Sci Total Environ 912:169421. https://doi.org/10.1016/j.scitotenv.2023.169421

Reichstein M, Camps-Valls G, Stevens B, Jung M, Denzler J, Carvalhais N (2019) Deep learning and process understanding for data-driven Earth system science. Nature 566(7743):195–204

Riediger J, Breckling B, Nuske RS, Schröder W (2014) Will climate change increase irrigation requirements in agriculture of Central Europe? A simulation study for Northern Germany. Environ Sci Eur 26(1):1–13

Roth SK, Lindroth RL (1995) Elevated atmospheric CO2: effects on phytochemistry, insect performance and insect-parasitoid interactions. Glob Chang Biol 1(3):173–182

Schneider T, Lan S, Stuart A, Teixeira J (2017) Earth system modeling 2.0: a blueprint for models that learn from observations and targeted high-resolution simulations. Geophys Res Lett 44(24):12–396

Schroth G, Ruf F (2014) Farmer strategies for tree crop diversification in the humid tropics. A review. Agron Sustain Dev 34(1):139–154

Singh SV, Chaturvedi S, Nayak P, Dhyani VC, Nandipamu TMK, Singh DK, Gudapaty P, Mathyam P, Srinivasarao K, Gavindaraju K (2023) Carbon offset potential of biochar based straw management under rice – wheat system along Indo-Gangetic Plains of India. Sci Total Environ 897:165176. https://doi.org/10.1016/j.scitotenv.2023.165176

Skendžić S, Zovko M, Živković IP, Lešić V, Lemić D (2021) The impact of climate change on agricultural insect pests. Insects 12(5):440

Sloat LL, Davis SJ, Gerber JS, Moore FC, Ray DK, West PC, Mueller ND (2020) Climate adaptation by crop migration. Nat Commun 11(1):1243

Stott PA, Gillett NP, Hegerl GC, Karoly DJ, Stone DA, Zhang X, Zwiers F (2010) Detection and attribution of climate change: a regional perspective. Wiley Interdiscip Rev Clim Chang 1(2):192–211

Sun Y, Zhang J, Wang H, Wang L, Li H (2019) Identifying optimal water and nitrogen inputs for high efficiency and low environment impacts of a greenhouse summer cucumber with a model method. Agric Water Manag 212:23–34

Thiery W, Visser AJ, Fischer EM, Hauser M, Hirsch AL, Lawrence DM, Lejeune Q, Davin EL, Seneviratne SI (2020) Warming of hot extremes alleviated by expanding irrigation. Nat Commun 11(1):290

Trenberth KE, Dai A, Van Der Schrier G, Jones PD, Barichivich J, Briffa KR, Sheffield J (2014) Global warming and changes in drought. Nat Clim Chang 4(1):17–22

United Nations (2017) World population prospects 2017. DESA/Population Division. UN. https://esa.un.org/unpd/wpp/

Wang L, Huang J, Luo Y, Yao Y, Zhao Z (2015) Changes in extremely hot summers over the global land area under various warming targets. PLoS One 10(6):e0130660

Woznicki SA, Pouyan Nejadhashemi A, Parsinejad M (2015) Climate change and irrigation demand: uncertainty and adaptation. J Hydrol Reg Stud 3:247–264

Chapter 2
Building Food Security and Climate Resilience: The Promise of Climate Smart Pulse Production Technologies

Bishal Mukherjee, Manish Kumar Naskar, Shaon Chakraborty, Suman Dutta, and Tanuj Kumar Mandal

Abstract As unpredictable weather patterns pose threats to farming communities globally, this chapter goes deep into how pulse crops offer a glimmer of hope. This has been made possible through experiences garnered in the field and innovation research on why chickpeas, lentils, and pigeon peas deserve special attention as we enter this new changing climate. While integrating tales of victorious farmers adopting state-of-the-art pulse-growing technology into scientific pieces on the novel nature of resilience in extreme-weather conditions, these crops feed a community while improving soil health owing to their native nitrogen-fixation abilities. These solutions are grounded in practice at all levels through financial and practical implementation possibilities for even marginal farmers. This chapter sees how investment into climate-smart pulse production brings together a change within food systems coupled with supporting successful farm communities on times of high uncertainty.

Keywords Food Security · Climate Resilience · Pulse Crops · Resource Use Efficiency · Environmental Sustainability

B. Mukherjee (✉) · T. K. Mandal
Department of Agronomy, School of Agriculture and Allied Sciences, The Neotia University, Sarisa, West Bengal, India

M. K. Naskar
Department of Agricultural Meteorology & Physics, Bidhan Chandra Krishi Viswavidyalaya, Mohanpur, West Bengal, India

S. Chakraborty
Krishi Vigyan Kendra, Deoghar, Jharkhand, India

S. Dutta
Department of Genetics and Plant Breeding, School of Agriculture and Rural Development, Ramkrishna Mission Vivekananda Educational and Research Institute, Belur Math, Howrah, West Bengal, India

B. Pramanick et al. (eds.), *Climate-Smart Agricultural Technologies*, Sustainability Sciences in Asia and Africa, https://doi.org/10.1007/978-981-96-7699-6_2

2.1 Introduction

Agriculture and allied activities are facing the most threatening challenges because of global climate change and extreme weather fluctuations round the globe. As climate change has a direct impact on food security, it is one of the most discussed words of the last decade. ‘Climate change’ depicts long-term variability in average weather patterns for a region over a longer period of time. Industrialization, urbanization, deforestation, mechanized agriculture, change in land use pattern, etc. are the major anthropogenic activities responsible for the occurrence of such climatic variabilities (Bera 2021). Increase in atmospheric concentration of carbon dioxide (CO_2) and global warming are the most harmful effects of climate change. Based on the projections, a notable warming of the climate system by 0.74 °C has occurred between 1906 and 2005 (IPCC 2007). IPCC has also predicted that global average temperature would rise by 2.8 °C during the twenty-first century, with best guess increases ranging from 1.8 to 4 °C. Parry et al. (2004) predicted the effects of projected climate change during the period between 2010 and 2060 on global food production and the risk of hunger enhanced in different parts of the world. Seasonal rainfall patterns are another feature of climate change that significantly impacts agroecosystems. Increasing frequency of extreme weather events such as cyclones, heat waves, cold waves, frosts, hailstorms, and hailstorms over short periods adversely affects crop performance (Bhadra et al. 2022).

In recent centuries, impulsive activities like fossil fuel burning, deforestation, pollution, and increasing intensive agriculture have led to an unprecedented situation where we have to worry regarding climate change. Coming to the present scenario, the atmospheric CO_2 concentration has increased from 280 to 395 ppm, CH_4 concentration rises from 715 to 1882 ppb, and N_2O concentration from 227 to 323 ppb for the year 1750 and 2012 (IPCC 2007). The rise in CO_2 concentration can be attributed to land use change and fossil fuel use whereas the release of methane and nitrous oxide is primarily related to agriculture. Besides all these greenhouse gases, water vapour is also considered a silent driver of climate change. As the air gets warmer due to high GHG concentration, it can hold more moisture. Water vapour acts as an insulator and traps the heat which amplifies the greenhouse effect. It has been studied that for a rise in 1 °C temperature, water vapour can increase up to 7%. After several modelling studies, IPCC has shown that the rise in greenhouse gases is correlated with 0.74 °C temperature elevation in the global climate system during the time period between 1906 and 2005. As per the projections of IPCC, the mean temperature in India is estimated to increase up to 1.7 °C in *kharif* (July–October) and up to 3.2 °C during *rabi* (November–March) season, while mean rainfall is expected to rise by 10% by 2070 (Gupta 2011). Since the family income of a huge population in the Indian context is heavily dependent on agriculture, such a scenario originating from climate change presents a huge impact on the country's food and feed economy. The development of a policy framework to implement mitigation and adaptation strategies is required on urgent basis to address the arising challenges of climate change.

2.2 Impacts of Climate Change on Agriculture

Climate change is an escalating global concern that harbours profound ramifications for diverse sectors, with agriculture being among the most susceptible. As our planet experiences unprecedented shifts in weather patterns and atmospheric conditions, the repercussions on agricultural systems and food production are becoming increasingly apparent. This introduction will delve into the multifaceted impacts of climate change on agriculture and allied activities, exploring the formidable challenges it poses to crop yields, water resources, livestock, and the overall sustainability of our food systems.

2.2.1 Impacts on Crop Productivity

Research conducted by Malhi et al. (2021) has highlighted the detrimental effects of climate change on crop productivity. Increasing temperatures, alterations in precipitation patterns, and escalated extreme weather events hamper crop yields, decreased nutritional value, and compromised food quality. These impacts demand enhanced agricultural techniques, such as precision farming, improved irrigation systems, and integrated pest management, to mitigate losses and maintain productivity.

2.2.2 Alterations in Pest and Disease Dynamics

Scientists, such as Masood et al. (2022), have examined the intricate relationship between climate change and pest dynamics. Rise in temperatures and changing precipitation patterns create favourable conditions for pests and diseases, resulting in increased infestations and the spread of pathogens. Advanced monitoring systems, biocontrol methods, and pest-resistant crop varieties offer promising avenues to safeguard crops and reduce reliance on chemical interventions.

2.2.3 Water Availability and Irrigation Challenges

Water scarcity, intensified by climate change, is a pressing concern for agricultural systems. Research by Sikka et al. (2022) emphasizes the need to optimize water management strategies, improve irrigation efficiency, and explore alternative water sources. Adoption of precision irrigation technologies, water recycling, and watershed management approaches can mitigate water scarcity impacts, ensuring sustainable water use in agriculture.

2.2.4 Impacts on Soil Health and Nutrient Cycling

Climate change affects soil health and nutrient cycling processes, with implications for agricultural productivity. Recent research has shown that elevated carbon dioxide levels, a key driver of climate change, can alter microbial population in the soil, affecting nutrient availability and cycling. For instance, Dubey et al. (2019) revealed in a study published in Nature Climate Change that increased temperatures and changing precipitation patterns can accelerate soil organic matter decomposition, reducing soil fertility and nutrient retention. These findings highlight the need for sustainable soil management practices to mitigate the deleterious impacts of climate change on soil health.

2.2.5 Livestock and Animal Husbandry

Climate change poses significant challenges to livestock production and animal husbandry. Scientific investigations by Bogale and Erena (2022) have highlighted the detrimental effects of heat stress, reduced forage availability, and altered disease dynamics on livestock health and productivity. Innovative solutions, including improved breeding practices, optimized feed formulations, and climate-controlled housing, are critical for enhancing livestock resilience and minimizing economic losses.

2.3 A Compendium on Climate Smart Agricultural Practices

Climate change presents an ominous global predicament, wielding potent threats to food security, rural livelihoods, and ecological integrity. In light of these challenges, the concept of Climate-Smart Agriculture (CSA) is an intricate and innovative approach to combatting the far-reaching impacts of climate change. Climate-Smart Agriculture (CSA) represents an encompassing strategy that holistically fosters the sustainable augmentation of productivity, fortifies resilience to climate change impacts, mitigates greenhouse gas emissions whenever feasible, and propels the attainment of food security and developmental objectives (FAO 2013). By empowering agricultural practitioners with effective tools and knowledge, CSA empowers them to adeptly address the challenges posed by climate change (Lipper et al. 2014; Maitra et al. 2023). This approach seeks to elevate productivity and livelihoods in a sustainable manner that safeguards against deforestation, while simultaneously adapting to climate change and endeavouring to minimize greenhouse gas emissions wherever possible. Climate-Smart Agriculture (CSA) assumes a pivotal role in effectively addressing the impacts of climate change (CC) and variability (CCV), notably by enhancing resilience during prolonged and intensified dry seasons while simultaneously mitigating pressures on natural forests (Nyasimi et al. 2017). An illustrative example lies in the integration of trees within agroforestry systems, which bolster the ability of seasonal crops to withstand drought, thereby safeguarding farmers' food security by averting complete crop failure.

2.4 An Overview of Pulse Production Technologies for Enhancing Productivity

Pulses are also an important component of Indian agricultural economy next to food grains and oilseeds in terms of acreage, production, and economic value (Choudhary et al. 2009). Pulses are a cheap, high-quality source of vegetable protein and dietary fibre. They are also gluten-free, low in fat, cholesterol, and glycaemic, as well as low in fat, cholesterol, and glycaemic index. Apart from supplying essential amino acids and proteins, they also maintain soil fertility by fixing nitrogen biologically in their nodules through a symbiotic relationship with Rhizobium bacteria. The projected food legume requirement of the country by the year 2030 is 32 million tonnes demanding an annual growth rate of 4.2% (IPGA 2018). India makes 1.5% of pulses production in the world, especially occupies third position in chickpea production and fourth in lentil production (Uysal and Subaþý 2016). The important pulse crops cultivated in India are chickpea, pigeon pea, mung bean, urd bean, lentil, and field pea (Ali and Gupta 2012).

The productivity of pulses in India is on the much lower side than in other pulse-producer countries of the globe. India has to depend on foreign countries to some extent for the import of pulses to meet demand in the country. This is not always only because of the inefficiency of the local pulse producers but also lack of dissemination of proper production technologies to the farmers. Unavailability of certified quality seeds of high-yielding varieties, biotic and abiotic stresses, improper sowing time, injudicious use of fertilizers and micronutrients, high labour requirement at several critical stages of life cycle, lack of small processing units or milling facility, and lack in marketing facilities are some major constraints related to pulse crop production. To overcome these, the following production technologies have been identified which are mentioned below (Pooniya et al. 2015):

- Selection of suitable varieties/cultivars of legumes
- Proper tillage management of pulses
- Maintenance of optimum planting geometry
- Optimum sowing time for legumes
- Optimization of proper seed rate and spacing
- Ensuring crop-specific nutrient management
- Adequate water management in pulses
- Proper implementation of plant protection measures
- Area expansion for pulses in non-traditional areas and rice fallow lands

- Proper analysis of technological gap and ensuring attractive market price by the Govt
- Initiatives regarding post-harvest management and storage facilities of pulse seeds

2.5 Climate Resilient Agriculture and Pulse Production

2.5.1 Benefits of Pulse Crops: Need of the Hour

Pulses can be broadly indicated as low-water requirement crop. They require relatively lower tillage operations sustaining soil health. They are suitable for intercropping and relay cropping doubling farmers' income. Pulse crops are capable of fixing atmospheric nitrogen through symbiosis developed with Rhizobium. Faba bean, chickpea, lentil, and field pea biological nitrogen fixation is relatively higher, ranging from 69 to 106 kg N ha^{-1}. Field pea is well known for its steady symbiosis with rhizobium, indicating stable nitrogen fixation (Hossain et al. 2016). Incorporation of pulse crop in monotonous cropping system enhances soil aggregation, increases organic carbon (%), and naturally revives the essential elements. In most cases, pulses being short-duration crop with low water requirement can easily escape drought through utilizing soil water storage. Besides their importance as a crop incorporated in the cropping system for maintaining soil health, they pose high nutrition value also. They are the primary source of plant-based proteins.

2.5.2 Challenges of Food Insecurity and Climate Change in Relation to Pulse Production

Pulses are wide adaptive crop that can thrive well in arid to semi-arid region but are highly sensitive to sudden weather aberrations. In the Indian subcontinent, pulses serve as major protein source for the majority of the population. Major climatic constraints related to pulse production should be addressed properly. Cold and heat stress situations can adversely affect the chickpea, lentil, green gram, and cow pea yield (Khattak et al. 2009; Ahmed et al. 1992; Sita et al. 2017, 2018; Patel and Hall 1990). Deformation of reproductive organ, delayed flowering, and excess vegetative growth are one of the prominent symptoms of temperature stress. Moisture stress can negatively impact pulse production (Navkiran et al. 2014; Bisen et al. 1980; Pandey et al. 1984). Under the context of climate change, the occurrence of frequent thunderstorm (Nor 'wester) in the winter season significantly hampers pulse productivity. To mitigate such problems wide adaptive pulse varieties are in demand. Incorporation of weather impact assessment during varietal evaluation should be maintained strictly.

2.5.3 *Approaches to Combat Climate Change Impacts on Pulse Production*

In response to the grave predicament, an array of approaches has been devised to effectively combat the deleterious consequences of climate change on crop production. These multifaceted strategies encompass an expansive spectrum, ranging from pioneering technological innovations to well-orchestrated policy interventions and sustainable farming practices. By adopting a comprehensive and diverse suite of methodologies, we not only endeavour to ameliorate the adverse effects of climate change, but also fortify pulse production with resilience.

2.5.3.1 Conservation Tillage

Conservation tillage is primarily aimed at minimizing tillage operations to preserve soil structure and sustain ground cover through the application of mulch, such as stubble. These practices effectively mitigate water runoff and enhance soil infiltration. Conservation tillage is commonly practised in arid land systems. Similarly, the adoption of conservation practices like zero or minimum tillage practices with mulching exerts a discernible positive impact on the productivity of pulses, particularly chickpea, in the Middle Indo Gangetic Plains, especially during seasons characterized by low post-rainy season precipitation (Mishra et al. 2012). Varieties exhibit differential responses under conditions of water adequacy or deficit. For instance, among the chickpea genotypes, the small-seeded and semi-spreading 'JG 16' genotype demonstrated superior productivity and water-use efficiency under conditions of sufficient rabi rainfall, whereas 'KWR 108' appeared to exhibit enhanced performance under situations of limited rainfall (Mishra et al. 2012). Thus, judicious selection of pulse (and other crop) varieties becomes valuable when considering appropriate land configuration for planting based on the climatic characteristics of the specific location (Singh et al. 2015). In the prevailing agro-ecosystem of the Indo-Gangetic Plains (IGP), where the discernible impacts of extreme weather events, particularly rainfall, are more pronounced, raised bed planting could present a viable alternative to other land configurations for a remunerative pigeon pea-wheat system (Singh et al. 2015). Similarly, in certain soils, deep tillage has proven highly beneficial in ameliorating soil moisture storage, particularly in the presence of compacted soils or hardpans.

2.5.3.2 Maintenance of Adequate Soil Organic Matter

In the face of shifting climate conditions, the soil organic carbon (SOC) is facing significant challenges. Employing innovative agricultural methods and embracing suggested management practices hold immense promise in effectively capturing carbon within agricultural soils. In low-input agroecosystems, soil organic carbon

(SOC) offers significant benefits, such as nutrient retention and storage, increased buffering capacity, improved soil aggregation, enhanced moisture retention, and increased cation-exchange capacity. These advantages are particularly valuable for pulse production, which typically occurs in rainfed regions. Maintaining optimal levels of organic matter in soils has a positive impact on water and nutrient retention, making it highly beneficial for pulse cultivation. Additionally, incorporating and preserving SOC contributes to enhanced soil structure, texture, and tilth, while activating a diverse range of beneficial microorganisms (Bagrecha et al. 2024). It is evident that SOC plays a vital role in maintaining soil health. Research has demonstrated that integrating pulses into maize-based systems and implementing organic nutrient management practices result in higher levels of organic carbon sequestration and improved soil health in the Inceptisols of the Indo-Gangetic plains of India (Venkatesh et al. 2013).

2.5.3.3 Reduction of Greenhouse Gas Emissions

Numerous reports have underscored the potential of organic farming in mitigating the release of greenhouse gases (GHGs). Within the realm of pulse production, the adoption of organic methods holds the promise of augmenting soil organic matter levels through the incorporation of composted organic manures and the cultivation of cover crops (Suri et al. 2012). Consequently, these systems diminish reliance on synthetic fertilizers, thereby reducing gas emissions associated with their production and transportation. By incorporating pulses into crop rotation, the necessity for fertilizer inputs is diminished, as pulses possess the capability to self-supply nitrogen and contribute nitrogen to subsequent crops (Lemke et al. 2007). This not only aids in lowering GHG emissions but also presents an opportunity to synergistically integrate diverse elements of organic agriculture with other sustainable farming systems, such as conservation tillage, to yield supplementary climate change mitigation benefits.

2.5.3.4 Application of Biofertilizers

The utilization of specific biofertilizers, such as arbuscular mycorrhizal (AM) fungi, has demonstrated significant advancements in water-use efficiency (11–24%) during rainfed pea cultivation (Kumar et al. 2016). This technology, in addition to enhancing overall nutrient utilization, particularly phosphorus, is characterized by its simplicity, convenience, affordability, and environmental compatibility. The AM fungi achieve this by extending the root system into the soil through branching hyphae, thereby expanding the exploration area for extracting water from deeper layers. However, there is an urgent necessity for further research in this realm and the generation of an exhaustive database. Similarly, leguminous crops require reduced external nitrogen inputs as a substantial portion of their nitrogen requirement is fulfilled through biological nitrogen fixation. Hence, it is vital to encourage

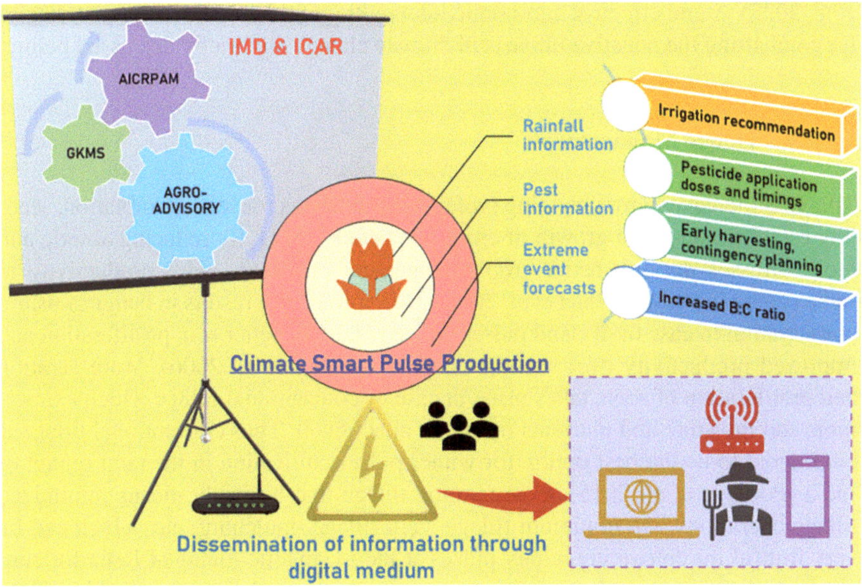

Fig. 2.1 Schematic diagram on climate smart pulse production

farmers to adopt agronomic practices that facilitate nitrogen fixation, specifically through seed treatment with Rhizobium strains tailored to the specific crop (Fig. 2.1).

2.5.4 Agronomic Interventions and Crop Husbandry Strategies in Mitigating Climate Change Impacts in Pulses

Most of the pulse crops in India are usually grown in low fertile soils and stressful environmental conditions. As a crop of dryland farming in the Indian subcontinent, pulses are mainly grown in marginal areas with low rainfall without irrigation or as rotation crops after cereal crops grown in the kharif season to replenish soil moisture (Bahl 2015; Jena et al. 2022). The degraded yield scenario for pulse crops grown in rainfed areas of peninsular India, on the other hand, is frequently caused by high temperatures during the crop growth period, resulting in drought, especially if the growing period of long-duration pulse crops is likely to be reduced under these unfavourable conditions. So, under such risk-prone environmental situations, the emerging climatic changes are likely to create an impact on decreasing pulse production which is a severe threat for sustainability maintenance in the Indo-Gangetic Plains. There is an urgent need to disseminate newest technologies among pulse growers to mitigate the growing threat and to sustain the potential of crop production under unfavourable and unpredicted situations (Bahl 2015; Hossain et al. 2022). Some effective agronomic interventions and crop husbandry strategies for combatting the negative impacts of climate change are briefly discussed below:

2.5.4.1 Tillage Management

Tillage is ineluctable for obtaining preferable conditions for seed germination, seedling establishment, and growth of crops (Das et al. 2014). By reducing abiotic and biotic stresses through better crop management practices, pulse productivity in India can be increased (Rana et al. 2016). Deep ploughing results in better moisture conservation in case of dryland pulses which results in better root proliferation and improved productivity over shallow cultivation (Vadi et al. 2006). More aeration and proliferation of roots takes place in case of conventional tillage which extracts more soil moisture and nutrients from unit area of soil. Thus conventional tillage is considered to be the best option for winter pulses cultivation in the tarai region of India (Pooniya et al. 2015). Conservation tillage that basically means minimized tillage operations like minimum tillage, zero tillage, mulching, etc. which can be very fruitful for cultivating winter pulses at Indo-Gangetic Plains of India for conserving soil structure and maintaining ground cover by mulch, such as stubble (Bera 2021; Islam et al. 2023). These practices increase water infiltration rate by reducing runoff. In the same way that rice can be successfully grown under no tillage, chickpeas can be planted in a sandy loam soil using no tillage or ploughing followed by planking. When the post-rainy season precipitation is low, it is possible to achieve yields similar to those obtained when deep ploughing (40 cm) with a disc plough (Mishra et al. 2012). Thus ensuring proper tillage management under climate-changing scenario can be one of the beneficial option for enhancing pulse productivity.

2.5.4.2 Adoption of Diversified Practices

Crop diversification is an effective approach for risk reduction in farming under changing climatic scenario. The practice of mixed cropping or intercropping is a successful method of crop variegation that involves growing multiple crops in a variety of configurations (Basu et al. 2016). Alternate land use systems like agri-horticultural system, sylvo-pastoral system, and alley cropping can also stabilize pulse production as these systems can withstand climatic extremes by proper nourishment of soil (FAO 2016). Varietal diversification of pulses which may be based on phenological, genetical, or flowering characteristics can also be helpful in a better probability. Crop damage can be minimized when the crop is sown to several varieties of different flowering dates (Basu et al. 2016). Planting in situ may be prohibited by higher rainfall intensities forecast during cropping season, but contingency planning could compensate for this loss in productivity (Praharaj et al. 2015).

2.5.4.3 Development and Popularization of Climate-Specific Crop Varieties

Various kinds of improved varieties of different pulse crops are already available in the market having different characteristic features like higher yield, shorter duration, synchronous maturity, short stature, disease-pest resistance, etc., suitable for variable agro-climatic conditions. Development and popularization of such special trait-possessing cultivars can replace less productive and problematic varieties based on climatic scenario of particular regions (Gudadhe et al. 2018). For example, many pigeon pea varieties like 'TGT 501', 'Bennur Local', 'JKM 7', 'BDN 2008–12', 'MAL 13', 'Bahar', 'GRG 815', 'BSMR 853', etc. have been identified having higher osmotic adjustment under drought (Basu et al. 2016). An Indian Council of Agriculture Research and International Crops Research Institute for the Semi-Arid Tropics (ICRISAT) joint venture has developed a new variety of hybrid pigeon pea, offering a revolutionary pulse opportunity (Saxena and Nadarajan 2010). There seems to be an urgent need regarding revisit of gene banks of pulse crops in search of distinctive traits required for climate change.

2.5.4.4 Ensuring Optimum Planting Time and Sowing Depth

Non-monetary inputs such as planting time are crucial to crop growth, phenological development, insect-pest dynamics, and weed dynamics. Shifting of the monsoon cycle (Sahana et al. 2015) and increasing post-monsoon rainfall (Krishnakumar et al. 2009) have made winter agriculture vulnerable across all over India. To buy the sowing time, short/medium duration rice varieties need to be cultivated during *kharif* season because short/medium duration rice will evacuate the land by the end of October to early November. Switching to such varieties eases the incorporation of winter legumes in the cropping sequence after *kharif* rice by allowing timely sowing. Thus, terminal heat and drought stress as well as productivity issues can be minimized (Pooniya et al. 2015). Sowing of summer mung bean in the first fortnight of March recorded a higher yield as compared to the last week of March (Patel et al. 2003). Type of crop/cultivar, growing season, soil moisture, soil texture, and more importantly seed size of the respective pulse crops are the key factors to be considered for optimum sowing depth (Dass et al. 1997).

2.5.4.5 Opting Proper Method of Sowing

Among different agronomic practices, the proper method of sowing is considered to be one of the most efficient to improve pulse productivity. An ideal canopy development and planting geometry can be helpful in better utilization of available resources (Basu et al. 2016). Pulse crops are generally sown in flat beds using the broadcasting technique for sowing. Raised/ridge-furrow planting techniques have been found to be very effective at draining excess water from crop root zones and increasing yields by 25–30% over flat beds in kharif pulses (Das et al. 2014). Raised bed planting can provide higher grain yields using lesser amounts of irrigation water for pigeon pea and mungbean (Dhindwal et al. 2006).

2.5.4.6 Optimum Seed Rate and Plant Population

Seed rate in pulse crop generally depends on climatic factors, purity of seed material as well as crop growth duration. Primarily, plant population desired per unit of land area determines the seed/seedling rate (Prasad 2012). Optimum seed rate for higher yield of green gram is 37.5 kg/ha during summer and 30 kg/ha during the rainy season in a study conducted in Punjab (Sekhon et al. 2004). Appropriate planting geometry in pulses ensures optimum plant population which ultimately helps in harnessing better solar radiation that is converted into higher yield. Moreover, maintaining proper plant population reduces the competition between the plants for moisture and nutrients. There is a preference for narrow spacing for short-duration pulse varieties, but a preference for wider spacing for long-duration varieties (Pooniya et al. 2015). Duc to prolonged vegetative growth, warm temperatures, and copious branching, kharif-sown crops generally require wider spacing and fewer plants than summer-sown crops (Prasad 2012). Therefore, farmers must maintain a proper spacing of 20–30 cm × 10–15 cm for pulses in order to obtain a higher yield and become self-sufficient in production.

2.5.4.7 Balanced Fertilization

As a result of fertilizer management, the correct amount of nutrients are added at the right time through an appropriate method in order to maximize crop productivity and soil fertility while minimizing nutrient losses (Dass et al. 2014). Pulse crops have the capability to fix biological N_2 which can help to meet 70–80% of their N_2 requirement. So, the application of a small dose of 15–25 kg/ha of N_2 along with 20–60 kg/ha P_2O_5 can be enough to fulfil the demand of pulse crops based on climatic preference (Basu et al. 2016). In pulse-growing regions of different states, there is a widespread deficiency of certain micronutrients like zinc and secondary nutrients, especially sulphur. The use of nutrient fertilizers is necessary in eastern and northeastern states with acidic soils (DAC and GoI 2012). The inclusion of different micronutrients like Zn, B, Mo, and Fe has resulted in significant improvement in pulse productivity across different regions. As a result, balanced nutrition promotes better nutrient-use efficiency and reduces the use of synthetic fertilizers, resulting in lower global warming emissions.

2.5.4.8 Improved Water Management Practices

Due to the rapid increase in population, huge intensification of agricultural activities generally increases the requirement of groundwater day by day. Thus, not only water-saving planning is needed but also innovative irrigation practices need to be adopted. Therefore, water-saving irrigation methods can increase yields, increase water and nutrient-use efficiency, and improve the area covered by irrigation by scientifically scheduling irrigation, estimating the amount of water to apply, and deploying water-saving irrigation techniques (DAC and GoI 2012). In comparison with traditional farmers' practices, furrow-irrigated seeding in northwest India yielded higher yields and resulted in 16–20% water savings (Lumpkin and Sayre 2009). Adoption of sprinkler irrigation can also be very beneficial in saving irrigation water under changing climatic scenario. It may be beneficial to increase pulse productivity by micro-irrigating at critical stages, typically at flowering. Higher water-use efficiency and in turn water economy can be ensured by micro-irrigation techniques (Kumar et al. 2014a, b). Under water scarcity conditions, supply of life-saving irrigation can improve crop yield under critical stages like pre-flowering pod formation. Furthermore, in rainfed areas, the study of antitranspirants can be a new area of research in increasing the water use efficiency of pulse crops (Pooniya et al. 2015).

2.5.4.9 Introduction of New Cropping System and Intercropping Pattern

New options regarding new cropping system introduction as per the climatic condition of a particular zone can enhance the productivity of legumes. Introduction of summer season pulses like green gram, blackgram, and cowpea in irrigated areas after harvesting of *rabi* season crops may be beneficial for the particular region. Replacement of upland crops like rice, sorghum, maize, pearl-millet, and introduction of short-duration pulses in those regions of eastern and southern states can be a new option (Pooniya et al. 2015). It is also possible to increase productivity while increasing income by intercropping pigeon peas, green gram, and black gram with sorghum, pearl millet, maize, cotton, groundnut, soybean, etc. In the Indian subcontinent, chickpea + mustard is an advantageous intercropping system (Arya et al. 2007). Gudadhe et al. (2015) recommended that cotton-chickpea can be a good sequence for ensuring higher return to farmers.

2.5.4.10 Improved Crop Management Practices

Agronomic practices play a crucial role in determining the productivity of pulses and can be influenced by factors such as tillage, sowing time, planting method, ridge-planting during the kharif/rainy season, crop geometry, plant population, nutrient and water management, seed treatment, weed management, and plant protection. Zonal Agricultural Research Stations continue to conduct various research trials and develop crop-specific recommendations for diversified agro-climatic conditions. It is important to disseminate these recommendations to farmers for adoption. Implementing crop-specific agronomic practices holds immense potential for increasing pulse productivity, especially in water-stressed regions facing changing climatic conditions. For instance, a common practice involves placing fertilizers in furrows below the seed to address limited soil moisture and ensure nutrient availability to plants. Furthermore, the balanced use of organic and inorganic fertilizers improves soil moisture retention and enhances drought tolerance. On the other hand, conservation agriculture systems offer promising solutions to mitigate the challenges posed by climate change. A study on rabi crops, such as lentil, field pea, faba bean, Lathyrus, and chickpea, comparing zero-tillage and conventional tillage after rice harvest, revealed comparable performance for all pulses under both tillage practices. Similarly, chickpea performance in zero-tillage system after rice showed no statistically significant difference compared to conventional tillage, although retaining rice residue on the surface proved advantageous in the zero-tillage system (Kumar et al. 2012).

2.5.5 Breeding Strategies and Genetic Improvement in Developing New Varieties and Mitigation of Abiotic and Biotic Stress in Pulses

Crop output must increase to fulfil the increasing global need for food. However, due to the changing environmental conditions, this effort is challenging. As a result of climate change, Earth will become warmer and drier (FAO 2019). Over the past ten years, the global agriculture industry has incurred over US $29 billion in losses due to drought (Varshney et al. 2021). Freshwater availability is expected to have declined by up to 50% by 2050 as a result of intensifying climate change, perhaps resulting in a two-fold increase in water demand for agriculture. For the sake of food security, investments in this sector are urgently required, particularly for the development of high-yielding crops with resiliency to changing climatic conditions (Trnka et al. 2019). The development of drought-tolerant crop types has been facilitated by significant improvements in traditional breeding programs (Kumar et al. 2014a; Noaman et al. 2007). However, this labour-intensive process usually takes many years to complete the early stages of analysing phenotypes and genotypes. Conservative breeding ignores the genetic diversity of adaptive traits that support yield, running the risk of selecting only the alleles that are favourable in all studied situations (Tardieu et al. 2018). The alleles chosen in this manner are uncommon compared to alleles whose effects rely on the environment (Millet et al. 2016). On the other hand, in some environmental circumstances, context-dependent trait optimization may maximize the positive effects on yield. Realizing their importance, a great deal of research has gone into enhancing their genetics, turning them from so-called 'orphan legumes' into widespread crops. Using conventional plant breeding methods, more than 3800 enhanced varieties of various pulse crops have been developed globally (Pratap et al. 2022). These types produce more grains and have longer, more resilient crops that also provide superior nutrition. However, in contrast to the dramatic increases in grain production between 1961 and 2017, the average pulse yields have only slightly improved (Kumar et al. 2020a, b). Productivity benefits in pulses have been noticed when especially considered along with the noticeably shorter duration of the enhanced kinds, which results in increased cropping intensity. Additionally, features that give resistance to substantial biotic and abiotic stressors, herbicide tolerance, larger seeds, and enhanced nutritional quality have all been linked to genetic improvements. As a result, production and productivity increased in the major countries that produce pulses. For instance, India has seen the greatest increase in mung bean production over the past 15 years, followed by chickpea (125%), urdbean (90%), pigeon pea (51%), and lentil (34%). Notably, breeding in the majority of pulses has continued to be limited to the use of genetic variation present within the primary gene pool, resulting in a limited range of cultivars (Pratap et al. 2022). The second-largest pulse crop globally is the chickpea (*Cicer arietinum* L.), which is raised by smallholder growers in more than 50 nations. In recent years, the chickpea has seen the development of the genome sequence information (Varshney et al. 2019), millions of single nucleotide polymorphism (SNP) polymorphisms (Thudi et al. 2016; Varshney et al. 2019), and affordable genotyping platforms like low- to high-density SNP arrays (Roorkiwal et al. 2020). Additionally, QTLs and markers connected to biotic and abiotic stresses were discovered, enabling the marker-assisted breeding of improved cultivars. Novel genetic loci were identified associated with chickpea phosphorus acquisition and utilization efficiency as well as root morphological characteristics using genome-wide association mapping (Thudi et al. 2016). The MutMap technique is based on BSA analysis of mutant progenies acquired in F_2 population and aims to identify mutant genes for evaluating gene function (Etherington et al. 2014). In doing so, it directly targets the causal SNPs produced during mutagenesis that are in charge of phenotypic behaviour (Tribhuvan et al. 2018). It depends on the cross between the mutant and its wild type. Using the MutMap method, markers for early blooming and improved seed size in chickpea were developed (Manchikatla et al. 2021). According to numerous circumstances, the breeding cycle typically lasts 7–10 years for the production of a new cultivar (Kumar et al. 2020b). By increasing the number of plant generations per year, speed breeding can enhance genetic benefits in agricultural improvement programs (Chiurugwi et al. 2019). Fang et al. (2021) showed how to breed soybeans at a lower cost using speed breeding by combining a fresh-seeding technique, an off-site nursery, and marker-assisted selection.

In comparison to one generation produced using conventional procedures, they could produce at least four using the aforesaid combination in a year. In an Australian cultivar of desi chickpea, Croser et al. (2021) showed the efficiency of cooperative breeding efforts for the deployment of novel breeding technologies for producing and quickly introducing imidazolinone group 2 herbicide tolerance. Although minor pulses are locally produced and consumed, there is typically no material interchange between countries unless farmers are already producing and consuming these small pulses (Ahlawat et al. 2016). The faba bean (*Vicia faba* L.), a small pulse, is the fourth most significant cool-season legume eaten regionally in Asia, America, and Mediterranean nations. The lack of genomic information for this crop is mostly caused by the inherent challenges of assembling and annotating its large 13 Gb genome (Pratap et al. 2022). There are closely connected QTLs available for both biotic and abiotic stress tolerance that might be exploited to boost the effectiveness of the selection procedure. The challenges and opportunities of increasing cowpea yield to guarantee food and nutritional security, environmental sustainability, and sustainability were examined (Omomowo and Babalola 2021). They paid particular attention to Africa and spoke in detail about suggestions for achieving extremely high productivity increases. The demand for plant-based proteins has recently dramatically expanded, even in those countries that have not historically consumed many pulses. This necessitates a concentration on techniques that will vastly increase the output of pulses and enhance their nutritious value (Pratap et al. 2022). It is also evident that small legumes of strategic importance continue to lag behind other pulses and require research funding for the generation of fresh genomic data and the application of molecular technologies. Extensive research is required to determine the specific genetic advantages of pulses with regard to yield and nutrition features in order to develop approaches for focused breeding (Pratap et al. 2022).

2.5.6 New Options in Pulse-Based Intercropping and Cropping System Across the Globe

Because pulses provide nutrient-rich, energizing food, feed, and forage, they are a crucial component of an effective agroecosystem (Singh et al. 2009). In a system-based approach, crops and cropping patterns are selected for ecological services, economic viability, energy-rich diverse nutritional food, and climate resilience. This category of food crops can be incorporated into rainfed agroecosystems due to its innate hardiness, minimal input requirements, and exceptional ability to thrive successfully under current challenging climatic conditions (Kumar et al. 2023). In a variety of agroecological settings, pulses can be cultivated as catch crops, ratoon crops, and green manuring crops under sequential cropping or mono-cropping. Pulses are included in the intercropping system as a mixed intercrop, a row intercrop (such as pearl millet and mung bean), and a relay intercrop (such as mung bean). Relay cropping (paira/utera) is used to better utilize the remaining soil moisture in humid north-eastern India and dry central and coastal southern India. In relay cropping, rice and pulses like mung bean, urdbean and lentil work well since they allow for double cropping and system sustainability. Crop diversification and the system intensification (SI) of already-existing non-pulse crops were made possible in northern parts of India due to the introduction of early-maturing mung bean, pigeon pea, and urd bean varieties. Fodder pulses like French bean, mung bean, urd bean, and rice bean not only serve as good nutritious food on difficult slopes, but they also make excellent cover crops (Kumar et al. 2023).

Numerous conventional rainfed agroecosystems have either been modified or swapped by numerous new pulse-based agroecosystems in response to concerns about sustainability and marketing opportunities in rainfed areas. The current state of resource degradation in agriculture and rising nutrient demand increases the standing of pulses in agroecosystem. Therefore, pulses must be a part of the primary rainfed agricultural systems. However, there isn't much room for expansion of the land devoted to pulses given the severe competition they face from cereals and cash crops. However, four to six million hectares can readily be devoted to pulses by intercropping them with additional crops (Praharaj and Blaise 2016). This approach is also consistent with SI and other cutting-edge systems. In particular, dryland or rainfed farming, short growing seasons, summer farming, short monsoon period, marginal and/or degraded soils, and underdeveloped ecoregions are factors that contribute to the importance of pulses for food and nutritional security. Pulses were grown in 150 nations worldwide in 1961, and there were 171 in 2019 (FAOSTAT 2022). Pulses are planted throughout South Asia as inter-/mixed crops, sequential crops, monocrops, and relay crops to meet the regional needs for nourishing food and income (Adarsh et al. 2019). The customary summer season is being changed in some areas, including the Indo-Gangetic plains of India and the Great Plains of North America. Since the 1980s, pulses have grown in importance as a cash crop in Australia. During the summer, faba beans are cultivated from 20 to 40 S, while during the winter, chickpeas are grown from 10 to 40 S (south) (Lal 2016). Pulses have been grown on the African continent since antiquity and will stay to be a part of their rainfed agricultural schemes, which are vulnerable to water scarcity. By minimizing continuous submersion and breaking the chain of insects and pests, pulses under cereal-based cropping systems play a crucial role in increasing system productivity, soil physical qualities, and soil fertility status (Kumar et al. 2023). By fixing atmospheric nitrogen and providing fixed nitrogen to the subsequent grain, pulses in rotation contribute to the cropping system sustainability. This aids in foraging mineral N, encouraging nitrogen cycling deep in the soil, aiding weed suppression, improving mycorrhizal colonization, and enhancing SOC stock (Jakhar et al. 2017). Such crop category helps to create jobs, provide nutritious, well-balanced food, and promote the socioeconomic circumstances of marginal farmers (Adarsh et al. 2019). Component crops contest for remaining moisture, nutrients, soil, and light in sequential cropping. Because of their lingering effect, pulses help subsequent non-legume crops to utilize less nitrogen. Depending on the crop and the environment, pulse residues range from 20 to 80 kg N per ha and account for close to 70% of biologically fixed N (Giller 2001). According to Kaur et al. (2018), with sequential cropping, the previous pulse crop contributes 18–70 kg N ha^{-1} to the soil on average. Following non-legume crops can still use this N, which lowers their need for fertilizer by up to 25–30% (Garg 2009). For example, producing cowpea and mung beans before pearl millet in the previous season provides the pearl millet crop with about 60 kg N (Ghosh et al. 2007). In contrast, pulses donate between 20% and 40% of the N needed by wheat in the pulse-wheat sequential cropping. Due to the larger variability in N fixation, it is difficult to estimate the quantity of fixed N transported to the succeeding cereal crop. The biologically fixed N also improves fertilizer utilization, which in turn lowers the negative environmental effects of chemical fertilization. This is accomplished by reducing chemical N inputs into the cycle (Kumar et al. 2023). Pulses in the cropping system help to lessen bulk density while enhancing physical soil conditions including aggregate stability and soil structure. Due to the existence of fungus, aggregate stability is a significant indicator of long-term improvement of soil quality in a pulse-based cropping system. To create a stable soil aggregate and further enhance soil structure, the fungi release a glycoprotein named glomalin that traps solid-oxide-oxygen-ion conducting membrane (SOM), plant detritus, and soil minerals (Singh et al. 2009). Additionally, soil bulk density responds well to management modifications in the long-term pulse-based cropping scheme (Wang et al. 2020). Legumes also stimulate earthworm activity, which along with the root channels increases soil porosity, encourages aeration, upsurges water-holding capacity, and permits water to penetrate deeper into the subsoil layers (Kumar et al. 2020a, b). Groundwater consumption is significantly impacted by pulses. In the rice-wheat system, for example, substituting pigeon peas for lowland rice reduces the pace of groundwater aquifer depletion, boosts water use efficiency, and therefore encourages efficient water use in the era of water crises made worse by climate change. In the cereal-based system, growing pulses also affect the chemical characteristics of the soil and enhance soil health (Wang et al. 2020). Pulses minimize reliance on soil N and fix atmospheric N, which lowers soil pH (Singh et al. 2009). The elimination of fallow land by the growing of a pulse crop and the addition of additional C to the soil through the use of a pulse-inclusive system have the combined impact of improving soil C stock (Ghosh et al. 2020). Following decomposition, the biomass-N added by these leftovers becomes accessible to the succeeding cereal crops. A sizeable amount of fixed nitrogen is maintained in the roots and nodules after the crop is harvested. According to Herridge et al. (2008), this fixed N makes up around 30% of the cumulative N in lentils and peas and about 50% of the cumulative N in chickpeas. Although less biomass is produced by pulses than by grains, it is of greater quality. They also assist succeeding non-pulse crops in producing more biomass by providing N advantages that have a positive impact on soil productivity (Liu et al. 2019). Since the root exudates support to mobilize the soil soluble P, green manuring of pulse-based wastes increases the amount of soil P that is available in the agroecosystem while simultaneously preserving N.

Crop intensification, which can reduce the current restrictions, adapt well, and provide an ideal yield, is a basic principle for sustainability of rainfed agroecosystem (Lehman et al. 2015). Due to their naturally low water requirement, pulses can withstand high temperatures and drought stress. Pulses may therefore effectively use the available water through their deeper root structure, whether they are grown as a mono-crop, inter-/mixed-crop, or a subsequent crop in rotation in a rainfed agroecosystem (Liu et al. 2019). Additionally, by effectively covering the soil and lowering the likelihood of water loss due to evaporation, leaching, percolation, run-off, salinization of soil, leaching of nutrients, and soil erosion, these crops conserve scarce soil moisture. This category of crops also produces nutrient-rich biomass, which could increase SOC levels and sustain a healthy soil biological communal (Lehman et al. 2015). Pulses fit well in low-fertility rainfed areas as well because of their low nutrient requirements, capacity to use biologically fixed N for self-use, and ability to leave some residual N for the use of partner crops in inter-/mixed-crop or subsequent crop in rotation system (Sheoran et al. 2021). In these low-fertile rainfed settings, soil health can be restored, the concentration of micronutrients in the edible sections is enhanced, and the N and C footprints of the agroecosystems are decreased (Lal 2016). It is further envisioned that the cultivation of pulses in a rainfed environment could usher in the next green revolution, strengthening ecosystem amenities and enhancing production sustainability in the face of new health crises, climate change, and the depletion of natural resources (Kumar et al. 2023). As a result, it is crucial to work towards sustainable production of pulse in rainfed regions by expanding cultivable land and implementing the best management techniques in line with local soil and environmental conditions.

2.5.7 Agro-Advisory and Pulse Production

In India, agro-advisories are mainly prepared under two government schemes, one funded solely by ICAR namely All India Coordinated Research Project on Agrometeorology (AICRPAM) (Bal et al. 2023) and another one is Gramin Krishi Mausam Sewa (GKMS) which is funded by IMD in close collaboration with ICAR (Singh et al. 2023). There are 23 AICRPAM centres nationwide. GKMS comprised 130 Agromet Field Units (AMFUs) distributed among the SAUs, ICAR, IITs and 199 District Agromet Units (DAMUs) under the umbrella of Krishi Vigyan Kendras (KVKs) guided by ICAR. With this gigantic network precise observed weather is collected and forecast is given in various categories moderated with the help of satellite observations. Crop-specific advisory, contingent crop planning is issued by Agrometeorologists twice a week (every Tuesday and Friday). Farmers involved in pulse cultivation mainly need to focus only on the rainfall forecast issued. A clear indication of the probable amount of rainfall (mm) in the upcoming week is given for each block. Farmers are benefitted from this agro-advisory services as reported by Rajegowda et al. (2008), Majumder et al. (2020), Dupdal et al. (2020), and Mahato et al. (2022) across different locations.

2.6 Successful Implementation of Climate-Resilient Pulse Production Technologies Across The World

2.6.1 Summarizing the Key Findings Resulting in Success of Initiatives Related to Climate-Resilient Pulse Production Technologies

The progress achieved in pulse research is a result of dedicated efforts to discover and utilize genes associated with important traits such as early maturity, disease resistance, Orobanche tolerance, and micronutrient content found in landraces and wild relatives of crops (Sallam and Ul-Allah 2019). Traditionally, chickpea is planted in spring. However, by advancing the planting time to November and taking advantage of winter rainfall, the yield of winter chickpea can be significantly increased. Aschochyta blight, a highly destructive disease, poses a major threat to winter-grown chickpea. However, through the identification of new sources of resistance and the application of gene pyramiding techniques, breeding programs have successfully enhanced aschochyta resistance in chickpea (Kumar et al. 2019).

Crop improvement research and technology have various impacts, including increased yield, reduced production costs, technology spill-over, and improved yield stability. When assessing the effects of such research on a specific crop, factors such as the development of new varieties, traits, and practices are considered, along with their impact on yield and other relevant measures (Kumara et al. 2014). For instance, the development of photo-thermo-insensitive and short-duration varieties in major pulse crops like chickpea, pigeon pea, and green gram has created opportunities to cultivate them in new niches and seasons, potentially increasing cropping intensity. Certain short-duration pulses such as greengram, blackgram, and cowpea can be introduced into existing cropping systems without displacing the current crops (Ali and Kumar 2007).

The adoption of early maturing varieties has significantly enhanced the area and productivity of Chickpea in short-season environments in the southern states of India. In India, Chickpea cultivation occurs in diverse agro-ecological conditions, ranging from the cooler and longer growing seasons in the north (32°N) that last 140–170 days, to the warmer and shorter seasons in the south (10°N) with crop maturity in just 90–100 days (Chaturvedi et al. 2014). Early maturing varieties have proven valuable in escaping terminal drought and heat stresses, facilitating the expansion of Chickpea cultivation in the south, where the growing season is limited. Among the major pulses in northern India, Chickpea has experienced the highest decline, with a 63% reduction in area from 4.98 million ha in 1971 to 1.85 million ha in 2010, primarily due to its replacement by cereals, particularly wheat (Gowda et al. 2013).

2.6.2 Initiatives on Food Security, Climate Resilience, and Other Related Outcomes

After green revolution, malnutrition is identified as a human disaster and indicates unequal distribution of resources in the society. The National Food Security Mission (NFSM) and National Mission on Sustainable Agriculture (NMSA) are one of the two pioneer initiatives adopted by the Government of India combating climate change and uneven distribution of resources. The schemes ensure the successful production and distribution of food grains as well as legumes fulfilling the Indian population's demand. Under the context of climate change, with high rainfall variability agro-advisories play crucial role in climate-smart pulse production. Incorporation of high-yielding climate-resilient varieties under conservation tillage system can perform outstandingly under normal rainfed situation. This can fulfil our goal of sustainability without hampering our potential crop output, i.e., yield. Crop insurance policies can change the course of Indian farming through minimizing unprecedented weather and pest-disease infestation crop losses. However, more awareness and smoothness regarding insurance schemes are required for easy implementation at the root level.

2.7 Conclusion

Agriculture plays a pivotal role in the overall economic scenario and rural livelihoods of India. Due to severe climatic changes as well as many kinds of biotic and abiotic stresses, crop production is expected to suffer much more in the coming decades. Pulses being an important part of food nutrition, there is an urgent need to match improved pulse production to the increasing population through intensification of sustainable crop management strategies. Pulses are actually grown in resource-poor and harsh conditions with limited facilities in our country. Pulses are among the neglected crops since Green Revolution in India. As a result, the productivity of the pulses in India is very low and we have to import legumes from foreign countries to meet our demand.

We need to reform our future strategies by adopting different technological interventions to popularize food legumes in traditional and non-traditional areas where pulse cultivation can be a beneficial option for the poor farming community. Pulse research in India led by IIPR, Kanpur with support from other International Agricultural Institutes like ICARDA is creating a lot of opportunities for smallholder farming community. Govt. of India is also taking several initiatives and policies for remunerative and sustainable pulse production throughout the country. The improved production technologies and strategies for legumes discussed here will definitely enlighten agricultural professionals and policymakers to enhance their capabilities.

References

Adarsh S, Jacob J, Giffy T (2019) Role of pulses in cropping systems: a review. Agric Rev 40(3):185–191

Ahlawat IPS, Sharma P, Singh U (2016) Production, demand and import of pulses in India. Indian J Agron 61(4):33–41

Ahmed FE, Hall AE, DeMason DA (1992) Heat injury during floral development in cowpea (Vigna unguiculata, Fabaceae). Am J Bot 79(7):784–791

Ali M, Gupta S (2012) Carrying capacity of Indian agriculture: pulse crops. Curr Sci 102(6):874–881

Ali M, Kumar S (2007) Promising technologies for improving pulse production. Indian Farm 57(4):15–18

Arya RL, Varshney JG, Kumar L (2007) Effect of integrated nutrient application in chickpea + mustard intercropping system in the semi-arid tropics of North India. Commun Soil Sci Plant Anal 38:229–240

Bagrecha S, Nithinkumar K, Saikia N, Meena RS, Singh SV (2024) Microbial dynamics and carbon stability under biochar-amended soils. In: Biochar production for green economy. Academic Press, Elsevier, London, pp 285–309

Bahl PN (2015) Climate change and pulses: approaches to combat its impact. Agric Res. https://doi.org/10.1007/s40003-015-0163-9

Bal SK, Chandran MS, Rao AS, Manikandan N, Rao BR (2023) Coordinated research on agrometeorology: India perspective. J Agrometeorol 25(2):215–223

Basu PS, Singh U, Kumar A, Praharaj CS, Shivran RK (2016) Climate change and its mitigation strategies in pulses production. Indian J Agron 61(4th IAC Special issue):S71–S82

Bera A (2021) Impact of climate change on pulse production and its mitigation strategies. Asian J Adv Agric Res 15(2):14–28

Bhadra P, Maitra S, Shankar T, Hossain A, Praharaj S, Aftab T (2022) Climate change impact on plants: plant responses and adaptations. In: Roychoudhury A, Aftab T (eds) Plant perspectives to global climate changes. Academic Press, London, pp 1–24

Bisen CR, Tomar SS, Shivamurthy RB, Kashyap ML (1980) Response of lentil (*Lens esculenta*) to rhizobium inoculation and fertilization under different moisture regimes. Mysore J Agric Sci 14(4):483–486

Bogale GA, Erena ZB (2022) Drought vulnerability and impacts of climate change on livestock production and productivity in different agro-ecological zones of Ethiopia. J Appl Anim Res 50(1):471–489

Chaturvedi SK, Mishra N, Gaur PM (2014) An overview of chickpea breeding programs in India. The timeless chickpea Big hearts pulse within pods perspectives: the timeless chickpea big hearts pulse within pods. J Int Legume Soc 3:50–52

Chiurugwi T, Kemp S, Powell W, Hickey LT (2019) Speed breeding orphan crops. Theor Appl Genet 132:607–616

Choudhary AK, Yadav DS, Singh A (2009) Technological and extension yield gaps in oilseeds in Mandi district of Himachal Pradesh. Indian J Soil Conserv 37(3):224–229

Croser J, Mao D, Dron N, Michelmore S, McMurray L, Preston C, Bruce D, Ogbonnaya FC, Ribalta FM, Hayes J, Lichtenzveig J (2021) Evidence for the application of emerging technologies to accelerate crop improvement–a collaborative pipeline to introgress herbicide tolerance into chickpea. Front Plant Sci 12:2839

DAC, GoI (2012) Report of expert group on pulses. Department of Agriculture and Cooperation, Government of India, Ministry of Agriculture, New Delhi, pp 1–148

Das TK, Choudhary AK, Sepat S, Vyas AK, Das A, Bana RS, Pooniya V (2014) Conservation agriculture: a sustainable alternative to enhance agricultural productivity and resources-use efficiency. Technical Extension Folder, Indian Agricultural Research Institute, New Delhi

Dass A, Kharwara PC, Rana SS (1997) Response of gram varieties to sowing dates and phosphorus level under on-farm conditions. Himachal J Agric Res 23(1 and 2):112–115

Dass A, Suri VK, Choudhary AK (2014) Site-specific nutrient management approaches for enhanced nutrient-use efficiency in agricultural crops. Res Rev J Crop Sci Technol 3(3):1–6
Dhindwal AS, Hooda IS, Malik RK, Kumar S (2006) Water productivity of furrow-irrigated rainy season pulses planted on raised beds. Indian J Agron 51(1):49–53
Dubey A, Malla MA, Khan F, Chowdhary K, Yadav S, Kumar A et al (2019) Soil microbiome: a key player for conservation of soil health under changing climate. Biodivers Conserv 28:2405–2429
Dupdal R, Manjunatha BL, Dhakar R, Patil SL (2020) Perception and economic impact of agromet advisory services: a case study of Thrissur AICRPAM Centre of Kerala State. Indian J Ext Educ 56(3):10–16
Etherington GJ, Monaghan J, Zipfel C, MacLean D (2014) Mapping mutations in plant genomes with the user-friendly web application CandiSNP. Plant Methods 10(1):1–11
Fang Y, Wang L, Sapey E, Fu S, Wu T, Zeng H, Sun X, Qian S, Khan MAA, Yuan S, Wu C (2021) Speed-breeding system in soybean: integrating off-site generation advancement, fresh seeding, and marker-assisted selection. Front Plant Sci 12:717077
FAO (2013) Climate-smart agriculture sourcebook. Food and Agriculture Organization of the United Nations (FAO), Rome. Retrieved from http://www.fao.org/publications
FAO (2016) Pulses and climate change. Food and Agriculture Organization of the United Nations. Online: http://www.fao.org/3/a-i5426e.pdf
FAO (2019) Proactive approaches to drought preparedness—where are we now and where do we go from here? Food and Agriculture Organization of The United Nations. Rome, Italy.
FAOSTAT (2022) Rome, Italy: Statistics Division, Environment Statistics Team. Available at: https://www.fao.org/faostat/en/#home. Accessed on: May 23, 2022
Garg N (2009) Symbiotic nitrogen fixation in legume nodules: process and signalling: a review. Sustain Agric 27:519–531
Ghosh PK, Bandyopadhyay KK, Wanjari RH, Manna MC, Misra AK, Mohanty M, Rao AS (2007) Legume effect for enhancing productivity and nutrient use-efficiency in major cropping systems–an Indian perspective: a review. J Sustain Agric 30(1):59–86
Ghosh PK, Hazra KK, Venkatesh MS, Praharaj CS, Kumar N, Nath CP, Singh U, Singh SS (2020) Grain legume inclusion in cereal–cereal rotation increased base crop productivity in the long run. Exp Agric 56(1):142–158
Giller KE (2001) Forage legumes in pastures and leys. In: Nitrogen fixation in tropical cropping systems. CABI, Wallingford, pp 187–211
Gowda CLL, Srinivasan S, Gaur PM, Saxena KB (2013) Enhancing the productivity and production of pulses in India. In: Shetty PK, Ayyappan S, Swaminathan MS (eds) Climate change sustainable food security. National Institute of Advanced Studies, Bangalore and Indian Council of Agricultural Research, New Delhi, pp 145–159. ISBN: 978-81-87663-76-8
Gudadhe N, Dhonde MB, Hirwe NA (2015) Effect of integrated nutrient management on soil properties under cotton-chickpea cropping sequence in vertisols of Deccan plateau of India. Indian J Agric Res 49(3):207–214
Gudadhe N, Raut AA, Thanki JD, Bodke PS, Imade SR (2018) Agronomic technologies for enhancing pulse production ahead of climate change (Chapter 2). In: Climate risks management: sustainable pulse production. Astral International Private Limited, New Delhi, pp 27–52
Gupta HS (2011) Climate change and Indian agriculture: impacts, mitigation and adaptation. X agricultural science congress on soil, plant and animal health for enhanced and sustained agricultural productivity. ICAR, New Delhi, pp 73–81
Herridge DF, Peoples MB, Boddey RM (2008) Global inputs of biological nitrogen fixation in agricultural systems. Plant Soil 311:1–18
Hossain Z, Wang X, Hamel C, Knight JD, Morrison MJ, Gan Y (2016) Biological nitrogen fixation by pulse crops on semiarid Canadian prairies. Can J Plant Sci 97(1):119–131
Hossain A, Maitra S, Ahmed S, Mitra B, Ahmad Z, Garai S, Mondal M, Adeel M, Shankar T, Meena RS (2022) Legumes for nutrient management in the cropping system. In: Meena RS,

Kumar S (eds) Advances in legumes for sustainable intensification. Academic Press, London, pp 93–112

IPCC (2007) In: Solomon S, Qin D, Manning M, Chen Z, Marquis M, Averyt KB, Tignor M, Miller HL (eds) Climate change 2007: the physical science basis. Contribution of Working Group I to the fourth assessment report of the Intergovernmental Panel on Climate Change. Cambridge University Press, Cambridge/New York

IPGA (2018) Pulse sector in India. http://ipga.co.in/pulse-growth-in-indian-region/. Accessed 10 May 2018

Islam MA, Sarkar D, Alam MR, Jahangir MMR, Ali MO, Sarker D, Hossain MF, Sarker A, Gaber A, Maitra S, Hossain A (2023) Legumes in conservation agriculture: a sustainable approach in rice-based ecology of the Eastern Indo-Gangetic Plain of South Asia-an overview. Technol Agron 3:3. https://doi.org/10.48130/TIA-2023-0003

Jakhar SR, Kumar S, Jangir CK, Meena RS (2017) The role of mycorrhizal relationship in sustainable manner towards plant growth and soil fertility. Indian J Agric Allied Sci 3(4):19–24

Jena J, Maitra S, Hossain A, Pramanick B, Gitari HI, Praharaj S, Shankar T, Palai JB, Rathore A, Mandal TK, Jatav HS (2022) Role of legumes in cropping systems for soil ecosystem improvement. In: Jatav HS (ed) Ecosystem services. Nova Science Publishers, Inc, Hauppauge, NY, pp 1–21

Kaur R, Shivay YS, Singh G, Virk HK, Sen S, Rajni R (2018) Increasing area under pulses and soil quality enhancement in pulse-based cropping systems–retrospect and prospects. Indian J Agric Sci 88(1):10–21

Krishnakumar KN, Rao GP, Gopakumar CS (2009) Rainfall trends in twentieth century over Kerala, India. Atmospheric environment 43(11):1940–1944

Khattak GSS, Saeed I, Muhammad T (2009) "Flowers shedding under high temperature in mung bean" (*Vigna radiata* (L.) Wilczek). Pak J Bot 41(1):35–39

Kumar N, Singh MK, Ghosh MK, Venkatesh MS, Hazra KK, Nadarajan N (2012) Resource conservation technology in pulse based cropping systems. Indian Institute of Pulses Research, Kanpur, pp 1–32

Kumar A, Dixit S, Ram T, Yadav RB, Mishra KK, Mandal NP (2014a) Breeding high-yielding drought-tolerant rice: genetic variations and conventional and molecular approaches. J Exp Bot 65(21):6265–6278

Kumar A, Choudhary AK, Suri VK, Bana RS, Pooniya V, Singh U (2014b) Site specific water management for sustainable agriculture. In: Water management in agriculture. Jaya Publishing House, Delhi, pp 327–336

Kumar A, Choudhary AK, Suri VK (2016) Influence of AM fungi, inorganic phosphorus and irrigation regimes on plant water relations and soil physical properties in okra (*Abelmoschus esculentus* l.) – pea (*Pisum sativum* L.) cropping system in Himalayan acid Alfisol. J Plant Nutr 39(5):666–682. https://doi.org/10.1080/01904167.2015.1087030

Kumar J, Choudhary AK, Gupta DS, Kumar S (2019) Towards exploitation of adaptive traits for climate resilient smart pulses. Int J Mol Sci 20(12):2971

Kumar R, Sharma P, Gupta RK, Kumar S, Sharma MMM, Singh S, Pradhan G (2020a) Earthworms for eco-friendly resource efficient agriculture. In: Resources use efficiency in agriculture. Springer, Singapore, pp 47–84

Kumar S, Gupta P, Choukri H, Siddique KH (2020b) Efficient breeding of pulse crops. In: Accelerated plant breeding, vol 3: food legumes. Cham, Switzerland, pp 1–30

Kumar S, Gopinath KA, Sheoran S, Meena RS, Srinivasarao C, Bedwal S, Jangir CK, Mrunalini K, Jat R, Praharaj CS (2023) Pulse-based cropping systems for soil health restoration, resources conservation, and nutritional and environmental security in rainfed agroecosystems. Front Microbiol 13:1041124

Kumara CD, Bantilan C, Rajalaxmi A, Reddy BVS, Borikar ST, Kumar AA, Singh NP, Shyam DM (2014) Development and diffusion of sorghum improved cultivars in India: impact on growth and stability in yield. Working paper series no. 50, Patancheru 502 324. International Crops Research Institute for the Semi-Arid Tropics, Andhra Pradesh, p 92

Lal R (2016) Soil health and carbon management. Food Energy Secur 5(4):212–222
Lehman RM, Cambardella CA, Stott DE, Acosta-Martinez V, Manter DK, Buyer JS, Maul JE, Smith JL, Collins HP, Halvorson JJ, Kremer RJ (2015) Understanding and enhancing soil biological health: the solution for reversing soil degradation. Sustainability 7(1):988–1027
Lemke RL, Zhong Z, Campbell CA, Zentner R (2007) Can pulse crops play a role in mitigating greenhouse gases from North American agriculture? J Agron 99:1719–1725
Lipper L, Thornton P, Campbell BM, Baedeker T, Braimoh A, Bwalya M (2014) Climate-smart agriculture for food security. Nat Clim Chang 4:1068–1072
Liu K, Blackshaw RE, Johnson EN, Hossain Z, Hamel C, St-Arnaud M, Gan Y (2019) Lentil enhances the productivity and stability of oilseed-cereal cropping systems across different environments. Eur J Agron 105:24–31
Lumpkin AT, Sayre K (2009) Enhancing resource productivity and efficiency through conservation agriculture. In: Proceedings of 4th World Congress on Conservation Agriculture Innovations for Improving Efficiency, Equity and Environment, New Delhi, 4–7 Feb 2009, pp 4–9
Mahato S, Mandal S, Rahman FH (2022) Study on economic benefit by following agromet advisory services received from District Agromet Unit in the Lateritic Belt of Birbhum. International Journal of Agriculture Science 7:102–107
Maitra S, Sahoo U, Sairam M, Gitari HI, Rezaei-Chiyaneh E, Battaglia ML, Hossain A (2023) Cultivating sustainability: a comprehensive review on intercropping in a changing climate. Res Crop 24(4):702–715. https://doi.org/10.31830/2348-7542.2023.ROC-1020
Majumder D, Roy R, Rahman F, Rudra B (2020) Impact assessment of block-level agro advisories for saving input cost of farmers under old Gangetic plains of West Bengal: a case study in Malda. Curr J Appl Sci Technol 39(20):86–96
Malhi GS, Kaur M, Kaushik P (2021) Impact of climate change on agriculture and its mitigation strategies: a review. Sustainability 13(3):1318
Manchikatla PK, Kalavikatte D, Mallikarjuna BP, Palakurthi R, Khan AW, Jha UC, Bajaj P, Singam P, Chitikineni A, Varshney RK, Thudi M (2021) MutMap approach enables rapid identification of candidate genes and development of markers associated with early flowering and enhanced seed size in chickpea (Cicer arietinum L.). Front Plant Sci 12:688694
Masood N, Akram R, Fatima M, Mubeen M, Hussain S, Shakeel M et al (2022) Insect pest management under climate change. In: Building climate resilience in agriculture: theory, practice and future perspective, pp 225–237. https://doi.org/10.1007/978-3-030-79408-8_15
Millet EJ, Welcker C, Kruijer W, Negro S, Coupel-Ledru A, Nicolas SD, Laborde J, Bauland C, Praud S, Ranc N, Presterl T (2016) Genome-wide analysis of yield in Europe: allelic effects vary with drought and heat scenarios. Plant Physiol 172(2):749–764
Mishra JP, Praharaj CS, Singh KK, Kumar N (2012) Impact of conservation practices on crop water use and productivity in chickpea under middle indo-Gangetic plains. J Food Legumes 25(1):41–44
Navkiran R, Jagmeet K, Sarvjeet S, Inderjit S (2014) Growth and yield in chickpea (Cicer arietinum L.) genotypes in response to water stress. Afr J Agric Res 9(11):982–992
Noaman MM, Ahmed IA, El-Sayed AA, Abo-El-Enin RA, El-Gamal AS, El-Sherbiny AM, Asaad FA, El-Hag AA, Moustafa KA, El-Bawab AM, El-Moselhy MA (2007) Registration of 'Giza 2000' drought-tolerant six-rowed Barley for rainfed and new reclaimed areas in Egypt. Crop Sci 47(1):440–440
Nyasimi M, Kimeli P, Sayula G, Radeny M, Kinyangi J, Mungai C (2017) Adoption and dissemination pathways for climate-smart agriculture technologies and practices for climate-resilient livelihoods in Lushoto, Northeast Tanzania. Climate 5(3):63
Omomowo OI, Babalola OO (2021) Constraints and prospects of improving cowpea productivity to ensure food, nutritional security and environmental sustainability. Frontiers in Plant Science 12:751731
Pandey RK, Herrera WAT, Villegas AN, Pendleton JW (1984) Drought response of grain legumes under irrigation gradient: III. Plant growth 1. Agron J 76(4):557–560

Parry ML, Rosenzweig C, Iglesias A, Livermore M, Fischer G (2004) Effects of climate change on global food production under SRES emissions and socio-economic scenarios. Glob Environ Chang 14:53–67

Patel PN, Hall AE (1990) Genotypic variation and classification of cowpea for reproductive responses to high temperature under long photoperiods. Crop Sci 30(3):614–621

Patel JJ, Mevada KD, Chotaliya RL (2003) Response of summer mungbean to date of sowing and level of fertilizers. Indian J Pulses Res 16(2):122–124

Pooniya V, Choudhary AK, Dass A, Bana RS, Rana KS, Rana DS, Tyagi VK, Puniya MM (2015) Improved crop management practices for sustainable pulse production: an Indian perspective. Indian J Agric Sci 85(6):747–758

Praharaj CS, Blaise D (2016) Intercropping: an approach for area expansion of pulses. Indian J Agron 61(4th IAC Special issue):S113–S121

Praharaj CS, Kumar N, Singh U, Singh SS, Singh J (2015) Transplanting in pigeon pea – a contingency measure for realizing higher productivity in Eastern Plains. J Food Legumes 28(1):34–39

Prasad R (2012) Textbook of field crops production-food grain crops, Vol I, Indian Council of Agricultural Research, New Delhi, pp 248–319

Pratap A, Kumar S, Polowick PL, Blair MW, Baum M (2022) Accelerating genetic gains in pulses. Front Plant Sci 13:879377

Rajegowda MB, Janardhanagowda NA, Jagadeesha N, Babu BT (2008) Influence of agromet advisory services on economic impact of crops. J Agrometeorol 10:215–218

Rana DS, Dass A, Rajanna GA, Kaur R (2016) Biotic and Abiotic stress management in pulses. Indian J Agron 61(4th IAC Special issue):S238–S248

Roorkiwal M, Bharadwaj C, Barmukh R, Dixit GP, Thudi M, Gaur PM, Chaturvedi SK, Fikre A, Hamwieh A, Kumar S, Sachdeva S (2020) Integrating genomics for chickpea improvement: achievements and opportunities. Theor Appl Genet 133:1703–1720

Sahana AS, Ghosh S, Ganguly A, Murtugudde R (2015) Shift in Indian summer monsoon onset during 1976/1977. Environmental Research Letters 10(5):054006

Sallam A, Ul-Allah S (2019) Genomics-aided breeding for climate-smart traits in faba bean. In: Genomic designing of climate-smart pulse crops. Springer, Cham, pp 359–395

Saxena KB, Nadarajan N (2010) Prospects of pigeon pea hybrids in Indian agriculture. Electron J Plant Breed 1(4):1107–1117

Sekhon HS, Singh G, Sharma P, Sharma P (2004) Agronomic management of mung bean grown under different environments. Improving income and nutrition by incorporating mung bean in cereal fallows in the Indo-Gangetic Plains of South Asia. DFID Mung bean Project for 2002–2004. In: Proceedings of the final workshop and planning meeting, Punjab Agricultural University, Ludhiana, India, 27–31 May 2004, pp 82–103

Sheoran S, Kumar S, Kumar P, Meena RS, Rakshit, S (2021) Nitrogen fixation in maize: breeding opportunities. Theor Appl Genet 134:1263–1280

Sikka AK, Alam MF, Mandave V (2022) Agricultural water management practices to improve the climate resilience of irrigated agriculture in India. Irrig Drain 71:7–26

Singh KK, Ali M, Venkatesh MS (2009) Pulses in cropping systems. Technical bulletin. IIPR, Kanpur, p 47

Singh U, Praharaj CS, Singh SS, Kumar N (2015) Influence of crop establishment practices and genotypes in pigeon pea–wheat system under IGP of India. J Food Legumes 28(4):315–319

Singh KK, Ghosh K, Bhan SC, Singh P, Vishnoi L, Balasubramanian R et al (2023) Decision support system for digitally climate informed services to farmers in India. J Agrometeorol 25(2):205–214

Sita K, Sehgal A, Kumar J, Kumar S, Singh S, Siddique KH, Nayyar H (2017) Identification of high-temperature tolerant lentil (Lens culinaris Medik.) genotypes through leaf and pollen traits. Front Plant Sci 8:744

Sita K, Sehgal A, Bhandari K, Kumar J, Kumar S, Singh S et al (2018) Impact of heat stress during seed filling on seed quality and seed yield in lentil (*Lens culinaris* Medikus) genotypes. J Sci Food Agric 98(13):5134–5141

Suri VK, Kumar A, Choudhary AK (2012) Soil health management through carbon sequestration under changing climatic scenario. Lead paper, ICLDBT International Symposium published during September 2012 in Progressive Agriculture 11(Conf. issue), pp 29–42

Tardieu F, Simonneau T, Muller B (2018) The physiological basis of drought tolerance in crop plants: a scenario-dependent probabilistic approach. Annu Rev Plant Biol 69:733–759

Thudi M, Chitikineni A, Liu X, He W, Roorkiwal M, Yang W, Jian J, Doddamani D, Gaur PM, Rathore A, Samineni S (2016) Recent breeding programs enhanced genetic diversity in both desi and kabuli varieties of chickpea (Cicer arietinum L.). Sci Rep 6(1):1–10

Tribhuvan KU, Kumar K, Sevanthi AM, Gaikwad K (2018) MutMap: a versatile tool for identification of mutant loci and mapping of genes. Indian J Plant Physiol 23:612–621

Trnka M, Feng S, Semenov MA, Olesen JE, Kersebaum KC, Rötter RP, Semerádová D, Klem K, Huang W, Ruiz-Ramos M, Hlavinka P (2019) Mitigation efforts will not fully alleviate the increase in water scarcity occurrence probability in wheat-producing areas. Sci Adv 5(9):eaau2406

Uysal O, Subaþý OS (2016) A study on determination of strategy for pulse industry through SWOT and SOR analyses in Mersin Province. Turkish J Agric Food Sci Technol 2(6):256–259

Vadi HD, Kachot NA, Polara JV, Sekh MA, Kikani VL (2006) Effect of tillage and mulching on yield and yield attributing characters of pigeon pea. Adv Plant Sci 19(2):497–499

Varshney RK, Thudi M, Roorkiwal M, He W, Upadhyaya HD, Yang W, Bajaj P, Cubry P, Rathore A, Jian J, Doddamani D (2019) Resequencing of 429 chickpea accessions from 45 countries provides insights into genome diversity, domestication and agronomic traits. Nat Genet 51(5):857–864

Varshney RK, Barmukh R, Roorkiwal M, Qi Y, Kholova J, Tuberosa R, Reynolds MP, Tardieu F, Siddique KH (2021) Breeding custom-designed crops for improved drought adaptation. Adv Genet 2(3):e202100017

Venkatesh MS, Hazra KK, Ghosh PK, Praharaj CS, Kumar N (2013) Long-term effect of pulses and nutrient management on soil carbon sequestration in Indo-Gangetic plains of India. Can J Soil Sci 93:127–136

Wang L, Zhao Y, Al-Kaisi M, Yang J, Chen Y, Sui P (2020) Effects of seven diversified crop rotations on selected soil health indicators and wheat productivity. Agronomy 10(2):235

Chapter 3
Climate-Resilient Technologies for Food Crop Production

Urjashi Bhattacharya, Augustina Saha, Manish Kumar Naskar, Pijush Das, Snigdha Samanta, and Maha Awjan Alreshidi

Abstract Increased frequencies of extreme weather events are projected by climate change and are among the greatest threats to global food crop production. This is due to increased temperatures, irregular patterns of rainfall, and soil degradation that have rendered farmers' work uncertain and unreliable in many regions. Urgent steps must be taken to create farm systems resilient to environmental shocks that ensure food for an increasingly expanding population. This chapter will focus on climate-resilient technologies and the role they have played in this respect. Advanced developments like crop varieties with resistance to drought stress, precision irrigation systems, and soil health management practices all form part of its productivity and sustainability contributions. This chapter will provide the practical tools that bridge traditional knowledge into modern science while providing practical ways to alter production in food crops with changing climatic conditions.

Keywords Crop Planning · Nutrient Management · Precision Water Management · Smart Plant Protection · Remote Sensing

U. Bhattacharya · P. Das · S. Samanta
Department of Agronomy, School of Agriculture and Allied Sciences, The Neotia University, Sarisa, West Bengal, India

A. Saha
Department of Agronomy, School of Agriculture Sciences, Sister Nivedita University, Newtown, Kolkata, West Bengal, India

M. K. Naskar
Department of Agricultural Meteorology and Physics, Bidhan Chandra Krishi Viswavidyalaya, Mohanpur, West Bengal, India

M. A. Alreshidi (✉)
Department of Chemistry, University of Ha'il, Ha'il, Saudi Arabia
e-mail: m.alrashedi@uoh.edu.sa

B. Pramanick et al. (eds.), *Climate-Smart Agricultural Technologies*, Sustainability Sciences in Asia and Africa, https://doi.org/10.1007/978-981-96-7699-6_3

3.1 Introduction

Agriculture and weather are extremely interrelated. Food crop production around the world depends on the availability of water, nutrients, and energy mostly sufficed by favorable weather conditions. A slight change in any of the climatic parameters is said to have a huge impact on food security. According to IPCC special report (2023), there is a projected global warming of 1.5 °C and even more, which is likely to affect livelihood and agriculture throughout the world, if necessary, actions are not taken to mitigate the temperature below 1.5 °C. The global surface temperature was found to be 1.09 °C higher in 2011–2020 than in 1850–1900 and it is expected to be even more in the upcoming decades if GHG emissions are not curtailed (IPCC 2023). Extreme weather events, like drought, heatwaves, flood, cold waves, etc. are compounding and making communities especially over Africa, Asia, Central and South America, Small Islands, and the Arctic vulnerable to such changes. Until now, though agricultural productivity has increased globally, its pace has slowed down over the last 50 years due to climate change, negatively affecting food production in mid and low-latitude regions (IPCC 2023). This has resulted in the shortening of crop growing seasons, incessant and uneven distribution of rainfalls, and changes to water temperature and quality, hampering food and nutritional security. A recent multi-model yield study of the four most important crop groups (wheat, rice, oilseeds, and coarse grains which account for 70% of total area cultivated globally) under the highest level of global warming suggests an overall yield decrease of 17% by 2050, if necessary, climate action is not taken (FAO 2015). Variability in yield in various locations around the world would be triggered mainly by CO_2 levels, ozone concentrations, rising temperature, and biotic stresses viz. weeds and pests. In simulating future crop yields, researchers are indicating a decline by the second half of the twenty-first century, which is to be observed mainly in tropical zones of the world. IFPRI reported that rice yield can be reduced by 10–15% due to climate change (Li 2018). Similarly, various precipitation models suggested maize production in the USA be reduced by 20–50% in 2050 under current global warming scenarios (Leng and Huang 2017). A 2% rise in temperature and a 20% decline in rainfall will cause a 10% yield reduction in maize in Africa (Lobell and Burke Marshall 2010). Wheat production throughout the world is projected to be reduced by 17–23%, in the latter half of the century if current climate change prevails (Wang et al. 2018). An increase in temperature actually accelerates the respiration rates of plants and carbon metabolism resulting in yield reduction. High levels of carbon dioxide affect the nutritional value of crops like wheat, soybean, and rice by lowering the concentration of proteins and minerals.

The global population will hit 9.8 billion by 2050, and food production needs to be upscaled by 75–100% in order to feed the growing population. But with the vulnerability of climate change in the backdrop, it is quite impossible to meet the needs if climate-resilient production technologies are not harbored. Climate-smart technologies are improved agricultural practices for diverse agroecosystems to combat the advent of climate change so as to build resilience under extreme weather events for sustainable yield in the long run (Pramanick et al. 2022, 2023). Crop contingency plans, building soil resilience, climate-smart: water management, nutrient management, weed management, plant protection measures, the proper intervention of agro-advisories, digital automation, and capacity building by linking various stakeholders (like scientists, policymakers, governments, and farmers) are need of the hour for mitigating climate change in food production level (Laik et al. 2021; Dey et al. 2022). The main focus of climate-resilient technologies is to incorporate carbon dioxide in soils and plant systems so that they act as sinks. Numerous researchers had brought forth various kinds of climate-resilient technologies over the years as mitigation for climate change. This chapter will focus on such climate-smart technologies.

3.2 Contingency Crop Planning

Rainfall behavior is inconsistent and unpredictable in dry farming regions. The variations in rainfall behavior that are frequently experienced in dry regions include inadequate and uneven distribution of rainfall, early and delayed onset of monsoon, early withdrawal, and prolonged dry spells during the rainy season. These aberrant rainfall conditions often lead to poor crop performance and/or complete crop failures. Therefore, suitable adjustments in crop management practices are required to minimize such adverse effects of anomalous rainfall behavior. Contingency crop planning entails these management choices.

Contingency crop planning refers to implementing an alternative plan with a crop or cultivar choices in accordance with the actual rainfall situations and soil conditions in a specific location. In rainfed areas, early sowing of crops with the start of monsoon is the best-bet practice that results in greater realizable yield. Real-time contingency crop planning contributes a major part in global food and fodder security when it is followed timely and effective (Srinivasrao et al. 2010). The salient achievements of real-time contingency measures are mentioned below (Table 3.1).

3.3 Soil Resilience and Climate-Smart Nutrient Management

Soil resilience is the "Capacity of the soil to recover its functional and structural integrity after a disturbance; where this integrity can be considered as soil capacity to perform essential soil functions" (Karlen et al. 1997). The IPCC Special Report on Climate Change and Land (IPCC 2019) addresses how a changing climate will increasingly affect how we use the land. On a holistic human ecosystem scale, soil represents the nexus of water, energy, and food. If we rely on the soil for food, energy, water, health, and well-being, the impacts of climate change will exacerbate any pressures of current land management practices.

Table 3.1 Achievements of real-time contingency crop planning/management approaches

Type of contingency	Issues/problems	Contingent crop plan to be followed
Inadequate and uneven distribution of rainfall	(a) Shortage of moisture for crop growth and development (b) Reduced crop yields (c) Excess moisture when crops do not have the requirement and when crops require it lacks	(a) Selection of low water requiring crops like finger millet, pearl millet, green gram, black gram, sunflower for cultivation (b) Short-duration crop varieties of rice (MTU 1010), green gram (Varsha), black gram (PU 30) to be incorporated in the system (c) Provision of supplemental irrigation as and when required by the crop (d) Absolute control of weeds to reduce competition with crops for moisture and nutrients
Early onset of monsoon/Late season drought	(a) Shortage of moisture at later growth stages of kharif crops (b) Delay in sowing of rabi crops (c) Improper germination of crop seeds	(a) Cultivate crops like baby corn, finger millet, pearl millet, sesamum, soybean, etc. (b) Crop diversification with vegetables like brinjal, okra, frenchbean, etc. (c) Creation of conservation furrows between two crop rows of pigeon pea in pigeon pea + finger millet intercropping system (d) Formation of farm ponds for in-situ water harvesting
Delayed onset of monsoon/Early season drought	(a) Short sowing window due to limited soil moisture (b) Delay in seedling raising in rice nursery (c) Late sowing of crops resulting in poor yields (d) Poor crop establishment and growth	(a) Select alternate crops like finger millet, pearl millet, foxtail millet, cowpea, green gram, castor (b) Seedling raising through community nurseries and transplanting of rice seedlings with the arrival of rain (c) Pre-sowing irrigation in pulses for seed sowing and initial crop establishment (d) Mechanization in sowing to place wheat seeds in the moist zone of soil (e) Timely sowing of groundnut crop seeds through tractor-drawn groundnut planter (f) Grow low-water-requiring crops in place of high-water-requiring crops (g) Formation of farm ponds for in-situ water harvesting (h) Soaking of seeds before sowing
Early cessation of rains/Terminal drought	(a) Unfavorable conditions for seed sowing of rabi crops (b) Moisture stress during flowering and maturity stages (c) Improper grain filling and pod development (d) Reduced crop yields	(a) Selection of short duration variety of rabi crops (wheat, mustard, sunflower, etc.) for cultivation (b) Selection of low water requiring crops like finger millet, pearl millet, green gram, black gram, sunflower, etc. (c) Minimize soil moisture depletion through reduced plant population (d) Supplemental irrigation at critical crop growth stages of castor and cumin (e) Protective and micro-irrigation techniques for increased water use efficiency in vegetable crops (f) Timely sowing and proper placing of wheat seed at moist soil zone with seed drill

Prolonged dry spells/Mid-season drought	(a) Very low plant population (b) Moisture stress in post-rainy season (c) Total crop failure	(a) Re-sowing of finger millet crop seeds (var. GPU 28) when dry spell occurs within 10 days after sowing (b) Thinning of plant population when mild moisture stress occurs at 30–35 days after sowing (c) Cutting of sorghum and pearl millet as fodder and ratooning to be followed during severe moisture stress (d) Practice relay cropping of castor in between the gaps of cotton crop (e) Creation of conservation furrows between two crop rows of pigeon pea in pigeon pea + finger millet intercropping system (f) Spray of 2% urea solution is effective for the crops like castor, pigeon pea and groundnut that experienced a prolonged dry spell after sowing (g) Alternate cropping system of pearl millet + pigeon pea (5:1) with early sowing of pearl millet (h) Provision of life-saving irrigation during most critical stages of crop growth
Excess rainfall	(a) Delay in sowing of kharif crops (b) Water logging situation (c) Death of rice seedling in nursery due to submergence	(a) Selection of short-duration kharif crop varieties of green gram, black gram, cowpea, horse gram for delayed sowing (b) Direct sowing of pre-germinated seeds of rice with drum seeder in the puddled field (c) Cutting of bunds in rice nursery to avoid water submergence (d) Earthing up in maize following proper drainage facilitates better performance of crop and higher yield (e) Digging of trenches at the lower side of the field for effective drainage of excess moisture

Climate-smart nutrient management is key to the enhancement of vertical productivity to meet the food demand of the ever-increasing world population under changing climate scenario. Overall plant nutrition process can be impacted by climate change at multiple scales. An increase in temperature creates variation in water regimes which impacts the availability of plant nutrients. Elevated levels of CO_2 in the climate change scenario will also impact plant nutrient uptake. Therefore, it is probably time to think about how appropriate nutrient management can help us to reduce GHG emissions from agriculture, thereby contributing to the mitigation of climate change (Kar et al. 2021). Precision nutrient management tools play a crucial role in modern agriculture by enabling farmers to optimize fertilizer application and monitor crop health effectively. Climate-smart nutrient management technologies depicted in Figs. 3.1 and 3.2 and discussed below help in making informed decisions on nutrient requirements while minimizing environmental impact.

3.3.1 Leaf Color Chart (LCC)

The leaf color chart (LCC) is an innovative and cost-effective precise nutrient management tool containing different shades of green color which can be useful in comparing with the leaf color and applying the required amount of fertilizer dosages. The use of LCC helps in determining the nitrogen demand of the crop and guides the right time of fertilizer nitrogen application which in turn prevents unwanted nitrogen losses and their serious impacts on the ecosystem.

3.3.2 Soil Plant Analysis Development (SPAD)

The SPAD chlorophyll meter is a quick and non-destructive tool that provides a quick and on-field direct reading of the chlorophyll content of any leaf in a plant and indirectly assesses the proportional parameter of leaf, plant nitrogen status, and finally, grain yield. It helps in providing the required dosage of nitrogen to the crop thereby reducing the nitrogen losses in crop cultivation.

Fig. 3.1 Climate-smart nutrient management technologies

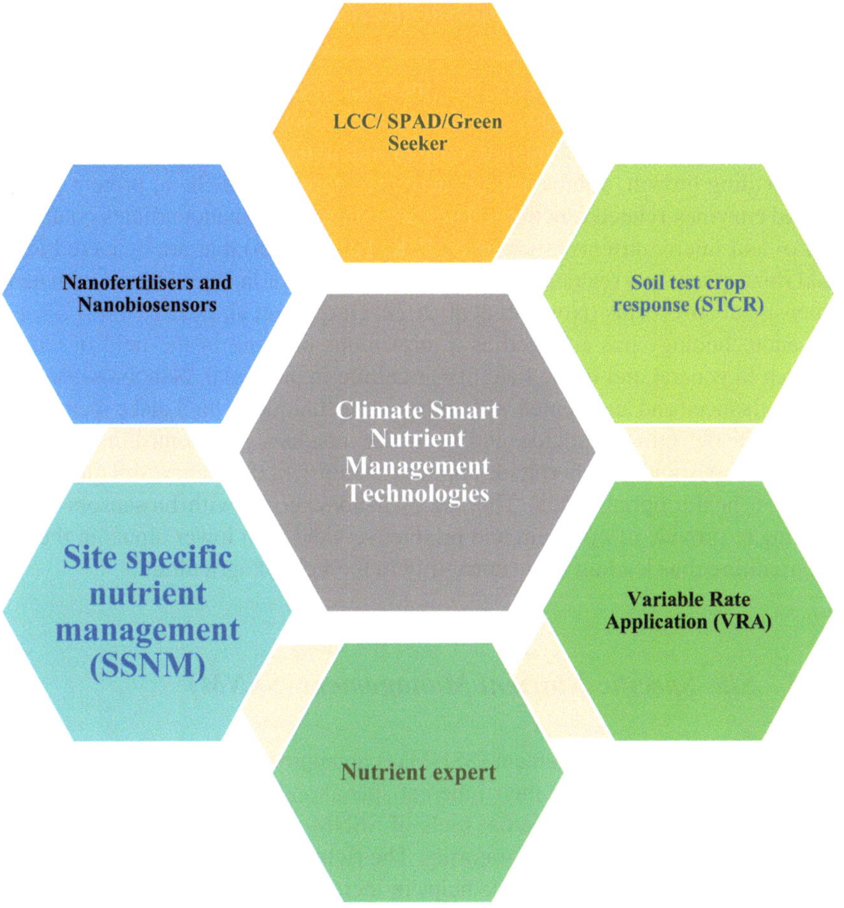

Fig. 3.2 Climate-smart nutrient management technologies

3.3.3 *Green Seeker*

Green seeker is an integrated optical sensor based on variable rate application and mapping system which is used for various purposes in precision farming like measuring crop nitrogen deficiencies, prediction of yield potential, crop conditions, measuring application of fertilizer at variable rates, and control chemical applications. Research proved that NDVI readings can note the effects of balanced fertilizer applications which can be utilized as a tool for monitoring and tracking the nutrient application and nutrient management effects.

3.3.4 Nanotechnology: Nanofertilizers and Nanobiosensors

Nanotechnology-based fertilizers, i.e., nanofertilizers, are very popular among researchers and fertilizer industries due to their controlled and targeted delivery mechanism and can able to improve crop productivity by enhancing seed germination, seedling growth, photosynthetic activity, energy metabolism, protein synthesis, and enzymes-related functions. They are composed of nanoparticles containing macro- and micronutrients (such as N, P, K, Fe, and Mn) that are delivered to the plant rhizosphere in a regulated manner bonded alone or in combination with nano-dimensional adsorbents (Nongbet et al. 2022; Durgude et al. 2022).

Nanotechnology has emerged as a miraculous blessing in the field of various research in general and in the field of agriculture in particular. Nanobiosensors are highly advanced and augmented with improved techniques which make them highly efficient in detecting pesticide and herbicide residues and controlling fertilizer application, pesticide and herbicide application, nutrient deficiency, soil quality, and good soil health optimization. The use of nanomaterials with biosensors is now enabling to provide of optimum and productive yield with lower inputs in the case of agriculture thus leading to sustainability in the field of agriculture.

3.3.5 Site-Specific Nutrient Management (SSNM)

Site-specific nutrient management (SSNM) is an approach to supplying plants with nutrients to optimally match their inherent spatial and temporal needs for supplemental nutrients by using different tools of SSNM such as remote sensing, GPS, GIS systems, VRT, and yield monitoring. The right amount of fertilizer quantity at the right time and at the right place helps in increasing the yield and production of the crop and also reducing the losses of nutrients with the excess application of fertilizers. SSNM helps in attaining the yield target and also helps in better usage of existing soil nutrients and fulfilling of nutrients with site-specific nutrient management practices (Shankar et al. 2020).

3.3.6 Nutrient Expert

Nutrient Expert is a computer-based automation tool that provides fertilizer recommendations to farmers in the absence of soil testing using the principle of the 4Rs of nutrient stewardship and site-specific nutrient management into a fertilizer recommendation. SSNM-Nutrient Expert software is convenient to manage daily field activities and useful for small farm-holding farmers to solve their crop and farm issues through information supplied by the system (Pooniya et al. 2015; Pramanick et al. 2022).

3.3.7 Soil Test Crop Response (STCR)

The soil test crop response (STCR) approach is an innovative, predictable, and one of the most scientific approaches to nutrient application for crops by using the soil test values and targeted yield equations. Recommendations based on the STCR correlation concept are developed by considering the contribution of nutrients from the soil, manures, fertilizers, and combined use of soil and plant analysis, which provide information on real balance between applied nutrients and available nutrients of soil through various field trials in a different location (Sharma et al. 2016).

3.3.8 Variable Rate Application (VRA)

Variable rate applications (VRA) in precision agriculture are the most site-specific nutrient management tool which focused on automated applications of materials in specific landscapes. VRA includes drones, satellites, artificial intelligence (AI), and hyperspectral imaging. The method of applying the material is based on data collected from sensors, maps, and GPS. There are proficiencies in various rate applicator technologies which include fertilizers, chemicals, seeds, etc., all of which help to optimize crop production. It is useful in managing and mapping cropping areas and planning better cropping patterns and farming systems (Javadi 2022).

3.4 Climate-Smart Water Management

Approximately 85% of water drawn from rivers, streams, lakes, and aquifers is being used in agriculture in many developing countries like India. Global warming has caused alterations in the hydrologic cycle resulting in water scarcity in many parts of the world. Mid-latitudes and lower latitudes arid and semiarid areas practicing rainfed agriculture would be the first ones to get affected by water shortage (Wu et al. 2020). 45% of worldwide food production comes from 20% of irrigated areas throughout the world. So, any kind of water shortage will surely affect agriculture at all levels. Climate change is the main driver of variations in rainfall, water infiltration, and groundwater recharge. A rise in temperature causes higher crop evapotranspiration. Together with the rise in population and the continuous threat from extreme weather events climate-smart water management practices are imperative. Meta-analysis of different crop models under various climate scenarios has concluded that location-specific climate-smart water technologies followed individually or on combined levels will improve yields by 7–15% and conserve water by 20–50% when compared to conditions without the mitigation strategies (Knox et al. 2016). Any kind of practices exacerbating the water use efficiency of crops are said to be climate-smart water management practices. Some of the interventions are discussed below (Table 3.2).

Table 3.2 Climate-smart approaches for efficient water management

Interventions	Details
1. Rain water harvesting (RWH)	RWH refers to the collection and storage of excess runoff in structures or infiltration in aquifers for future use for maximizing agricultural production. At the farm level, agronomic measures include contour cultivation, deep tillage, mulching, contour farming, furrow irrigation, ridges and furrow method of cultivation, etc. Whereas, conservation ditches, contour trenches, contour bunds, graded bunds, bench terraces, etc. fall under engineering measures. RWH is the viable adaptation strategy to sustain agriculture in arid, semiarid, and subhumid areas where rainfall is very erratic. *Advantages*: Restore degraded dryland and make it feasible for agricultural purpose, reduces runoff, and holds it in micro-catchments Slows down topsoil erosion, especially in hilly areas Some of RWH structures in India are: guls and khuls for hilly regions, inundation channels in West Bengal for irrigation, etc.
2. Floating agriculture	This an important mitigation strategy in flood prone and waterlogged areas. In this technique, crops are planted in a floating soil-less bed supported by bamboo poles and made up of composted organic matter which might include water hyacinth, rice stubbles topped with quickly decomposing materials like water-rots (APEIS and RIPSO 2004). Major nutrients like nitrogen, phosphorus, and potassium along with other micronutrients will be easily available from these beds, so application of fertilizers is not required. It is been widely practiced in Bangladesh where arable lands are waterlogged for most of the times during monsoon season. *Advantages*: Waterlogged lands turned into arable ones. Floating beds after the harvest of crops can be re-used as organic fertilizer in the main field or as fertilizer in the next year's floating rafts. Controls aquatic invasive weeds viz. water hyacinth, and use it in a more productive way.
3. Laser land leveling	This is a technology that involves cutting and filling of soil in high-lying and low-lying areas of the field respectively in order to have a uniform and smooth slope of 0–0.2% by laser-guided levelers. Uneven agricultural lands cause moisture scarcity in high-lying areas and waterlogging conditions in low-lying areas of the field thereby wasting irrigation water by 30%. Improper field design leads to poor germination of seedlings and nonuniform crop stands consequently increasing crop yield. On the pretext of shortage of water resources, this is an efficient resource conservation technology, that has proved to improve crop canopy structures and finally crop yield. In Indo-Gangetic plains in rice-wheat system it is said to lower irrigation water losses by 10–25% (Kahlown et al. 2000). *Advantages*: Leveled lands reduce the use of seeds, chemicals, water, fertilizers. Precision water management Increases water use efficiency, surface irrigation efficiency, application efficiency lowers irrigation timing. Uniform soil moisture distribution This is being adopted worldwide in countries like Japan, USA, Australia, India, etc. In India states like Uttar Pradesh, Madhya Pradesh, Punjab, Haryana, etc. have been practicing this technique.

4. Phytorid technology	This is a technology for treating and reusing wastewater originating from industries, household, and agriculture. Declining freshwater resources can be conserved for irrigation purpose if this technology is adopted. It resembles a wetland, where aquatic plants like cattails (*Typha* spp.), Elephant grass (*Pennisetum purpureum*), *Cannas* sp., *Nerium* sp., etc. are grown which filtrates the wastewater by absorbing nutrients from it. This system is a soilless system consisting of three zones: (i) intel zone—consisting of stones and crushed bricks, (ii) treatment zone—having the same media as of intel zone along with plant species, (iii) outlet zone—zone where treated water resides. The wastewater passes through all the zones to be used for other purposes. It is been developed by the Council of Scientific and Industrial Research-National Environmental Engineering Research Institute (CSIR-NEERI), Nagpur. *Advantages*: Mitigating and reusing polluted water for productive purposes like agriculture. Creating wetland habitats. Energy saving, cost saving green technology.
5. Deficit irrigation	In arid regions and areas where rainfall is insufficient, this technique can be used. It is a regulated irrigation approach where water is provided to plants based on three principles: (a) critical growth stage-based irrigation (when irrigation is applied only in the critical growth stages of crop), (b) partial root zone-based irrigation (alternate drying and wetting of root zone when half of the root zone is dried while the other one is wetted), (c) subsurface drip irrigation (water is applied to plants at subsurface so that it gets absorbed by capillary movement) for optimal growth by enhancing water use efficiency. *Advantages*: Saves 50% water requirement of crops (Abdelraouf 2016).
6. Micro-irrigation system	It is a low-pressure irrigation water distribution system, in which water is applied as droplets on the surface, below the surface or from emitters suspended above the ground. This includes drip and sprinkler systems. It is being used worldwide in countries like Israel, South Africa, Russia, Spain, Italy, Saudi Arabia, the USA, Brazil, India, etc. India occupies the highest area under micro-irrigation (7.73 mha) among the South Asian countries with a potential of >42 mha in the coming days (ICFA 2017). *Advantages*: Scientists have reported from across the globe under various crop, soil, and agro-ecological conditions sprinkler system can save up to 30–70% of irrigation water (Bandyopadhyay et al. 2010). Drip irrigation system lowers nutrient loss, soil erosion, and saves about 50% of water in most cereal crops (Wang et al. 2016). This system can be clearly used as an alternative in areas facing acute water crisis.

(continued)

Table 3.2 (continued)

Interventions	Details
7. IoT-based smart irrigation systems	IoT (Internet of Things)-based smart irrigation systems is a new climate-smart farmer-friendly technology that integrates artificial intelligence, sensors, and wireless communication with irrigation controls. This irrigation system is based on real-time weather and soil data, so that irrigation is adjusted by knowing the exact condition of the field. *Advantages*: Sensors used in this system allow the farmers to understand their crop and plan irrigation at the correct timing with the correct amount of water (Ayaz et al. 2019)
8. Application of remote sensing and GIS irrigation scheduling	Remote sensors using visible, infrared, and near-infrared region of electromagnetic spectrum are generally used for scheduling irrigation. The remote sensing tools along with the ground data obtained from GIS create maps spatially and temporally, focusing on the availability of groundwater resources and crop water consumption (Al-Bakri et al. 2023). It provides information on vulnerable areas suffering from water scarcity and identifies periods of high-water demand so as to schedule irrigation accordingly. *Advantages*: It can map crop evapotranspiration over a huge area with minimum climatic and ground data. The sensors measure soil moisture data, climate-based data, and plant moisture stress data to monitor irrigation scheduling for a region.
9. Hydrogel	This is a super absorbent polymer (SAP) with hydrophilic properties absorbing water to about hundred times its dry weight without dissolving and releasing it when exposed to mechanical stress. Generally, proteins, lignocellulosic materials, and other materials obtained from ecofriendly and biodegradable sources like crop residues can be used for hydrogel preparation. It is a feasible climate-smart water management technology for dry areas. When irrigation (rainwater or artificial irrigation) is given, it absorbs water and reduces its loss via evaporation or drainage. When soil dries up it slowly releases the retained water to the surrounding soil by the process of diffusion and keeps it moist for a longer period of time (Oladosu et al. 2022) *Advantages*: It improves soil structure and water infiltration, in drought-prone areas. It reduces drought stress in plants by minimizing reactive oxygen species (ROS). Ex-Pusa Hydrogel also known as wonder gel has been developed by IARI and has been reported to increase the growth and yield of plants by 10–25% in drought areas.

3.5 Climate-Smart Plant Protection Measures

Climate change is an important contributing factor with regard to the abundance and distribution of species. This climate change is affecting the outbreak potential of insect pests and diseases in a vast majority of crops, across all geographical regions (FAO 2013). Data states around 40% of the world's food supply is already lost due to pests (Heeb et al. 2019). Hence, reducing pest impact is an important facet to maintain global food security.

Change in climate has a profound impact on population dynamics and distribution of pests so much so that it can ultimately lead to the extinction of species (Walther et al. 2002). Sudden changes in temperature, rainfall, or climate in general can affect pest populations in any of the following ways. Fluctuation in diversity pattern and abundance of insect pests

- Increased overwintering insects
- Variations in geographical distribution of insect pests
- Imbalance in synchrony between insect pests and their host crops
- Increased incidence of vector-transmitted plant diseases
- Changes in insect biotypes
- Extinction of species
- Impact in the activity and relative abundance of natural enemies
- Introduction and spread of invasive pest species
- Reduced efficacy of crop protection technologies

India has recorded a declining trend in agricultural productivity due to fluctuation in temperature (Aggarwal 2008), frequently occurring natural calamities, and outbreak of insect pest and diseases. Yield losses from diseases/plant pathogens have been recorded between 20% and 40% in important food crops, regardless of the adaptation of various crop protection measures (Savary et al. 2019). The same problems are likely to get worse as a result of changing climatic conditions which shall further put forth a tremendous task to attain the ultimate goal of food security.

Therefore, immediate action on multilevel is required across all geographical regions, especially pest and disease-related yield losses due to climate change as this arena has received very little importance in comparison to human/animal health affected by changing climate.

Across the globe, new invasive species are appearing, expedited by such extreme unprecedented climatic conditions. In the absence of effective management systems, the newly invasive species have the full potential to become major pests in these new areas. The ongoing climate change directly affects the relative abundance of different species of insect and not all species are able to acclimatize to the changes hence, lost in due course of time. Moreover, changes in climatic variables have eventually led to increased frequency and outbreak of insect-pests. Upsurge in temperature increases the severity of disease caused by pathogens of crops such as cereals, oilseed, and potatoes (Marco et al. 2015). Unprecedented weather events result in an undesirable impact on the existing interactions between crops, pests, and diseases. Strong air currents can be a medium for transporting disease agents like spores or overwintering/hibernating insects to different areas which can add to the existing problems in those areas (Rosenzweig et al. 2001).

Keeping in mind the current climatic change and the possible risk it poses to the entire agricultural system, it has become more important than ever to adapt practices which shall focus on crop protection and maintain food security across the globe. The model of Climate-smart Pest Management (CSPM) is a very recent and innovative approach which combines a set of interdisciplinary strategies vital for primary production to adapt, with its supporting functions, for example, research and extension. Implementing this concept will result in more effective management of the threat posed by new and existing crop pests to agriculture, and ultimately increase overall food security in times of climate change. CSPM contributes in more ways than one by improving the overall greenhouse gas (GHG) balance (Heeb et al. 2019).

Additionally, climate monitoring, combined with pest risk forecasting as carried out by the Agrhymet Regional Centre (Agrhymet 2013), farmers can proactively identify and implement specific pest prevention practices in order to prevent the occurrence of expected pest problem. For instance, delineating the influence of spring weather pattern on the population of *Helicoverpa armigera* (Hübner) (Lepidoptera: Noctuidae) and other insect species implies that spring weather observation can be further used to improve forecasting of insect pest abundance and improve implementing pest management strategies (Gu et al. 2018). Reports indicate aphid population on wheat and other crops was negatively affected by rainfall and application of sprinkler irrigation (Chander 1998). Moreover, Masters et al. in 1998 carried out innovative manipulation of local climate to inspect how warmer winters with either wetter or drier summers would affect the homopteran insects of the grasslands.

FAW (Fall Armyworm), an alien invasive pest, is reported to cause havoc in India across all agro-climatic zones and it has been a tedious task to manage this pest. It has widened its host range wherein earlier it fed only on maize, now reports of FAW feeding on various other crops including rice, sugarcane is also surfacing. Certain heat shock proteins have been reported to be found in different vertebrates and invertebrates including FAW which enables these invasive species to thrive across extensive thermal regime (Samanta et al. 2021). Thermal tolerance is considered to be an important factor for successful establishment and spread of any insect species particularly in the era of climate change. Gene silencing has a huge role to play in if exploited for silencing the genes which are responsible for thermal adaptability of these pests and also resistance to different insecticides. RNA interference has been widely used in insects such as whitefly for its effective management (Grover et al. 2019). This technique is highly species-specific and environmentally friendly unlike other management strategies. The potential of RNAi has been well documented in a wide range of insect, which includes fruit fly (*Drosophila melanogaster*), the flour beetle (*Tribolium castaneum*; Whyard et al. 2009), silkworm (*Bombyx mori*; Bellés 2009), the kissing bug (Taracena et al. 2015), the striped rice borer *Chilo suppressalis* (Hui et al. 2011), and the list continues.

Ongoing climate changes are expected to affect pathogenesis and survivability rates resulting in slow but vital changes in the impact of diseases on agricultural crops or crops in general. However, the extent of impact varies from different geographical regions (Elad and Pertot 2014). Furthermore, temperature and moisture affect disease development by altering the susceptibility of the host to infection and thus increasing the level of symptom expression. The ultimate goal in such times of unprecedented climatic conditions is to develop coping techniques to adjust with the changes in the pathosystems. Establishment of mixed populations of different plant species is one important technique to deal with the risk of the monoculture-stimulated pathogen spread across large extents. Sturrock et al. (2011) recommended enhancement of tree species richness across different landscapes. This helped minimize the adverse effect of pathogen activity when there was ambiguity regarding the effects of climate on forest ecosystems.

Relocation of crops and genotypes from their traditional areas of cultivation to new locations around the world has occurred quite commonly. In this era of climate change, the deliberate movement of genotypes into areas where the climate is believed to be favorable for their persistence is absolutely necessary. However, one major drawback to this would be the unintended introduction of new pathogens into areas (Elad and Pertot 2014). The ultimate solution in the perspective of crop protection to climate change is breeding for the desired characteristics associated with tolerance to disease and pest resistance. Biochar is the solid product obtained from thermal degradation of biomass in anaerobic condition. The major plus point of Biochar is that its production can be considered as a climate change mitigation strategy. Additionally, Biochar when added to soil assists in a long-term sequestering agent of atmospheric CO_2 (Lehmann 2007). Thus, the constructive effects of Biochar on crop growth under conditions of intensive cropping can help plants cope with the effects of climate change to a greater extent. Remarkably, earlier reports demonstrate that the use of Biochar in soil results in the induction of systemic resistance to foliar pathogens, for instance, *Botrytis cinerea* (gray mold) in tomato, *Podosphaera aphanis* in strawberry, and *Oidiopsis sicula* (powdery mildew of pepper and tomato) (Graber et al. 2010).

Microbial communities have been reported to play a vital role in maintaining plant health status. There is an urgent need to promote microbial communities which will ensure good plant health, soil health and also increase the resistance of plants against different diseases and pest. They further help in combating temperature, salinity, and other abiotic stresses also. Some plant growth-promoting microorganisms ultimately reduce the use of chemical pesticides, thus reducing agricultural pollution and increasing crop productivity. Management of plant diseases in the era of climate change has necessitated the use of biocontrol agents in crop ecosystem. For example, *Simplicillium lamellicola* has shown great efficiency in managing various diseases including leaf rust, late blight, and powdery mildew in tomato, barley, and wheat respectively (Le Dang et al. 2014). Moreover, rice brown sheath rot disease which is caused by *Pseudomonas fuscovaginae* can be effectively controlled by the rhizobacteria *Bacillus amyloliquefaciens*. Microbial interactions with a wheat crop play a pivotal role in disease suppression. *Pseudomonas fluorescens* HC1-07 suppresses the causal organisms of root rot of wheat (Dubey et al. 2020).

The present climatic condition calls for an urgent interdisciplinary effort for crop protection and ultimately maintenance of crop yield. This is one of the key global challenges of the twenty-first century. The aim is to maximize agricultural output while restoring the land and environment quality. There is a further need for an in-depth understanding of complex system between plant, pest/diseases, and the environment in general. With the help of the integration of different approaches, we can indeed improve food security in the context of climate change.

3.6 Climate-Smart Weed Management Practices

Weeds can affect food production in agricultural systems, decreasing the product quality and productivity due to the competition for natural resources (Fig. 3.3). Recent development of weed control technologies can promote higher levels of food production, lower the amount of inputs needed and reduce environmental damage, invariably bringing us closer to more sustainable agricultural systems. Currently, weed management in agricultural systems branches out into two distinct directions corresponding to different approaches. On the one hand, is the widespread use of synthetic herbicides, while on the other, weed control is widely based on mechanical, cultural, and physical methods.

Climate-smart agriculture (CSA)-based weed management practices are evolving as a viable and sustainable alternative for better resources conservation and yield stability as CSA-based management practices rely on a wide range of indicators, i.e., tillage, crop establishment, crop residue retention, crop diversification, and precise water and nutrient management, along with the use of information and communication tools (ICTs) for timely implementation of crop practices.

3.6.1 Crop Residue Management Through Mechanized Cultivation

Mulching crop residues might be a compelling reason for the observed reduction in weed density and biomass. It helps in weeds suppression by imposing a physical barrier to emerging seeds, restricting the growth of germinated ones by limiting light availability and by potential allelopathic effects (Malik et al. 2018). Use of

Fig. 3.3 Climate-smart weed management technologies

high-precision, advanced farm machines such as zero-till seed drills, slit drills, happy seeder, tractor-operated sprayers, combine harvesters, and super straw management systems (SMSs), etc. can retain surface residues which decrease and buffer soil temperature, thus the germination of some weed species is reduced due to the reduced fluctuations in soil temperature.

3.6.2 *Precise Water Management*

Precise water management through subsurface drip irrigation (SDI) is a promising technology for altering weed density and weed suppression effects. The positive effect of SDI might be due to adjusting water and fertilizer placement. Since under ZT weed seeds accumulate near and/or on the soil surface, weeds germination is dramatically reduced because SDI leaves the few top cms of the soil dry and delivers water and nutrients where it will most benefit crops, not weeds (Shrestha et al. 2007).

3.6.3 *Application of UAV and Remote Sensing Technique for Weed Management in Agricultural Crops*

Smart farming technologies, such as smart sensors, remote sensing, air vehicles, satellites, the Internet of Things (IoT) technology, etc., are becoming increasingly common in modern agriculture to assist in optimizing agricultural production and minimizing the wastes and costs. The use of these emerging technologies in agriculture, such as drones, provides the ability to face many major or minor challenges; irrigation, crop monitoring, soil, and field analysis. Unmanned aerial vehicles (UAVs) can be highly valuable, since they allow for site-specific weed management (SSWM), an improved weed management approach for the highly efficient and environmentally safe control of weed populations, enabling precise and continuous monitoring and mapping of weed infestation (Esposito et al. 2021).

Remote sensing methods, such as thermal infrared (IR) sensors and light detection and ranging (LiDAR) technology, attained astonishing results in measuring vegetation canopy temperatures and heights in order to evaluate the biomass, chlorophyll, and nitrogen (N) contents at a discrete time. Additionally, remote sensing has been broadly employed to detect and map weeds in the agricultural sector.

3.7 Intervention of Agro-advisories

The National Centre for Medium Range Weather Forecasting introduced the weekly Agromet Advisory Service (AAS) in 1988, aiming to provide effective agrometeorological information to the farming community. With a wide network of 107 centers across the country, this service has been instrumental in minimizing losses caused by weather aberrations. The Advisory Services plays a vital role in agriculture by offering valuable information for all stages of agricultural operations based on weather forecasting. Real-time and accurate weather forecasts are essential for making informed decisions, mitigating risks, and optimizing farming practices. Weather is a key factor that determines the success or failure of crop production, and timely forecasts can help overcome losses caused by unpredictable weather situations. The utility of weather forecasts depends on their reliability and applicability at a micro-level. Farmers rely on agrometeorological information to make decisions on variety selection, input utilization, crop management, protection practices, and harvesting. Spatial and temporal changes in important weather parameters, including rainfall, temperature, wind speed and direction, cloud cover, and humidity, influence these decisions. The crop protection advisory services heavily rely on real-time availability of quality regional short to medium-range weather forecasts and tested epidemiological models. Providing farmers with risk-related information at least 7–10 days ahead of likely disease or pest occurrences allows them to implement necessary mitigation strategies, such as pesticide spraying, to minimize the impact.

Earlier approaches focused on the effects of single meteorological variables on hosts, pathogens, or their interactions under controlled conditions. However, advancements have been made with weather-based location-specific forewarning models. These models, utilizing techniques like stepwise regression, discriminant function analysis, and artificial neural networks, have successfully validated major diseases and pests, enabling the issuance of timely advisories for farmers (Fig. 3.4).

The agro-advisories in India are provided through Subject Matter Specialists, Technical Officers, Agrometeorologists involved in Krishi Vigyan Kendra, State Agriculture University, Regional Research Station (RRS) by Gramin Krishi Mausam Sewa Project (GKMS) and All India Coordinated Research Project on Agrometeorology (AICRPAM). The agro-advisory is provided based on the 5 days forecast provided by the National Centre for Medium Range Weather Forecasting (NCMRWF). GKMS is funded by the India Meteorological Department (IMD) in close collaboration with the Indian Council of Agricultural Sciences (ICAR) while AICRP is funded by the Central Research Institute for Dryland Agriculture (CRIDA) under the guidance of ICAR. The IMD advisory comes with level-wise (National AAS Bulletin, State AAS Bulletin, District AAS Bulletin and Block Level AAS Bulletin) bulletin prepared across the stations issued for micro to macro-level policymakers.

The agro-advisories are helping the farming community by providing hazard information 1 week earlier, this helps minimize yield loss and reduction in agricultural input costs. The advisories reported average increased economic benefit of 31.4%, 24.7%, 16.2% and 20.6% in finger millet, pigeon pea, field bean, and tomato respectively in the eastern dry tract of Karnataka (Rajegowda et al. 2008), 25% in rice-wheat cropping system and 28% in maize cultivation in Lower Gangetic Plains of West Bengal (Majumder et al. 2020) and 19–31% hike in Benefit-cost ratio in Red Lateritic zone of West Bengal (Mahato et al. 2022).

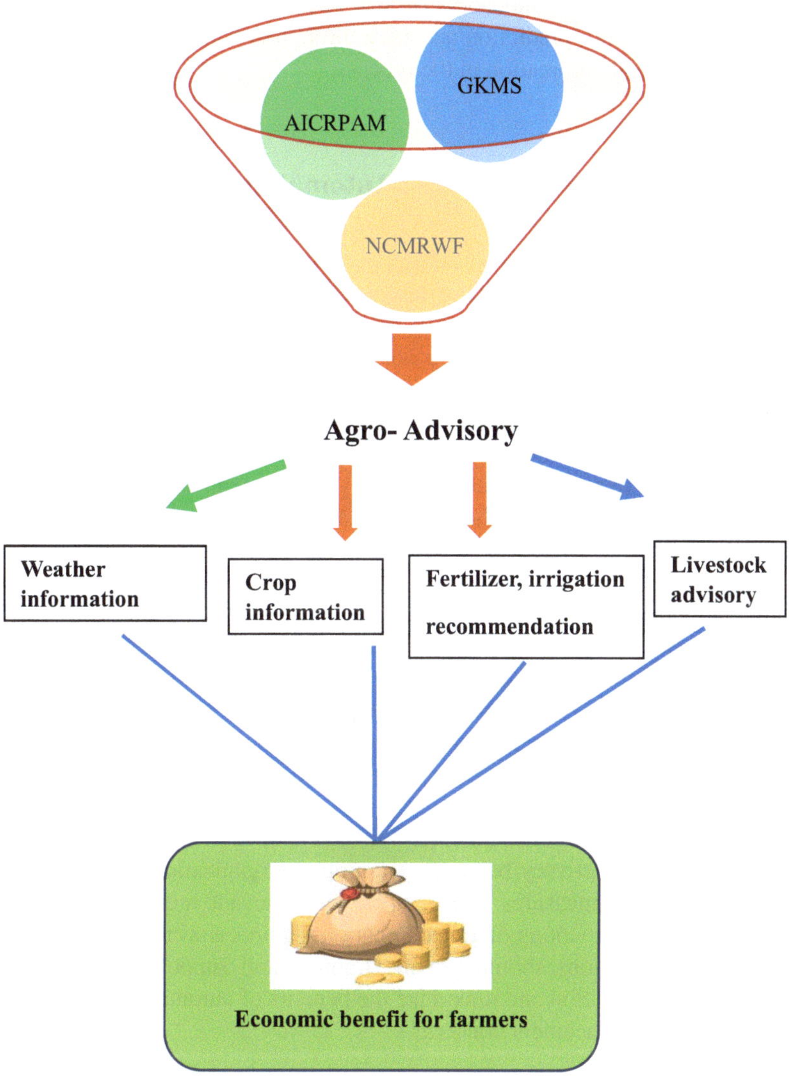

Fig. 3.4 Schematic diagram of typical work mechanism of Agro-advisories

A study from Karnataka depicted that on average, across different seasons, the usability of rainfall forecasts was highest during the winter season (75–93%). It was followed by the pre-monsoon season, where the usability percentages were 73–76%. However, during the southwest monsoon season, the usability of rainfall forecasts was comparatively lower at 45–71% (Sridhara et al. 2014).

Similar results were found by Das et al. (2022) where the prediction accuracy of rainfall is way lower in the monsoon season than in other seasons. However it is not a big issue where the AAS farming community nearly gained 20% extra income from the services. The agro-advisory services are bliss to the farmers. More specific efforts lie in enhancing prediction accuracy through enhancing model results and application of digital automation in the process.

3.8 Resilience Through Digital Automation

Digital automation plays a crucial role in enhancing resilience in various domains by streamlining processes, optimizing resource utilization, and improving decision-making. Resilience is the ability to withstand, adapt, and recover from disturbances or shocks. In today's rapidly changing world, organizations and systems need to be resilient to effectively navigate uncertainties and disruptions. It encompasses technologies such as artificial intelligence, machine learning, robotics, and Internet of Things (IoT), which has emerged as a powerful enabler of resilience (Misra et al. 2020). Digital automation in weather recording and subsequent bulletin preparation can be done. This will reduce human error and time. This allows organizations (IMD, ICAR, States) to respond swiftly and effectively to disruptions, reducing downtime and minimizing the impact of extreme weather events. Through advanced analytics and predictive modeling, digital automation enables optimal resource allocation. This capability enhances resilience by ensuring efficient utilization of resources, minimizing waste, and enabling proactive decision-making. It provides real-time data and analytics, empowering decision-makers with timely and accurate insights. This data-driven decision-making enhances resilience by enabling organizations to make informed choices in response to changing conditions or unexpected events.

As digital automation relies on extensive data collection and processing, ensuring data security and privacy becomes paramount. Organizations must implement robust cybersecurity measures and comply with relevant regulations to address these concerns. Implementing digital automation requires workforce adaptation and upskilling. Organizations should provide training and support to employees to ensure a smooth transition and maximize the benefits of automation while mitigating potential job displacement concerns.

3.8.1 Automation in Weather Observations Recording

The first step involves installation of weather sensors or instruments at the desired locations to measure various weather parameters such as temperature, humidity, wind speed, rainfall, etc. The sensors require proper calibration and positioned for accurate measurements. Then they are connected with microcontroller board capable of reading sensor data. We can use appropriate protocols (e.g., RS-232, USB, Ethernet) or wireless technologies (e.g., Wi-Fi, LoRa, GSM) to interface with the sensors and retrieve data. Depending on the distance and location, wired connections, wireless networks, or satellite communication are used in data transmission. Ensuring data integrity and security during transmission to prevent data loss from unauthorized access should be checked thoroughly. Thereafter we can use software like R, Python, SQL, etc. to access the data, perform statistical analysis, and create visuals.

3.8.2 Automation in Weather Bulletin Creation

Automating the weather bulletin involves collecting weather data from various sources, processing and analyzing the data, and generating a formatted weather report. We can access the data online through provided APIs using scrapping tools like 'rvest', 'httr' and weather downloaders like 'weatherData', 'rnoaa'. With the help of 'dyplr', 'tidyverse' we can process and analyze the data. A template can be designed with the input from the environment for the weather bulletin using R Markdown. To automate the process, we can use a task scheduler like 'cron' (for Unix-based systems) or the 'taskscheduleR' package (for Windows). We can periodically run the R script to obtain the bulletin in just one click.

3.9 Conclusion and Way Forward

Planned and successful adaptations of climate resilient technologies like contingency crop planning; climate-smart nutrient management, water management, plant protection measures and weed management; intervention of agro-advisories, etc. at the field level, require various stakeholders to join their hands at macro and micro level. Proper interventions of institutions and policies are essential for small-scale farmers to produce food specially to combat risks and vulnerabilities posed by climate change scenarios. Since 83% of the farmers worldwide are small-scale farmers, residing in mid and low latitudes of Africa and Asia (Fanzo 2017) mostly, policies regarding access and availability to timely financial and informational support for farmers at the national and international level are necessary for transitioning into climate resilient practices. Incorporating extensive participatory approach and collective action among farmers is another way to avert climate change in the longer run. Institutions and organizations need to promote region-specific resilient strategies. Location-specific traditional farming practices like soil and water conservation, etc. which might have been abandoned, need to be encouraged more under government schemes. For building social and community resilience to climate vulnerabilities, perpetuation and exchange of information regarding adaptation practices is important. Various weather simulation models, yield response models along with real time weather data act as early surveillance and warning system for managing any climatic risks. Coordination for disseminating such information from local to international level is an important component of risk analysis. Strategies for natural resource management as well as maintaining food security need to be implemented. Plans like NAPA (National Adaptation Plan of Action) developed by least developed countries aim at identifying short-term country-specific urgent needs, and vulnerable sectors to meet food security and economy. While NAP frame worked by UNFCCC pinpoints medium-term and long-term adaptation plans. UNDP and FAO support the least-developed countries in this (UNFCCC 2010). NAP plans need to be more strengthened by national and international cooperation. Some of the approaches which can be adopted worldwide for better resilience are listed below (Table 3.3):

Vigorous location-specific R&D activities along with dissemination of climate-smart practices are crucial to meet the SDGs (sustainable development goals). In the face of threatening climate change, immediate multiscale and multisector actions are called for in order to neutralize its cascading effects on agricultural ecosystems to maintain food security. Especially the low- and medium-income countries that are more prone to climate damage need international support to regularly carry out climate risk assessments and streamline their mitigation strategy accordingly.

Table 3.3 Approaches to be integrated worldwide as necessary

S. No.	Approach	Details	Country
1.	Citizen Science Approach	This approach allows farmers to verify the use of ICT (Information and Communication Technology) and other climate-resilient technologies as solutions for climate change.	Africa
2.	Highlighting participation of women in NAP	Disseminating knowledge about the effects of climate change on agriculture and food security among women officials of parliament and strengthening women SHGs (self-help groups).	Nepal
3.	Climate Field Schools	It empowers farmers about climate data for their better decision-making, It allows farmers to interact directly and regularly with providers of such information. Field visits and other institutions are mandatory in this approach along with hands-on training.	Indonesia
4.	Flood mapping and insurance	Deploying weather simulation models and high-tech satellite images for mapping purpose in order to have fast price compensations to farmers.	Bangladesh
5.	National Initiative on Climate-Resilient Agriculture	Enabling Indian agriculture to combat climate change through strategic research and technological interventions. Strategic research includes climate-smart natural resource management while technological interventions cover participatory validation of location-specific climate-smart approaches in vulnerable districts.	India
6.	Data-driven Agronomy improving rice productivity	In 2014, project involved CIAT (International Centre for Tropical Agriculture) scientists collecting 10 years of crop data and weather data to study the climatic extremes limiting rice yield by 5–6 tons/ha, in the rice-growing areas of the country. The farmers were advised in advance about the onset of drought year in which they didn't grow rice and averted USD 3.6 million. Rainfall was delayed again in the following season, and the farmers adjusted their planting dates by 2 months and got a remunerative yield.	Colombia

References

Abdelraouf RE (2016) Effect of partial root zone drying and deficit irrigation techniques for saving water and improving productivity of potato. Int J ChemTech Res 9(8):170–177

Aggarwal PK (2008) Climate change and Indian Agriculture: impacts, adaptation and mitigation. Indian J Agric Sci 78:911–919

Agrhymet (2013) Centre Regional Agrhymet—Nos Produits. Int J Curr Microbiol Appl Sci 8:2500–2505. http://www.agrhymet.ne/produit.html. Accessed 16 May 2018. Agriculture

Al-Bakri JT, D'Urso G, Calera A, Abdalhaq E, Altarawneh M, Margane A (2023) Remote sensing for agricultural water management in Jordan. Remote Sens 15:235. https://doi.org/10.3390/rs15010235

APEIS, RIPSO (2004) Floating Agriculture in the flood-pr one or submerged areas in Bangladesh (Southern regions of Bangladesh). APEIS and RIPSO, Bangladesh

Ayaz M, Ammad-Uddin M, Sharif Z, Mansour A, Aggoune eL-H M (2019) Internet-of-Things (IoT)-based smart agriculture: toward making the fields talk. IEEE Access 7:129551–129583. https://doi.org/10.1109/ACCESS.2019.2932609

Bandyopadhyay KK, Misra AK, Ghosh PK, Hati KM, Mandal KG, Moahnty M (2010) Effect of irrigation and nitrogen application methods on input use efficiency of wheat under limited water supply in a Vertisol of Central India. Irrig Sci 28:285–299. https://doi.org/10.1007/s00271-009-0190-z

Bellés X (2009) Beyond Drosophila: RNAi in vivo and functional genomics in insects. Annu Rev Entomol 55:111–128. https://doi.org/10.1146/annurev-ento-112408-085301

Chander S (1998) Infestation of root and foliage/ear head aphids on wheat in relation to predators. Indian J Agric Sci 68:754–755

Das H, Sial P, Mishra PJ, Dash SR (2022) Usability analysis of weather forecast and impact assessment of weather based agro advisories on rice and groundnut. Int J Agric Environ Biotechnol 15(1):95–99

Dey P, Mahapatra BS, Pramanick B, Pyne S, Pandit P (2022) Optimization of seed rate and nutrient management levels can reduce lodging damage and improve yield, quality and energetics of subtropical flax. Biomass Bioenergy 157:106355. https://doi.org/10.1016/j.biombioe.2022.106355

Dubey RK, Tripathi V, Prabha R, Chaurasia R, Singh DP, Rao CS, El-Keblawy A, Abhilash PC (2020) Belowground microbial communities: key players for soil and environmental sustainability. In: Easter (ed) Unravelling the soil microbiome: perspectives for environmental sustainability, pp 5–22. https://doi.org/10.1007/978-3-030-15516-2_2

Durgude SA, Ram S, Kumar R, Singh SV, Singh V, Durgude AG, Pramanick B, Maitra S, Gaber A, Hossain A (2022) Synthesis of mesoporous silica and graphene-based FeO and ZnO nanocomposites for nutritional biofortification and sustained the productivity of rice (*Oryza sativa* L.). J Nanomater 2022:13. https://doi.org/10.1155/2022/5120307. Article ID 5120307

Elad Y, Pertot I (2014) Climate change impacts on plant pathogens and plant diseases. J Crop Improv 28(1):99–139

Esposito M, Crimaldi M, Cirillo V, Sarghini F (2021) Drone and sensor technology for sustainable weed management: a review. Chem Biol Technol Agric 8:18

Fanzo J (2017) From big to small: the significance of smallholder farms in the global food system. Lancet Planet Health 1(1):e15–e16. https://doi.org/10.1016/s2542-5196(17)30011-6

FAO (2013) International standard for phytosanitary measure 5. FAO, Rome. https://www.ippc.int/en/publications/622/. Accessed 15 May 2018

FAO (2015) Climate change and food security: risks and responses. FAO. ISBN 978-92-5-108998-9. https://www.fao.org/3/i5188e/I5188E.pdf

Graber ER, Harel YM, Kolton M, Cytryn E, Silber A, David DR, Tsechansky L, Borenshtein M, Elad Y (2010) Biochar impact on development and productivity of pepper and tomato grown in fertigated soilless media. Plant Soil 337:481–496

Grover S, Jindal V, Banta G, Taning C, Nji T, Smagghe G, Christiaens O (2019) Potential of RNA interference in the study and management of the whitefly, *Bemisia tabaci*. Arch Insect Biochem Physiol 100(2):21522

Gu S, Han P, Ye Z, Perkins L, Li J, Wang H, Zalucki MP, Lu Z (2018) Climate change favours a destructive agricultural pest in temperate regions: late spring cold matters. J Pest Sci 91:1191–1198. https://doi.org/10.1007/s10340-018-1011-z

Heeb L, Jenner E, Cock MJW (2019) Climate-smart pest management: building resilience of farms and landscapes to changing pest threats. J Pest Sci 92(3):951–969

Hui X-M, Yang L-W, He G-L, Yang Q-P, Han Z-J, Li F (2011) RNA interference of ace1 and ace2 in Chilo suppressalis reveals their different contributions to motor ability and larval growth. Insect Mol Biol 20(4):507–518. https://doi.org/10.1111/j.1365-2583.2011.01081.x

ICFA (2017) Micro-irrigation market updates. Indian Council of Food and Agriculture, New Delhi. https://www.icfa.org.in/assets/doc/reports/1499354-micro-irrigation

IPCC (2019) Special report on climate change, desertification, land degradation, sustainable land management, food security, and greenhouse gas fluxes in terrestrial ecosystems (SRCCL). IPCC (Report); IPCC, Geneva. https://www.ipcc.ch/srccl/

IPCC (2023) AR6 synthesis report: climate change 2023. IPCC. https://www.ipcc.ch/sr15/

Javadi A (2022) Variable rate application in precision agriculture. Glob J Agric Health Sci 11:1000125

Kahlown MA, Raoof A, Hanif M (2000) Rice yield as affected by plant densities. Mona Experimental Project Bhalwal, Report No. 238

Kar S, Pramanick B, Brahmachari K, Saha G, Mahapatra BS, Saha A, Kumar A (2021) Exploring the best tillage option in rice based diversified cropping systems in alluvial soil of eastern India. Soil Tillage Res 205:104761. https://doi.org/10.1016/j.still.2020.104761

Karlen DL, Mausbach MJ, Doran JW, Cline RG, Harris RF, Schuman GE (1997) Soil quality: a concept, definition, and framework for evaluation. Soil Sci Soc Am J 61:4–10

Knox J, Daccache A, Hess T, Haro D (2016) Meta-analysis of climate impacts and uncertainty on crop yields in Europe. Environ Res Lett 11, 113004

Laik R, Singh SK, Pramanick B, Kumari V, Nath D, Dessoky ES, Attia AO, Hassan MM, Hossain A (2021) Improved method of boron fertilization in rice (*Oryza sativa* L.)–mustard (*Brassica juncea* L.) cropping system in upland calcareous soils. Sustain For 13:5037. https://doi.org/10.3390/su13095037

Le Dang Q, Shin TS, Park MS, Choi YH, Choi GJ, Jang KS, Kim IS, Kim J-C (2014) Antimicrobial activities of novel mannosyl lipids isolated from the biocontrol fungus Simplicillium lamellicola BCP against phytopathogenic bacteria. J Agric Food Chem 62(2014):3363–3370

Lehmann J (2007) A handful of carbon. Nature 447:143–144

Leng G, Huang M (2017) Crop yield response to climate change varies with crop spatial distribution pattern. Sci Rep 7:1463

Li M (2018) Climate change to adversely impact grain production in China by 2030. IFPRI, Washington, DC

Lobell DB, Burke Marshall B (2010) On the use of statistical models to predict crop yield responses to climate change. Agric For Meteorol 150:1443–1452

Mahato S, Mandal S, Rahman FH (2022) Study on economic benefit by following Agromet Advisory Services received from District Agromet Unit in the Lateritic Belt of Birbhum. Int J Agric Sci 7:102–107

Majumder D, Roy R, Rahman FH, Rudra BC (2020) Impact assessment of block-level agro advisories for saving input cost of farmers under old Gangetic plains of West Bengal: a case study in Malda. Curr J Appl Sci Technol 39(20):86–96

Malik RK, Kumar V, McDonald A (2018) Conservation agriculture-based resource-conserving practices and weed management in the rice-wheat cropping systems of the Indo-Gangetic Plains. Indian J Weed Sci 50:218

Marco B, Lamichhane JR, Booij K, Boonekamp P, Desneux N, Huber L, Kudsk P, Langrell SRH, Ratnadass A, Ricci P, Jean-Leouis S, Messean A (2015) Research and development priorities in

the face of climate change and rapidly evolving pests. In: Lichtfouse E (ed) Sustainable agriculture reviews, vol 17. Springer, Cham, pp 1–27. https://doi.org/10.1007/978-3-319-16742-8_1
Masters GJ, Brown VK, Clarke IP, Whittaker JB, Hollier JA (1998) Direct and indirect effects of climate change on insect herbivores: Auchenorrhyncha (Homoptera). Ecol Entomol 23(1):45–52
Misra NN, Dixit Y, Al-Mallahi A, Bhullar MS, Upadhyay R, Martynenko A (2020) IoT, big data, and artificial intelligence in agriculture and food industry. IEEE Internet Things J 9(9):6305–6324
Nongbet A, Mishra AK, Mohanta YK, Mahanta S, Ray MK, Khan M, Baek K-H, Chakrabartty I (2022) Nano fertilizers: a smart and sustainable attribute to modern agriculture. Plants 11:2587
Oladosu Y, Rafii MY, Arolu F, Chukwu SC, Salisu MA, Fagbohun IK, Muftaudeen TK, Swaray S, Haliru BS (2022) Superabsorbent polymer hydrogels for sustainable agriculture: a review. Horticulturae 8(7):605. https://doi.org/10.3390/horticulturae8070605
Pooniya V, Jat SL, Choudhary AK, Singh AK, Parihar CM, Bana RS, Swarnalakshmi K, Rana KS (2015) Nutrient Expert assisted site-specific-nutrient-management: an alternative precision fertilization technology for maize-wheat cropping system in South-Asian Indo-Gangetic Plains. Indian J Agric Sci 85(8):996–1002
Pramanick B, Kumar M, Naik BM, Kumar M, Singh SK, Maitra S, Naik BSSS, Rajput VD, Minkina T (2022) Long-term conservation tillage and precision nutrient management in maize–wheat cropping system: effect on soil properties, crop production, and economics. Agronomy 12:2766. https://doi.org/10.3390/agronomy12112766
Pramanick B, Mahapatra BS, Datta D, Dey P, Singh SP, Kumar A, Paramanik B, Awasthi N (2023) An innovative approach to improve oil production and quality of mustard (*Brassica juncea* L.) with multi-nutrient-rich polyhalite. Heliyon 9(3):e13997. https://doi.org/10.1016/j.heliyon.2023.e13997
Rajegowda MB, Janardhanagowda NA, Jagadeesha N, Babu BT (2008) Influence of agromet advisory services on economic impact of crops. J Agrometeorol 10:215–218
Rosenzweig C, Iglesias A, Yang XB, Epstein PR, Chivian E (2001) Implications of climate change for U.S. agriculture: extreme weather events, plant diseases and pests. Center for Health and the Global Environment, Harvard Medical School, Cambridge
Samanta S, Barman M, Chakraborty S, Banerjee A, Tarafdar J (2021) Involvement of small heat shock proteins (sHsps) in developmental stages of fall armyworm, Spodoptera frugiperda and its expression pattern under abiotic stress condition. Heliyon 7(4):e06906
Savary S, Willocquet L, Pethybridge SJ, Esker P, McRoberts N, Nelson A (2019) The global burden of pathogens and pests on major food crops. Nat Ecol Evol 3:430–439
Shankar T, Banerjee M, Malik GC, Dutta S (2020) Effect of nutrient management on growth, productivity and economics in rice based cropping system in lateritic soil of West Bengal. Int J Chem Stud 8(2):2694–2698
Sharma VK, Pandey RN, Kumar S, Chobhe KA, Chandra S (2016) Soil test crop response-based fertilizers recommendations under integrated nutrient management for higher productivity of pearl millet (*Pennisetum glaucum)* and (*Triticum aestivum)* wheat under long term experiment. Indian J Agric Sci 86:1076–1081
Shrestha A, Mitchell JP, Lanini WT (2007) Subsurface drip irrigation as a weed management tool for conventional and conservation tillage Tomato (Lycopersicon esculentum Mill.) production in semi-arid agroecosystems. J Sustain Agric 31:37–41
Sridhara S, Gopakkali P, Somashekarappa PR, Pradeep S, Agnihotri G (2014) Validation, usability and assessment of economic impact of agro advisories for southern transitional zone of Karnataka. J Agrometeorol 16:214–218
Srinivasrao C, Venkateswarlu B, Dixit S, Veeraiah R, Rammohan S, Reddy BS, Kundu S, Devi KG (2010) Implementation of contingency crop planning for drought in a tribal village in Andhra Pradesh: impact on food and fodder security and livelihoods. Indian J Dryland Agric Res Dev 25(1):23–30
Sturrock RN, Frankel SJ, Brown AV, Hennon PE, Kliejunas JT, Lewis KJ, Worrall JJ, Woods AJ (2011) Climate change and forest diseases. Plant Pathol 60:133–149

Taracena ML, Oliveira PL, Almendares O, Umaña C, Lowenberger C, Dotson EM, Paiva-Silva GO, Pennington PM (2015) Genetically modifying the insect gut microbiota to control chagas disease vectors through systemic RNAi. PLoS Negl Trop Dis 9(2):e0003358. https://doi.org/10.1371/journal.pntd.0003358

UNFCCC (2010) Handbook on vulnerability and adaptation assessment. Consultative group of experts on national communications from parties not included in Annex I to the Convention

Walther G-R, Post E, Convey P, Menzel A, Parmesan C, Beebee TJC, Fromentin J-M, Hoegh-Guldberg O, Bairlein F (2002) Ecological responses to recent climate change. Nature 416:389395

Wang YP, Li XG, Fu T, Wang L, Turner NC, Siddique KHM, Li F-M (2016) Multi-site assessment of the effects of plastic-film mulch on the soil organic carbon balance in semiarid areas of China. Agric For Meteorol 228(2016):42–51

Wang J, Vanga SK, Saxena R, Orsat V, Raghavan VGS (2018) Effect of climate change on the yield of cereal crops: a review. Climate 6(2):41. https://doi.org/10.3390/cli6020041

Whyard S, Singh AD, Wong S (2009) Ingested double-stranded RNAs can act as species-specific insecticides. Insect Biochem Mol Biol 39(11):824–832. https://doi.org/10.1016/j.ibmb.2009.09.007

Wu W-Y, Lo M-H, Wada Y, Famiglietti JS, Reager JT, Yeh PJ-F, Ducharne A, Yang Z-L (2020) Divergent effects of climate change on future groundwater availability in key mid-latitude aquifers. Nat Commun 11:3710. https://doi.org/10.1038/s41467-020-17581-y

Chapter 4
Impact of Climate Change and Abiotic Stresses on Food Crop Production

Shivani Ranjan, Smruti Ranjan Padhan, Sanjay Kumar, Sumit Sow, Dhirendra Kumar Roy, Ritwik Sahoo, Harun Gitari, and Souvik Sadhu

Abstract Climate change is a term that describes the long-term change in the statistical distribution of weather patterns. It leads to several biotic and abiotic stresses that have negative effect on crop production. The negative impacts of climate change get intensified by overexploitation of natural resources mainly in agriculture. Abiotic factors such as high and low temperatures, soil salinity, drought, and other stresses lead to approximately 60% yield loss, with interrelated effects on morphology, physiology, biochemistry, and molecular changes in crop. The negative impact of climate change on global crop production and food security is a major concern for researchers. Hence, there is a need to focus on the planned adaptation and mitigation strategies to enhance the resilience of agriculture.

Keywords Abiotic stress · Climate change · Heat stress · Water stress · Crop management

---

S. Ranjan (✉) · S. Sow · D. K. Roy
Department of Agronomy, Dr. Rajendra Prasad Central Agricultural University, Pusa, Bihar, India

S. R. Padhan
Division of Agronomy, ICAR-Indian Agricultural Research Institute, Pusa Campus, New Delhi, India

S. Kumar
Department of Agronomy, Bihar Agricultural University, Sabour, Bihar, India

R. Sahoo
Department of Plant Pathology, Uttar Banga Krishi Viswavidyalaya, Pundibari, Cooch Behar, West Bengal, India

H. Gitari
School of Agriculture and Environmental Sciences, Department of Agricultural Science and Technology, Kenyatta University, Nairobi, Kenya

S. Sadhu
Department of Soil Science and Agricultural Chemistry, Bihar Agricultural University, Sabour, Bihar, India

B. Pramanick et al. (eds.), *Climate-Smart Agricultural Technologies*, Sustainability Sciences in Asia and Africa, https://doi.org/10.1007/978-981-96-7699-6_4

4.1 Introduction

The adverse effects of devastating environmental changes have severely impacted agricultural production. Abiotic stresses in both direct as well as indirect ways are continuously affecting crop productivity due to unpredicted changes in environmental conditions. The concentration of carbon dioxide (CO_2) in the atmosphere is expected to reach 400 μmol^{-1} due to deforestation, excessive use of fossil fuels, etc., and will double to 800 μmol^{-1} by the end of the twenty-first century (Vaughan et al. 2018). Climate change and environmental variation have resulted in an increasing number of stress episodes that affect agricultural crops. Climate-smart crop varieties must be developed to address these constraints and ensure food security. Crop experiences stress, such as waterlogging, drought, heat, cold, flooding, and salinity during the growth period (Fig. 4.1). At the end of the twenty-first century, the average temperature of the earth is expected to increase by 2–4.5 °C. According to IPCC-2014, the period between the nineteenth and twenty-first centuries has experienced the most warming. Extreme precipitation leads to floods that cause destruction, while the absence or scarcity of rainfall over an extended period results in drought stress. Industrialization is a significant contributor to global climate change (Dar et al. 2020). There will be an increase in the frequency of weather disasters which will ultimately disturb the entire global ecosystem. Ziska et al. (2016) have

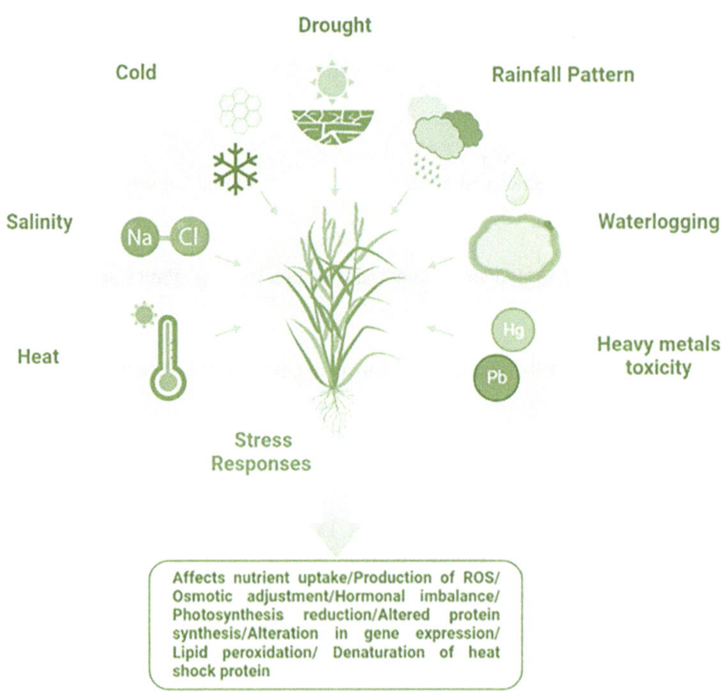

Fig. 4.1 Crop response to different abiotic stresses

found that susceptibility to climate change is more in developed countries as compared to developing ones.

According to Giannini et al. (2017), maintaining sustainability along with food security are the two major issues of the twenty-first century. The global population is supposed to reach 9.7 billion in the year 2050 and the requirement for food will be escalated by about 85% (Rosenzweig et al. 2013). The increased frequency of drought and high-intensity rainfall, salinity stress, temperature fluctuation, and insect pest attacks lead to a reduction in crop productivity (Datta et al. 2024). Therefore, there is an immediate requirement to know and understand the nature as well as the severity of diverse extreme weather conditions and their impact on agriculture. This chapter emphasizes the effect of abiotic stresses induced by climate change on the production of food crops, as well as approaches to cope with them.

Crop growth and yield are affected by various abiotic stresses such as floods, droughts, and extreme temperatures. Farmers must be informed regarding the appropriate management options to deal with these stresses (Pan et al. 2020). As a result of the burning of crop residues, forest fires, and rapid industrialization there is an increase in the concentration of aerosols, which has led to a decrease in direct incoming solar radiation and an increase in diffused radiation. This event may reduce the total radiation, which may change combined weather patterns. Depending on the extent of damage to the total cultivated area, abiotic stresses have a major impact on crop production.

4.2 Crop Response to Climate Change-Induced Abiotic Stresses

4.2.1 Heat Stress

Crop stress can be caused by exposure to high or low temperatures, with different crop species responding differently at different stages of their growth. Heat waves can be defined as the period of abnormally hot weather which lasts for 3 or more days. It can increase plant respiration more than photosynthesis thereby reducing net carbon gain. Higher temperatures affect the function of Rubisco and electron transport in the chloroplast, leading to a decrease in photosynthetic capacity. Moreover, higher temperatures or increased atmospheric evaporative demand can cause stomatal closure, which reduces CO_2 flux into leaves and further reduces photosynthesis. Rising temperature can cause a significant decrease in wheat yield, with a 10% reduction per 1 °C rise in temperature. In many annual crops, temperatures over 36 °C can decrease seed set, particularly during the critical period of 1–3 days around flowering. Heat stress can also reduce pollen viability in maize and induce grain sterility in rice.

Burning fossil fuels and deforestation increases carbon dioxide emissions, which indirectly increase temperature, and cause heat stress in crops. Heat stress can have severe consequences on plant development, leading to yield loss by affecting growth, photosynthesis, and pollination. Heat shock proteins denature, reactive oxygen species (ROS) production occurs, and enzyme activities change under heat stress, reducing photosynthetic activity. However, heat stress can have an advantageous effect on the cell division cycle, which can improve yield.

4.2.2 Cold and Frost Stress

Cold and frost stresses include chilling injury that occurs in the temperature range of 0–20 °C and frost injury which occurs at temperatures less than 0 °C. These affect crop growth, net photosynthesis, respiration, and yield (Takahashi et al. 2018). Crop yield losses due to these stresses range from 15% to 50% in wheat, 30% in gram, and 50% to 60% in mustard. The extent of the injury depends on the crop type and growth stage. Frost-affected plants have a water-soaked and scorched appearance due to the rupture of cell walls caused by ice crystal formation. Frost and cold injury in plants are caused by accelerated ROS production resulting from freezing, which leads to lipid peroxidation.

4.2.3 Drought

Insufficient soil moisture or excessive water loss through evaporation and transpiration caused by weather conditions leads to drought stress (Mukherjee et al. 2018). This leads to a decrease in soil moisture required for plant growth and development, resulting in reduced biomass, growth rate, and ultimately yield. In this era of climate change, the severity of drought severity has increased and over 50% of arable land is expected to be affected by 2050 (Vurukonda et al. 2016). Table 4.1 summarizes the adverse effects of drought stress on various crops. Drought stress affects plant functions by altering plant turgor and water potential, morphology, physiology, and molecular traits. In crops, water serves as a carrier for nutrient transport to roots. Nutrients such as calcium (Ca), silicon (Si), magnesium (Mg), and nitrate (NO_3^-) whose availability depends on diffusion and mass flow are reduced due to drought stress (Selvakumar et al. 2011).

During drought stress, plant structure and physiological processes are affected due to the negative interaction between ROS and abscisic acid (ABA) which ultimately inhibits plant growth (Jones 2015). Drought causes the accumulation of ABA that generates H_2O_2 in plants. In order to ensure water availability in the aerial parts of the crop, roots sense water availability and transmit signals through root to shoot to decrease water loss and turgidity of guard cells which causes closure of the stomata (Menezes-Silva et al. 2017).

ROS production results from electron leakage during different pathways of electron transport, causing membrane disruption and lipid peroxidation in plants (Jan

Table 4.1 Effect of drought stress on different crop species

Crop	Stage	Effect	References
Triticum aestivum	Grain filling	Reduced grain yield (34%) and weight (27%)	Saeidi and Abdoli (2015)
Cicer arietinum	Vegetative and anthesis	Reduced total chlorophyll, stomatal CO_2 concentration and grain yield (45–66%)	Mafakheri et al. (2010)
Oryza sativa	Germination and seedling growth stage	Decreased root and shoot length, dry weight, activity of catalase (CAT), peroxidase and superoxide dismutase (SOD) increases	Lum et al. (2014)
Zea mays	Germination and early growth stage	Reduced germination percentage, and seed vigour	Khodarahmpour (2011)
Sesamum indicum	Germination and seedling stage	Decreased germination percentage, and dry weight	Bahrami et al. (2012)
Brassica napus	Flowering and pod-filling stage	Decreased chlorophyll content, grain yield and number of pods per plant	Din et al. (2011)
Vigna unguiculata	Vegetative growth stage and flowering stage	Decreased growth attributes and chlorophyll content	Ndiso et al. (2016)

et al. 2016). Partial stomatal closure caused by drought stress decreases the rate of photosynthesis. Moreover, reduced water availability leads to shrinkage of the cell, and increased viscosity of cellular contents which damages the proteins and enzymes responsible for photosynthesis. Partial stomatal closure also reduces the entry of CO_2 resulting in the production of H_2O_2 in peroxisomes and mitochondria via the process of photorespiration, eventually limiting photosynthesis (Dixit et al. 2017).

4.2.4 Change in Rainfall Pattern

Global warming has a negative effect on rainfall patterns. A rise in temperature results in storms and dry weather conditions in some areas. Increasing air temperature increases the capacity of air to hold moisture, which results in an increase in atmospheric moisture. The uneven pattern of rainfall poses a considerable challenge to agriculture. The variability of weather patterns makes forecasting rainfall more difficult than temperatures. Areas that are presently wet are expected to become even wetter, but it is not easy to determine the extent of the influence and how it will affect the local ecosystem. Because of these altered precipitation patterns, plants, and soil can be adversely affected, mainly during intense rainfall events. Many plants cannot survive in drought-affected areas due to the absence of optimum moisture in the soil ultimately leading to increased vulnerability to disease and fire.

4.2.5 *Waterlogging and Salinity Stress*

According to Schmidhuber and Tubiello (2007), waterlogging, moisture stress, and salinity are the three that can negatively affect crop production. Waterlogging is aggravated by flooding caused either due to snowmelt or rainfall events (Espeland and Kettenring 2018). Accumulation of water-soluble sodium chloride (NaCl), sodium carbonate (Na_2CO_3), or partially water-soluble calcium chloride ($CaCl_2$) leads to salinity stress. The adverse impact of salinity stress on different crops is presented in Table 4.2. On the basis of adaptive mechanisms towards saline soil, plants are classified as glycophytes or halophytes. Glycophytes are unable to tolerate high levels of salinity and eventually die, while halophytes can withstand high salt concentrations due to enhanced tolerance mechanisms. Salinity leads to osmotic stress restricting water uptake, decreased growth, and eventually reduction in productivity (Shrivastava and Kumar 2015). It can occur at any stage of crop growth. Hence, it is important to understand and find potential solution for reducing the detrimental effect of waterlogging and salinity on crop production for ensuring food security (Morris and Brewin 2014; Pramanick et al. 2023).

Table 4.2 Effect of different types of salinity stress on various crops

Crop	Concentration of NaCl (mM) and type of stress	Effect	References
Oryza sativa	210 (Moderate)	Decreased dry and fresh weight, leaf relative water and starch content	Amirjani (2011)
Glycine max	30–110 (Low)	Decreased plant height, root and shoot weight and leaf number	Dolatabadian et al. (2011)
Solanum tuberosum	30–50 (Low)	Reduced tuber yield and root length	Akhtar et al. (2015)
Zea mays	30–110 (Low)	Inhibition of root and shoot length	Turan et al. (2010)
Vicia faba	60–240 (Moderate)	Reduced plant height, osmotic potential and total chlorophyll	Qados (2011)
Phaseolus vulgaris	50–210 (Low)	Decreased dry matter content in root and shoot and photosynthetic pigments	Taïbi et al. (2016)

4.3 Impact of Climate Change and Abiotic Stresses on Food Security

4.3.1 Effect on Some Major Crops

4.3.1.1 Rice

Rice cultivation thrives in areas with abundant rainfall and low labour costs. However, extreme abiotic factors, such as floods, frost damage, osmotic stress, low and high temperatures, heavy rainfall, and salinity, pose a serious threat to rice yield. High temperatures negatively affect rice yield, with even a brief period of elevated temperature during anthesis resulting in a significant number of sterile grains. Moreover, increased temperatures in conjunction with rising CO_2 levels reduce the carbon sink capacity of the grain (Kar et al. 2021). Extremely cold temperatures also impede rice yield, as the growth of the seedling is highly susceptible to chilling injury, which can increase ROS and cause cellular damage. Moisture stress affects rice at the molecular, biochemical, and physiological levels, altering gene expression, accumulating osmoprotectants, and reducing transpiration and photosynthesis. Rice production systems in semi-arid and arid regions are particularly at risk of drought as they heavily rely on water supply, owing to the diversity of rice ecosystems, its semi-aquatic phylogenetic origin, and growing conditions.

4.3.1.2 Sugarcane

The growth and development of sugarcane are largely dependent on climate, as it influences several biochemical processes. Photosynthetic efficiency in sugarcane increases linearly with temperature ranging between 8 and 34 °C. Climate change will affect precipitation, temperature, pollination, and pest populations which will ultimately reduce sugarcane yield. High temperature increases the incidence of smut and other diseases. Change in rainfall patterns during the growing season can impede growth and increase weed infestation. Elevated levels of carbon dioxide will directly affect stomatal aperture and photosynthesis.

4.3.1.3 Maize

Maize is a highly valuable food crop that is cultivated across approximately 150 million hectares worldwide due to its adaptability to various soil types and environmental conditions (Chen et al. 2017). Temperature rises are also predicted to result in reduced maize yields during the growing season. While high rainfall may negatively impact maize in the short term, it can have a positive long-term effect on its yield (Shakoor et al. 2017). Moderate increases in precipitation during summer season, such as 50 mm, can increase yield by 5–10%, compensating for some of the adverse effects of rising temperatures (Shiriga et al. 2014).

Researchers have reported that extreme temperatures will significantly decrease maize yield (10%) in the next 20 years (Lashkari et al. 2012).

4.3.1.4 Wheat

The most economic portion in wheat is the spike and the number of grains per spike depends both on the genetic makeup as well as environmental conditions (Lopes et al. 2015). An increase in the concentration of CO_2 is positively correlated with yield since it results in increased photosynthesis, leading to higher output (Kosová et al. 2013). Aggarwal and Sinha (1993) found that wheat grain yield significantly increased at a CO_2 concentration of 425 ppm with no temperature rise. However, higher temperatures have negative effects on crops, increasing the rate of evapotranspiration, water loss due to stomatal opening, and sometimes even causing plant death in severe cases (Johansson et al. 2020). Higher temperatures can reduce kernel weight by approximately 15–25% due to an increase in the rate of grain filling, which shortens the grain filling duration (Lopes et al. 2015).

4.4 Strategies to Combat Climate Change and Abiotic Stress

Abiotic decreases crop productivity and endangers the global food safety. Therefore, it is the need of the hour to adopt effective strategies to deal with crop abiotic stress on crops (Tao and Zhang 2010). Strategies to mitigate climate change and abiotic stress can be broadly divided into breeding approaches and agronomic or crop phenotypic management approaches.

4.4.1 Breeding Strategies

Breeding for abiotic stress tolerance involves selecting plants with desirable traits such as drought tolerance, heat tolerance, salinity tolerance, and flood tolerance. Crop ideotype refers to the set of desirable traits that are selected and optimized for a particular crop in a specific environment. By modifying crop ideotype, breeders can develop cultivars that are better suited to cope with changing environmental conditions (Marcinkowski and Piniewski 2018). For example, ideotypes can be developed with traits such as deep root systems for improved drought tolerance, short stature to resist lodging during high winds, or enhanced photosynthetic efficiency for improved growth in low light conditions. Several genes have been identified that can help plants tolerate abiotic stress (Feng et al. 2019). By transferring this gene to other plants, it is possible to improve their tolerance to drought stress.

Similarly, the HKT1 gene in rice has been found to regulate the salt uptake (Garriga et al. 2017). By transferring this gene to other crops, it is possible to improve their tolerance to saline soils. The following are some of the breeding strategies used to develop crops with abiotic stress tolerance:

4.4.1.1 Conventional Breeding Methods

In conventional breeding, plants with desirable qualities are chosen and crossed to create offspring with enhanced traits. For many years, this technique has been utilized to create crops with higher yields, disease resistance, and quality (Oladosu et al. 2019). By choosing plants with desired qualities like drought tolerance, heat tolerance, salinity tolerance, and flood tolerance, conventional breeding can be utilized to generate crops with abiotic stress resistance (Yijun et al. 2022).

4.4.1.2 Genetic Engineering

An effective tool for reducing abiotic stress on crops is genetic engineering involving the introduction of genes that give crops stress resistance (Naz et al. 2023). Crops have been successfully engineered with genes encoding for proteins that can withstand abiotic stress, including aquaporins, and osmo-protective enzymes (Bhattacharya et al. 2019). The modification of existing genes to improve their expression is another aspect of genetic engineering. For instance, it has been demonstrated that overexpressing the genes for the enzymes CAT and SOD can lessen oxidative stress in plants (Sharma et al. 2021).

4.4.1.3 Genomic Selection

Genomic selection is a breeding technique that involves selecting plants with desirable traits based on their genomic data. It chooses plants with the required features based on their genomic information after identifying genomic areas linked with particular traits using high-throughput genotyping and phenotyping techniques (Singh et al. 2018). This method is faster and more accurate than conventional breeding and with the use of this technique, crops have been developed that are more resilient to abiotic challenges like salinity, heat, and drought.

4.4.1.4 Transgenic Breeding

Transgenic breeding involves introducing foreign genes from other plants or organisms into the genome of the target plant in order to produce offspring with desirable traits (Bhatnagar-Mathur et al. 2008). Usually, the foreign genes come from different plant species or even organisms that are not plants but yet have the desired properties. Transgenic breeding aims to produce crops that are more resistant and tolerant of challenging environmental circumstances, such as salinity, high heat, and drought (Cominelli and Tonelli 2010). Agrobacterium-mediated transformation, electroporation, and gene gun are some of the techniques used in the transgenic breeding process to locate the desired gene, isolate it, and then introduce it into the crop genome (Anwar and Kim 2020). The crop can express the desired feature once the foreign gene has been incorporated into the genome of the plant.

4.4.1.5 Genome Editing

Genome editing, commonly referred to as gene editing, is a potent technology that enables researchers to precisely pinpoint alterations to the DNA (Ansari et al. 2020; Maharajan et al. 2022). By enabling scientists to make genetic adjustments that can strengthen capacity of crop to withstand abiotic stress and fend off the consequences of climate change, it has the potential to revolutionize plant breeding. Molecular scissors, like the CRISPR-Cas9 system, are used in genome editing to cut the DNA at a specific region and insert desired alterations (Razzaq et al. 2022). This can involve removing, substituting, or introducing particular genes, which can change the features of the plant and increase its resistance to environmental stressors (Wang et al. 2022a).

4.4.2 *Agronomic Approaches*

There are several agronomic strategies available to help combat these challenges and promote resilient and adaptive agricultural practices.

4.4.2.1 Soil Management Techniques

In order to prevent climate change and abiotic stress on crops, soil management is essential. By implementing soil management techniques including conservation tillage, crop rotation, cover crops, and integrated nutrient management, the resistance of crop to abiotic stress can be increased. In order to decrease soil erosion and maintain soil moisture, conservation tillage includes minimizing soil disturbance during cultivating (Wang et al. 2022a, b; Pramanick et al. 2022). Crop rotation aids in halting the spread of disease and pest cycles and preserves soil fertility (Rodriguez-Ortega et al. 2017). Increased soil organic matter and improved soil health are three benefits of using cover crops (Das et al. 2021). Utilizing both organic and inorganic fertilizers in a balanced way to increase soil fertility while lowering greenhouse gas emissions is known as integrated nutrient management (Rubio et al. 2010).

4.4.2.2 Appropriate Nutrient Management

It plays a crucial role in mitigating abiotic stress on crops. An adequate and balanced nutrient supply can help plants to better tolerate and recover from various abiotic stresses. For instance, the application of micronutrients like zinc, iron, silicon, and manganese can enhance plant tolerance to drought stress (Thakral et al. 2021; Laik et al. 2021). Nutrient management practices such as precision fertilization, use of slow-release fertilizers, and application of organic fertilizers can improve nutrient-use efficiency and reduce environmental pollution (Dhaliwal et al. 2022; Dey et al. 2022).

4.4.2.3 Foliar Spray Techniques

Foliar spraying is an agricultural technique in which a fertilizer solution is sprayed directly onto the leaves of plants. It works well to protect crops from abiotic stress, especially during crucial growth stages. Crops' ability to absorb nutrients more effectively encourages growth and development, which increases their ability to withstand abiotic stressors including salinity, heat, and drought (Malko et al. 2022). Additionally, foliar spraying with particular plant hormones can improve crops' ability to withstand stress (Thakral et al. 2021). The timing, frequency, and concentration of the solution administered, as well as the stage of crop growth, all have a significant role in foliar spray effectiveness (Peng and Guan 2021). To guarantee maximal benefit and prevent potential bad consequences like phytotoxicity, careful attention should be paid to these elements (De Pinto et al. 2020).

4.4.2.4 Crop Diversification

By lowering reliance on a single crop and boosting the resilience of agricultural systems, crop diversity helps battle climate change and abiotic stress. Crop diversification entails rotating different crops through various seasons or growing many crops with various features in the same field. This lessens the likelihood that crops will fail owing to weather variations, pests, or illnesses (Cortinovis et al. 2020). Additionally, diverse cropping systems can improve pest control, nutrient cycling, and soil fertility. By fostering local food production, maintaining biodiversity, and providing farmers with additional income, crop diversification can also support sustainable development (Kar et al. 2021).

4.4.2.5 Climate Resilient Crops

Crops that have been developed to endure extreme weather conditions including drought, flooding, and high temperatures are said to be climate-resilient. These crops are created using breeding approaches that emphasize qualities including disease resistance, heat tolerance, and water use efficiency. Climate-resilient crops can retain productivity even in the face of shifting climatic conditions by adopting these characteristics. In areas where climate fluctuation is a significant concern, this can help to ensure farmers' food security and means of subsistence. By using less inputs like water, herbicides, and fertilizers, all of which have the potential to harm the environment, climate-resilient crops can also help to mitigate the effects of climate change. Traits of plants which are required to survive in various abiotic stress conditions are depicted in Table 4.3. Some of the important characteristics of climate resiliency are early maturing crop cultivar, higher drought index; fitting to several cropping systems, photo insensitivity, etc.

4.4.2.6 Use of Plant Hormones

Plant hormones can be defined as chemical substances that occur naturally and play an important role in regulating plant growth and development. Several environmental stresses, such as salt, drought, and extremely high or low temperatures have been overcome by these hormones. Abscisic acid (ABA) and salicylic acid (SA) are important plant hormones that play significant role in growth, development, and stress responses (Zahid et al. 2023). Exogenous application of plant hormones can enhance abiotic stress tolerance in crops. For example, exogenous application of ABA can enhance drought tolerance in maize and wheat through closing stomata,

Table 4.3 Specific traits required by plants to survive and thrive in various abiotic stress conditions

Abiotic stress	Plant characteristics needed for different abiotic stress tolerance	Examples of plants with these traits	References
Drought	Deep root system, developmental plasticity, higher root-shoot ratio, efficient water uptake, thick cuticle, water storage, reduced transpiration	Succulents, desert shrubs, xerophytes	Hammer et al. (2009) and Spollen et al. (2008)
Heat	Heat tolerance proteins, increased antioxidant activity, improved membrane stability, heat shock proteins	Maize, tomato, rice	Bita and Gerats (2013)
Cold	Ability to maintain cellular function at low temperatures, ability to recover from freezing, tolerance of ice formation	Winter wheat, rye, barley	Pan et al. (1994)
Salinity	Salt exclusion or compartmentalization, osmotic adjustment, ion homeostasis, water-use efficiency	Quinoa, Spartina grass, salt-tolerant rice	James et al. (2006)
Heavy metals	Chelation, exclusion, reduction, compartmentalization, and tolerance to high metal concentrations	Sunflower, Indian mustard	Sedlacko et al. (2020)
Acid soils	Root adaptation to aluminium and low pH, phosphorus uptake ability	Common bean, wheat, soybean	Zhu and Shen (2023)
Flooding	Tolerance to anoxia, aerating roots, and other physiological adaptations	Rice, wheat, maize	Xu et al. (2006)

lowering transpiration rates, and promoting root growth in turn strengthening plants in conserving water. SA application has been shown to enhance tolerance to heavy metal toxicity in plants. Gibberellic acid can help plants grow and develop, including the lengthening of stems and the formation of fruits, and can lessen the effects of abiotic stresses. It has also been demonstrated that additional plant hormones, such as auxin and cytokinin, enhance crop growth and stress resistance (Peleg and Blumwald 2011). Plant hormones can help farmers increase crop output and adjust to shifting environmental conditions in a sustainable and affordable manner.

4.4.2.7 Use of Plant Growth Promoting Rhizobacteria (PGPR)

Various soil bacteria are known to colonize the roots of plants and stimulate their growth and development. These are called as Plant Growth Promoting Rhizobacteria (PGPR). Furthermore, they can increase the susceptibility of crops to abiotic stresses, such as drought, salinity, and heavy metal toxicity (Goswami and Suresh 2020). PGPR has the potential to improve soil health and fertility, nutrient uptake, and defence against pathogenic microorganisms. It can also induce systemic resistance in plants that helps plants to survive during abiotic stresses. Kaushal and Wani (2016) reported that *Bacillus amyloliquefaciens* (PGPR) enhances drought tolerance in maize. PGPR has gained considerable attention in recent years as a sustainable and environmentally friendly approach to improve crop productivity and resilience to climate change.

4.4.2.8 Use of Nanoparticles

Nanoparticles are small particles with unique physicochemical properties that can be used to confer abiotic stress tolerance in crops. Nanoparticles have been used to enhance plant growth and development, improve nutrient uptake, and increase stress tolerance in crops (Jamil et al. 2023). Kareem et al. (2022) observed that nano-sized zinc oxide particles can improve drought tolerance of maize plants by enhancing antioxidant enzyme activity and regulating water use efficiency.

4.5 Conclusion

Abiotic stresses lead to decrease in growth and chlorophyll content in a variety of plant species. Moreover, salinity stress decreases dry matter, chlorophyll content, and phenolic content of crops. The occurrence of salts in a concentration of around 400 mM causes increased oxidative stress and reduced osmotic potential in crops, affecting lipid peroxidation and electrolyte leakage. Researchers have recognized a number of phytoregulators, such as phytohormones, osmolytes, and biostimulants, that can improve plant tolerance to several abiotic stresses through protein, proline, and antioxidant metabolism. In order to adapt and acquire tolerance, crops respond to these changes in different morphologically, structurally, physiologically, biochemically, and molecularly ways. These stresses may cause alterations in the soil-plant-atmospheric continuum (SPAC), resulting in reduction in yields. As a result of the continued effects of climate change on these stresses, especially low or high temperature and drought or flooding in various locations, suitable adaptation strategies must be adapted to meet future demands. Climate-resilient agricultural practises, which include the continual management of natural and man-made resources, may greatly boost production and produce quality, and are critical for ensuring sustainability in the face of climate change.

References

Aggarwal PK, Sinha SK (1993) Effect of probable increase in carbon dioxide and temperature on wheat yields in India. J Agric Meteorol 48(5):811–814

Akhtar SS, Andersen MN, Liu F (2015) Biochar mitigates salinity stress in potato. J Agron Crop Sci 201(5):368–378

Amirjani MR (2011) Effect of salinity stress on growth, sugar content, pigments and enzyme activity of rice. Int J Bot 7(1):73–81

Ansari WA, Chandanshive SU, Bhatt V, Nadaf AB, Vats S, Katara JL, Sonah H, Deshmukh R (2020) Genome editing in cereals: approaches, applications and challenges. Int J Mol Sci 21(11):4040

Anwar A, Kim J-K (2020) Transgenic breeding approaches for improving abiotic stress tolerance: recent progress and future perspectives. Int J Mol Sci 21(8):2695

Bahrami H, Razmjoo J, Ostadi Jafari A (2012) Effect of drought stress on germination and seedling growth of sesame cultivars (*Sesamum indicum* L.). Int J AgriSci 2(5):423–428

Bhatnagar-Mathur P, Vadez V, Sharma KK (2008) Transgenic approaches for abiotic stress tolerance in plants: retrospect and prospects. Plant Cell Rep 27:411–424

Bhattacharya S, Dhar S, Banerjee A, Ray S (2019) Structural, functional, and evolutionary analysis of late embryogenesis abundant proteins (LEA) in Triticum aestivum: a detailed molecular level biochemistry using in silico approach. Comput Biol Chem 82:9–24

Bita CE, Gerats T (2013) Plant tolerance to high temperature in a changing environment: scientific fundamentals and production of heat stress-tolerant crops. Front Plant Sci 4:273

Chen J, Chopra R, Hayes C, Morris G, Marla S, Burke J, Xin Z, Burow G (2017) Genome-wide association study of developing leaves' heat tolerance during vegetative growth stages in a Sorghum Association Panel. Plant Genome 10(2):2016–2009

Cominelli E, Tonelli C (2010) Transgenic crops coping with water scarcity. New Biotechnol 27(5):473–477

Cortinovis G, Di Vittori V, Bellucci E, Bitocchi E, Papa R (2020) Adaptation to novel environments during crop diversification. Curr Opin Plant Biol 56:203–217

Dar AA, Chen J, Shad A, Pan X, Yao J, Bin-Jumah M, Allam AA, Huo Z, Zhu F, Wang Z (2020) A combined experimental and computational study on the oxidative degradation of bromophenols by Fe(VI) and the formation of self-coupling products. Environ Pollut 258:113678. https://doi.org/10.1016/j.envpol.2019.113678

Das A, Rangappa K, Basavaraj S, Dey U, Haloi M, Layek J, Idapuganti RG et al (2021) Conservation tillage and nutrient management practices in summer rice (*Oryza sativa* L.) favoured root growth and phenotypic plasticity of succeeding winter pea (*Pisum sativum* L.) under eastern Himalayas, India. Heliyon 7(5):e07078

Datta D, Ghosh S, Das D, Singh SV, Mazumdar SP, Mandal S, Singh Y (2024) Biochar imparting abiotic stress resilience. In: Biochar production for green economy. Academic Press, Elsevier, pp 227–248. https://doi.org/10.1016/B978-0-443-15506-2.00021-3

De Pinto A, Cenacchi N, Kwon H-Y, Koo J, Dunston S (2020) Climate smart agriculture and global food-crop production. PLoS One 15(4):e0231764

Dey P, Mahapatra BS, Pramanick B, Pyne S, Pandit P (2022) Optimization of seed rate and nutrient management levels can reduce lodging damage and improve yield, quality and energetics of subtropical flax. Biomass Bioenergy 157:106355. https://doi.org/10.1016/j.biombioe.2022.106355

Dhaliwal SS, Sharma V, Shukla AK (2022) Impact of micronutrients in mitigation of abiotic stresses in soils and plants—a progressive step toward crop security and nutritional quality. Adv Agron 173:1–78

Din J, Khan SU, Ali I, Gurmani AR (2011) Physiological and agronomic response of canola varieties to drought stress. J Anim Plant Sci 21(1):78–82

Dixit S, Singh A, Sandhu N, Bhandari A, Vikram P, Kumar A (2017) Combining drought and submergence tolerance in rice: marker-assisted breeding and QTL combination effects. Mol Breed 37:1–12

Dolatabadian A, Sanavy SAMM, Ghanati F (2011) Effect of salinity on growth, xylem structure and anatomical characteristics of soybean. Not Sci Biol 3(1):41–45

Espeland EK, Kettenring KM (2018) Strategic plant choices can alleviate climate change impacts: a review. J Environ Manag 222:316–324

Feng W, Li J, Long S, Wei S (2019) A DREB1 gene from zoysiagrass enhances Arabidopsis tolerance to temperature stresses without growth inhibition. Plant Sci 278:20–31

Garriga M, Raddatz N, Véry A-A, Sentenac H, Rubio-Meléndez ME, González W, Dreyer I (2017) Cloning and functional characterization of HKT1 and AKT1 genes of Fragaria spp.—relationship to plant response to salt stress. J Plant Physiol 210:9–17

Giannini TC, Costa WF, Cordeiro GD, Imperatriz-Fonseca VL, Saraiva AM, Biesmeijer J, Garibaldi LA (2017) Projected climate change threatens pollinators and crop production in Brazil. PLoS One 12(8):e0182274. https://doi.org/10.1371/journal.pone.0182274

Goswami M, Suresh DEKA (2020) Plant growth-promoting rhizobacteria—alleviators of abiotic stresses in soil: a review. Pedosphere 30(1):40–61

Hammer GL, Dong Z, McLean G, Doherty A, Messina C, Schussler J, Zinselmeier C, Paszkiewicz S, Cooper M (2009) Can changes in canopy and/or root system architecture explain historical maize yield trends in the US corn belt? Crop Sci 49(1):299–312

James RA, Davenport RJ, Munns R (2006) Physiological characterization of two genes for Na+ exclusion in durum wheat, Nax1 and Nax2. Plant Physiol 142(4):1537–1547

Jamil A, Umair M, Rizwan A, Ahmed R (2023) Molecular basis of abiotic stress alleviation by nanoparticles. In: Sustainable plant nutrition. Academic Press, pp 355–380

Jan SA, Shinwari ZK, Rabbani MA (2016) Morpho-biochemical evaluation of Brassica rapa subspecies for salt tolerance. Genetika 48(1):323–338

Johansson E, Henriksson T, Prieto-Linde ML, Andersson S, Ashraf R, Rahmatov M (2020) Diverse wheat-alien introgression lines as a basis for durable resistance and quality characteristics in bread wheat. Front Plant Sci 11:1067

Jones AM (2015) A new look at stress: abscisic acid patterns and dynamics at high-resolution. New Phytol 210(1):38–44. https://doi.org/10.1111/nph.13552

Kar S, Pramanick B, Brahmachari K, Goutam Saha BS, Mahapatra AS, Kumar A (2021) Exploring the best tillage option in rice based diversified cropping systems in alluvial soil of eastern India. Soil Tillage Res 205:104761

Kareem HA, Hassan MU, Zain M, Irshad A, Shakoor N, Saleem S, Niu J et al (2022) Nanosized zinc oxide (n-ZnO) particles pretreatment to alfalfa seedlings alleviate heat-induced morpho-physiological and ultrastructural damages. Environ Pollut 303:119069

Kaushal M, Wani SP (2016) Plant-growth-promoting rhizobacteria: drought stress alleviators to ameliorate crop production in drylands. Ann Microbiol 66:35–42

Khodarahmpour Z (2011) Effect of drought stress induced by polyethylene glycol (PEG) on germination indices in corn (Zea mays L.) hybrids. Afr J Biotechnol 10(79):18222–18227

Kosová K, Vítámvás P, Planchon S, Renaut J, Vanková R, Prášil IT (2013) Proteome analysis of cold response in spring and winter wheat (Triticum aestivum) crowns reveals similarities in stress adaptation and differences in regulatory processes between the growth habits. J Proteome Res 12(11):4830–4845

Laik R, Singh SK, Pramanick B, Kumari V, Nath D, Dessoky ES, Attia AO, Hassan MM, Hossain A (2021) Improved method of boron fertilization in rice (*Oryza sativa* L.)–mustard (*Brassica juncea* L.) cropping system in upland calcareous soils. Sustainability 13:5037. https://doi.org/10.3390/su13095037

Lashkari A, Alizadeh A, Rezaei EE, Bannayan M (2012) Mitigation of climate change impacts on maize productivity in northeast of Iran: a simulation study. Mitig Adapt Strateg Glob Chang 17:1–16

Lopes MS, El-Basyoni I, Baenziger PS, Singh S, Royo C, Ozbek K, Aktas H et al (2015) Exploiting genetic diversity from landraces in wheat breeding for adaptation to climate change. J Exp Bot 66(12):3477–3486

Lum MS, Hanafi MM, Rafii YM, Akmar ASN (2014) Effect of drought stress on growth, proline and antioxidant enzyme activities of upland rice. JAPS: J Anim Plant Sci 24:5

Mafakheri A, Siosemardeh AF, Bahramnejad B, Struik PC, Sohrabi Y (2010) Effect of drought stress on yield, proline and chlorophyll contents in three chickpea cultivars. Aust J Crop Sci 4(8):580–585

Maharajan T, Ajeesh Krishna TP, Rakkammal K, Ceasar SA, Ramesh M (2022) Application of CRISPR/Cas system in cereal improvement for biotic and abiotic stress tolerance. Planta 256(6):106

Malko MM, Khanzada A, Wang X, Samo A, Li Q, Jiang D, Cai J (2022) Chemical treatment refines drought tolerance in wheat and its implications in changing climate: a review. Plant Stress 6:100118

Marcinkowski P, Piniewski M (2018) Effect of climate change on sowing and harvest dates of spring barley and maize in Poland. Int Agrophys 32:2

Menezes-Silva PE, Sanglard LM, Ávila RT, Morais LE, Martins SC, Nobres P, Patreze CM et al (2017) Photosynthetic and metabolic acclimation to repeated drought events play key roles in drought tolerance in coffee. J Exp Bot 68(15):4309–4322. https://doi.org/10.1093/jxb/erx211

Morris J, Brewin P (2014) The impact of seasonal flooding on agriculture: the spring 2012 floods in Somerset, England. J Flood Risk Manag 7(2):128–140

Mukherjee S, Mishra A, Trenberth KE (2018) Climate change and drought: a perspective on drought indices. Curr Clim Chang Rep 4(2):145–163. https://doi.org/10.1007/s40641-018-0098-x

Naz M, Akhtar K, Khan A, Zaib S, Tariq M, Raza MA, Zhou J, Dai Z, Daolin D (2023) Genetic engineering strategies for regulation of phytohormones in plants exposed to biotic and abiotic stresses. In: Phytohormones and stress responsive secondary metabolites. Academic Press, pp 265–274

Ndiso JB, Chemining'wa GN, Olubayo FM, Saha HM (2016) Effect of drought stress on canopy temperature, growth and yield performance of cowpea varieties. Int J Plant Soil Sci 9(3):1–12

Oladosu Y, Rafii MY, Samuel C, Fatai A, Magaji U, Kareem I, Kamarudin ZS, Muhammad I'i, Kolapo K (2019) Drought resistance in rice from conventional to molecular breeding: a review. Int J Mol Sci 20(14):3519

Pan A, Hayes PM, Chen F, Chen THH, Blake T, Wright S, Karsai I, Bedö Z (1994) Genetic analysis of the components of winterhardiness in barley (*Hordeum vulgare* L.). Theor Appl Genet 89:900–910

Pan X, Wei J, Ruijuan Q, Shiping X, Chen J, Al-Basher G, Li C, Shad A, Dar AA, Wang Z (2020) Alumina-mediated photocatalytic degradation of hexachlorobenzene in aqueous system: kinetics and mechanism. Chemosphere 257:127256. https://doi.org/10.1016/j.chemosphere.2020.127256

Peleg Z, Blumwald E (2011) Hormone balance and abiotic stress tolerance in crop plants. Curr Opin Plant Biol 14(3):290–295

Peng B, Guan K (2021) Harmonizing climate-smart and sustainable agriculture. Nat Food 2(11):853–854

Pramanick B, Kumar M, Naik BM, Kumar M, Singh SK, Maitra S, Naik BSSS, Rajput VD, Minkina T (2022) Long-term conservation tillage and precision nutrient management in maize–wheat cropping system: effect on soil properties, crop production, and economics. Agronomy 12:2766. https://doi.org/10.3390/agronomy12112766

Pramanick B, Dubey R, Kesarwani A, Bera A, Bhutia KL, Kumar M, Maitra S (2023) Advancement in mitigating the effects of waterlogging stress in wheat. In: Khan MK, Pandey A, Hamurcu M, Gupta OP, Gezgin S (eds) Abiotic stresses in wheat, 1st edn. Academic Press, Cambridge, MA, Elsevier, RELX, Elsevier Inc. ISBN: 9780323953689. https://doi.org/10.1016/B978-0-323-95368-9.00013-8

Qados AMSA (2011) Effect of salt stress on plant growth and metabolism of bean plant Vicia faba (L.). J Saudi Soc Agric Sci 10(1):7–15

Razzaq MK, Akhter M, Ahmad RM, Cheema KL, Hina A, Karikari B, Raza G, Xing G, Gai J, Khurshid M (2022) CRISPR-Cas9 based stress tolerance: new hope for abiotic stress tolerance in chickpea (Cicer arietinum). Mol Biol Rep 49(9):8977–8985

Rodriguez-Ortega WM, Martinez V, Rivero RM, Camara-Zapata JM, Mestre T, Garcia-Sanchez F (2017) Use of a smart irrigation system to study the effects of irrigation management on the agronomic and physiological responses of tomato plants grown under different temperatures regimes. Agric Water Manag 183:158–168

Rosenzweig C, Elliott J, Deryng D, Ruane AC, Müller C, Arneth A, Boote KJ et al (2013) Assessing agricultural risks of climate change in the 21st century in a global ridded crop model intercomparison. Proc Natl Acad Sci 111(9):3268–3273. https://doi.org/10.1073/pnas.1222463110

Saeidi M, Abdoli M (2015) Effect of drought stress during grain filling on yield and its components, gas exchange variables, and some physiological traits of wheat cultivars. J Agric Sci Technol 17(4):885–898

Schmidhuber J, Tubiello FN (2007) Global food security under climate change. Proc Natl Acad Sci 104(50):19703–19708

Sedlacko EM, Chaparro JM, Heuberger AL, Cath TY, Higgins CP (2020) Effect of produced water treatment technologies on irrigation-induced metal and salt accumulation in wheat (*Triticum aestivum*) and sunflower (Helianthus annuus). Sci Total Environ 740:140003

Selvakumar G, Panneerselvam P, Ganeshamurthy AN (2011) Bacterial mediated alleviation of abiotic stress in crops. In: Bacteria in agrobiology: stress management, pp 205–224. https://doi.org/10.1007/978-3-662-45795-5_10

Shakoor U, Rashid M, Saboor A, Khurshid N, Husnain Z, Rehman A (2017) Maize production response to climate change in Pakistan: a time series assessment. Sarhad J Agric 33:2

Sharma A, Kalia R, Sareen S, Kumari A, Jandyal S, Katnoria JK, Bhardwaj R, Nagpal AK (2021) Genetic engineering of legumes for abiotic stress tolerance. In: Abiotic stress and legumes. Academic Press, pp 371–393

Shiriga K, Sharma R, Kumar K, Yadav SK, Hossain F, Thirunavukkarasu N (2014) Genome-wide identification and expression pattern of drought-responsive members of the NAC family in maize. Meta Gene 2:407–417

Shrivastava P, Kumar R (2015) Soil salinity: a serious environmental issue and plant growth promoting bacteria as one of the tools for its alleviation. Saudi J Biol Sci 22(2):123–131

Singh B, Mishra S, Bohra A, Joshi R, Siddique KHM (2018) Crop phenomics for abiotic stress tolerance in crop plants. In: Biochemical, physiological and molecular avenues for combating abiotic stress tolerance in plants. Academic Press, pp 277–296

Spollen WG, Tao W, Valliyodan B, Chen K, Hejlek LG, Kim J-J, LeNoble ME et al (2008) Spatial distribution of transcript changes in the maize primary root elongation zone at low water potential. BMC Plant Biol 8:1–15

Taïbi K, Taïbi F, Abderrahim LA, Ennajah A, Belkhodja M, Mulet JM (2016) Effect of salt stress on growth, chlorophyll content, lipid peroxidation and antioxidant defence systems in Phaseolus vulgaris L. S Afr J Bot 105:306–312

Takahashi F, Kuromori T, Sato H, Shinozaki K (2018) Regulatory gene networks in drought stress responses and resistance in plants. Adv Exp Med Biol:189–214. https://doi.org/10.1007/978-981-13-1244-1_11

Tao F, Zhang Z (2010) Adaptation of maize production to climate change in North China Plain: quantify the relative contributions of adaptation options. Eur J Agron 33(2):103–116

Thakral V, Bhat JA, Kumar N, Myaka B, Sudhakaran S, Patil G, Sonah H, Shivaraj SM, Deshmukh R (2021) Role of silicon under contrasting biotic and abiotic stress conditions provides benefits for climate smart cropping. Environ Exp Bot 189:104545

Turan MA, Elkarim AHA, Taban N, Taban S (2010) Effect of salt stress on growth and ion distribution and accumulation in shoot and root of maize plant. Afr J Agric Res 5(7):584–588

Vaughan NE, Gough C, Mander S, Littleton EW, Welfle A, Gernaat DE, Van Vuuren DP (2018) Evaluating the use of biomass energy with carbon capture and storage in low emission scenarios. Environ Res Lett 13(4): 044014

Vurukonda SS, Vardharajula S, Shrivastava M, SkZ A (2016) Enhancement of drought stress tolerance in crops by plant growth promoting rhizobacteria. Microbiol Res 184:13–24. https://doi.org/10.1016/j.micres.2015.12.003

Wang M, Dungait JAJ, Wei X, Ge T, Hou R, Ouyang Z, Zhang F, Tian J (2022a) Long-term warming increased microbial carbon use efficiency and turnover rate under conservation tillage system. Soil Biol Biochem 172:108770

Wang Y, Zafar N, Ali Q, Manghwar H, Guanying Wang LY, Ding X et al (2022b) CRISPR/Cas genome editing technologies for plant improvement against biotic and abiotic stresses: advances, limitations, and future perspectives. Cells 11(23):3928

Xu K, Xia X, Fukao T, Canlas P, Maghirang-Rodriguez R, Heuer S, Ismail AM, Bailey-Serres J, Ronald PC, Mackill DJ (2006) Sub1A is an ethylene-response-factor-like gene that confers submergence tolerance to rice. Nature 442(7103):705–708

Yijun G, Zhiming X, Jianing G, Qian Z, Rasheed A, Hussain MI, Ali I et al (2022) The intervention of classical and molecular breeding approaches to enhance flooding stress tolerance in soybean–an review. Front Plant Sci 13

Zahid A, Razzaq AA, Munawar M, Ramzan M, Almutairi BO, Almutairi MH (2023) Foliar spray of salicylic acid and ascorbic acid ameliorates the biochemical compounds in hybrid chillies. J King Saud Univ-Sci 35(5):102660

Zhu XF, Shen RF (2023) Towards sustainable use of acidic soils: deciphering aluminum-resistant mechanisms in plants. Fundam Res 4:1533

Ziska L, Crimmins A, Auclair A, DeGrasse S, Garofalo JF, Khan AS, Loladze I et al (2016) Ch. 7: food safety, nutrition, and distribution. In: The impacts of climate change on human health in the United States: a scientific assessment. https://doi.org/10.7930/j0zp4417

Chapter 5
Precision Agriculture Tools, Techniques, and Future Directions for Climate Resilience

Sourajit Dey, Nishadini Widanagamage, Somdatta Achar, Udit Debangshi, Sindhu Palla, Jiyung Kim, and Gaurav Jha

Abstract The increasing frequency of extreme weather events and rising temperatures have significantly impacted global crop production. Precision agriculture technologies such as remote sensing, proximal sensing, variable rate applications, and autonomy in farming provide solutions to mitigate these uncertainties. Precision tools can optimize the use of seeding rate, fertilizers, pesticides, and irrigation, improving efficiency and resilience. These tools can provide real-time, continuous data on soil conditions, crop health, and resource use efficiency. Detecting nutrient deficiencies, water stress, and disease infestation enables a farmer for targeted interventions at right time, right spot, right resources, right amount simply by using right data from the sensors. Recent developments in autonomous operations have modernized the fields of farm machinery and mechatronics for efficient functionality and optimized use of energy to manage operations. These farm data layers derived from these technologies can be normalized to efficient management zones for an optimized decision-making during extreme weather, soil, or crop variabilities.

Keywords Precision agriculture · Remote sensing · Variable rate applications · Resource optimization · Decision support systems

---

S. Dey (✉) · N. Widanagamage · U. Debangshi · S. Palla · J. Kim · G. Jha
Department of Agronomy, Kansas State University, Manhattan, KS, USA
e-mail: sdey@ksu.edu

S. Achar
Department of Agronomy & Horticulture, University of Nebraska-Lincoln, Lincoln, NE, USA

B. Pramanick et al. (eds.), *Climate-Smart Agricultural Technologies*, Sustainability Sciences in Asia and Africa, https://doi.org/10.1007/978-981-96-7699-6_5

5.1 Role of Technology in Mitigating Climate Change for Crop Production

5.1.1 Impact of Climate Variability on Agronomic Production

Climate change is a global challenge impacting the qualitative and quantitative shifts in cropping systems. It refers to alterations in the average values of meteorological parameters like precipitation and temperature, calculated over an extended period. The average global temperature is projected to rise by 2 °C by 2100 and 4.2 °C by 2400 (Pachauri et al. 2014). Global warming poses a severe threat to world food security but limiting warming up to 1.5 °C would reduce vulnerability in 76% of developing countries compared to 2 °C (Betts et al. 2018). Evidence suggests that approximately 9% of cropland expansion in developing nations over the last two decades is due to severe to extreme drought conditions, demanding increased area to compensate for yield and nutritive losses (Zaveri et al. 2020). By 2050, agricultural production requires a 60% increase than the demand from 2005 to 2007 levels, with a 77% increase in developing nations and a 24% increase in developed countries (Alexandratos and Bruinsma 2012). Global temperature changes and rainfall abnormalities significantly affect crop yields, varying by crop type, location, and magnitude of environmental changes. Warming of 2 °C is likely to reduce wheat, rice, and corn production in both temperate and tropical regions (Challinor et al. 2014). Tropical regions are more affected by climate change as the crops are closer to its high temperature optima, experiencing heat stress during elevated temperatures. Drought-prone areas across the world may see crop yield reductions of over 50% by 2050 and nearly 90% by 2100 (Li et al. 2009). Weather extremities in critical growth stages of crop increase the yield reduction. Crop yield losses can increase food prices and negatively impact global agriculture, with a projected 0.3% annual GDP loss by 2100 (Stevanović et al. 2016).

5.1.2 Influence of Greenhouse Gas Emissions on Crop Production

Since 1750, the concentrations of greenhouse gases (GHGs), including methane (CH_4), carbon dioxide (CO_2), and nitrous oxide (N_2O), have risen by 150%, 40%, and 20%, respectively (Pachauri et al. 2014). Globally, the annual GHG emission is approximately 50 billion tons measured in CO_2 equivalents (Ritchie and Roser 2024). The top sectors that produce GHGs are energy sector including electricity, heat, transportation, manufacture, and fuel combustion (75.6% emissions), agriculture including livestock and crop cultivation (11.6%), industrial processes of chemicals, cement, and more (6.1%) (Ge et al. 2020). These activities substantially raise GHG concentrations in the atmosphere, which intensifies the greenhouse effect and affects the climate. As agriculture is strongly dependent on climate variables, any disruptions in the climate factors can greatly affect crop yield and productivity. For example, temperature has a direct effect on crop physiology, and any temperature alterations put constraints on factors like crop growth, development, and yield. So, weather extremes like high temperature, rainfall abnormality, severe drought, and heat wave increase aridity, and decrease soil moisture which affects crop production.

5.1.3 Technological Adaptions to Address Climate Uncertainty

Variable rate application (VRA) is a precision agriculture method that involves modifying the application rate of inputs like fertilizers, herbicides, pesticides, irrigation, and seeds in various parts of a field according to their specific requirements, considering no need for changing rate settings on equipment manually or making multiple passes over an area. VRA utilizes data from multiple sources, such as sensors, satellite imagery, soil maps, and yield data, to generate a customized prescription map that directs the application of inputs to specific areas within the field; resulting in a more efficient system and minimizing potential environmental impacts. VRA can optimize crop yields, minimize input waste, and reduce costs, making it an effective and efficient tool to mitigate climate uncertainty. The VRA provides crops with nutrients at the right time, right amount, and right place, resulting in increased nutrient use efficiency. It prevents environmental problems such as water pollution by reducing the risk of nutrient runoff. In traditional irrigation, the irrigation dose was based on the crop's evapotranspiration. In contrast, precision irrigation systems rely on soil moisture sensors that monitor volumetric soil-water content using multi-depth capacitance sensors. It can potentially lead to water savings of up to 20–25% (Sharma et al. 2024). VRA helps to reduce the environmental footprint from agriculture by eliminating the over-application of chemicals. Targeted pesticide application to the parts of the field where it's most needed reduces chemical usage, helps in managing pest infestations, and protects beneficial insects. Additionally, the precise application of inputs helps preserve biodiversity by keeping the disruptions at a minimum to nearby ecosystems. VRA also optimizes farmland yield potential, decreasing the need to convert native forests and prairies into agricultural land.

Precision soil fertility management integrates VRA data layers for precise fertilizer application prescription maps to improve the accuracy of nutrient applications vs. use efficiency. It improves the overall effectiveness of resources, reduces excessive usage of agrochemicals, increases nutrient efficiency, and mitigates soil degradation. Some essential elements of VRA, that eventually help in the prevention of soil degradation are Grid Soil Sampling. Grid soil sampling ranging from 1 sample per 1 acre to 2.5 acres is widely used for assessing soil nutrient levels and to understand variability of soil properties within the field.

Variable rate irrigation can optimize irrigation application in crop fields according to its in-field variability of irrigation requirement and ensures prevention of over-watering, under-watering, and runoff, which ultimately maintains soil health and ameliorates GHG emissions. Proper utilization of irrigation water through VRA distributing the right amount of water at the right intervals to the crops satisfies crop water needs, and also reduces GHG emission. The VRA technology reduces GHG indirectly due to the reduced need to pump energy powered by fossil fuel or electricity. The GHG emission is correlated with the amount of nitrogenous (N) fertilizer and irrigation applied to the soil. Moreover, N fertilizer accompanied by accurate irrigation rate (also known as fertigation) transports nutrients directly to crop rhizosphere, reduces the substrate concentration of soil microorganisms in the process of nitrification and denitrification, thus reducing nitrous oxide emissions (Fig. 5.1). Fertigation also reduces evapotranspiration and improves the N and water use efficiency (Ding et al. 2022). Application of irrigation with modified application methods (such as subsurface irrigation, drip irrigation), duration, and amount can successfully decrease GHG emissions. Andrews et al. (2022) mentioned a result from a GHG emission study in forage cropping system that emission of major greenhouse gases (CO_2 and N_2O) can be reduced by up to 62% through adoption of subsurface drip irrigation (Fig. 5.1).

Variable rate seeding ensures optimal plant populations, and reduces competition among plants for nutrients and water, resulting in healthier crops, ultimately increasing crop production. Overall, VRA is proven as both profitable and advantageous, as it enhances the efficiency of farm inputs, protects water quality, and most

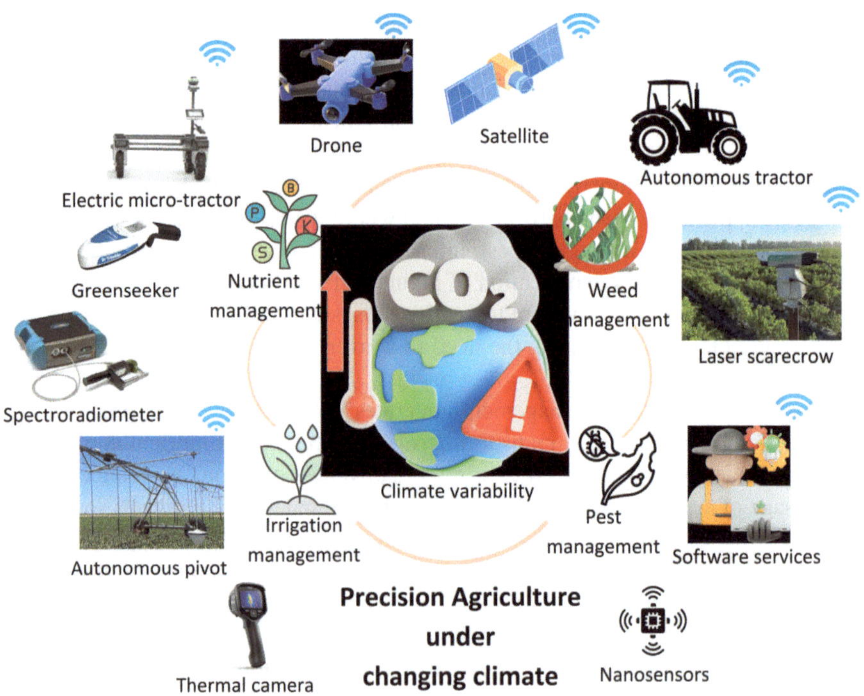

Fig. 5.1 Use of precision agriculture tools under changing climate

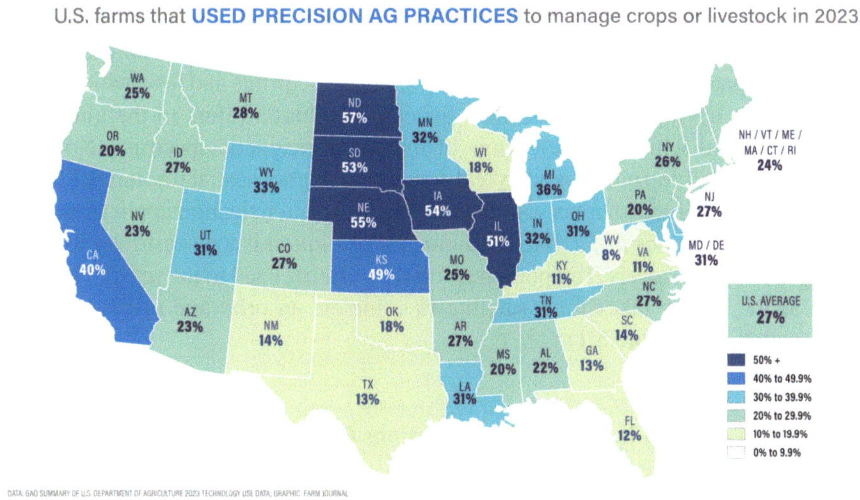

Fig. 5.2 Statistics of agricultural farm adaption to precision agriculture technologies in the United States in 2023 (New Report Details Top States for Precision Ag Technology 2024)

importantly maintains or improves crop yield and quality. VRA simplifies data collection and data analysis, provides best management decisions like optimum input application rate, best management practices to adopt, and these ensure best utilization of available land resources, and enhanced productivity. As per reports, a farm in Iowa had a 20% increase in corn yield and a 15% reduction in fertilizer use by adopting variable rate technologies (Saud 2024) (Fig. 5.2).

5.2 Precision Farming Tools for Resilient Crop Production

5.2.1 Remote Sensing for Crop Assessment and Production

Over nearly six decades, extensive research has demonstrated that a vast amount of crucial agricultural information can be remotely obtained through aircraft and satellite-based sensor systems. These systems can be combined with technological advancements such as Global Navigation Satellite System (GNSS) receivers, microcomputers, geographic information systems (GIS) software, yield monitors, simulation models, and remote sensing technology that has the potential to revolutionize agricultural practices. This integration allows growers to manage their lands with precision, optimizing resources and employing advanced farming techniques to boost productivity and sustainability (Berry et al. 2003).

The synergy of these technologies facilitates more informed decision-making, enhancing crops, water, and soil management while improving overall efficiency and environmental sustainability in agricultural systems. Sensors used for remote sensing vary based on their spatial, spectral, radiometric, and temporal resolution. Spatial resolution is defined by the pixel size representing the area on the ground, with high spatial resolution sensors having smaller footprints and low spatial resolution sensors having larger footprints. Spectral resolution refers to the number of bands captured within a given range of the electromagnetic spectrum (Nowatzki et al. 2004). For crop biomass and yield estimation, higher spatial resolution (1–3 m) is typically required, whereas variable rate fertilizer and irrigation applications may only need resolutions of 5–10 m. Radiometric resolution is the sensitivity to subtle changes in energy levels. Temporal resolution is often related to the sensor platform, such as the time it takes for a satellite to complete an orbit and revisit the same observation area (Table 5.1).

5.2.1.1 Vegetation Indices and Applications for Crop Management

Vegetation indices are mathematical formulas that combine reflectance measurements from different spectral bands to generate values that help assess various vegetation properties, such as crop growth, vigor, biomass, and chlorophyll content

Table 5.1 Some of the active satellite sensors and their spatial and temporal resolutions used in precision agricultural applications

Satellite name	Country	Started year	Temporal resolution (days)	Spatial resolution (m)[a]
AVHRR	Unites States	1979	1	1100
Landsat 7	United States	1999	16	60
Terra MODIS	United States	1999	1–2	250–1000
Terra-ASTER	United States	2000	16	TIR—90 SWIR—30 VNIR—15
Aqua MODIS	United States	2002	1–2	250–1000
GeoEye-1	United States	2008	2.1–8.3	1.65
RapidEye	Germany	2008	1–5.5	6.5
Worldview-2	United States	2009	1.1	1.4
Spot-6	France	2012	1	6
Landsat 8	United States	2013	16	100
Worldview-3	United States	2014	<1	1.24
Spot-7	France	2014	1	6
Sentinel-1	European countries	2014	1–3	5–40
TripleSat	China	2015	1	3.2
Sentinel-2	European countries	2015	2–5	2 NIR—60 Red edge and SWIR—20 VNIR—10
ECOSTRESS-PHyTIR	United States	2018	1–5	38 (in-track) × 69 (cross-track)

[a]*TIR* thermal infrared, *NIR* near-infrared, *SWIR* short-wave infrared, *VNIR* visible and near-infrared

(McKinnon and Hoff 2017). The amount of solar radiation reflected by plants is influenced by their chemical and morphological characteristics. Factors such as plant type, water content, and canopy structure affect the light reflected in each spectral band differently. Scientists have commonly used reflected light measurements in the ultraviolet, visible, and near- and mid-infrared regions of the spectrum to create various vegetation indices. These indices provide valuable insights into plant structure and conditions related to growth, photosynthesis, and vigor. By mapping these indices, researchers can understand the spatiotemporal variability in crop conditions, which is essential for precision agriculture. This understanding allows for more efficient and targeted agricultural practices, leading to better crop management and optimized resource use.

Plant pigments absorb light strongly in the visible part of the spectrum, except in the green region. This is why healthy plants appear green to our eyes. Several commonly used vegetation indices, such as the normalized difference vegetation index and green normalized difference vegetation index, take advantage of the fact that plant reflectance is low in the blue and red regions of the spectrum but peaks in the green region. However, in the near-infrared (NIR) part of the spectrum, plants do not absorb light strongly. Instead, it reflects a significant amount of NIR light, especially if it is green and healthy. This difference in light absorption and reflection between the visible and NIR regions is the basis for the vegetation indices, which help in monitoring plant health and vigor (Table 5.2).

5.2.1.2 Remote Sensing for Spatiotemporal Estimates in Crop Protection and Input Management

Remote sensing applications have revolutionized precision agriculture by providing farmers with detailed, real-time data about their crops and soil. By utilizing satellite imagery and aerial sensors, farmers can monitor crop health, assess soil conditions, and optimize the use of resources such as water, fertilizers, and pesticides. These technologies enable precise management of agricultural practices, leading to increased yields, reduced environmental impact, and more efficient use of inputs. Remote sensing tools offer valuable insights into crop growth patterns, disease outbreaks, and nutrient deficiencies, allowing for timely and targeted interventions.

Weed Management

Weeds are a significant management challenge for growers as they compete with crops for essential resources like water, nutrients, and light, often leading to a reduction in crop yield, quality, and return on investment. Traditional weed management practices involving uniform application of herbicides are inefficient and increase the risk of pesticide runoff, leading to environmental contamination (Davis and Pradolin 2016). Remote sensing technology integrated with energy optimization tools and machine vision has advanced weed control within crop fields, facilitating site-specific weed management like spot spraying, spectral differences and identification, and electrocution (Weiss et al. 2020). Weeds can be distinguished from crops

Table 5.2 Some of the widely used remote sensing vegetation indices and their applications

Vegetation index	Applications	Range and interpretation
Normalized difference vegetation index (NDVI)	Biomass, yield, nitrogen management, water stress, breeding	Ranges from −1 to +1. Higher values indicate healthy, dense vegetation; lower positive values represent stressed vegetation; positive near zero values indicate bare soil; negative values indicate clouds and water.
Green normalized difference vegetation index (GNDVI)	Biomass, yield, water stress	Ranges from −1 to +1. Higher values indicate healthier vegetation; positive values closer to zero indicates bare soil; negative values represent clouds, water, manmade structures.
Red edge normalized difference vegetation index (RENDVI)	Yield, nitrogen management, water stress, irrigation	Ranges from −1 to +1. Higher values towards 0.7 indicate healthier vegetation.
Normalized difference red edge (NDRE)	Biomass, yield, nitrogen management	Ranges from −1 to +1. Negative and near zero values indicate bare soil; positive values less than 0.6 immature or unhealthy crops; values more than 0.6 represent healthy and mature crops.
Soil adjusted vegetation index (SAVI)	Biomass, yield, nitrogen uptake and management, water stress, disease identification	Corrects for soil brightness influences, making it more effective in areas with sparse vegetation or exposed soil.
Modified soil adjusted vegetation index (MSAVI)	Biomass, yield, nitrogen management	Similar to SAVI but adjusted for more accurate measurements in areas with varying soil and vegetation coverage. Negative to 0.2 values indicate Bare soil and values over 0.6 represent dense vegetation.
Normalized water index (NWI)	Yield, soil moisture	Higher values indicate more water presence; used for soil moisture estimation and water stress detection.
Enhanced vegetation index (EVI)	Biomass, disease management	Sensitive to canopy structure and less affected by soil background and atmospheric noise; high values indicate dense, healthy vegetation.
Normalized difference water index (NDWI)	Vegetation water content	High values represent high or sufficient water content in vegetation, while low values indicate water stress or dry conditions.
Transformed soil adjusted vegetative index (TSAVI)	Yield, water stress	Adjusts for both soil background and canopy cover, improving vegetation health estimates in areas with sparse vegetation or exposed soil.
Difference vegetation index (DVI)	Yield, leaf area index, disease management	Higher values indicate more leaf area and biomass; useful in assessing canopy cover and crop health.

(continued)

Table 5.2 (continued)

Vegetation index	Applications	Range and interpretation
Red edge difference vegetation index (REDVI)	Biomass, yield, nitrogen management	Sensitive to nitrogen and chlorophyll content, particularly useful for detecting plant stress in early growth stages.
Ratio vegetation index (RVI)	Biomass, yield	Ratio-based index where higher values represent healthier and more vigorous vegetation.
Chlorophyll index (CI)	Chlorophyll content	Used to assess chlorophyll levels, a key indicator of plant health and growth potential; higher values signify healthier vegetation.
Water balance index (WBI)	Irrigation management	Higher values indicate a better water balance in vegetation, showing less water stress and sufficient irrigation levels.
Triangular vegetation index (TVI)	Disease identification and management	Detects differences in canopy structure due to disease; higher values indicate healthier vegetation.
Shortwave infrared water stress index (SIWSI)	Crop water stress	Identifies water stress in crops by measuring moisture levels; higher values indicate less stress, while lower values indicate more stress.

by their unique spectral signatures, which are related to differences in their phenological or morphological attributes. Machine vision techniques can identify weed seedlings based on leaf shapes and structural configurations of the adaxial surfaces of the leaf. In recent decades, machine learning approaches have emerged as highly accurate and efficient methods for image classification in weed mapping (Partel et al. 2019). Using sentinel-2 satellite images for identifying *Cirsium arvense* in grain fields has shown 6–46% potential herbicide savings (Rasmussen et al. 2021). A study has shown that pixel-level classification of weed species is achievable using satellite imagery (Rist et al. 2019). These approaches help to avoid herbicide application in weed-free areas, reducing herbicide usage and potential environmental pollution. Remote sensing also helps to determine the optimal rate of herbicide application.

Disease and Insect-Pest Management

Early detection and mapping of plant diseases can help contain their spread and minimize production losses. Traditional field scouting is time-consuming, labor-intensive, and prone to human error. Additionally, this method may struggle to detect diseases in their early stages, when symptoms are not yet fully visible, or in cases where symptoms are not outwardly apparent until it is too late to act (Sladojevic et al. 2016). Mapping the spatial extent and severity of disease spread is also challenging with conventional field scouting. Remote sensing offers a more efficient way to monitor diseases, especially in their early stages, when signs are difficult to detect through field scouting. Various remote sensing techniques, including multispectral, hyperspectral, thermal, and fluorescence imaging, have been used to identify diseases in different crops (Mahlein 2016). Compared to traditional vegetation indices like the NDVI, developing disease-specific spectral indices can improve the accuracy of disease detection and differentiation under real-world conditions. Higher accuracy in disease detection during critical crop growth stages, such as flowering, allows for the timely implementation of effective management plans, thereby avoiding significant yield loss. Furthermore, using disease-specific spectral indices instead of typical vegetation indices can simplify disease detection methods, enhance system efficacy, and reduce computational demands (Liang et al. 2019).

Nutrient Management

Efficient nutrient management is a challenge in production agriculture. Timely and precise fertilizer application is crucial to optimize crop growth and yields while minimizing environmental damage from nutrient runoff into groundwater and surface water. Crop nutrient requirements vary with crop type, variety, and growth stage. The demand fluctuates spatially and temporally due to factors like soil type, management practices, topography, weather, and hydrology (Hendricks et al. 2019). Spatial estimates of crop nutrient status using remote sensing tools can enhance nutrient use efficiency, maintain or increase crop yields, and reduce nutrient losses. Remote sensing offers in-situ and site-specific diagnostic methods that can detect abiotic stresses like nutrient deficiencies early in the season, while preventing yield, quality, and economic losses. These applications that depend on spectral signatures or specific wavelengths commonly correspond to proxy estimations of soil nitrogen, phosphorous contents, and organic matter creating spatial maps for site-specific management (Kawamura et al. 2021). Yet, remote sensing monitoring of symptoms of trace nutrient deficiencies (such as iron, manganese, zinc, and copper) is challenging because they rarely occur uniformly across a field and must be distinguished from background variations in canopy density.

Irrigation Management

The timing and rate of irrigation are crucial factors in reducing crop water stress and achieving optimal crop growth and yield. Traditional agricultural practices often fail to account for field variability, typically applying a uniform irrigation rate across the entire field. However, remote sensing data can identify spatiotemporal variability in soil, canopy growth, and weather patterns. One or a combination of these spatiotemporal estimates for variability can facilitate variable rate irrigation. By adjusting irrigation rates to match field conditions, water stress from extreme wet or dry areas can be mitigated, ensuring uniformly high yields and minimizing water and nutrient losses (Smith and Munoz 2002). Globally, remote sensing data in optical, thermal, and microwave bands have been utilized to estimate soil moisture (Zhu et al. 2017; Babaeian et al. 2018). Despite varying complexity and the amount of required meteorological and crop-specific data, all these indices are based on plant temperatures measured remotely using infrared thermometers or thermal imaging devices. The extent to which plant temperatures vary due to soil water availability depends on atmospheric evaporative demand and crop-specific transpiration characteristics.

Precision agriculture employs several non-invasive thermal indices to assess whether plants meet atmospheric transpiration demands and to infer plant water status. These indices, including the stress degree day, crop water stress index, and water deficit index, have been researched across more than 40 different crops such as wheat, corn, cotton, sorghum, soybean, and sunflower. These strategies involve monitoring plant water status, measuring evapotranspiration rates, and estimating crop coefficients. Studies have demonstrated that thermal infrared (3–14 μm range) is more sensitive to acute water stress than reflectance in visible, near-infrared (0.7–1.5 μm range), or short-wave infrared (1.5–3 μm range) wavelengths. However, reflective spectral fractions and vegetation indices respond to plant water status, particularly when it affects canopy architecture (such as wilting or leaf rolling) or results in chronic water stress that hinders growth, reduces green leaf area index, or alters senescence rates. Thermal plant water stress indices are valuable tools for irrigation scheduling, especially in regions where supplemental water is necessary for optimal crop growth. These indices are based on the relationship between the temperature of the plant canopy and the surrounding air, which reflects the plant's water status. When the atmospheric evaporative demand is high, plants need to transpire more water to cool themselves, which is facilitated by stomatal conductance. If soil moisture is sufficient, stomata remain open, leading to a lower canopy temperature (cold pixel). However, when water is scarce, stomata close to conserve water, causing the canopy temperature to rise (hot pixel), indicating water stress. By correlating evapotranspiration rate, which is driven by both stomatal conductance and atmospheric conditions, to the thermal index, farmers can predict when plants will begin to experience water stress, providing sufficient lead time to irrigate and prevent yield loss (Romero-Trigueros et al. 2019).

5.2.2 Proximal Sensing for In-Situ and Groundtruth Measurements

Proximal sensing is defined as the close monitoring of plants or soils to understand the growing conditions by signals obtained from sensors. These attributes define the strength of data in facilitating quantitative plant phenotyping. Instruments are placed within 2 m of the object in proximal sensing (Gebbers and Adamchuk 2010). These sensors are used to measure variability within soil, crops, and environment. Proximal sensors can be divided into two types: "on-the-go" and "stop-and-go" based on the deployment of the sensors (Adamchuk et al. 2018). On-the-go sensors are mobile sensors that can move across the field with farm machinery like planters, spraying machines, and combine harvester to collect a large amount of information while moving. To measure real-time soil and crop properties, as well as crop yield, variable rate technology employs on-the-go soil sensors, either in conjunction with predefined algorithms or machine learning models (real-time or online applications) or a geographic coordinating system (map-based or offline applications) (Andrade et al. 2023). For example, yield monitor which is used for gathering georeferenced data on crop yield and other crop characteristics including crop moisture, nutrient status, estimating profitability, and developing management zones (Fulton et al. 2018). Stop-and-go sensors are stationary sensors which are developed to provide detailed information at a specific field location. For example, apparent soil electrical conductivity measuring system, time domain reflectometer, and soil color sensor.

5.2.2.1 Proximal Soil Sensing

Proximal soil sensing technologies allow for real-time and non-destructive assessment of soil characteristics such as fertility level, moisture level, and soil electrical conductivity to provide an immediate feedback to farmers and growers (Nichols 2011). These sensors can reduce laboratory-based soil analytical costs and save time while minimizing the risk of soil sample deterioration and contamination during transit. It also allows for high-resolution spatial mapping of soil variability across the field.

Electrical Resistivity Tomography (ERT) and Electromagnetic Induction (EMI)

ERT and EMI are geophysical measurement techniques capable of measuring soil electrical conductivity up to several meters of soil depth. Electromagnetic induction can collect apparent electrical conductivity without physical contact, whereas electrical resistivity sensors require electrodes. ERT measures apparent electrical conductivity (ERa, ohm m) by injecting current into the soil through electrodes. ERT sensors, such as the 4-point light 10 W system (LIPPMANN, Schaufling, Bavaria, Germany), measure the voltage between sensors and calculate the ECa (Paz et al. 2024). EMI is a non-invasive and faster method for measuring electrical conductivity (ECa, dS m^{-1}). For example, the EM38 instrument (Geonics Limited, Mississauga, Ontario, Canada) consists of a transmitter and receiver coil separated by 1 m, and it measures the ECa using forward modeling. These methods are also useful for measuring soil cation exchange capacity, soil salinity, and sodality, which are all directly related to soil ECa. It is also useful for calculating soil organic matter, which is affected by soil salinity, clay percentage, and mineralogy. In a study conducted by Huang et al. (2017), a liner model was fitted between organic matter content and ECa to improve understanding of soil quality.

Time and Frequency Domain Reflectometry

The dielectric properties of soil are used by time and frequency domain reflectometry sensors to determine the water content. A key component of time domain reflectometry is the generation of a series of extremely precise electrical pulses that travel via an electromagnetic wave generator (waveguide) and transmission line. The duration of time it takes for electrical pulses to pass through soil and return to the waveguides is measured by this device. The dielectric constant of soil determines the impedance along the waveguides. The volumetric water content of the soil determines the dielectric constant of the soil. Therefore, soil water can be measured using the reflected measurement. The dielectric permittivity is estimated using a frequency domain reflectometer, which is a type of capacitance sensor or electrical impedance spectroscopy that is used to determine the soil water content (He et al. 2021). This device determines the soil dielectric constant by examining the frequency response of a signal that is transmitted through the soil, as opposed to time-domain reflectometry. The water content is determined by adjusting for frequency variations.

Ground Penetrating Radar

Ground penetrating radar (GPR) is a time-dependent radar system that can provide high-resolution 2D or 3D radar images of the subsurface with short pulses of high and ultra-high frequency, ranging from 10 MHz to 3 GHz (Wang and Birken 2022). This recording antenna attached in the system monitors the magnitude of temporal change in reflected energy over time as it encounters a layer with a different dielectric permittivity (Adamchuk et al. 2018; Sun and Di 2021). It is an efficient tool for assessing differences in water content between soil layers as soil dielectric permittivity is affected by water content. So, by measuring soil dielectric permittivity GPR can monitor soil moisture level. This system can also attach in center-pivot irrigation system to transform it into an intelligent, AI-driven robotic platform capable of optimizing irrigation application. For example, Autonomous pivot ltd. is an innovative Israel-based company that involves a variable rate technology that can modify the rate of irrigation as it moves across the field. This intelligent robotic platform can capture 300 soil moisture readings per rotation without disturbing the soil which enables autonomous irrigation based on point values. This technology combines a ground penetrating radar to continuously monitor crop health and soil conditions. The application of radar sensors in irrigation has several advantages, including real-time based irrigation and fertilization recommendation to maximize the resource use efficiency. It helps farmers and growers to prevent potential crop loss and disease outbreak based on their high-resolution camera. This system has potential for sustainable agriculture as it is conserving valuable resources like water, nutrients and increasing crop yield and lowering cost of cultivation. Based on a report, autonomous pivot can cut irrigation needs by 15% which can indirectly cut energy consumption and cost (finder.startupnationcentral.org).

Optical Reflectance Sensor

Optical reflectance sensors are used to assess the soil reflecting ability based on these absorption properties. It includes visible (400–750 nm), near-infrared (750–2500 nm), mid-infrared (2500–25,000 nm), and ultraviolet (100–400 nm). These sensors are effectively used in various soil properties such as mineral composition, clay content, soil color, moisture, organic matter, pH, and cation exchange capacity (CEC) (Mouazen et al. 2010; Tekin et al. 2013). Visible radiation or combined with UV (Vis-UV) is used to measure the inorganic minerals such as iron oxides. NixSpectro 2 sensor (Nix sensor Ltd., Hamilton, Ontario, Canada) (400–750 nm) is a colorimetric sensor which cost 349 USD, is operated by bluetooth system, and takes only 5 seconds to measure soil iron content with a coefficient of determination ranging from 0.79 to 0.81 (Jha et al. 2021). This sensor is also capable of predicting soil organic carbon. According to a report, NixSpectro 2 sensor can predict soil organic carbon with the coefficient of determination value of 0.70 (Mukhopadhyay et al. 2020). Dissolved organic matter is a key indicator for soil quality assessment. Vis-UV spectral radiation is used in estimating the soil-dissolved organic matter with a coefficient of determination exceeding 0.95 (Zhang et al. 2022). Another sensor is the ASD FieldSpec 4 Spectroradiometer (Analytical Spectral Devices Inc., Boulder, CO, USA) which is a Vis-NIR-SWIR (350–2500 nm) sensor used in measuring various properties of soil (Carmon and Ben-Dor 2017; Vasava et al. 2019). It has a fine spectral resolution (1–2 nm) which can provide detailed biophysical information.

5.2.2.2 Proximal Crop Sensing

Proximal crop sensing technology utilizes advanced sensors such as multispectral, hyperspectral, and thermal to gather detailed information on crop nutrient status, water content, biomass, and growth stages (Chakhvashvili et al. 2022; Mezera et al. 2021). Hyperspectral sensors capture reflectance data across hundreds of narrow spectral bands, enabling precise detection of crop health and stress indicators. For example, hyperspectral data often reveal key physiological features such as chlorophyll content, which can be assessed in the red absorption region (typically around 660–680 nm), and water status, identifiable through specific water absorption bands (around 970, 1450, and 1950 nm) (Clevers and Kooistra 2011).

A typical hyperspectral reflectance curve of a healthy plant shows low reflectance in the visible spectrum (400–700 nm), particularly in the red region (660–680 nm) where chlorophyll absorbs light. Higher reflectance is observed in the near-infrared region (700–1300 nm), which is sensitive to leaf structure and biomass. Water content can be detected through strong absorption bands at 970 and 1450 nm, indicating water stress or hydration levels in plants (Thenkabail et al. 2000). The graph below illustrates a typical hyperspectral reflectance curve from corn plants, highlighting key bands related to chlorophyll and water absorption (Fig. 5.3). These bands are critical for monitoring plant health and making informed management decisions about irrigation and fertilization.

Multispectral Proximal Crop Sensing

Multispectral-based proximal crop sensing uses sensors to capture data across multiple specific wavelengths in the electromagnetic spectrum. This technology allows for detailed analysis of various crop biophysical parameters by measuring different wavelengths as it is sensitive to different aspects of crop physiology. For example, the amount of chlorophyll in the palisade layer of a leaf controls the amount of visible light (400–720 nm) reflected. In contrast, the structure of the mesophyll tissues impacts the reflectance of the NIR electromagnetic spectrum (720–1300 nm) (Singh et al. 2016). Stressed plants are more susceptible to chlorophyll loss so spectral reflectance readings in the red region increase while NIR decreases in stressed canopy. Therefore, the NIR to red ratio is generally more sensitive to stressed plants. Optical sensors, operating within the visible and near-infrared spectrum, have the potential to provide valuable insights into the nutritional and health conditions of plants. Conversely, Light Detection and Ranging (LiDAR) sensors can be employed to evaluate and quantify the shape and size of the canopy. The Multiplex device, developed by Force-A, utilizes leaf fluorescence as a measurement method and has

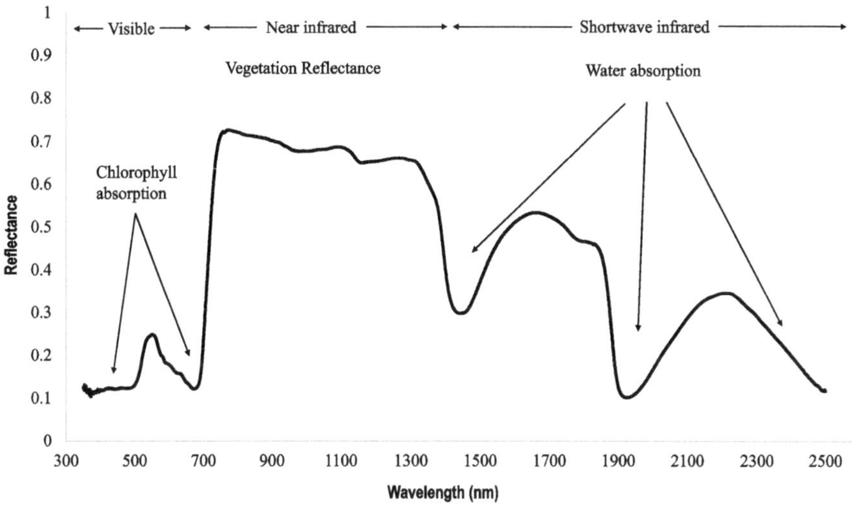

Fig. 5.3 Typical hyperspectral reflectance curve of a healthy crop

been brought to market. Multiplex was designed to estimate various pigments, including plant chlorophyll, by directly measuring fluorescence on leaves, fruits, or different sections of the canopy. In the context of precision agriculture, it is crucial to have swift and dependable methods for measuring nitrogen content in plants. This is essential for monitoring plant growth and effectively optimizing crop production. Crop nitrogen stress can be examined by a wide range of multispectral sensors such as ACS-210 Crop Circle (Holland Scientific, Inc., Lincoln, Nebraska, USA) (450–880 nm), SPAD-502 Chlorophyll Meter (Spectrum Technologies, Inc., Plainland, Illinois, USA), and GreenSeeker (Trimble, Inc., Sunnyvale, California, USA). ACS-210 Crop Circle sensor generates modulated light in the visible and NIR regions of the electromagnetic spectrum and measures canopy reflectance with visible (590 ± 5.5 nm, VIS_{590}) and NIR detectors (880 ± 10 nm, NIR_{880}) (Roberts et al. 2009). On the other hand, GreenSeeker canopy sensor is a commercially available and widely used active optical sensor that calculates NDVI by accounting red (650 ± 10 nm) and NIR (770 ± 15 nm) wavebands. Experimental result showed that optical sensor-based nitrogen management resulted in a 15% increase in nitrogen use efficiency (Bijay-Singh et al. 2011). The SPAD-502 is a portable non-destructive device that rapidly estimates the chlorophyll content in sample leaves. To obtain precise measurements of chlorophyll content in a non-destructive leaf sample, two LED lamps emitting red light at 650 nm and infrared light at 940 nm are used. The leaf sample is exposed to both red and infrared radiation. A fraction of light is assimilated, while the remaining light passes through the leaf and is captured by a silicon photodiode sensor. The intensity of light detected is inversely related to the concentration of chlorophyll in the leaf sample. The leaf chlorophyll concentration is quantified using a scale ranging from 0 to 99.9, with values expressed in arbitrary units. Nonetheless, due to its small sampling area, the SPAD sensor is highly sensitive to variations in leaf texture, veins, and imperfections. Consequently, obtaining a dependable reading that can be compared across different situations often requires some expertise and multiple measurements.

Thermal Proximal Crop Sensing

Thermography is a popular tool to assess plant water use and is commonly used in the field to detect plant water stress as plant temperature is influenced by transpiration rate. Along with that, plant response to biotic and abiotic stress causes biophysical and biochemical changes that lead to reducing biomass and chlorophyll (Adla et al. 2020). This can indirectly change the temporal pattern of thermal properties of crops and can be measured by thermal infrared cameras. Infrared thermometer measures the energy emitted by the surface. Based on this surface energy temperature is calculated using Stefan's law which states that, $E = \varepsilon\sigma T^4$, where ε is the emissivity of the object, σ is the Stefan-Boltzmann constant, and T is the absolute temperature. Examples include FLIR One Pro (Teledyne FLIR, Inc., Arlington, Virginia, USA) with an optical resolution of 640 × 480, spectral range of 8–14 μm, which can be used to measure crop temperature pattern. This spectral range falls within the long-wave infrared region, which is sensitive to thermal radiation emitted by objects. The camera operates within an ambient temperature range of 0–35 °C. In agriculture, the FLIR E8 Pro is used to detect temperature variations in crops, which can indicate water stress, disease, or other physiological changes. By providing a visual representation of temperature differences, the FLIR E8 Pro helps users quickly identify and diagnose potential issues. Another type of proximal sensor is porometer, such as the SC-1 Leaf Porometer (Decagon Devices, Inc., Pullman, Washington, USA) used to measure leaf stomatal conductance. This sensor operates by quantifying the rate of vapor movement from the leaf by stomates. Additionally, it provides information regarding actively transpiring leaves and ceased transpired leaves (metergroup.com). This instrument is also helpful for salt stress detection. Other than these, foliar wetness sensors can evaluate its moisture levels in crop plants, thereby helping to mitigate the risk of pathogens and diseases.

Moreover, in contemporary irrigation scheduling, there is a growing inclination toward prioritizing plant-oriented sensors. These sensors are widely used and dependable for the continuous assessment of plant water status. Such sensors encompass leaf turgor sensors, which measure leaf turgor pressure, an indicator closely linked to the plant's water status; sap flow sensors, which offer insights into plant transpiration rates; trunk dendrometers, which monitor changes in trunk size over time, reflecting the plant's hydration levels; and linear variable differential transformer fruit gauges.

5.3 Mechatronics Advances in Agriculture

The fusion of mechanical engineering, electronic components, and computer science has created the cutting-edge discipline termed as mechatronics which has the power of revolutionizing crop production (Cintra Faria and Barbalho 2023).

Producers can get valuable insights related to crop health, and early biotic and abiotic stress detection with the help of mechatronics. These technologies enhance precision and safety which leads to economic efficiency and stable productivity. Mechatronics-based machinery should have a control system to implement automation, actuators to translate electrical signals into physical movement, and sensors for position detection, imaging, and/or light detection. Mechatronics often use analog and digital data obtained from sensors to support automated operations and decision-making. The use of intelligent control systems such as artificial intelligence (AI) and machine learning algorithms enhances the working efficiency of mechatronics. To address climate change, mechatronics systems integrating with precision tools and technologies embedded with AI can ensure better crop production as it works efficiently in a cost-effective way.

Uncrewed ground vehicles (UGVs) equipped with advanced navigation, sensing, and location technologies are a valuable addition in farming (Bechtsis et al. 2019). The global agricultural UGV market value for 2023 is 2290 million USD which is expected to reach 3595.1 million USD by the end of the year 2030 (Valuates Reports 2024). To achieve full autonomy, UGVs have incorporated additional technologies, such as obstacle detection and safety systems to protect humans, animals, and other robots. The integration of big data, IoT, and cloud computing will enhance decision-making processes in smart farming (Gonzalez-De-Santos et al. 2020). Efforts to develop UGVs are split between automating traditional tractors and creating purpose-built mobile platforms, both of which are essential for the future of precision agriculture.

A case study reported by Mammarella et al. (2022) shows the cooperative use of unmanned systems in agriculture, specifically within a vineyard in the Barolo region of Piedmont, Italy. It showcases the integration of UGVs alongside uncrewed aerial vehicles (UAVs) to perform tasks such as remote sensing, mapping, and biopesticide distribution. The study introduces a decentralized, multi-phase approach where UAVs first create a 3D model of the vineyard, which then guides UGVs and rotary-wing UAVs in executing in-field operations. The UGV, designed with a four-wheel steering configuration, autonomously navigates the vineyard rows, handling tasks like scouting, bio-pesticide spraying, and shredding, even on sloped terrains. Tailored control algorithms optimize wheel velocity to prevent slippage and enhance traction on wet soils, with the RRT* algorithm ensuring efficient path planning. Preliminary simulations demonstrate the UGV's ability to follow planned trajectories, avoid obstacles, and maintain stability during complex maneuvers, highlighting the potential for improved autonomous navigation and task efficiency in challenging agricultural environments.

Another case study was reported in a pear orchard in the Pinggu district of Beijing, China (Jiang et al. 2022). It addressed working efficiency, environmental sustainability, water saving, and carbon dioxide reduction while using UAVs, UGVs, and conventional spray gun method for pesticide spray. Results showed that UGV had the highest deposit coverage (60.3%) of pesticide on both the adaxial and abaxial side of leaves than UAV (3.4%) and conventional method (34.9%) which was significant at $p < 0.05$. The working efficiency was highest in UAV, but effectiveness was lower than UGV. Study showed that intelligent techniques can have carbon dioxide emission reduction and water-saving potential.

The evolution of farm equipment and farming technology has been tremendous since mankind started farming. At 5500 BC, Sumerian farmers invented wooden plow where forked sticks were used to plow the field. In 2010, John Deere company started using iTEC Pro (Intelligent Total Equipment Control) guidance system for end turns in tractors based on global positioning technology (GPS). In contrast, the company upgraded the real-time kinetics by using Starfire receivers which taps into multiple satellite constellations such as GPS and Globalnaya Navigazionnaya Sputnikovaya Sistema (GLONASS) satellites, and BeiDou and Galileo constellations. The labor shortage is hitting the agricultural sector hard, but autonomous farming vehicles offer a solution by enabling farms to maintain productivity. These autonomous systems allow farm operators to continue working at necessary levels, even with smaller teams, by taking over repetitive, physical tasks. Autonomous tractors can operate around the clock, even in the dark, extending working hours throughout the year. This frees up skilled workers to focus on essential activities like planning for planting seasons, equipment maintenance, and managing production. By automating farming processes, these technologies enhance farm productivity and profitability, delivering fresher, faster, and more sustainable produce with fewer errors, leading to more consistent profit returns. These tractors are designed to navigate fields independently, processing and calculating their own position and speed while avoiding obstacles like people, animals, and objects. They can be fully autonomous or monitored remotely, with a single operator often overseeing a fleet from a nearby location. Autonomous tractors work by utilizing integrated systems, computers, and processors that convert electrical impulses into commands for the tractor to follow. Each tractor is equipped with a set of interconnected capabilities and emergency failsafe systems, including a remote emergency stop feature for safety. For navigation, they rely on GPS-enabled cameras, radars, and often LiDAR sensors. A sensor suite, which includes LiDAR, cameras, and robotic side cameras, is mounted on the tractor to monitor the environment and feed data into the main system for seamless operation. The advantages of autonomous tractors include improved farm productivity and profitability (Lowenberg-DeBoer et al. 2021), enhanced efficiency and time savings (Moorehead et al. 2012), reduced operating costs, increased agility in obstacle detection (Lindhorst 2019) and continuous adjustments, minimized environmental impacts, enhanced data collection and analytics, greater scalability and customization, alleviation of labor shortages, and environment-friendly farming practices. A Swedish dairy farm reported reduced energy consumption by 58% and greenhouse gas emissions by 92% for electric autonomous tractor compared to a diesel engine tractor (Engström and Lagnelöv 2018). The disadvantage of autonomous tractors includes a significant initial investment of at least $500,000 to buy the system. List of the autonomous tractor models used worldwide are John Deere 8R 410, CNH Industrial New Holland T7.315, Fendt 1000 Vario, Monarch Tractor MK-V Electric, etc. Other than high initial investment, the need for high-speed internet and other farm infrastructure, the possibility of interference with automation, vulnerability to cyber-attacks are drawbacks of automation in tractors.

Agricultural robots are revolutionizing modern agriculture by automizing agricultural activities such as planting, weeding, pest control, harvesting, and crop health monitoring. The global robotics market in agriculture was 13.4 billion USD in 2023 which has a projection to grow up to 86.5 billion USD by 2033 (Market.US 2023). Agricultural robots are equipped with sensors including cameras, LiDAR, GPS, and so on which are integrated with AI. Robots can be powered by electricity generated by renewable energy sources like solar or water which reduces fuel costs such as FarmDroid FD20, Solix, and Aigro UP. It also creates consistency, uniformity, and predictability, and reduces labor and maintenance. There are also agricultural robots which can provide real-time updates through cloud-based tools, and user-friendly application programming interfaces which help in decision-making. Agricultural robots operate on fields with reduced soil compaction. Some of the leading robotics and automation companies are listed in Table 5.3.

The crop yield is dependable on many factors such as climatic and soil variability, nutrient availability, disease and pest incidence, crop management strategies, and so on. To understand crop yield variability, yield maps could help in decision-making on crop management and resource allocation. The data for yield mapping can be collected from the yield monitors attached to the harvesters. The GPS sensors and yield monitoring systems are attached to the combined harvesters that help to create yield maps on a spatial scale. These maps are created with farm management information software such as FarmTRX, FarmingIT, Yield Editor, etc. An advantage of yield map is that it helps in finding the crop management problems and identifying ways to improve the practices for the next season. The limitation of yield mapping is that it visualizes values of certain range with color schemes. This implies that the number of ranges will impact the visualization of maps. Commonly, five to seven ranges are recommended, and green-yellow-red color schemes are used for yield mapping. Yield maps are used in nutrient removal maps creation which is

Table 5.3 List of some agricultural robots used for different agricultural operations

Agricultural robot	Company	Agricultural operations
Aigro UP	Aigro	Weeding and mowing
Amiga	Farm-ng	Planting, spraying, weeding
FarmDroid FD20	FarmDroid	Planting and weeding
FFRobot	FFRobotics	Harvesting
Harvest Croo	Harvest Croo Robotics	Harvesting
Laser weeder	Terra Robotics	Weeding
Laser weeder	Carbon Robotics	Weeding
Oz	Naio Technologies	Hoeing, planting, creating furrows, weeding
Sharpshooter smart sprayer	Verdant Robotics	Weeding and spraying
Solix	Solinftec	Weeding and pest control

useful for producers to understand the soil nutrient status of the field after harvesting. The idea is to map the amounts of nitrogen, phosphorus, and potassium that are lost from the soil for each unit of grain production. These maps can help in variable rate fertilizer application.

The yield mapping helps in creating management zones of a field. A management zone is a sub-part of a field area that possesses a relatively homogeneous combination of yield-limiting factors. Management zone delineation can help in understanding the management practices needed to be executed in a field for higher production. This could help in crop variety selection, seeding rate, fertilization, weed management, diseases, and pest infestation. Other than yield data layer, management zones can be created considering data layers such as satellite imagery, soil analysis, topography, historical yield data, etc. Some of the commonly used software for management zone creation are FastMapping, GeoPard, Management zone analyst, and Field monitor.

These yield maps and management zones help in applying variable rates as required. Variable rate technology (VRT) is a precision agriculture technology which uses maps for differential rate application. This map-based approach identifies where and how much of a given treatment such as fertilizer, pesticide or seeding rate should be applied. This needs the installation of a control system that can read maps to autonomously adjust the application rate. However, the fundamental functionality of existing applicators is not affected. Other than using maps to create variable rate treatments, real-time sensing can be used in VRT (Sharda et al. 2018). Vegetation indices like NDVI which is based on crop reflectance of near-infrared and red spectrum region can be used to understand crop health which helps in variable fertilizer or pesticide application in real time. Thermal infrared spectrum can help to sense the water stress, and the variable irrigation technology could help to manage it. The list of some vendors in VRT applications includes John Deere, Trimble Inc., AGCO corporation, Topcon corporation, and CNH industrial.

5.4 Future Innovations and Emerging Technologies

Agriculture has gone through the implementation of a lot of innovations in recent years. Farmers have embraced the effectiveness and advancement in precision technologies and welcome future innovations in farming. In precision agriculture, nanotechnology is utilized in nano-based smart delivery system and nano-biosensors that help producers in making decisions in crop and resource management to enhance productivity. The new concept of nanotechnology-based precision agriculture employs nano-biofertilizers with neuro-receptors and nano-chips to strengthen nutrient recycling process, increase resource conservation, and meet crop yield targets. There are different types of nanosensors with various functions used in precision agriculture. Near-infrared fluorescent hydrogen peroxide (H_2O_2) nanosensors can be used as early stress detection in crop plants as H_2O_2 is a key signaling molecule related to plant stress (Wu et al. 2020). Nanosensor-based electric nose and tongue (e-Nose and e-Tongue, respectively) are reliable devices with high accuracy and sensitivity with broad application in crop production. These devices can be portable or on-field installed with real-time sensing and advanced data analysis techniques to early detection of disease and pest and quality control to reduce the impact of climate change (Fundurulic et al. 2023; Hu et al. 2019). To obtain real-time data on crop health, soil and weather condition, wireless nanosensors are installed in the fields. Wireless nanosensor network can be used for real-time soil fertilization, disease detection, and smart pest control (Shafi et al. 2019). Electrochemical nanosensors can detect specific ions in the soil which is effective for plant nutrition through an ion-selective electrode (Cheong et al. 2021). A biometric fiber-based electrochemical biosensor can be inserted into the plant tissues to monitor the changes in solute content in real-time (Coppedè et al. 2017). The changes in solute content can be related to the plant stress, and monitoring through these electrochemical sensors the crop management could be for better production.

Big data analytics helps in anticipating trends, potential issues, and useful insights for crop production. Big data consists of 3Vs, i.e., volume (like exponential growth of collected data), velocity (speed of real-time in-field data collection through sensors), and variety (such as large number of data sources and formats) (Shekhar et al. 2017). It integrates datasets from diverse sources such as satellite imagery, geospatial data, meteorological data, crop data, soil data, etc., and transforms it into actionable intelligence (San Emeterio de la Parte et al. 2023). Through the use of predictive algorithms and machine learning, big data analytics can suggest specific interventions to optimize crop production. This holistic approach serves as a decision support system for farmers to make management decisions, enhance productivity, and ensure sustainable agricultural practices.

The concept of utilizing machine learning and big data in agriculture focuses on advanced technologies to assess and interpret big agricultural datasets. This approach enables researchers, producers, and agronomists to make informed decisions that can increase crop yield and quality. A major aspect is the emphasis on real-time data collection, accuracy, and timely information with minimal errors. Integration of historical trial data with current datasets allows us to predict and strengthen the database for more robust data analysis. The real-time visualization of analyzed data simplifies interpretation, facilitating better decision-making for stakeholders.

Some key examples of emerging farm software include Agmatix that provides a platform combining historical and current trial data, enabling better decision-making. Climate FieldView uses machine learning to analyze weather, soil, and field conditions, and also uses prediction modeling as a decision-support tool for farmers. Granular helps farmers optimize operations through data-driven decisions. FarmLogs provides real-time crop monitoring and planning tools. Taranis combines aerial imagery with machine learning to detect crop issues. TrialWerx captures canopy, climate, and soil data season-long from large-scale on-field trials and quantifies the crop response by integrating results across locations and seasons. Ceres Imaging focuses on crop health and water use. These platforms exemplify how technology is transforming agriculture by improving crop management and promoting sustainability.

Birds can damage a significant portion of the economical part of a crop. It is reported that a direct loss of over 150 million USD is experienced in the United States for bird damage annually (Animal and Plant Health Inspection Service 2024). The use of a new precision tool, Laser scarecrows is getting popular to deter the birds in agricultural fields. Researchers from the University of Florida and the University of Rhode Island in the United States have investigated the effectiveness of laser scarecrows to reduce bird menace using laser lights. Experiment showed that the laser scarecrows were able to reduce the damage by more than 20% in sweet corn in Florida (Manz et al. 2024). An average of 46% of bird damage was reduced by using laser devices with a wavelength of 532 nm ± 5 in 3 years long experiment in sweet corn in Rhode Island (Brown and Brown 2021). The mechanism behind it is that the birds perceive laser beams in the green spectrum as physical objects and when the laser is directed toward the birds, it triggers their intrinsic flight response. Vendors sell two types of laser bird deterrents, namely, automated and manual.

5.5 Summary

Climate change poses significant challenges to crop production through temperature extremities, erratic precipitation, and frequent droughts. By 2050, agricultural output must increase by 60% to meet demand, with extreme climate events potentially reducing crop yields by up to 50%. Technological interventions like variable rate application offer solutions to mitigate these effects by optimizing the use of fertilizers, pesticides, and irrigation based on field-specific data. Precision farming tools, such as remote sensing, provide real-time data on crop health, soil conditions, and resource use, significantly improving agricultural management. Remote sensing utilizes satellites and aerial sensors to capture spatial, spectral, radiometric, and temporal data, which can be used to identify nutrient deficiencies, water stress, weed distribution, crop health, and early disease detection. Proximal soil sensing technologies like electrical resistivity tomography and electromagnetic induction measure soil electrical conductivity, while ground penetrating radar monitors moisture levels. Crop sensing technologies including hyperspectral and multispectral sensors help in monitoring plant health, nitrogen levels, and water status. Thermal sensors detect plant water stress by measuring surface temperature variations, and leaf turgor and sap flow sensors aid in irrigation scheduling by tracking plant water status. Mechatronics such as UGV and agricultural robots offer improved crop monitoring, pesticide application, and autonomous operations, contributing to efficiency and accuracy. Technologies like yield mapping and management zone creation help to optimize resource allocation and crop management. VRT and precision tools like nanosensors and big data analytics further refine decision-making for better crop production and resource utilization. Emerging innovations in precision agriculture, including autonomous tractors and nanotechnology, promise to improve productivity, reduce climate uncertainty impact, and address labor shortages, all while offering significant advancements in real-time data analysis and automation.

References

Adamchuk V, Ji W, Viscarra Rossel R, Gebbers R, Tremblay N, Shannon DK, Clay D, Kitchen NR (2018) Proximal soil and plant sensing. In: Precision agriculture basics. American Society of Agronomy, Inc./Crop Science Society of America/Soil Science Society of America/ACSESS Publications, Madison, pp 123–144. https://doi.org/10.2134/precisionagbasics.2016.0093

Adla S, Rai NK, Karumanchi SH, Tripathi S, Disse M, Pande S (2020) Laboratory calibration and performance evaluation of low-cost capacitive and very low-cost resistive soil moisture sensors. Sensors 20(2):363. https://doi.org/10.3390/s20020363

Alexandratos N, Bruinsma J (2012) World agriculture towards 2030/2050: the 2012 revision. https://ageconsearch.umn.edu/record/288998/?v=pdf

Andrews HM, Homyak PM, Oikawa PY, Wang J, Jenerette GD (2022) Water-conscious management strategies reduce per-yield irrigation and soil emissions of CO2, N2O, and NO in high-temperature forage cropping systems. Agriculture, Ecosystems and Environment, 332, 107944

Andrade R, Silva SHG, Benedet L, Mancini M, Lima GJ, Nascimento K, Amaral FHC, Silva DRG, Ottoni MV, Carneiro MAC, Curi N (2023) Proximal sensing provides clean, fast, and accurate quality control of organic and mineral fertilizers. Environ Res 236:116753. https://doi.org/10.1016/j.envres.2023.116753

Animal and Plant Health Inspection Service (2024) NWRC research areas: bird damage. Retrieved July 10, 2024, from https://www.aphis.usda.gov/national-wildlife-programs/nwrc/research-areas/bird-damage

Babaeian E, Sadeghi M, Franz TE, Jones S, Tuller M (2018) Mapping soil moisture with the OPtical TRApezoid Model (OPTRAM) based on long-term MODIS observations. Remote Sens Environ 211:425–440

Bechtsis D, Moisiadis V, Tsolakis N, Vlachos D, Bochtis D (2019) Unmanned ground vehicles in precision farming services: an integrated emulation modelling approach. In: Salampasis M, Bournaris T (eds) Information and communication technologies in modern agricultural development, vol 953. Springer, Cham, pp 177–190. https://doi.org/10.1007/978-3-030-12998-9_13

Berry JK, Detgado JA, Khosla R, Pierce FJ (2003) Precision conservation for environmental sustainability. J Soil Water Conserv 58(6):332–339

Betts RA, Alfieri L, Bradshaw C, Caesar J, Feyen L, Friedlingstein P, Gohar L, Koutroulis A, Lewis K, Morfopoulos C, Papadimitriou L, Richardson KJ, Tsanis I, Wyser K (2018) Changes in climate extremes, fresh water availability and vulnerability to food insecurity projected at 1.5°C and 2°C global warming with a higher-resolution global climate model. Philos Trans R Soc A Math Phys Eng Sci 376(2119):20160452. https://doi.org/10.1098/rsta.2016.0452

Bijay-Singh, Sharma RK, Jaspreet-Kaur, Jat ML, Martin KL, Yadvinder-Singh, Varinderpal-Singh, Chandna P, Choudhary OP, Gupta RK, Thind HS, Jagmohan-Singh, Uppal HS, Khurana HS, Ajay-Kumar, Uppal RK, Vashistha M, Raun WR, Gupta R (2011) Assessment of the nitrogen management strategy using an optical sensor for irrigated wheat. Agron Sustain Dev 31(3):589–603. https://doi.org/10.1007/s13593-011-0005-5

Brown RN, Brown DH (2021) Robotic laser scarecrows: a tool for controlling bird damage in sweet corn. Crop Prot 146:105652. https://doi.org/10.1016/j.cropro.2021.105652

Carmon N, Ben-Dor E (2017) An advanced analytical approach for spectral-based modelling of soil properties. Int J Emerg Technol Adv Eng 7:90–97

Chakhvashvili E, Siegmann B, Muller O, Verrelst J, Bendig J, Kraska T, Rascher U (2022) Retrieval of crop variables from proximal multispectral UAV image data using PROSAIL in maize canopy. Remote Sens 14(5):Article 5. https://doi.org/10.3390/rs14051247

Challinor AJ, Watson J, Lobell DB, Howden SM, Smith DR, Chhetri N (2014) A meta-analysis of crop yield under climate change and adaptation. Nat Clim Chang 4(4):287–291

Cheong YH, Ge L, Lisak G (2021) Highly reproducible solid contact ion selective electrodes: emerging opportunities for potentiometry – a review. Anal Chim Acta 1162:338304. https://doi.org/10.1016/j.aca.2021.338304

Cintra Faria AC, Barbalho SCM (2023) Mechatronics: a study on its scientific constitution and association with innovative products. Appl Syst Innov 6(4):72

Clevers JG, Kooistra L (2011) Using hyperspectral remote sensing data for retrieving canopy chlorophyll and nitrogen content. IEEE J Sel Top Appl Earth Obs Remote Sens 5(2):574–583

Coppedè N, Janni M, Bettelli M, Maida CL, Gentile F, Villani M, Ruotolo R, Iannotta S, Marmiroli N, Marmiroli M, Zappettini A (2017) An in vivo biosensing, biomimetic electrochemical transistor with applications in plant science and precision farming. Sci Rep 7(1):16195. https://doi.org/10.1038/s41598-017-16217-4

Davis AM, Pradolin J (2016) Precision herbicide application technologies to decrease herbicide losses in furrow irrigation outflows in a northeastern Australian cropping system. J Agric Food Chem 64(20):4021–4028. https://doi.org/10.1021/acs.jafc.5b04987

Ding W, Chang N, Zhang J, Li G, Zhang J, Ju X, Zhang G, Li H (2022) Optimized fertigation mitigates N2O and NO emissions and enhances NH3 volatilizations in an intensified greenhouse vegetable system. Agric Water Manag 272:107797

Engström J, Lagnelöv O (2018) An autonomous electric powered tractor—simulation of all operations on a Swedish dairy farm. J Agric Sci Technol A 8(3):182–187. https://doi.org/10.17265/2161-6256/2018.03.006

Fulton J, Hawkins E, Taylor R, Franzen A (2018) Yield monitoring and mapping. In: Kent Shannon D, Clay DE, Kitchen NR (eds) ASA, CSSA, and SSSA books. American Society of Agronomy and Soil Science Society of America, Madison, pp 63–77. https://doi.org/10.2134/precisionagbasics.2016.0089

Fundurulic A, Faria JMS, Inácio ML (2023) Advances in electronic nose sensors for plant disease and pest detection. Eng Proc 48(1):Article 1. https://doi.org/10.3390/CSAC2023-14890

Ge M, Friedrich J, Vigna L (2020) 4 Charts explain greenhouse gas emissions by countries and sectors. https://www.wri.org/insights/4-charts-explain-greenhouse-gas-emissions-countries-and-sectors

Gebbers R, Adamchuk VI (2010) Precision agriculture and food security. Science 327(5967):828–831. https://doi.org/10.1126/science.1183899

Gonzalez-De-Santos P, Fernández R, Sepúlveda D, Navas E, Armada M, Gonzalez-De-Santos P, Fernández R, Sepúlveda D, Navas E, Armada M (2020) Unmanned ground vehicles for smart farms. In: Agronomy—climate change & food security. IntechOpen. https://doi.org/10.5772/intechopen.90683

He H, Turner NC, Aogu K, Dyck M, Feng H, Si B, Wang J, Lv J (2021) Time and frequency domain reflectometry for the measurement of tree stem water content: a review, evaluation, and future perspectives. Agric For Meteorol 306:108442

Hendricks GS, Shukla S, Roka FM, Sishodia RP, Obreza TA, Hochmuth GJ, Colee J (2019) Economic and environmental consequences of overfertilization under extreme weather conditions. J Soil Water Conserv 74(2):160–171

Hu W, Wan L, Jian Y, Ren C, Jin K, Su X, Bai X, Haick H, Yao M, Wu W (2019) Electronic noses: from advanced materials to sensors aided with data processing. Adv Mater Technol 4(2):1800488. https://doi.org/10.1002/admt.201800488

Huang J, Pedrera-Parrilla A, Vanderlinden K, Taguas EV, Gómez JA, Triantafilis J (2017) Potential to map depth-specific soil organic matter content across an olive grove using quasi-2d and quasi-3d inversion of DUALEM-21 data. Catena 152:207–217

Jha G, Sihi D, Dari B, Kaur H, Nocco M, Ulery A, Lombard K (2021) Rapid and inexpensive assessment of soil total iron using Nix Pro color sensor. Agric Environ Lett 6:e20050. https://doi.org/10.1002/ael2.20050

Jiang Y, He X, Song J, Liu Y, Wang C, Li T, Qi P, Yu C, Chen F (2022) Comprehensive assessment of intelligent unmanned vehicle techniques in pesticide application: a case study in pear orchard. Front Plant Sci 13:959429. https://doi.org/10.3389/fpls.2022.959429

Kawamura K, Nishigaki T, Tsujimoto Y, Andriamananjara A, Rabenaribo M, Asai H, Rakotoson T, Razafimbelo T (2021) Exploring relevant wavelength regions for estimating soil total carbon contents of rice fields in Madagascar from Vis-NIR spectra with sequential application of backward interval PLS. Plant Prod Sci 24(1):1–14. https://doi.org/10.1080/1343943X.2020.1785898

Li Y, Ye W, Wang M, Yan X (2009) Climate change and drought: a risk assessment of crop-yield impacts. Clim Res 39(1):31–46

Liang Q, Xiang S, Hu Y, Coppola G, Zhang D, Sun W (2019) PD2SE-Net: computer-assisted plant disease diagnosis and severity estimation network. Comput Electron Agric 157:518–529

Lindhorst C (2019) Field obstacle identification for autonomous tractor applications. Dissertations and theses, Department of Biological Systems Engineering. https://digitalcommons.unl.edu/biosysengdiss/94

Lowenberg-DeBoer J, Franklin K, Behrendt K, Godwin R (2021) Economics of autonomous equipment for arable farms. Precis Agric 22(6):1992–2006. https://doi.org/10.1007/s11119-021-09822-x

Mahlein A-K (2016) Plant disease detection by imaging sensors – parallels and specific demands for precision agriculture and plant phenotyping. Plant Dis 100(2):241–251. https://doi.org/10.1094/PDIS-03-15-0340-FE

Mammarella M, Comba L, Biglia A, Dabbene F, Gay P (2022) Cooperation of unmanned systems for agricultural applications: a case study in a vineyard. Biosyst Eng 223:81–102. https://doi.org/10.1016/j.biosystemseng.2021.12.010

Manz ST, Sieving KE, Brown RN, Klug PE, Kluever BM (2024) Experimental assessment of laser scarecrows for reducing avian damage to sweet corn. Pest Manag Sci 80(3):1547–1556. https://doi.org/10.1002/ps.7888

Market.Us (2023) Agricultural Robot market size, share | CAGR of 20.5%. Retrieved September 3, 2024, from https://market.us/report/agricultural-robot-market/

McKinnon T, Hoff P (2017) Comparing RGB-based vegetation indices with NDVI for drone based agricultural sensing. AgribotixLLC AGBX021-17, pp 1–8

Mezera J, Lukas V, Horniaček I, Smutný V, Elbl J (2021) Comparison of proximal and remote sensing for the diagnosis of crop status in site-specific crop management. Sensors (Basel) 22(1):19. https://doi.org/10.3390/s22010019

Moorehead SJ, Wellington CK, Gilmore BJ, Vallespi C (2012) Automating orchards: a system of autonomous tractors for orchard maintenance. In: Proceedings of the IEEE international conference of intelligent robots and systems, workshop on agricultural robotics. http://www.cs.cmu.edu/afs/cs.cmu.edu/Web/People/cvalles/papers/automatingOrchards.pdf

Mouazen AM, Kuang B, De Baerdemaeker J, Ramon H (2010) Comparison among principal component, partial least squares and back propagation neural network analyses for accuracy of measurement of selected soil properties with visible and near infrared spectroscopy. Geoderma 158(1):23–31. https://doi.org/10.1016/j.geoderma.2010.03.001

Mukhopadhyay S, Chakraborty S, Bhadoria PBS, Li B, Weindorf DC (2020) Assessment of heavy metal and soil organic carbon by portable X-ray fluorescence spectrometry and NixPro™ sensor in landfill soils of India. Geoderma Reg 20:e00249. https://doi.org/10.1016/j.geodrs.2019.e00249

New Report Details Top States for Precision Ag Technology (2024) Retrieved September 8, 2024, from https://westernagnetwork.com/new-report-details-top-states-for-precision-ag-technology

Nichols S (2011) Review and evaluation of remote sensing methods for soil-moisture estimation. J Photonics Energy 2:028001. https://doi.org/10.1117/1.3534910

Nowatzki J, Andres R, Kyllo K (2004) Agricultural remote sensing basics. https://library.ndsu.edu/ir/bitstream/handle/10365/5408/ae1262.pdf?sequence=1

Pachauri RK, Allen MR, Barros VR, Broome J, Cramer W, Christ R, Church JA, Clarke L, Dahe Q, Dasgupta P (2014) Climate change 2014: synthesis report. Contribution of Working Groups I, II and III to the fifth assessment report of the Intergovernmental Panel on Climate Change. IPCC, Geneva. https://epic.awi.de/id/eprint/37530/

Partel V, Kakarla SC, Ampatzidis Y (2019) Development and evaluation of a low-cost and smart technology for precision weed management utilizing artificial intelligence. Comput Electron Agric 157:339–350

Paz MC, Castanheira NL, Paz AM, Gonçalves MC, Monteiro Santos F, Farzamian M (2024) Comparison of electromagnetic induction and electrical resistivity tomography in assessing soil salinity: insights from four plots with distinct soil salinity levels. Land 13(3):295

Rasmussen J, Azim S, Nielsen J (2021) Pre-harvest weed mapping of Cirsium arvense L. based on free satellite imagery – the importance of weed aggregation and image resolution. Eur J Agron 130:126373

Rist Y, Shendryk I, Diakogiannis F, Levick S (2019) Weed mapping using very high resolution satellite imagery and fully convolutional neural network. In: IGARSS 2019-2019 IEEE international geoscience and remote sensing symposium. IEEE, pp 9784–9787. https://ieeexplore.ieee.org/abstract/document/8900442/

Ritchie H, Roser M (2024) Sector by sector: where do global greenhouse gas emissions come from? Our World in Data. https://ourworldindata.org/ghg-emissions-by-sector

Roberts D, Adamchuk V, Ferguson R, Schepers JS (2009) Optimization of crop canopy sensor placement for measuring nitrogen status in corn. Agron J 101:140. https://doi.org/10.2134/agronj2008.0072x

Romero-Trigueros C, Bayona Gambín JM, Nortes Tortosa PA, Alarcón Cabañero JJ, Nicolás Nicolás E (2019) Determination of crop water stress index by infrared thermometry in grapefruit trees irrigated with saline reclaimed water combined with deficit irrigation. Remote Sens 11(7):757

San Emeterio de la Parte M, Martínez-Ortega J-F, Hernández Díaz V, Martínez NL (2023) Big Data and precision agriculture: a novel spatio-temporal semantic IoT data management framework for improved interoperability. J Big Data 10(1):52. https://doi.org/10.1186/s40537-023-00729-0

Saud D (2024) Boosting modern agriculture by variable rate technology (VRT). AgriERP Blog, 27 May 2024. https://agrierp.com/blog/variable-rate-technology-vrt/

Shafi U, Mumtaz R, García-Nieto J, Hassan SA, Zaidi SAR, Iqbal N (2019) Precision agriculture techniques and practices: from considerations to applications. Sensors 19(17):Article 17. https://doi.org/10.3390/s19173796

Sharda A, Franzen A, Clay DE, Luck JD (2018) Precision variable equipment. In: Kent Shannon D, Clay DE, Kitchen NR (eds) ASA, CSSA, and SSSA books. American Society of Agronomy and Soil Science Society of America, Madison, pp 155–168. https://doi.org/10.2134/precisionagbasics.2016.0094

Sharma A, Prakash A, Bhambota S, Kumar S (2024) Investigations of precision agriculture technologies with application to developing countries. Environ Dev Sustain:1–37. https://doi.org/10.1007/s10668-024-04572-y

Shekhar S, Schnable P, LeBauer D, Baylis K, VanderWaal K (2017) Agriculture big data (AgBD) challenges and opportunities from farm to table: a Midwest Big Data Hub community whitepaper. White Paper for the US National Institute of Food and Agriculture. https://www-users.cse.umn.edu/~shekhar/talk/2017/NIFA_whitepaper_AgBD_final_12_14_2017.pdf

Singh V, Kaur D, Singh B, Brar B, Kaur A (2016) Precision nutrient management: a review. Indian J Fertil 12:1–15

Sladojevic S, Arsenovic M, Anderla A, Culibrk D, Stefanovic D (2016) Deep neural networks based recognition of plant diseases by leaf image classification. Comput Intell Neurosci 2016:1–11. https://doi.org/10.1155/2016/3289801

Smith M, Munoz G (2002) Irrigation advisory services for effective water use: a review of experiences. Irrigation Advisory Services and Participatory Extension in Irrigation Management Workshop Organized by FAO-ICID, Montreal, Canada. https://www.academia.edu/download/86886382/irrigation_adv_serv.pdf

Stevanović M, Popp A, Lotze-Campen H, Dietrich JP, Müller C, Bonsch M, Schmitz C, Bodirsky BL, Humpenöder F, Weindl I (2016) The impact of high-end climate change on agricultural welfare. Sci Adv 2(8):e1501452. https://doi.org/10.1126/sciadv.1501452

Sun Z, Di L (2021) A review of smart irrigation decision support systems. In: 2021 9th international conference on agro-geoinformatics (agro-geoinformatics). IEEE, pp 1–4. https://doi.org/10.1109/Agro-Geoinformatics50104.2021.9530351

Tekin Y, Kuang B, Mouazen A (2013) Potential of on-line visible and near infrared spectroscopy for measurement of pH for deriving variable rate lime recommendations. Sensors 13(8):10177–10190. https://doi.org/10.3390/s130810177

Thenkabail PS, Smith RB, De Pauw E (2000) Hyperspectral vegetation indices and their relationships with agricultural crop characteristics. Remote Sens Environ 71(2):158–182

Valuates Reports (2024) Agricultural UGV market, report size, worth, revenue, growth, industry value, share 2024. Retrieved September 3, 2024, from https://reports.valuates.com/market-reports/QYRE-Auto-39T13392/global-agricultural-ugv

Vasava HB, Gupta A, Arora R, Das BS (2019) Assessment of soil texture from spectral reflectance data of bulk soil samples and their dry-sieved aggregate size fractions. Geoderma 337:914–926

Wang ML, Birken R (2022) Sensing solutions for assessing and monitoring roads. In: Lynch JP, Sohn H, Wang ML (eds) Sensor technologies for civil infrastructures, 2nd edn. Woodhead Publishing, Oxford, pp 299–330. https://doi.org/10.1016/B978-0-08-102706-6.00020-9

Weiss M, Jacob F, Duveiller G (2020) Remote sensing for agricultural applications: a meta-review. Remote Sens Environ 236:111402

Wu H, Nißler R, Morris V, Herrmann N, Hu P, Jeon S-J, Kruss S, Giraldo JP (2020) Monitoring plant health with near-infrared fluorescent H2O2 nanosensors. Nano Lett 20(4):2432–2442. https://doi.org/10.1021/acs.nanolett.9b05159

Zaveri E, Russ J, Damania R (2020) Rainfall anomalies are a significant driver of cropland expansion. Proc Natl Acad Sci 117(19):10225–10233. https://doi.org/10.1073/pnas.1910719117

Zhang F, Zhang W, Wu S, Fu X, Li S, Yue S (2022) Analysis of UV–Vis spectral characteristics and content estimation of soil DOM under mulching practices. Ecol Indic 138:108869. https://doi.org/10.1016/j.ecolind.2022.108869

Zhu W, Jia S, Lv A (2017) A universal T_s-VI triangle method for the continuous retrieval of evaporative fraction from MODIS products. J Geophys Res Atmos 122(19):10206–10227. https://doi.org/10.1002/2017JD026964

Chapter 6
Precision Nutrient Management for Reducing Environmental Impacts Due to Non-judicial Use of Inputs

Devi Shidayaichenbi, Shubham Singh, Sushmita Thokchom, Ashok Kumar Singh, Santosh Kumar Singh, Arati Ningombam, Guntamukkala Sekhar, and Marian Brestic

Abstract The rising human population brought about the need for the Green Revolution (GR) in the 1960s to meet the growing demand for food. The GR focused on using inorganic fertilizers, irrigation, pesticides, high-yielding variety (HYV) seeds, and field equipment to increase crop yields. However, the GR approach paid little attention to the sustainable management of agricultural inputs, leading to environmental degradation. The mismanagement of agricultural inputs released Green House Gases (GHGs) into the atmosphere, contributing to 30% of the total GHG emissions and causing climate change. This deterioration also affected soil and water quality, reducing crop productivity. To reclaim environmental degradation, mitigating climate change is crucial. Precision agriculture with modern techniques and technologies, such as chlorophyll meters, ground positioning systems (GPS), satellite imagery, nutrient management models, and integrated plant nutrient management (IPNM), offers a promising approach to mitigate climate change. These

D. Shidayaichenbi · A. K. Singh · S. K. Singh
Dr. Rajendra Prasad Central Agricultural University, Pusa, Samastipur, Bihar, India

S. Singh
College of Post Graduate Studies in Agricultural Science, (CAU-Imphal), Umiam, Meghalaya, India

S. Thokchom
Assam Agricultural University, Jorhat, Assam, India

A. Ningombam
ICAR Research Complex for North Eastern Hill Region, Manipur Centre, Lamphelpat, Manipur, India

G. Sekhar
Centurion University of Technology and Management, Gajapati, Odisha, India

M. Brestic (✉)
Department of Botany and Plant Physiology, Faculty of Agrobiology, Food, and Natural Resources, Czech University of Life Sciences Prague, Prague, Czechia
e-mail: marian.brestic@uniag.sk

B. Pramanick et al. (eds.), *Climate-Smart Agricultural Technologies*, Sustainability Sciences in Asia and Africa, https://doi.org/10.1007/978-981-96-7699-6_6

techniques support the 5Rs—"right time," "right place," "right manner," "right amount," and "right input"—increasing nutrient use efficiency and enhancing plant growth and development. Proper nutrient handling and vegetation maintenance can help alleviate environmental degradation without compromising crop productivity.

Keywords Green revolution · Climate change · Soil health · Precision agriculture · Sustainable environment

6.1 Introduction

The history of the Green Revolution (GR) dates back to the 1840s in the USA, where it continued until the 1930s. However, the modern era of GR started in Mexico from 1940 until 1965, where it was initially successful before going away in the 1970s due to mishandling and misutilization of chemical inputs. India also adopted the GR approach in the 1960s to meet the growing demand for food due to its ever-increasing population. However, the approach only focused on maximizing crop production with little regard for the sustainable environment, leading to stagnant crop productivity, soil health deterioration, and climate change. The country's food security and environment were at risk for future generations of limited land and natural resources. The adoption of completely organic and natural farming will also be unable to support the population instantly as it takes 8–10 years to sustain. It suggests the utilization of agricultural inputs of GR is a must concerning a sustainable environment without compromising crop productivity which became a challenging task.

The adoption of precision nutrient management (PNM) came into light and became a promising aspect of increasing crop production from limited natural resources without detrimental effects on society and the environment (Varinderpal-Singh et al. 2016). It includes various techniques, technologies, and management methods and practices based on the application of agricultural requirements or inputs depending on a specific crop and site which varies with different crops and locations as the soil properties vary from one region to another (Jin and Jiang 2002). The determination of soil properties using the soil sampling method from the specific site is time-consuming and a destructive process. Fortunately, technological advancements like chlorophyll meters, leaf color charts, and geospatial technologies such as ground positioning systems (GPS) and remote sensing (RS) have made PNM more accessible and user-friendly (Varinderpal-Singh et al. 2010). These technologies are available to all users, regardless of field size or user type. PNM enables the efficient management of agricultural inputs, reducing costs and regulating the environment while enhancing crop yield. Misuse and misapplication of agricultural inputs can lead to environmental degradation. However, the PNM approach ensures that these inputs are applied in exact amounts and at the right time using advanced technologies and scientific knowledge to mitigate environmental problems while still maximizing crop productivity.

6.2 Reason Behind the Climate Change

The cause of climate change is the disturbance or imbalance between the natural resources. It involves the unproportionate gas components in the atmosphere due to excess emission of GHGs from mass deforestation and non-judicious utilization of chemical inputs under Green Revolution (GR) in the soil. This renders global warming into climate change. The most potent gas for climate change is carbon dioxide (CO_2) contributing 400 ppm by 2022 in the atmosphere. The solution to reducing its concentration is through carbon sequestration which is the long-term trapping of atmospheric CO_2 in the soil by vegetation. However, the reduction of forest and arable land area owing to increasing population fails to balance the CO_2 concentration in the atmosphere. Moreover, the deterioration of soil health after the agricultural practices under GR malfunction to support vegetation. It shows a great challenge to mitigate climate change along with enhancing soil and crop productivity and sustaining the environment.

6.3 Green Revolution and Agricultural Inputs

6.3.1 Types of Agricultural Inputs

Agricultural inputs are essential external resources used in agriculture to enhance crop productivity and profitability. There are two prominent types of agricultural inputs: consumable and capital (Table 6.1). Consumable inputs include natural resources utilized directly by plants, on the other hand, capital inputs are high-tech and mechanical resources that are not directly consumed by plants but support in conditioning the soil for plant growth. The former inputs include supplying of plant with water and available nutrients, a medium, i.e., soil for root to anchorage, quality seeds for better growth and development of the plant and adaptation to the environment, plant protection measures, etc. The latter inputs involve the mechanical manipulation of soil using tractors and plows, supplying water to fields through

Table 6.1 Types and resources of agricultural inputs

Types of inputs	Agricultural resources
Consumable inputs	*Natural resources*: High-yielding variety (HYV) seeds, chemical fertilizers, pesticides, insecticides, soil, insect traps, crop residue, water, hay, etc.
Capital inputs	*High technology and mechanical resources*: Nylon netting, tractors, irrigation systems, plows, etc.

canals, drainage systems, and greenhouse, and mulching using nylon netting to maintain the soil moisture and temperature regimes. The combined adoption of the inputs for cultivation is far more profitable in terms of soil and crop health and productivity than individual adoption (Meja and Kebede 2022). Since the Green Revolution (GR), these inputs play a vital role in agriculture and adoption beyond the sustainable level by small- and large-scale farmers to enhance their agricultural production leading to an environmental challenge.

6.3.2 *Role of Green Revolution in Mishandling of Agricultural Inputs*

The Green Revolution (GR) was launched to maximize crop production from limited resources and meet the increasing food demands by utilizing a high quantity of agricultural inputs. In India, GR was initiated during 1965–1966 after the country faced the risk of famine (1947–1960) due to the increasing population, traditional farming methods, and limited arable lands that made insufficient food supply. International organizations such as the CIMMYT, Mexico, and IRRI, Philippines played a significant role in eradicating the risk of famine in the country by introducing high-yielding variety (HYV) seeds of wheat (Norin-10) and rice (IR-8), respectively (Eliazer Nelson et al. 2019).

These HYV seeds have the potential for heavy consumption of chemical fertilizers, pesticides, and insecticides, and require heavy irrigation. They also possess early maturation in a short duration which offers to maximize the cropping intensity per unit of land. The GR facilitates cropping intensity of 300% under a cropping system. Moreover, heavy farm equipment such as tractors, threshers, and chisels were also implemented for efficient crop production in the practices of GR which destruct the soil's physical properties (soil structure, water infiltration, bulk density, etc.). This overburdened the land against its capacity.

Even though the crop production was satisfactory under the GR and met the food demands of the country, the natural resources (soil, water, and air) have a threshold-bearing capacity, and due to the mass misutilization and excessive utilization of the agricultural inputs under the GR, they deteriorate the natural resources. Among the chemical fertilizers, the N-based fertilizers, i.e., urea, ammonia, and nitrate were widely used as compared with phosphorus (P) and potassium (K)-based fertilizers (Department of Fertilizers and Department of Agriculture 2017), rendering nutrient imbalances in the soil.

The agricultural inputs are a boon to the farmers when utilized optimally. However, the overemphasis on crop production under the GR resulted in their misutilization and mishandling, leading to the destruction of the eco-friendly environment. It is crucial to optimize the use of agricultural inputs to safeguard the environment and maintain long-term sustainability in agriculture (FAO 2011).

6.3.3 *Non-judicial Use of Agricultural Inputs on Climate Change*

Today, climate change is a pressing environmental crisis that is expected to be caused by both natural and anthropogenic activities. Among the latter, the over and misutilization of agricultural inputs especially chemical fertilizers, exploitation of forest land, conventional agriculture, and overpopulation are major concerns. Chemical fertilizers are of particular concern as they can contribute to the deterioration of soil health. While the goal of farmers is to maximize production from a unit of arable land—made possible under the Green Revolution—a lack of concern for sustainable practices has resulted in the heavy implementation of chemical fertilizers, especially N-based fertilizers, beyond the recommended dose. This has resulted in an imbalance of soil nutrients and negatively affected soil health. As soil health declines, it loses its capacity to support plant growth and development, leading to declining crop production and vegetation failure (Pramanick et al. 2024). This widespread decline in vegetation reduces the trapping of atmospheric carbon dioxide (CO_2), which is the most potent greenhouse gas (GHG). Vegetation traps atmospheric CO_2 for the photosynthetic process along with water and sunlight and helps maintain the atmospheric CO_2 level, mitigating climate change. The reduction in the vegetal population increases the CO_2 concentration in the atmosphere inducing climate change. The only solution to mitigate climate change is the increment of the vegetated area or crop area with judicial management of agricultural inputs to maintain soil health and promote a sustainable environment.

6.3.4 *Understanding the Environmental Impacts of Non-judicial Use of Inputs*

The global population has increased from 6 billion in 2000 to 8 billion in 2021 by an annual average rate of 1.2% causing an increase in demand for food which led to intensified agricultural production by more use of chemicals (UNDP 2021). To meet the needs of this increasing population, pesticide usage increased by 30%, and inorganic fertilizer usage for N, P, and K by 23%, 13%, and 56% respectively (Mahapatra 2022). But the benefits of these chemicals are offset by several negative impacts on the environment and health indicating that the use of pesticides and fertilizers worldwide is unsustainable (Table 6.2). The inappropriate and irregular use of inputs without considering the environmental impacts is referred to as non-judicial utilization. Overuse of pesticides and fertilizers causes pollution by contaminating natural resources, thereby disrupting ecosystems and wildlife. It can cause climate change through the emission of greenhouse gases from the synthesis of these fertilizers and pesticides. Furthermore, some fertilizers release nitrogen oxide while applied to the field adding to greenhouse gas emissions. Due to this soil fertility and productivity is degrading over time impacting food security. However, the severity

Table 6.2 Environmental impacts of non-judicial use of inputs

Environmental impacts	Description
Soil degradation	Toxic elements such as Cd, Ar, Hg, and Pb present in fertilizers degrade the soil quality by affecting the soil organisms.
Water pollution	Nitrogen and phosphoric fertilizer when applied in excess may contaminate the water bodies by the eutrophication process encouraging algal growth and affecting the aquatic organisms by forming hypoxic areas.
Air pollution	Excessive nutrients cause air pollution and affect human health adversely when volatilized and inhaled for example ammonia and dusts from manure.
Habitat destruction	Resistance developed by pathogens, insects, and weeds gradually affects the pest and vector management process resulting in higher costs of new approaches. Monoculture practices also affect the ecosystem's resilience.
Climate change	The process of production, transportation, and use of fertilizers emit greenhouse gases into the atmosphere which contribute around 40% of total agricultural GHG emissions contributing to climate change.
Human health impacts	Effects on the health of humans will be acute and long-term impacts such as respiratory issues, cancer, and immunological, neurological, and reproductive effects. Also, its exposure during pregnancy causes leukemia in infants.
Diversity loss	The use of excessive fertilizers reduces the population of non-target and beneficial organisms. It also affects pollination and natural pest control which are important for ecosystem services.

of these impacts varies depending on the type, intensity, and context of use. So, to promote a sustainable environment, these problems should need to be addressed.

6.3.5 Mitigation of Environmental Impacts Due to Non-judicial Use of Inputs

The environmental impacts are mainly caused by the residual effects, negligence use, and misutilization of agricultural inputs. It is no offense to utilize inputs in agriculture but should be only where it is needed and when it is required at an optimum level. The implementation of short shelf life of pesticides and insecticides and easily mineralizable and available fertilizers are necessary to be adopted. The soil is the natural buffer medium of all misutilization of agricultural inputs besides providing plant nutrition and root anchorage. However, it also holds a threshold level for buffering and can cause environmental impacts. Heavy farm machinery such as tractors, harvesters, threshers, etc., release GHGs and destroy soil structures, besides reducing labor costs and energy in agricultural practices. Modification of agricultural practices regarding the utilization of agricultural inputs is of major importance. The promotion of precision nutrient management (PNM) using its advanced techniques, technologies, and management practices can judicious the agricultural inputs for mitigating environmental impacts into a sustainable environment.

6.3.6 Precision Nutrient Management and Its Techniques

Precision nutrient management (PNM) is the prediction technique for the application of organic and inorganic fertilizers and soil and plant management practices depending on the 5Rs such as "right input" at the "right time" in the "right manner" and "right amount" at the "right place" for maintaining the soil health and eco-friendly environment with enhancing the crop productivity. It targets to synchronize the optimum availability of the plant requirements (nutrients and water) at its critical growth stages for its proper growth and development with the nutrient supplied by the soil under a particular environment leading to discarding excess utilization and improving the management of agricultural inputs. The application of fertilizers, either organic or inorganic, is for the plant to provide the nutrients, however, its effectiveness depends on soil condition, climatic factors, quality and type of cultivar, and management practices. Many researchers have shifted toward feeding the plant rather than feeding the soil without deteriorating the environmental factors under PNM using its advanced techniques, methods, technologies, and management practices (Fig. 6.1).

6.3.6.1 Soil Testing and Mapping

Soil testing is performed to determine the soil's composition, nutritional level, and pH. It is essential for attaining sustainable agriculture since it provides farmers with a more lucrative, environmentally friendly, and efficient agricultural plan. Furthermore, it aids in judicious fertilizer usage and appropriate soil management practices, which aid in crop productivity. The lab test is the most suited for soil testing since it provides accurate technology. Soil testing-based fertilizer recommendations are the true connection between agronomic research and farmer's fields (Singare et al. 2020).

Soil mapping is one of the most essential techniques in precision farming since it employs modern technology to provide exact information on soil conditions, crop condition, and weather (Lagacherie and McBratney 2007). It can readily assess the humidity and texture of the soil in a region to assist effective irrigation, as well as the pH value or soil conductivity to pick appropriate fertilizer. Malone et al. (2018) described how soil mapping is commonly achieved utilizing an optical sensor, an electromagnetic sensor, a mechanical sensor, and an electrochemical sensor. Nowadays, digital soil mapping is frequently utilized to build models for soil function and variability in the form of maps in order to provide efficient soil management using Soil Land Inference Model as a tool.

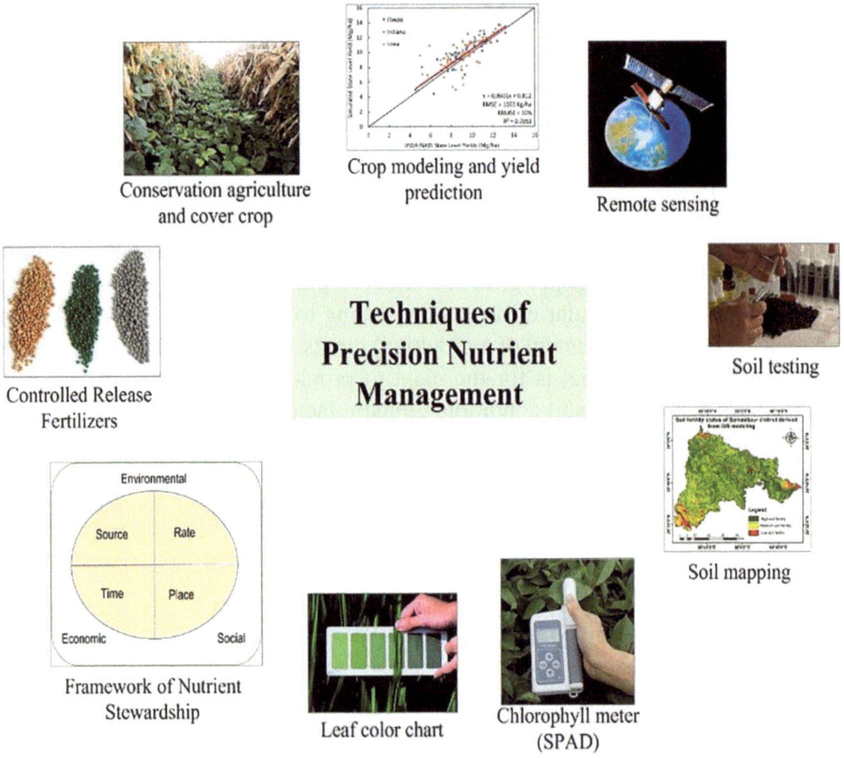

Fig. 6.1 Various techniques and technologies of PNM

6.3.6.2 Chlorophyll Meter and Leaf Color Chart

These techniques help to check the content of chlorophyll in the tissues of plants and they are basically diagnosing tools for nitrogen content in leaves which are directly linked to crop production. This chlorophyll meter works on a fixed limit threshold, and when the reading of chlorophyll drops down the limit, nitrogen-based fertilizers are needed to be applied. Minolta SPAD-502 is a commonly used hand-held chlorophyll meter that diagnoses in the standing crop synchronizing with the nitrogen content in the particular leave.

A leaf color chart (LCC) is a kind of plastic strip which has color chip hues that range from yellowish mild to dark green. It works by regular monitoring of the leaf color with the strip every 7–10 days till blooming starts. The color chip has a separate representation of nitrogen content and if LCC color drops down the limit, fertilizers of nitrogen-based are to be introduced.

6.3.6.3 Nutrient Stewardship

This framework promotes nutrition application by using the appropriate product, rate, time, and location.

(a) *Using the correct product*: Because it is a nutritional evaluation, it is critical to use the correct fertilizer product. When soil testing is unavailable, it may be used to guide site diagnostics and fertilizer application. DAP and MAP, for example, both include nutrients and vary in pH. DAP should not be used in soil with a high pH because it releases ammonia, which is harmful to seedlings.
(b) *Using the correct rate*: Using the correct and exact rate of fertilizer is essential. It is possible to do this by determining the yield in a field and determining the indigenous supply of soil nutrient ability so that we can determine how much nutrients may be found in the soil.
(c) *Applying fertilizer at the proper time*: It is critical to apply fertilizer at the proper time. It should be applied throughout the key periods of plant development (IPNI 2012; Pramanick et al. 2024). Phosphorous is sprayed during planting to make it accessible to growing plants and to assist reduce phosphorous fixation in the soil.
(d) *Proper fertilizer placement*: Proper fertilizer placement aids in nutrient absorption. When phosphorus is applied near the seed row in cereal crops, it may have a significant impact on the crop.

6.3.6.4 Controlled-Release Fertilizer

Controlled-release fertilizers (CRF) are required to achieve sustainable agriculture. Its mechanism lowers nutrient dissipation via leaching and volatilization while also providing enough nutrient-release designs that are appropriate for biochemical and physiological elements of plant growth (Jariwala et al. 2022). Certain polymers used as coating materials are now available in biodegradable form as bio-nanocomposites (Vejan et al. 2021). CRF also helps with nutrient release timing and may satisfy particular nutritional requirements during plant growth. Recently, scientists have attempted to create CRF materials from plant starch, rice straw, organic polymers, vegetable oil, and lignin in order to build CRF as a sustainable source (Singh et al. 2020; Zhang et al. 2020; Figueiredo et al. 2024). Because of technological advancements, CRF-based urea is designed utilizing liquefied bagasse, which is generated from bisphenol-A-diglycidyl ether and sugarcane residue. The bio-nano composites should be designed in such a way that they are heat and water-resistant so that they may be used in the soil. CRF aids in the reduction of nutrient loss such as nitrous oxide emissions and nitrate; prevents direct exposure of fertilizer granules to the soil; is less affected by environmental parameters; increases soil longevity; ability to maintain nitrogen level; GHG emission less; water retention; soil organic content increase; and optimum pH level inside the plants (Jariwala et al. 2022).

6.3.6.5 Cover Crops and Conservation Agriculture

Cover crops aid to prevent microbial habitat loss, fixed nutrient cycling, and soil organic carbon depletion, making food production more sustainable. They promote and facilitate WUE, reduce drought, and improve water filtering. It has a good impact on yield, land use sustainability, farming ease, revenue, and delivering long-term ecosystem benefits. They also boost SOM content and minimize soil erosion when fresh plant remains are applied to the soil. They are planted before the crop is planted to minimize allopathic effects from residues and to decrease nitrogen immobilization.

Conservation agriculture aids in water infiltration and wormhole improvement (Farooq and Siddique 2015). It employs scientific technology to promote crop productivity by reducing soil fertility, runoff, and soil erosion of topsoil, hence conserving moisture. It is commonly used in Bolivia, the USA, Chile, Argentina, Colombia, Finland, Kenya, China, the Falkland Islands, Morocco, Western Australia, the Malvinas, Zambia, and Kazakhstan (Friedrich et al. 2012).

6.3.6.6 Crop Modeling and Yield Prediction

Because of climate change, yield prediction has become critical to food security. It assists farmers in understanding management decisions, provides economic assistance, and aids in the prevention of famine. Crop modeling, which includes DSSAT, APSIM, R2WQM, WOFOST/SWAP, aids in the study of soil and plant physiology in order to build non-linear equations for forecasting yield and phenotypes (Eitzinger et al. 2004). All of the models aid in explaining the complicated relationship between environmental conditions and attributes at various stages of development (Hipólito et al. 2018). Many farmers attempt to predict production using his expectations. The yield forecast was discovered by analyzing the farmer's previous crop experience.

Crop modeling gives quantitative information on how a plant develops and interacts with its surroundings. They aid in the rapid simulation of places, situations, and therapies. Regular meteorological data, temperature, solar radiation, soil properties, rainfall, soil condition, crop management, and cultivar are all included in crop modeling.

6.3.6.7 Remote Sensing and Imaging

Remote sensing (RS) and imaging are techniques to observe the objects on the earth's surface through electromagnetic radiation (EMR) without direct contact. It provides geospatial data and photographs to commercial companies for mapping to monitor drought and vegetation health through GYURI, VCI, NDVI, and LAI (Sruthi and Aslam 2015). In agriculture, RS helps to monitor crop conditions for crop yield, humidity, and soil moisture for irrigation, insect pest management, and livestock condition. With all this information, it boosts maximum agriculture productivity. Optical sensing helps to navigate crop vigor, crop stress, and crop damage by visible wavelength. Through RS, crops can be managed timely and can identify crops affected by pests and weather. It helps to monitor SOM, soil pH, moisture content, and soil texture. It predicts crop productivity, canopy health, biomass, yield, vegetation density, and growth stage. It helps to navigate accurate and precise information such as irrigated cropland, water bodies, soil, and crop status. As RS can predict the conditions of weather, it will help in providing better crop needs and reduce unwanted costs. Crop mapping and precision farming by using remote sensing can help to increase crop yield and minimize environmental problems (Adhikary et al. 2022). RS gives a lot of advantages in climate change by finding the temporal stages of the atmosphere, oceans, and land. Yield and weather go hand in hand in agriculture; therefore this modern technology plays a pivotal role in predicting climate change.

6.3.6.8 Site-Specific Nutrient Management

The approach of Site-Specific Nutrient Management (SSNM) involves the use of organic and inorganic sources, depending on the temporal and spatial soil variability, cropping system, nutrient requirements, soil capacity for supplying nutrients, and the productive capacity of different crop varieties, as well as nutrient use efficiency (NUE) (Pramanick et al. 2024). This approach allows for feeding crops with suitable nutrients when needed by recognizing spatial availability. The application of nutrient sources varies with season and crop varieties. To ensure that N is applied at the proper time, avoiding unwanted waste, the most widely used technique is the leaf color chart (LCC). Additionally, integrated crop management techniques such as proper plant density, integrated pest management (IPM), quality seed, and proper water management are included (Dass et al. 2014). To achieve efficient SSNM, remote sensing, decision support systems (DSS), GIS, and other sensing devices are used to determine plant nutrient status, variable rate technology, and simulation models play a major role (Biswas et al. 2019).

6.3.6.9 Time of Application

It's crucial to apply nutrients during a crop's critical growth stages. Applying them at the wrong time can lead to nutrient loss through leaching, surface run-off, deep percolation, volatilization, and denitrification. The timing of nutrient application may also vary depending on the site. For instance, soils with low and high P-fixing potential should be fertilized before planting the crop, and sometimes P is applied in rotation in multiple crops. In sandy soil, it's advisable to split the application of fertilizers, especially N-fertilizers, as NO_3^{-2} is mobile and can be lost by leaching or denitrification. Splitting the application also helps reduce loss during low rainfall seasons by skipping the second application. However, in heavy rainfall areas, particularly in rice crops, it's recommended to place the fertilizer instead of splitting it.

6.3.7 *The Role of Technology in Precision Nutrient Management*

Technology has a critical role in achieving common objectives between environmental and economic goals. Recent advances indicate that the optimization of nutrients for crops can be achieved by PNM through the application of GIS, GPS, RS, variable rate technologies, and data analytics (Table 6.3). It helps farmers to improve

Table 6.3 Different precision farming equipment, their technology, description, and role in precision nutrient management

Equipment (Technology)	Description	Role
Soil sensors	Measure soil moisture, temperature, carbon, pH, and nutrient status for example IoT sensor by FASAL	Provides ground-level data and helps farmers decisions to apply nutrients more efficiently and accurately
Variable rate technology	Optimizes input use efficiency by modifying fertilizers, chemicals, or seeds based on GPS data	Help farmers apply fertilizer at the right time, and place, reducing nutrient waste and maximizing crop yields
Global positioning system	A spectral index map generated from satellite data shows farm health based on geo-referenced data.	Assist with the identification of precise field locations suitable for seed sowing, and agrochemicals application, manage water use effectively, and improve farm management practices.
Drones	Drones can perform real-time aerial monitoring and provide a bird's eye view of the farm when equipped with cameras and sensors.	Used for crop scouting, mapping, and monitoring crop health allowing for more targeted application of fertilizers
Automated irrigation system	Uses sensors and weather data to determine when and how much to water (Ex.- Ambit robotics, ARaas)	Provides the right amount of water at the right time, automated irrigation systems help farmers
Precision planters (GPS, sensors, software)	Planting machines that place seeds at precise depths and spacing	Ensures that crops are planted at optimal densities, maximizing nutrient uptake, and reducing the amount of fertilizer needed
Yield monitoring systems	Sensors that measure crop yields as the crop is harvested	Provides farmers with data on crop performance, allowing them to adjust nutrient management practices for future crops
Weather stations	Measures weather conditions such as temperature, rainfall, and humidity	Provides data for predicting plant growth and nutrient requirements, allowing more precise application of fertilizers
Mobile apps (software and data analytics)	Provides real-time information, digitization, and easy collection of field data.	Helps farmers make informed decisions, providing farmers with access to inputs, a market to sell, and a facility of insurance and credit.

their decision-making ability on nutrient management, resource allocation, and inputs for crop production. These technologies provide real-time data on soil and crop health enabling them to maximize crop yields and minimize the wastes. Several factors, including the seed crops to be sown, the size and shape of the growing area, the farmer's budget, and resources will need to be considered in order to determine specific equipment and technology that a farmer uses. For example, variable rate fertilization optimizes the fertilizer use efficiency by maintaining the over- and under-fertilization, and spatial and temporal variability enhances the nutrient assimilation in crops by standardizing site- and time-specific input application and developments of sensors for generating soil maps. To drive precision soil nutrient management and monitoring, cutting-edge technologies are currently being developed to facilitate intensive grid sampling in a cost-effective and timely manner.

6.4 Benefits of Precision Nutrient Management

Precision nutrient management (PNM) is a comprehensive solution that addresses the non-judicious use of agricultural inputs, which has led to deteriorated soil health and environmental degradation. It is not limited to the utilization of organic or inorganic inputs; rather, it focuses on the appropriate use of all agricultural inputs. PNM has a positive impact on the environment, society, economy, and crop health, and offers several benefits, such as:

- Reduction of the cost of nutrient inputs by avoiding the over-application of fertilizers
- Compatibility with different brands of equipment and machinery, which simplifies the process of proper time management, digital farm records, and tracing of crop productivity
- Provides environmental benefits by reducing the use of pesticides, chemical nutrients, and irrigation systems
- Aims to apply nutrients only when crops require them, leading to increased crop quality, optimized crop yield, better standard ability, and more sustainable field management
- Places nutrients where crops can easily use them, which maximizes nutrient uptake, reduces volatilization, and minimizes denitrification loss of N
- Meets societal expectations by reducing GHG emissions, mitigating climate change, and providing a reliable decision support system
- Improves NUE by monitoring crop conditions, detecting nutrient deficiencies, maintaining water retention, and balancing soil pH
- Reduces loss of soil habitat, enhances nutrient cycling, and helps reduce drought by supporting vegetation, thereby improving soil organic matter (SOM) content.

Overall, PNM is a valuable solution that can help to restore soil health, protect the environment, and sustainably improve agricultural productivity.

6.5 Precision Nutrient Management in a Sustainable Agriculture

The conversion of a degraded environment due to the non-judicial utilization and application of fertilizers into an eco-friendly environment will take a minimum of 8–10 years which will be unable to sustain the ever-increasing population. The non-judicial utilization of agricultural inputs affects plant nutrition and soil health leading to a reduction in plant biomass, and crop yields, and disturbing the environment too. Most of the side effects of the misutilization of fertilizers are unavoidable and uncontrollable. Precision nutrient management (PNM) is the solution for sustainable agriculture, enhancing crop productivity, and eco-friendly environment (Varallyay 2008). Sustainable agriculture is the maintenance of the ecosystem without any detrimental effects on the environment even with the utilization of its resources. By definition of sustainable agriculture, it is the utilization of agricultural inputs (organic and inorganic fertilizers) optimally with no harmful effects on the environment. It positively expects soil health, crop productivity, and an eco-friendly environment. PNM for sustainable agriculture improves the application of fertilizers depending on crop and site requirements, enhancing nutrient use efficiency (NUE), and reduction in nutrient loss due to surface run-off, denitrification, and volatilization. It involves studying the dynamics of soil nutrients and their transformation making it possible to judicious utilization of agricultural fertilizers. It also involves broadcasting its tools and techniques to the land users and educating them about sustainable agriculture.

6.6 Relationship Between Vegetation, Precision Nutrient Management, and Environment

For the survival of plants, it is essential to provide them with mineral nutrients, organic matter, soil, water, and air (Fig. 6.2). The 17 essential nutrients play a vital role in stimulating plant growth and development. Organic matter improves the physical properties of the soil and supports soil biodiversity, which ultimately enhances plant life. Soil serves as the medium for plants to grow and facilitates the availability of water and aeration. The soil microorganisms help to decompose and mineralize the plant unavailable form of nutrients from organic matter into easily available form. When all these components are in optimum condition, they provide a stable micro-environment for proper plant growth, and carbon sequestration, and promote a natural environment. However, anthropogenic activities under GR have led to disturbances in these components, causing environmental degradation. Overutilization and mishandling of agricultural inputs have released GHGs, leading to climate change. Mismanagement of irrigation systems and misuse of chemical fertilizers have caused soil degradation (salinity, alkalinity, erosion). Heavy runoff, percolation, and leaching from chemical fertilizers have also caused surface and

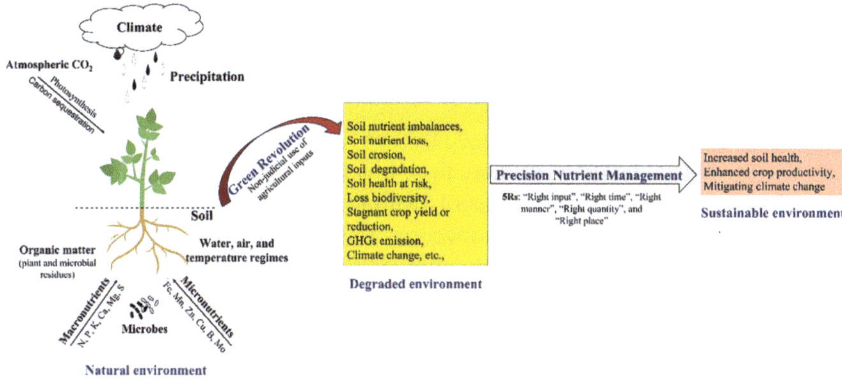

Fig. 6.2 Depicts the natural environment providing optimum plant growth with the natural resources which are then deteriorated due to the green revolution and shown that can be mitigated with precision nutrient management

groundwater pollution, degrading agricultural fields and surrounding areas. The approach of maximum crop production with maximum agricultural inputs has contributed to environmental degradation. Moreover, imbalance application of chemical fertilizers disregarding the recommended dose of fertilizer of N, P, and K is also a major issue under GR. Organic matter application is less concerned as an agricultural input under GR rendering poor soil aggregation leads to soil and nutrient loss. But, by implementing agricultural inputs concerning the 5Rs under PNM, environmental degradation can be mitigated, and approach sustainable agriculture. This approach can reduce the residual effects of chemical fertilizers on soil health and crop production. Consuming excessive doses of chemical inputs and irrigation can increase growth and production, but it can also deteriorate soil quality and lead to a decline in crop growth and stagnation in production. Therefore, optimum application of agricultural inputs depending on the crop and site requirement under PNM can sustain better crop growth without compromising crop production and mitigating climate change for a sustainable environment.

6.7 Challenges and Barriers to Implementing Precision Nutrient Management

Precision farming has already achieved remarkable growth in developed nations, especially in Europe and North America. But in India, it's relatively slow to understand and adopt these practices. The global smart agriculture market is increasing with a growth rate of 13.12% per year as compared to 10% per year for the Indian market. In 2019, the Indian market was valued at 57 million dollars, expected to reach approx. 1-billion-dollar mark in 2025 (UNDP 2021). So, it's important to minimize the challenges and maximize resource use efficiency. The adoption and implementation of this concept have been slower than initially thought because of several kinds of significant challenges. The most challenging issue is the average size of holdings among Indian farmers, as there are around 85% of the farmers belong to small and marginal categories. Therefore, the implementation is mainly in western India, i.e., Gujarat, Haryana, Punjab, and Rajasthan where 20% of the farmland has operational holdings of more than 4 ha as compared to the whole of India (Singh et al. 2016). The next problem is with the availability of real-time data on crop and soil conditions, if there is no accurate information it will affect the resource optimization. Further due to the high initial cost of implementation, it became difficult for the number of farmers and farm enterprises to adopt. Also, there is a reluctance of farmers toward this practice as they were more inclined to their traditional way of farming.

Some other challenges will be the lack of infrastructure, which will impact the availability and quality of data; climate variability, which will impact the prediction of crop and soil conditions; regulatory constraints, which will impact the widespread applicability and lastly, privacy concerns which will raise the ethical and legal issues. Besides these challenges, in order to efficiently implement such schemes there must be proper training and skill development activities among farmers and agricultural professionals. Government should also play a big role in making policies, investing in R&D, and the efficient dissemination of knowledge needed to overcome these challenges.

6.8 Success Stories and Case Studies of Precision Nutrient Management

The success of PNM is based on the development of Famer-centric low-cost technologies and innovations that cater to the needs of small as well as marginal farmers. In Western countries, PNM is widely adopted as compared to India because of its policies and incentives to farmers. They have tried and tested different techniques such as yield monitoring, fertility mapping, and variable fertility rates techniques to efficiently manage their nutrient applications and resulting in increased yields and less environmental impacts. For example, the USA started providing universal broadband access to their rural areas, financial assistance, and training to farmers for the widespread adoption of precision farming. In Europe, they are more focused on the environment and productivity. They included precision farming ideas in their Common Agricultural Policy for providing financial aid and farm advisory training services through different schemes for growers. Australians are the major adopter of precision farming. They have invested around 27 million dollars in the implementation and promotion of smart farming for better productivity and sustainability (Bah et al. 2012). They set up eight innovation hubs to promote research and started several pilot farm projects to test the efficiency of these technologies. Several start-ups, like Agrsmart, OneSoil, and Cropln, also come out and use satellite imagery under precision agriculture services for providing recommendations based on yield estimation and fertilizer application.

In India, the scenario is different as compared to Western nations. Out of 140 million hectares of net sown area, there is little scope to increase it. Our input application recommendations are based on composite samples and visual symptoms, but it has a wide variability in the field. It means there is an excessive application of inputs compared to recommended doses which reduces the efficiency and causes environmental degradation. Therefore, diversification of agriculture by incorporating this modern technique is necessary which can be achieved by precision farming tools to ensure the right input at the right time in the right amount in the right place by using the right techniques of precision farming management. Some of the flagship works in this field are mentioned here. ISRO in collaboration with top agricultural institutes started exploring the role of remote sensing in precision agriculture. They developed facilities in their Space Application Centres for studying the role of remote sensing and GIS-based technologies in mapping and forecasting. NABARD in collaboration with M S Swaminathan Research Foundation, Chennai studied the variable rate technology under precision farming in some districts of Tamil Nadu. IARI, New Delhi developed the methodology to do experiments related to precision farming in their institutes. Central Institute of Agricultural Engineering, Bhopal with collaborations with Project Directorates on farming system research also initiated variable rate application of inputs in different farming systems models. Tata Kishan Sansar set up by Tata Chemical Limited in northern and eastern India also helped farmers to adapt quickly to climate variabilities by providing recommendations based on remote sensing techniques for higher yields and enhanced farm incomes. In Odisha, a mobile-based custom advisory service named "Ama Krushi" for 17 crop types is helping farmers with fertilizer application, disease/pest monitoring, optimal planting, and nursery-based practices. IIT Bombay developed SoilSenseGo technology a portable moisture system for monitoring and irrigation recommendations. A non-profit Wadhwani AI in collaboration with Google.org developed a mobile app called "CottonAce" to provide advice to farmers for control of pink bollworms.

Furthermore, government policy and initiatives played a big role in the adoption of PNMs in India. Indian government by their flagship National Programme for Sustainable Agriculture provided financial and technical assistance to farmers with the adoption of PNM methods and technologies. In the next few years, precision farming could help Indian farmers to achieve the desired goals by using frontier technology without compromising the quality of the soil.

6.9 Future Prospects

To sustain livelihoods, the modern technique of precision nutrient management (PNM) is quite promoting nowadays which functions on the 5Rs. To date, many ideas and technologies are being launched under the PNM, but necessary to improve in its feasibility to make adopted on a large scale as it involves many tools and techniques which are unable to be afforded and understood by a lay farmer, especially in the developing countries like India. The implementation of PNM by joining hand in hand with private, public, and NGO sectors to provide the technical skills and knowledge on PNM, and subsidies to afford the technologies to the farmers would be a promising approach for avoiding any non-judicial utilization of agricultural inputs.

6.10 Conclusion

The concept of the green revolution without environmental concern led to the deterioration of natural resources such as soil and water, and climate change due to the non-judicial utilization of agricultural inputs. Various life forms will be at risk which depends on natural resources for its survival. The adoption of PNM is expected as a solution for reclaiming deteriorated natural resources and mitigating climate change through vegetation with proper handling of agricultural inputs using its advanced techniques, practices, and technologies that can sustain long-term food security. Moreover, it can increase nutrient use efficiency, and crop productivity, and mitigate environmental degradation by reducing the non-judicial utilization of agricultural inputs for a more sustainable future.

References

Adhikary S, Biswas B, Naskar MK, Mukherjee B, Singh AP, Atta K (2022) Remote sensing for agricultural applications. In: Arid environment-perspectives, challenges and management. IntechOpen

Bah A, Balasundram SK, Husni MHA (2012) Sensor technologies for precision soil nutrient management and monitoring. Am J Agric Biol Sci 7(1):43–49

Biswas S, Gawade BH, Singh P (2019) Site-specific nutrient management (SSNM) for increasing fertilizer use efficiency and agriculture profitability. Published on July 27. https://www.krishisewa.com/resource-management/1031

Dass A, Suri VK, Choudhary AK (2014) Site-specific nutrient management approaches for enhanced nutrient-use efficiency in agricultural crops. Res Rev J Crop Sci Technol 3(3):1–6. Accessed 9 Apr 2023

Department of Fertilizers and Department of Agriculture, Cooperation & Farmers Welfare (DAC&FW), India (2017)

Eitzinger J, Trnka M, Hösch J, Žalud Z, Dubrovský M (2004) Comparison of CERES, WOFOST and SWAP models in simulating soil water content during growing season under different soil conditions. Ecol Model 171:223–246

Eliazer Nelson ARL, Ravichandran K, Antony U (2019) The impact of the Green Revolution on indigenous crops of India. J Ethnic Foods 6(1):1–10. Accessed 11 May 2023

FAO (2011) The state of food and agriculture 2010-2011. FAO, UN, Rome

Farooq M, Siddique HMK (2015) Conservation agriculture: concepts, brief history, and impacts on agricultural systems. Springer, Cham

Figueiredo CC, Melo LCA, Silva CA, Fachini J, de Carneiro JSS, de Morais EG, Ndoung OCN, Singh SV, Nandipamu TMK (2024) Nutrient enriched and co-composted biochar: system productivity and environmental sustainability. In: Biochar production for green economy. Academic Press, Elsevier, London, pp 311–331. https://doi.org/10.1016/B978-0-443-15506-2.00010-9

Friedrich T, Derpsch R, Kassam AH (2012) Global overview of the spread of conservation agriculture. Field Action Sci Rep 6:1–7

Hipólito J, Boscolo D, Viana BF (2018) Landscape and crop management strategies to conserve pollination services and increase yields in tropical coffee farms. Agric Ecosyst Environ 256:218–225

IPNI (2012) 4R plant nutrition manual: a manual for improving the management of plant nutrition, metric version (Bruulsema TW, Fixen PE, Sulewski GD, eds.). International Plant Nutrition Institute, Norcross

Jariwala H, Santos RM, Lauzon JD, Dutta A, Chiang YW (2022) Controlled release fertilizers (CRFs) for climate-smart agriculture practices: a comprehensive review on release mechanism, materials, methods of preparation, and effect on environmental parameters. Environ Sci Pollut Res 36:53967–53995. https://doi.org/10.1007/s11356-022-20890-y

Jin J, Jiang C (2002) Spatial variability of soil nutrients and site-specific nutrient management in the PR China. Comput Electron Agric 36(2–3):165–172. https://www.sciencedirect.com/science/article/abs/pii/S0168169902000996

Lagacherie P, McBratney AB (2007) Spatial soil information systems and spatial soil inference systems: perspectives for digital soil mapping. In: Lagacherie P, McBratney AB, Voltz M (eds) Digital soil mapping: an introductory perspective. Elsevier, Amsterdam, pp 3–24

Mahapatra T (2022) Precision agriculture in India: a perspective. NAAS Newsl 22(3):1–4

Malone BP, Odgers NP, Stockmann U, Minasny B, McBratney AB (2018) Digital mapping of soil classes and continuous soil properties. In: McBratney AB, Minasny B, Stockmann U (eds) Pedometrics, progress in soil science. Springer, Cham, pp 373–413

Meja M, Kebede AK (2022) Impact of agricultural inputs use on productivity of major crops in southern Ethiopia. Res Sq:1–31. https://doi.org/10.21203/rs.3.rs-2062064/v1

Pramanick B, Choudhary S, Kumar M, Singh SK, Jha RK, Singh SK, Salmen SH, Ansari MJ, Hossain A (2024) Can site-specific nutrient management improve the productivity and resource use efficiency of climate-resilient finger millet in calcareous soils in India? Heliyon 10(12):e32774. https://doi.org/10.1016/j.heliyon.2024.e32774

Singare PB, Meshram NA, Jondhale AS, Kadam VS (2020) Need of soil testing for improvement of soil health and crop productivity. Agric Obs 1(2):48–51

Singh V, Kaur R, Singh B, Brar BS, Kaur A (2016) Precision nutrient management: a review. Indian J Fertil 12(11):1–15. https://www.researchgate.net/publication/310606305

Singh SV, Chaturvedi S, Dhyani VC, Govindaraju K (2020) Pyrolysis temperature influences the characteristics of rice straw and husk biochar and sorption/desorption behavior of their biourea composite. Bioresour Technol 314:123674. https://doi.org/10.1016/j.biortech.2020.123674

Sruthi S, Aslam MMA (2015) Agricultural drought analysis using the NDVI and land surface temperature data; a case study of Raichur district. Int Conf Water Resour Coast Ocean Eng 4:1252–1264

UNDP (2021) Annual report 2021. UNDP, New York, NY 10017, USA

Varinderpal-Singh, Bijay-Singh, Yadvinder-Singh et al (2010) Need based nitrogen management using the chlorophyll meter and leaf colour chart in rice and wheat in South Asia: a review. Nutr Cycl Agroecosyst 88:361–380. https://doi.org/10.1007/s10705-010-9363-7

Vejan P, Khadiran T, Abdulla R, Ahmad N (2021) Controlled release fertilizer: a review on developments, applications and potential in agriculture. J Control Release 339:321–334. https://doi.org/10.1016/j.jconrel.2021.10.003

Varallyay G (2008) Extreme soil moisture regime as limiting factor of the plants' water uptake. Cereal Research Communications 36, 3–6

Zhang M, Newton C, Grove J, Pritzker M, Loannidis M (2020) Design and assessment of a hybrid chemical engineering laboratory course with the incorporation of student-centred experiential learning. Educ Chem Eng 20:1–8. https://doi.org/10.1016/j.ece.2019.09.003

Chapter 7
Precision Water Management Under Dryland Region: Climate-Resilient Strategies

Sougata Roy, Sanjay Kumar Rathour, Anjali Mehta, Ritu Dwivedi, Surabhi, Aastika Pandey, Anamika Barman, Shreyas Bagrecha, and Shiv Vendra Singh

Abstract Water is a critical input for agricultural production systems. Increasing water productivity in field crops through precision water management (PWM) is essential for optimizing resource use. PWM aims to achieve maximum crop yield with optimal water utilization, enhancing productivity and crop quality through judicious irrigation. Proper crop monitoring related to applied irrigation helps growers manage their fields efficiently, saving water, time, and costs while improving yields compared to conventional systems. Analyzing crop-water requirements and applying irrigation based on these parameters is crucial. Geographic Information System (GIS) technology integrates spatial and temporal data, aiding practical managerial decisions for irrigation. Proper adoption of precision irrigation can save up to 25% more water than conventional practices. Agriculture, the largest water consumer, uses 70% of global water resources, with drylands producing over half of the world's food. In India, drylands encompass 68% of the net sown area, contributing significantly to crop production but facing freshwater scarcity, global warming, and erratic rainfall, impacting productivity. Despite efforts to improve irrigation

S. Roy (✉)
ICAR-Indian Agricultural Research Institute, New Delhi, India

International Potato Centre (CGIAR-CIP), Patna, Bihar, India

S. K. Rathour
G. B. Pant University of Agriculture and Technology, Pantnagar, Uttarakhand, India

A. Mehta · S. Bagrecha
ICAR-National Dairy Research Institute, Karnal, Haryana, India

R. Dwivedi · Surabhi
University of Agricultural Sciences, GKVK, Bangalore, Karnataka, India

A. Pandey · A. Barman
ICAR-Indian Agricultural Research Institute, New Delhi, India

S. V. Singh
ICAR-College of Agriculture, Rani Lakshmi Bai Central Agricultural University, Jhansi, Uttar Pradesh, India

B. Pramanick et al. (eds.), *Climate-Smart Agricultural Technologies*, Sustainability Sciences in Asia and Africa, https://doi.org/10.1007/978-981-96-7699-6_7

efficiency in India, overall efficiency remains between 35% and 40%. Advances in irrigation systems and technologies now enable site-specific application of water within fields, enhancing water use efficiency and reducing overall water usage for sustainable crop production, particularly in arid and semi-arid regions. This chapter explores recent developments in Variable-Rate Irrigation (VRI) technologies, including drip and central pivot irrigation systems. Effective precision water management can significantly boost crop yields, ensuring food security and contributing to sustainable development goals (SDG).

Keywords Water management · Central pivot system · Water footprint · Sustainable development goals · Variable rate technology

7.1 Introduction

Agriculture is the largest consumer of water, utilizing 70% of the total consumption (Qin et al. 2018). Over half of the world's food is produced on drylands, which represent 41.3% of global agricultural land (Chen et al. 2018a, b). In India, drylands encompass 177 districts and 68% of the net sown area (136.8 million hectares (Mha)). These regions account for 68% of the area used for non-food crops and 48% for food crops. Nearly half of India's rural labor force and 60% of its livestock are concentrated in arid zones (CRIDA 2015). Sustainable rainfed agriculture in these drylands is essential for global food supply due to increasing freshwater scarcity. However, global warming and erratic rainfall patterns are depleting water resources and limiting agricultural productivity in these areas (Qin et al. 2018). Agricultural cropping systems rely heavily on water resources for their survival, and the water requirements can vary across different areas within a field due to variations in soil characteristics like texture, topography, water-holding capacity, infiltration, and drainage rate. Consequently, the need for irrigation may differ from one zone to another within the same field. Conventional moving irrigation systems apply water at uniform rates, which can lead to some areas receiving too much water while others receive too little. Precision Water Management (PWM) aims to optimize water use for sustainable management by applying the right amount of quality water at the right time, place, and crop growth stage across the target area. Various strategies and technologies have been developed to support PWM, including computer-aided tools designed to enhance irrigation project management. Water is a critical input for crop growth and yield, influencing plant health and productivity. Both excess and deficit water conditions can cause significant stress to crops. Excess water can lead to root rot, which compromises root function and nutrient uptake. Conversely, water deficit conditions can cause wilting, reducing photosynthesis and potentially leading to plant death. Efficient water management is essential to maintain optimal soil moisture levels, ensuring crops receive adequate hydration without the risk of waterlogging or drought stress. Therefore, it is essential to provide crops with the optimal amount of water to maximize yield potential. Water quality, including factors such as salinity and pH, is vital for the effective growth of crops. This makes Precision Water Management (PWM) an essential practice, as it takes into account both the quality and quantity of water used (Raine et al. 2007).

7.1.1 Significance of Precision Water Management in Indian Context

Recent news reports from sources like the "Times of India" (2024) highlight the severe impact of drought in regions such as Maharashtra and Saurashtra (Gujarat), where crop failure and farmer suicides have become alarmingly common (Times of India 2024). To address these issues, the government has promoted drip irrigation and other water-saving techniques, as evidenced by headlines like "Farmers take to drip irrigation to cut costs and save water" and "Make every drop of water count for sustainable agriculture." Despite these efforts, there remains a significant gap between the created and utilized irrigation potential. By the end of the Eleventh Plan, only 87.86 Mha of the total 113.53 Mha of irrigation potential had been utilized, leaving a gap of 25.67 Mha (Central Water Commission 2015). The gap between IPC and IPU is widening due to declining irrigation efficiency. Factors such as inadequate irrigation infrastructure and shifts in agricultural practices, like the intensive rice-wheat cropping system, also contribute to this growing disparity. According to a Central Water Commission report from 2015, the overall irrigation efficiency of major irrigation projects in India ranges between 35% and 40%, which is low compared to global standards. Improving irrigation efficiency is essential to meet the increasing water demands for food, the environment, urban areas, and industry (Rockström et al. 2007).

The low productivity in rainfed agricultural regions is due to various constraints, including environmental factors (drought, flooding, soil compaction), socioeconomic conditions of farmers, and other agronomic management issues. Rockström et al. (2007) noted that India's actual crop yield is only 45% of its potential yield, suggesting that proper management in rainfed areas could significantly boost yields. Recent research by Bal et al. (2022) indicated that the effect of dry spell (DSI) on crop yield decreases linearly, with regression coefficients of 0.63 for groundnut and 0.545 for maize. To ensure food security and achieve potential yields, it is crucial to focus on efficient water management in agriculture (Figs. 7.1, 7.2, and 7.3).

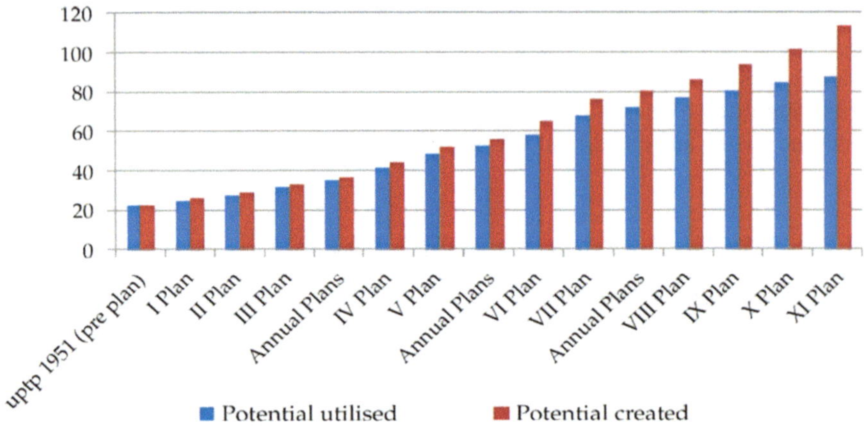

Fig. 7.1 Gap between irrigation potential created vs irrigation potential utilized. (Source: Central Water Commission (2015))

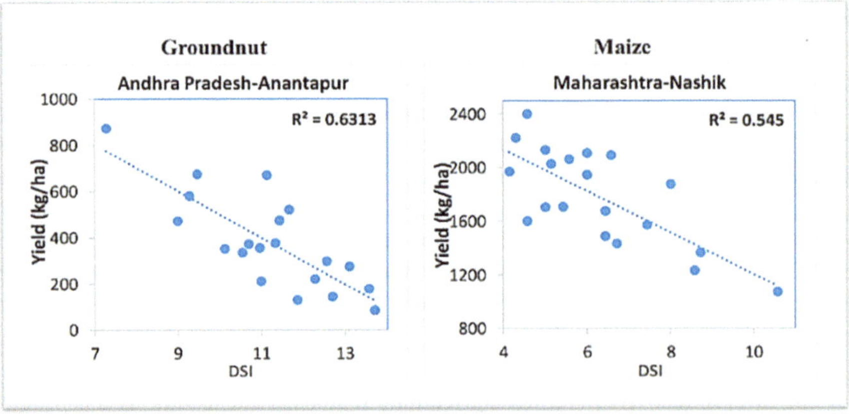

Fig. 7.2 Effect of dry spell on groundnut and maize yield. (Source: Bal et al. (2022))

7.2 Water and Dryland Scenario of India and Its Contribution to Indian Economy

India holds about 4% of the world's freshwater resources but must support around 17% of the global population. Despite this, it uses two to four times more water than countries like China and the USA to produce the same quantity of major agricultural products (NITI Aayog 2015–16). Annually, India receives approximately 4000 billion cubic meters (BCM) of water, mainly from rainfall and some snowfall, but the distribution is highly variable across regions and seasons, impacting water availability. Of the 4000 BCM of water received, around 1869 BCM are accessible as water resources, with only 1123 BCM (690 BCM from surface water and 433 BCM

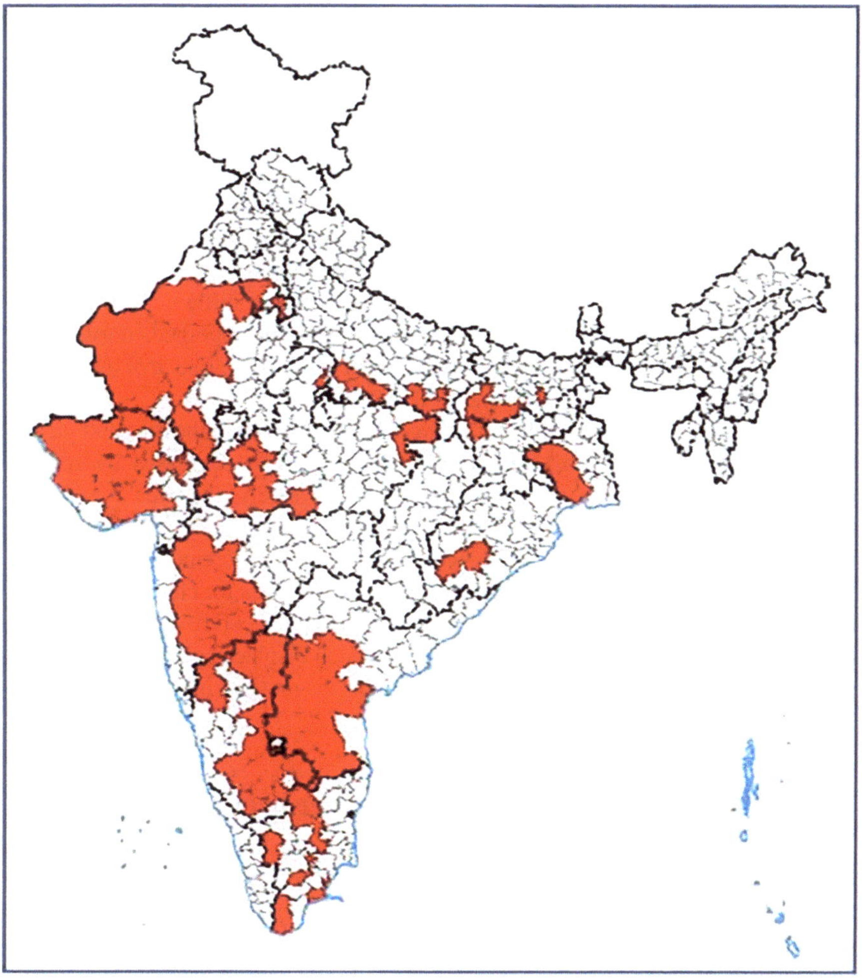

Fig. 7.3 Drought-affected area of India

from groundwater) being usable. In 2000, water demand in India was 634 BCM. Projections indicate that total water demand will rise to 814 BCM by 2025 and reach 1077 BCM by 2050. Per capita water availability in India has significantly declined, from 5000 cubic meters annually in 1950 to an estimated 1465 cubic meters by 2025, underscoring a sharp reduction in water resources for agriculture. In 2010, out of 781 BCM of total water withdrawals, around 91% (688 BCM) was utilized for agriculture, making it the largest consumer of freshwater in the country.

Globally, out of the 1.5 billion hectares (Bha) of cropland, 1.223 Bha (82%) is rainfed, contributing to 70% of the world's staple food (FAO 2016). In India, out of 143.8 Mha of arable land, 32% (43.8 Mha) is irrigated, and the remaining 68%

(100 Mha) is used for dry farming. Within this 100 Mha, 65.5 Mha is rainfed, and 34.5 Mha is dry farming (Drought Manual 2020). India's drought-prone areas, marked in red on the map, include almost all of Rajasthan (except Ganganagar district), the Kutch and Gujarat region, the rain shadow areas of the Western Ghats, parts of Tamil Nadu, Karnataka, and Andhra Pradesh, Bankura and Purulia districts of West Bengal, and the Bundelkhand region. The arid zone covers about 12% of India's geographical area, with 32 Mha of hot arid land spread across Rajasthan (61%), Gujarat (20%), Andhra Pradesh and Karnataka (10%), and Punjab and Haryana (9%). Dryland agriculture accounts for approximately 91% of the area for coarse cereals (sorghum, pearl millet, maize, and finger millet), 91% of pulses (chickpea and pigeon pea), 80% of oilseeds (groundnut, rapeseed, mustard, and soybean), and 65% of cotton. Additionally, about 50% of the rice area and 19% of the wheat area are rainfed (Reddy and Reddy 2019).

Over the past 25 years, significant changes have occurred in the area and yield of key crops in rainfed regions. The area for coarse cereals decreased by approximately 10.7 Mha, mainly affecting sorghum. Conversely, the area for oilseeds increased by 9.2 Mha, largely due to irrigated rapeseed, mustard, and soybean. The total area for pulses and cotton remained stable, although more cotton became irrigated, and there were regional shifts in crop areas. The area for chickpea decreased in the northern belt but increased in the central belt. This shift was driven by the expansion of the rice-wheat cropping system, which displaced chickpea, as well as pearl millet to a large extent and maize to a lesser extent (Drought Manual 2020).

7.2.1 Water Footprints and Water Management of Field Crops

The concept of the water footprint was introduced by Hoekstra in 2003 (Hoekstra 2015). It represents the total volume of water used to produce a product, typically measured in cubic meters per ton (m^3/ton). This includes the water used in all process steps involved in manufacturing the product. The water footprint for a specific geographic area, such as a province, state, basin, or river basin, is the cumulative water footprint of all processes occurring within that area (Fig. 7.4; Tables 7.1 and 7.2).

Agricultural activities consume more than 70% of freshwater resources, amounting to 1500 billion m^3 out of the total 2500 billion m^3 used annually for crop production and related enterprises (Goodwin and O'Connell 2015). However, about 40% of this water is lost in developing countries through evaporation or percolation beyond the root zone of plants (Panchard et al. 2007). In these regions, nearly half (48.8%) of the cultivated area benefits from assured irrigation, while the remainder relies on rainfed conditions, out of a total cultivated area of 140 Mha. As water availability per capita declines to around 1000–1100 m^3, countries risk falling into the water-stressed category, making efficient water management crucial to enhance water use efficiency and reduce production costs (UN DESA 2023). To address this, efforts over the past four decades have focused on micro-irrigation methods like

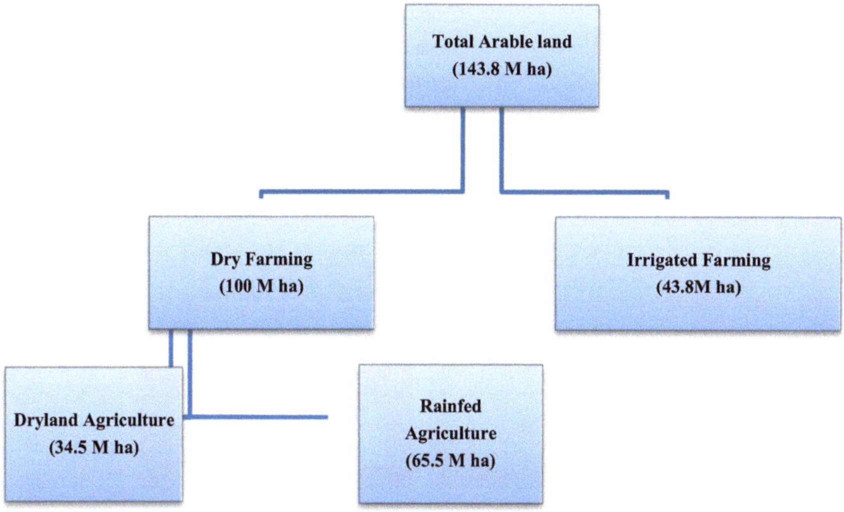

Fig. 7.4 Distribution of dryland area in India. (Source: Reddy and Reddy (2019))

Table 7.1 Types of water footprint (Hoekstra 2015)

Types of water footprint	Characteristics
Blue water footprint	This represents the volume of surface and groundwater consumed (evaporated) during the production of goods.
Green water footprint	This denotes the volume of rainwater used for the crop production and another use
Gray water footprint	This indicates the volume of freshwater required to dilute pollutants to meet existing water quality standards.

drip and sprinkler irrigation across various parts of the world. Despite these advancements, a universally perfect irrigation method suitable for all weather conditions, soil structures, and crop varieties remains elusive. In developing countries, many small and marginal farmers depend on rainfall for their crops due to the high costs of powered irrigation. Therefore, preventing water loss and misuse through the development of efficient irrigation scheduling systems is essential. These systems must deliver the right amount of water at the right time and place, optimizing irrigation efficiency (Fig. 7.5).

Agricultural production is intricately linked to water sources, quality, and quantity, with variations due to soil characteristics (topography, texture, water-holding capacity, and drainage), crop type (short or long duration, water requirements), and regional climate. Consequently, water needs differ across agro-climatic zones, and irrigation frequency varies between fields and crops. Inefficient water distribution can harm crops, increase cultivation costs, and deplete water resources. Given the insufficient natural precipitation in different zones, irrigation from canals, rivers, bore wells, and storage tanks becomes essential. Precision irrigation, therefore,

Table 7.2 Average water footprint (m^3/ton) of different crops (Lovarelli et al. 2016)

Crop type	Average water footprint (m^3/ton)
Sugar crops	~200
Vegetables	300
Roots and tubers	400
Fruits	1000
Cereals	1600

Fig. 7.5 17 Sustainable development goals of UN. (Source: Marathe (2019))

emerges as the best option to enhance water productivity and efficiency. The concept of precision water management (PWM) is critical for the judicious use of water, ensuring it is applied in the right quantity, at the right time, place, and crop growth stage. This approach supports the achievement of FAO's sustainable development goals (SDG6) by optimizing water productivity and use efficiency across various agricultural settings (Ortigara et al. 2018).

7.3 Precision Water Management and Technological Innovations

Precision water management traditionally involves applying precise amounts of water to crops at specific locations and times, uniformly across the field. However, it is best viewed as a management approach where crops grow and yields are maximized only if soil moisture remains high and water is readily available throughout the growing season. Precision water management evaluates crop water requirements and applies the optimal quantity of water at the right time, place, and manner to enhance water use efficiency and crop productivity.

Defined generally, precision water management means applying the precise quantity of water to crops at precise locations and times (Ahmad and Mahdi 2018). Other definitions include "applying water in the right place with the right amount" (Al-Karadsheh et al. 2002), and "the application of water to a given site in a volume and at a time needed for optimum crop production and profitability" (Camp et al. 2006). It also involves irrigation management based on crop needs, defining sub-areas of a field as management zones (King et al. 2006), and accurately applying water to meet specific plant requirements while minimizing environmental impact (Raine et al. 2007).

Irrigation water is critical in agriculture as crop yields often maximize when soil moisture remains high throughout the growth period, and yields generally increase linearly with water transpired by the crop. Water quality is also crucial under precision water management. The main challenge is to improve water use efficiency and sustainability, achievable by increasing crop water productivity, reducing water losses through soil evaporation, and increasing soil water storage within the plant rooting zone through better soil and water management practices at both farm and area-wide scales. Environmental inputs such as climate, water quality, and soil properties, along with management practices and crop requirements, are key in designing precision irrigation. The optimized outcomes of precision irrigation include improved soil and water availability, optimal salt distribution and storage, reduced deep drainage and runoff, and ultimately, optimized crop production and sustainable water and salt load to the catchment.

7.3.1 Tools and Technologies Used in Precision Water Management

To effectively gather and utilize information, those interested in precision farming should become acquainted with the available technological tools. These include hardware, software, and best practices (Figs. 7.6, 7.7, and 7.8).

Global Positioning System (GPS) Satellite-based navigation, known as GPS, was developed by the US Department of Defense in the 1970s for military use in positioning and timing. Today, GPS has wide-ranging applications, including commercial use, agriculture, robotics, and clock synchronization. GPS, using a satellite network records positional information (latitude, longitude, elevation) with an accuracy of 100 to 0.01 m (Lang 1992). In agriculture, GPS helps farmers identify precise field locations (soil type, pests, weeds, water holes, boundaries, obstructions) using an automated system with a DGPS, antenna, and receiver. GPS signals allow receivers to calculate positions, enabling targeted application of seeds, fertilizers,

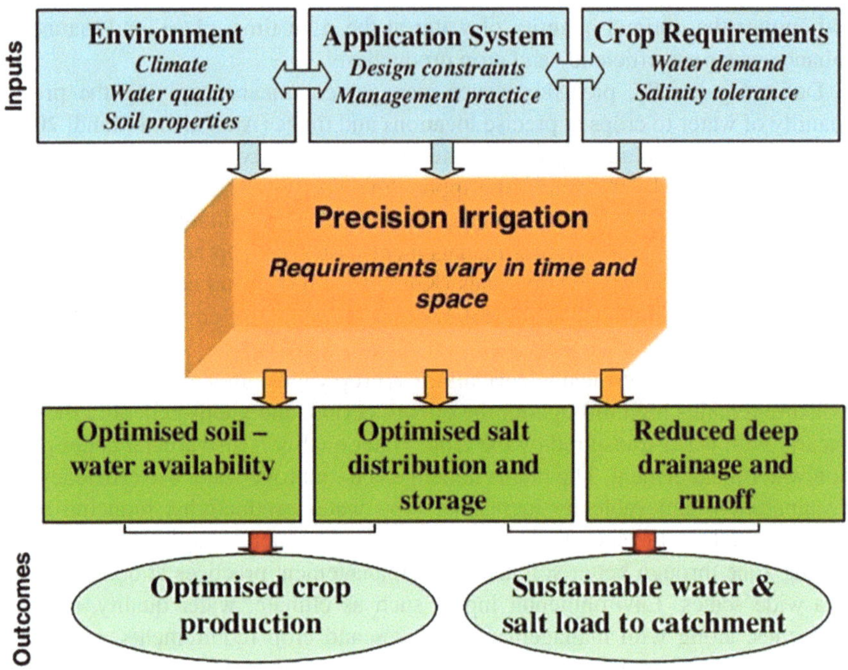

Fig. 7.6 Input and output of precision irrigation system. (Source: Smith et al. (2021))

pesticides, herbicides, and irrigation based on field performance and past inputs (Bill et al. 2012).

Geographic Information System (GIS) This system integrates hardware, software, and procedures to support compiling, storing, retrieving, and analyzing feature attributes and location data for map production. GIS consolidates diverse information for easy retrieval. Unlike traditional maps, computerized GIS maps contain multiple layers of data such as yield, soil surveys, rainfall, crops, soil nutrients, and pests. GIS uses statistical and spatial methods to analyze geographic features and data. A agricultural GIS database can detail field topography, soil types, surface and subsurface drainage, soil tests, irrigation, chemical usage, and crop yields. Analyzing this data helps understand the interrelationships affecting crops at specific locations (Bill et al. 2012).

Sensor Technologies Various technologies like electromagnetism, conductivity, photoelectricity, and ultrasound measure humidity, vegetation, temperature, texture, structure, physical characteristics, nutrient levels, and vapor. Remote sensing data differentiates crop species, locates stress conditions, identifies pests and weeds, and monitors drought, soil, and plant conditions. Sensors collect extensive data without lab analysis. Wireless sensor networks (WSNs) use distributed sensors for automatic,

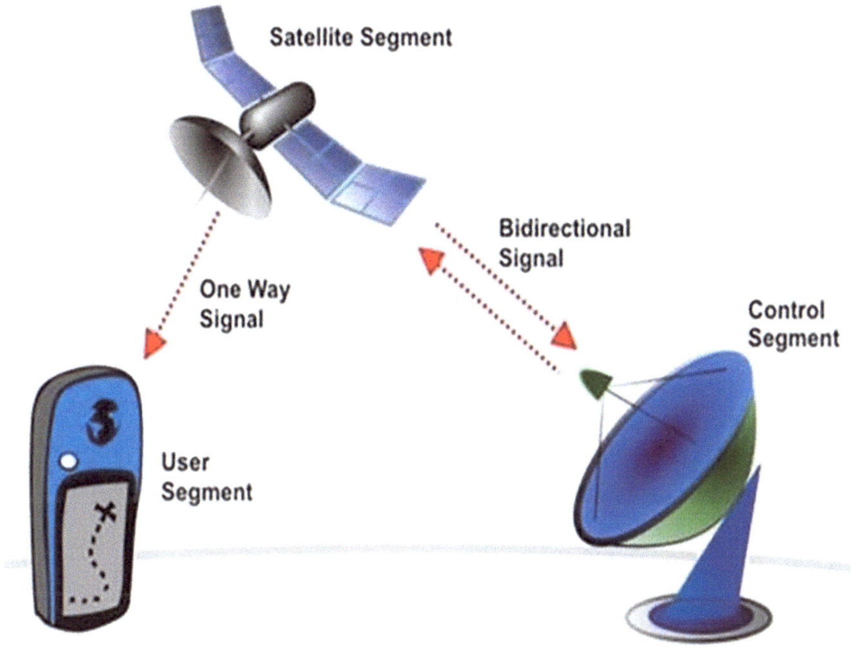

Fig. 7.7 Working of GPS. (Source: Ortigara et al. (2018))

wireless monitoring of weather, water status, and soil conditions. They are cost-effective and efficient compared to ground-based sensors, providing continuous, real-time measurements. WSNs employ sensor nodes across an area, each with wireless communication (Devadas et al. 2011), integrated with cellular and internet technologies for remote measurement, data transmission, and access. This approach ensures long-term soil parameter monitoring without disturbing sensors during field operations like plowing (Figs. 7.9, 7.10, 7.11, 7.12, 7.13, and 7.14).

Digital Soil Mapping Digital soil mapping involves creating and populating a soil database with geographic references. This is done at a specific resolution using field and lab observations, coupled with environmental data, to establish quantitative relationships. Efforts have been made to create digital soil maps (DSMs) that integrate soil type, weather patterns, and chemical composition in a GIS. These maps aim to illustrate the fertility of agricultural regions or predict their productivity. Notably, the Natural Resources Conservation Service (NRCS) has developed a prominent regional soil map (Söderström et al. 2016).

Software Precision agriculture often relies on specialized software for tasks like interfacing display controllers, mapping information layers, and analyzing farm data. Key functions include generating yield and soil maps, filtering data, and creating variable rate application maps for inputs like irrigation, fertilizer, and chemicals.

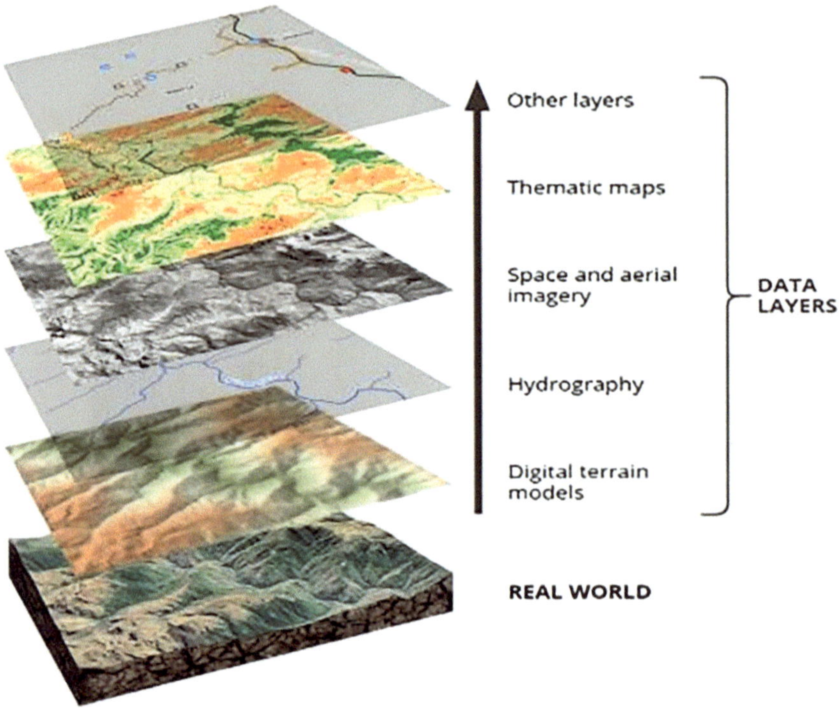

Fig. 7.8 GIS produced thematic layers. (Source: Kumar and Patel (2022))

These software solutions are crucial for efficient farm management in today's data-driven agricultural systems. They vary in complexity and cost, with some offering comprehensive mapping, statistical analysis, and record-keeping features. Challenges remain in data transfer between farmers, cooperatives, and consultants, and in effectively overlaying maps such as soil and yield data (Nikkilä et al. 2010).

Variable Rate Technologies Variable rate technologies (VRT) automate various farming tasks by adjusting input delivery rates based on soil type from soil maps. GIS-derived data directs processes like seeding, irrigation, fertilizer and pesticide application, and herbicide use, ensuring optimal variable-rate applications. VRT is widely adopted in the USA, with grid soil sampling enhancing traditional soil sampling by intensifying sampling frequency and enabling precise mapping using location-tagged data (Šarauskis et al. 2022).

Fig. 7.9 Soil moisture sensor. (Original Picture)

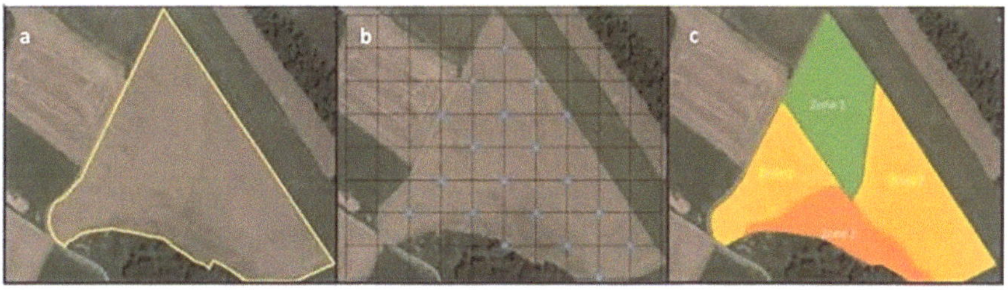

Fig. 7.10 A digital soil map of crop field. (Source: Devadas et al. (2011))

7.3.2 Steps in Variable Rate Irrigation/Precision Water Management

Hillel (1990) defines a well-managed irrigation system as one that efficiently distributes water spatially and temporally to maximize crop growth, yield, and economic returns. Given the site-specific and seasonal nature of irrigation conditions, there is no universal solution for optimal development and management. Figure depicts the framework for precision irrigation strategy. Initial spatial variability data is derived from soil maps available at agricultural planning offices, but these maps alone are insufficient for precision irrigation due to their broad-scale coverage. Next, in-field data is gathered using fast, non-destructive real-time sensors and surrogate properties like electrical conductivity (EC). This data informs soil sampling and correlation with specific properties (e.g., EC vs. available water capacity,

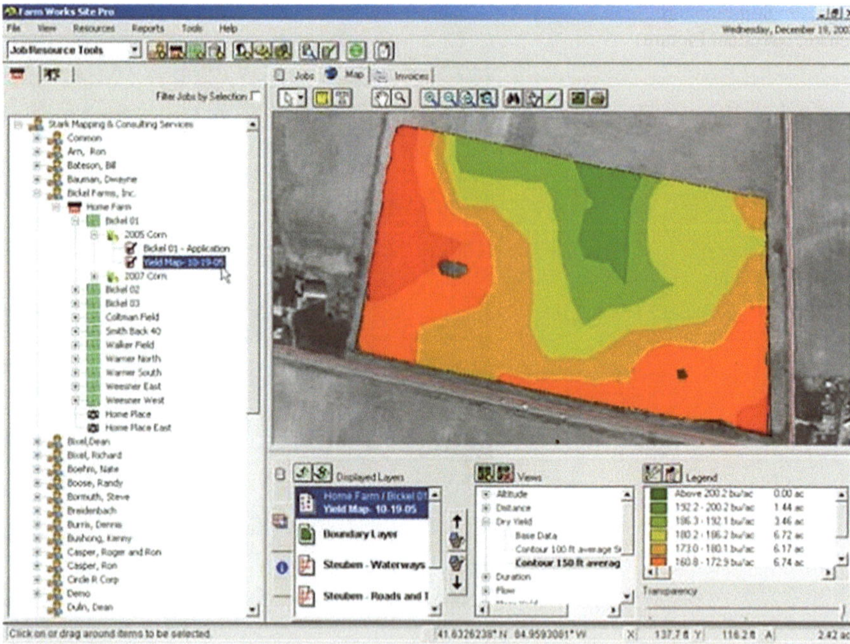

Fig. 7.11 An image processing software. (Original Picture)

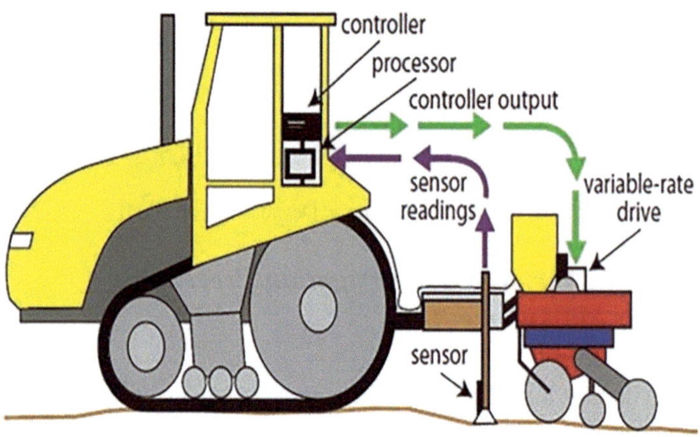

Fig. 7.12 VRT hardware in tractor. (Source: Ortigara et al. (2018))

AWC). Management zone maps (application maps) are then created to guide field activities, such as irrigation, specifying depths and locations (Chavez et al. 2010). Integration of variable rate technologies into existing machinery or new introductions is decided, involving evaluations of parameters like travel speed and discharge rate.

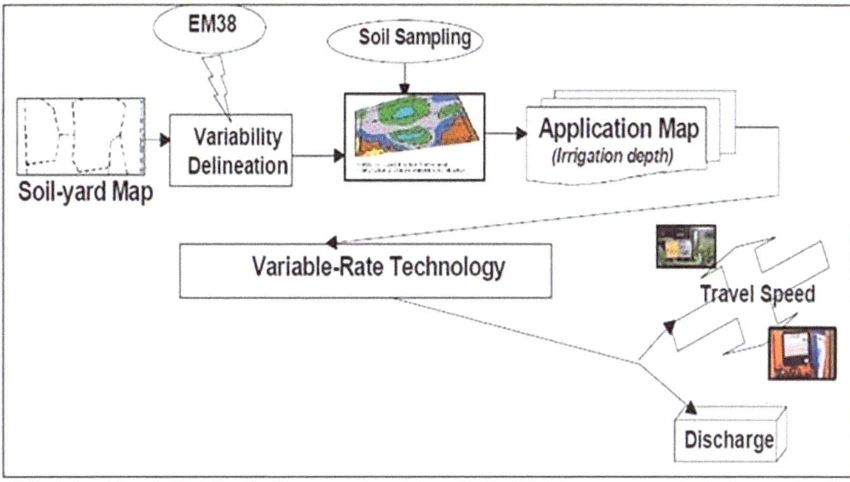

Fig. 7.13 Steps involved in precision irrigation application. (Source: Ortigara et al. (2018))

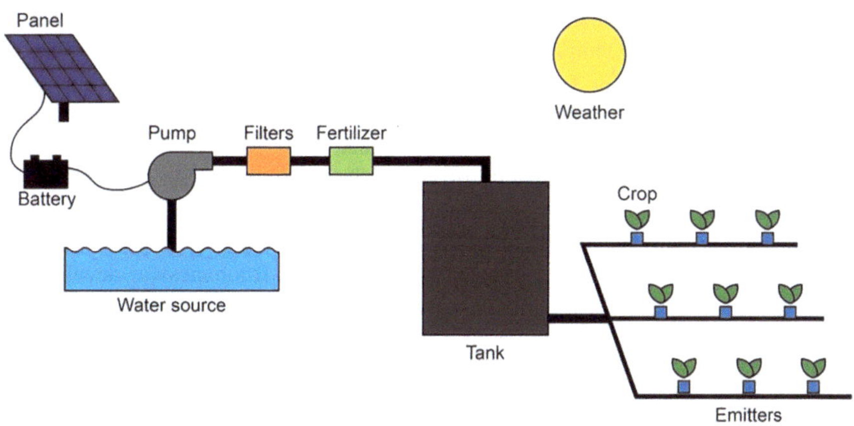

Fig. 7.14 A systematic diagram of precision drip irrigation system. (Source: Chen et al. (2018a, b))

7.4 Precision Irrigation Methods for Dryland Agriculture

Precision irrigation management, aided by data analytics, significantly enhances crop monitoring and water use efficiency. Advance tools now support growers in optimizing irrigation practices, thereby conserving water, saving time, reducing costs, and improving crop yields compared to traditional methods. Effective irrigation management relies on accurate data, such as weather forecasts, evaporation rates, soil moisture levels, and crop types. This information helps determine additional soil moisture needs, ensuring irrigation is precisely tailored to crop requirements.

Geographic Information Systems (GIS) integrate hardware and software for managing spatially referenced data, linking spatial information with descriptive details. GIS technology supports effective decision-making by synchronizing spatial and temporal data. It is particularly useful in rice farming for managing water resources through both simple and complex operations. GIS provides necessary numerical and graphical data for timely irrigation, enhancing water management efficiency. This approach minimizes costs and avoids over-irrigation, which can damage crops and increase cultivation expenses.

7.4.1 Precision Drip Irrigation System

Drip irrigation is a highly efficient precision irrigation method widely used in agriculture. It significantly reduces water losses compared to conventional irrigation methods. Water is delivered through a filter into drip pipes, also known as drip-tape or drip-line, which have tiny emitters that release water directly into the root zone at low pressure maintained by software. This system can be fully automated, minimizing human effort. Advanced moisture-sensitive sensors enable automation, providing real-time data on temperature, evaporation, humidity, and soil moisture to regulate water supply. The system's automation helps monitor and control the water flow accurately, ensuring the right amount is delivered to the crops. Water flow meters connected to sensors adjust the water quantity based on crop requirements. Additionally, analyzing the capacity of water discharge by the motor or irrigation pump ensures precise water application, enhancing water use efficiency and crop yield.

The ICAR-Indian Institute of Water Management in Bhubaneswar developed a sensor-based automated drip irrigation system for bananas, saving 25% water and increasing yield by 13% compared to manual drip irrigation. The All India Coordinated Research Programme on Irrigation Water Management (AICRP-IWM) in Bhubaneswar introduced precision irrigation techniques like drip fertigation, mini portable sprinkler systems, and alternate furrow irrigation. Studies showed drip irrigation significantly improves water use efficiency (WUE), with increases ranging from 14.3% in Chiplima, Odisha to 270.4% in Sri Ganganagar, Rajasthan. At Rahuri, Maharashtra, drip fertigation enhanced tomato WUE by 77% and nitrogen uptake by 811% (Singandhupe et al. 2003). Drip fertigation increased the productivity of bitter gourd, ladies' finger, and cowpea by 59.1%, 66.6%, and 141.6%, respectively, in a 2018 ICAR-IIWM study.

7.4.2 *Centre Pivot (Sprinkler) and Lateral Move Machines*

The central pivot system is a type of sprinkler irrigation method that distributes water through a rotating mechanism, where water is sprayed into the air and falls onto the crop effectively simulating natural rainfall. This process is accomplished by forcing water through small nozzles or orifices under pressure. The central pivot irrigation system, enhanced by GIS, GPS, and NDVI (Normalized Difference Vegetation Index) technology, offers a sophisticated approach to water management. GIS and GPS technologies optimize the system by mapping field layouts, tracking the movement of pivot units, and ensuring precise water application. NDVI, a remote sensing technique, monitors crop health and water stress levels by analyzing vegetation indices from satellite or drone imagery. This data helps in adjusting irrigation schedules and quantities based on real-time crop needs, reducing water waste and improving efficiency. By integrating these technologies, the central pivot system achieves uniform water distribution across large areas and diverse crops. It minimizes issues such as water stagnation and soil compaction while maintaining adequate air circulation for rapid seed germination. This approach not only enhances water use efficiency (WUE) but also adapts to varying field conditions and crop requirements. The use of GIS and NDVI further helps in addressing water scarcity challenges and supports sustainable agriculture by optimizing irrigation practices and reducing labor costs (Al-Ghobari and Dewidar 2021) (Fig. 7.15).

The field application of variable rate irrigation (VRI) through a central pivot system involves several key steps to optimize water use and enhance crop yield.

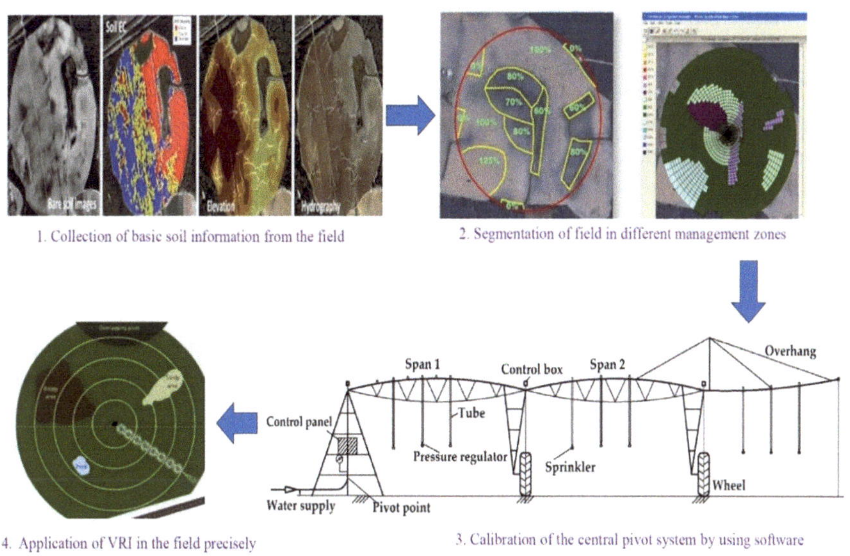

Fig. 7.15 Diagram of field-level implementation of central pivot irrigation system controlled by software. (Modified form of Ahmad and Mahdi (2018))

Initially, the field's location is determined using GPS technology, followed by remote sensing to capture detailed imagery. This data is processed with Geographic Information System (GIS) software to create thematic maps, highlighting soil conditions, electrical conductivity, elevation, and hydrogeographic features. The field is then segmented into management zones based on soil and crop characteristics, which can also be refined through grid soil sampling. Calibration of the VRI system's hardware follows, utilizing algorithms or programming software to ensure precise water delivery tailored to the varying needs of crops and soil conditions. The calibrated central pivot system then delivers water variably, aligning irrigation with crop requirements and soil conditions to optimize water usage and field management (Ahmad and Mahdi 2018).

A 2-year field study evaluated the potential of Site-Specific Irrigation Management (SSIM) to enhance potato yield, quality, and economic return compared to Conventional Uniform Irrigation Management (CUIM). Both irrigation methods used near real-time soil water content for scheduling, with similar water application rates. Although tuber yield under SSIM was 4% higher, the difference was not statistically significant. However, SSIM resulted in a 4–6% increase in yield per unit of water applied and an average gross income increase of $159/ha. Despite this, the income gain likely covers only half the cost of commercial SSIM technology. This improvement is attributed to an increase in the average water-holding capacity and enhanced water use efficiency (King et al. 2006).

7.5 Impact of Precision Water Management on Crop Yield and Sustainability

In Tamil Nadu, a state-sponsored NADP project successfully implemented precision irrigation in the semi-arid regions of Dharmapuri and Krishnagiri districts. Over 3 years, 750 acres were equipped with drip irrigation and fertigation, delivering nutrients precisely at critical crop stages. This resulted in a 60–80% increase in crop yields, with sugarcane yields especially improving by 90%, due to the precise application of water-soluble fertilizers to the root zone and maintenance of deal soil moisture regime of 60% and aeration of 40%. The project also achieved a 30–40% reduction in water and pesticide use, along with higher-quality produce. To address small farm sizes and lack of farmer compatibility, a cluster approach was adopted, grouping 20 farmers in 20-ha clusters for better logistics. Training and monitoring, guided by the Tamil Nadu Agricultural University (TNAU), ensured technology transfer and success, making the project a model for other farms in the state (NADP Projects 2008).

In Gujarat's Chandrala village, precision irrigation through micro-irrigation significantly boosted vegetable yields by 20–30% and increased mechanization, despite reduced groundwater levels. The village's consolidated land holdings facilitated the adoption of drip irrigation, leading to a shift towards labor-intensive crops like vegetables and cotton. Surprisingly, laborers supported mechanization as it improved working conditions and maintained employment through increased irrigated areas. However, farmers noted that there were no savings on fertilizer costs due to the lack of subsidies on water-soluble fertilizers used in drip irrigation (Bhamoriya and Mathew 2014). Similarly, in Chanvelly village, Andhra Pradesh, 70% of farmers adopted micro-irrigation after visits to government model farms, despite sufficient groundwater. Benefits included labor cost savings, expanded irrigated areas, reduced weed growth, and improved yields. The remaining farmers faced challenges due to a lack of awareness about subsidies and high initial costs (Bhamoriya and Mathew 2014).

Precision irrigation significantly enhances crop yields by tailoring water delivery to each crop's unique requirements (Mushtaq et al. 2024). In summer crops such as Maize, Moong (Green Gram), Urad (Black Gram), Sesame (Til), Groundnut, Sorghum (Jowar), Pearl Millet (Bajra) Tomato, Cucumber, Bitter Gourd, Pumpkin, Watermelon, Muskmelon, etc., where irrigation supplements up to 75% of the maximum water requirement, LEPA (Low Energy Precision Application) proves more advantageous than sprinkler systems. Subsurface drip irrigation (SDI) is a method of irrigation where water is delivered directly to the root zone of plants through a network of polyethylene tubing and emitters buried beneath the soil surface. This technique allows for precise water application, reducing evaporation and runoff, and improving water use efficiency. Under irrigation treatments in semi-arid regions providing less than 50% of the total water needed, SDI (Subsurface Drip Irrigation) achieves a significantly higher yield of 14% than LEPA, while LEPA outperforms sprinklers by 16%. LEPA continues to be an important water conservation tool in semi-arid regions facing declining irrigation availability. At irrigation levels above 50%, sprinkler systems yield slightly more than LEPA, with SDI yields being 7% higher than LEPA. The LEPA irrigation method is the catalyst for innovations in chemigation, no-till planting, and site-specific irrigation. (Bordovsky). Sprinkler irrigation, applied at 80% of crop evapotranspiration (ETC) alongside precise fertilizer management, results in significantly higher grain (2.63 t/ha) and biological yields (8.37 t/ha). This suggests that combining System of Crop Intensification (SCI) protocols with sensor-based precision fertilizer and irrigation control can enhance soybean yields and resource-use efficiency (Sachin et al. 2023). The System of Crop Intensification (SCI) is an agricultural methodology aimed at enhancing crop productivity, resource conservation, and sustainability across various crops by improving root growth and soil fertility while reducing inputs like water, fertilizers, and agrochemicals. It integrates precision irrigation, which uses technologies like soil moisture sensors and automated systems to optimize water use, and precision fertilizer management, employing sensor-based tools and variable rate technology to ensure efficient nutrient application. This holistic approach results in improved crop yields, water and nutrient use efficiency, and reduced environmental impact, addressing modern agricultural challenges such as climate change and resource constraints. In addition to improving crop yields, precision irrigation revolutionizes soil health management. By applying the optimal amount of water, it prevents soil from becoming too wet, reduces waste, maintains a loose and crumbly soil structure, and inhibits erosion and salt buildup. This approach enhances water absorption and prevents soil compaction, which can suffocate roots and essential bacteria. Loose soil allows for better movement of water, nutrients, and roots, promoting root growth. Early, frequent light watering can keep soil at field capacity for optimal root development, especially in loose areas (Coelho et al. 2003).

7.5.1 Barriers to Adoption of Precision Water Management

The adoption of precision water management (PWM) technologies, such as drip and sprinkler irrigation systems, faces significant socio-economic challenges in India.

Small Land Holdings Precision irrigation systems are more viable for large farms, typical in the USA and Europe. However, in India, 82% of farmers have less than 2 ha of land, making it challenging to adopt such systems. Drip and sprinkler systems require significant investment, unsuitable for small land sizes (Brahmanand and Singh 2022).

Lack of Investment Capacity Small and marginal farmers in India invest only 9% of their total assets in agriculture. The high initial cost of precision water management (PWM) systems is unaffordable for most, limiting adoption (Levidow et al. 2014).

Insufficient Infrastructure The rural areas lack the necessary infrastructure, such as computers and controllers, required for PWM. This infrastructure gap hinders the implementation of these advanced systems (Marathe 2019).

The adoption of Precision Water Management (PWM) systems faces several challenges. There is a shortage of technically qualified personnel to operate and maintain these complex systems, particularly in rural areas where farmers often lack the necessary skills. Additionally, the limited field demonstrations reduce farmers' awareness and understanding of PWM benefits. Furthermore, the development of micro-irrigation is uneven across Indian states, with Andhra Pradesh and Maharashtra leading in adoption, while many other states fall behind, hindering widespread implementation and effectiveness (Marathe 2019).

Addressing these barriers through targeted policies, infrastructure development, training programs, and extensive field demonstrations is essential to enhance the adoption of precision water management in India.

7.5.2 *Policy Recommendations and Research Needs*

To address these challenges, government policies should incentivize the formation of farmers' cooperatives, increase comprehension, application, and adoption of precision water management techniques and technologies, and support deeper research and development. Research and development should focus on developing cost-effective PWM technologies suitable for small and marginal farmers, creating robust and scalable supporting infrastructure in rural areas, and implementing training programs to develop local technical expertise (Smith et al. 2021). Additionally, research on adaptive cropping systems that align with the precision irrigation framework is essential (Kumar and Patel 2022). Extensive field demonstrations and pilot projects are needed to showcase the benefits and practical applications of PWM technologies, promoting their widespread adoption (Chen et al. 2018a, b). By addressing these areas, PWM can become a viable option for small and marginal farmers, enhancing agricultural sustainability and water use efficiency in India (Raza et al. 2023).

7.6 Conclusion

In the context of climate change, utilizing water resources judiciously is crucial to address scarcity and quality issues. Precision water management (PWM) enhances water and nutrient use efficiency, ensuring food security. In India, 68% of the total net sown area (136.8 Mha) is dry land, spread over 177 districts, with dry land crops accounting for 48% of food crop area and 68% of non-food crop area. Despite advances in some nations, precision irrigation is still primitive in India, facing challenges such as small landholding sizes, high investment needs, and field heterogeneity. Adoption can increase through cooperative farming models, national research projects, public-private partnerships, and mobile app-based digital solutions. This transition requires significant human and capital infrastructure, supported by government policies. To meet global food demands and achieve sustainable development goals (SDG 6), enhancing water use efficiency is vital. Precision water use in agriculture boosts production, water productivity, and crop quality. PWM aims to maximize yield with optimal water use, economizing water, saving time, reducing costs, and improving yields compared to conventional systems. Reliable information on weather, evaporation, soil moisture, and crop types is essential for determining soil moisture needs. GIS technology integrates spatial and temporal data for informed irrigation decisions. Accurate irrigation, applied only, when necessary, can save up to 25% of water compared to conventional practices, highlighting the importance of adopting precision irrigation.

References

Ahmad L, Mahdi SS (2018) Satellite farming. Springer, Basel

Al-Ghobari H, Dewidar AZ (2021) A comparative study of standard center pivot and growers-based modified center pivot for evaluating uniformity coefficient and water distribution. Agronomy 11(8):1675

Al-Karadsheh E, Sourell H, Krause R (2002) Precision irrigation: new strategy irrigation water management. In: Conference on International Agricultural Research for Development, Witzenhausen, Germany

Bal SK, Sandeep VM, Vijaya Kumar P (2022) Assessing impact of dry spells on the principal rain-fed crops in major dryland regions of India. Agric For Meteorol 313:108768

Bhamoriya V, Mathew S (2014) An analysis of resource conservation technology: a case of micro-irrigation system (drip irrigation). Final report of Centre for Management in Agriculture, Indian Institute of Management, Ahmedabad

Bill R, Nash E, Grenzdörffer G (2012) GIS in agriculture. In: Springer handbook of geographic information. Springer, Cham, pp 461–476

Brahmanand PS, Singh AK (2022) Precision irrigation water management-current status, scope and challenges. Indian J Fertil 18:372–380

Camp CR, Sadler EJ, Evans RG (2006) Precision water management: current realities, possibilities, and trends. In: Srinivasan A (ed) Handbook of precision agriculture. Food Products Press, Binghamton, pp 1–20

Central Water Commission (2015) Annual report CWC 2015–16. Retrieved from https://cwc.gov.in/sites/default/files/annual-report-cwc-2015-16.pdf

Chavez JL, Pierce FJ, Evans RG (2010) Compensating inherent linear move application errors using a variable rate irrigation system. Irrig Sci 28:203–210

Chen B, Liu E, Mei X, Yan C, Garré S (2018a) Modelling soil water dynamics in rainfed spring maize fields with plastic mulching. Agric Water Manag 198:19–27

Chen R, Zhao W, Liu Q (2018b) Field demonstrations of precision irrigation: lessons from pilot projects. Irrig Sci 36(4):313–327

Coelho MB, Villalobos FJ, Mateos L (2003) Modeling root growth and the soil–plant atmosphere continuum of cotton crops. Agric Water Manag 60(2):99–118

CRIDA (2015) VISION 2030. Central Research Institute for Dryland Agriculture, Santosh Nagar, Hyderabad, India

Devadas R, Jones SD, Fitzgerald GJ, McCauley I, Matthews BA, Perry EM, Kouzani AZ (2011) Development of a wireless sensor network for in-situ image validation for water and nitrogen management. Asian J Geoinform 11(2):1–11

Drought Manual (2020) Drought manual 2020. Retrieved from https://vedas.sac.gov.in/static/pdf/Drought%20Manual-2020.pdf

Food and Agriculture Organisation (FAO) (2016) Annual report 2016. Retrieved from https://www.fao.org/family-farming/detail/en/c/447860/

Goodwin I, O'Connell MG (2015) Drought water management: an Australian perspective. In: VIII international symposium on irrigation of horticultural crops. ISHS, Leuven, pp 219–232

Hillel D (1990) Role of irrigation in agricultural systems. Agronomy 30:5–30

Hoekstra AY (2015) The water footprint of industry. In: Assessing and measuring environmental impact and sustainability. Butterworth-Heinemann, Oxford, pp 221–254

King BA, Stark JC, Wall RW (2006) Comparison of site-specific and conventional uniform irrigation management for potatoes. Appl Eng Agric 22(5):677–688

Kumar S, Patel R (2022) Adaptive cropping systems for precision water management. Agric Water Manag 243:106426

Lang L (1992) GPS, GIS, remote sensing: an overview. Earth Observation Magazine, pp 23–26

Levidow L, Zaccaria D, Maia R, Vivas E, Todorovic M, Scardigno A (2014) Improving water-efficient irrigation: prospects and difficulties of innovative practices. Agric Water Manag 146:84–94

Lovarelli D, Bacenetti J, Fiala M (2016) Water footprint of crop productions: a review. Sci Total Environ 548:236–251

Marathe M (2019) Reimagining water infrastructure in its cultural specificity: case of Pune, India. Thesis

Mushtaq M, Ali H, Raza A, Maqbool S, Safdar M, Ahmed M, Sattar J (2024) Precision irrigation for sustainable agricultural productivity. In: Emerging technologies and marketing strategies for sustainable agriculture. IGI Global, Hershey, pp 184–208

NADP Projects (2008) Precision farming. Tamil Nadu Agricultural University. Retrieved from https://agritech.tnau.ac.in/govt_schemes_services/govt_serv_schems_nadp_tnau%20_precisionfarming.html

Nikkilä R, Seilonen I, Koskinen K (2010) Software architecture for farm management information systems in precision agriculture. Comput Electron Agric 70(2):328–336

NITI Aayog (2015–16) Annual report. Retrieved from https://www.niti.gov.in/sites/default/files/2018-12/AnnualReport_15-16-Eng.pdf

Ortigara ARC, Kay M, Uhlenbrook S (2018) A review of the SDG 6 synthesis report 2018 from an education, training, and research perspective. Water 10(10):1353

Panchard J, Prabhakar TV, Hubaux JP, Jamadagni HS (2007) Commonsense net: a wireless sensor network for resource-poor agriculture in the semiarid areas of developing countries. Inf Technol Int Dev 4(1):51

Qin S, Li S, Tang K, Hu K (2018) Can plastic mulch save water at night in irrigated croplands? J Hydrol 564:667–681

Raine SR, Meyer WS, Rassam DW, Hutson JL, Cook FJ (2007) Soil-water and solute movement under precision irrigation: knowledge gaps for managing sustainable root zones. Irrig Sci 26(1):91–100

Raza A, Khaliq A, Hu Y, Zubair N, Acharki S, Zubair M, Syed NR, Ahmad F, Iqbal S, Elbeltagi A (2023) Water resources and irrigation management using GIS and remote sensing techniques: case of Multan district (Pakistan). In: Pande CB, Kumar M, Kushwaha NL (eds) Surface and groundwater resources development and management in semi-arid region, Springer hydrogeology. Springer, Cham, pp 137–156. https://doi.org/10.1007/978-3-031-29394-8_8

Reddy TY, Reddy GH (2019) Principles of agronomy. Kalyani Publishers, Ludhiana

Rockström J, Lannerstad M, Falkenmark M (2007) Assessing the water challenge of a new green revolution in developing countries. Proc Natl Acad Sci 104(16):6253–6260. https://doi.org/10.1073/pnas.0605739104

Sachin KS, Dass A, Dhar S, Rajanna GA, Singh T, Sudhishri S, Devi AD (2023) Sensor-based precision nutrient and irrigation management enhances the physiological performance, water productivity, and yield of soybean under system of crop intensification. Front Plant Sci 14:Article 1282217. https://doi.org/10.3389/fpls.2023.1282217

Šarauskis E, Kazlauskas M, Naujokienė V, Bručienė I, Steponavičius D, Romaneckas K, Jasinskas A (2022) Variable rate seeding in precision agriculture: recent advances and future perspectives. Agriculture 12(2):Article 305

Singandhupe RB, Rao GGSN, Patil NG, Brahmanand PS (2003) Fertigation studies and irrigation scheduling in drip irrigation system in tomato crop (*Lycopersicon esculentum* L.). Eur J Agron 19(2):327–340

Smith L, Johnson M, Lee T (2021) Affordable technologies for small-scale precision agriculture. Precis Agric 22(1):85–103

Söderström M, Sohlenius G, Rodhe L, Piikki K (2016) Adaptation of regional digital soil mapping for precision agriculture. Precis Agric 17:588–607

Times of India (2024) 143 in 152 days: Amravati is new farm suicide capital of Maha. The Times of India. https://timesofindia.indiatimes.com/city/nagpur/143-in-152-days-amravati-is-new-farm-suicide-capital-of-maha/articleshow/111121270.cms

UN DESA (2023) The sustainable development goals report 2023: special edition – July 2023. UN DESA, New York. https://unstats.un.org/sdgs/report/2023/

Chapter 8
Climate-Smart Rice Production Technologies for Improving Productivity and Resilience

Debabrata Nath, Malay K. Bhowmick, Sunil Kumar, Panneerselvam Peramaiyan, Sudhanshu Singh, Biswajit Pramanick, Ranjan Laik, and Vandana Kumari

Abstract Rice is a vital commodity in South Asian countries, where resource-poor farmers mostly rely on it for their sustenance, nutrition, and livelihood. Injudicious use of agricultural inputs along with inappropriate management practices have been vastly impacting the quality of the natural resources and contributing to the climatic variability that not only adversely affects rice productivity, but also poses a significant threat to overall food security. It is very crucial to improve adaptive capacity of farmers and ensure food security while reducing environmental footprints. Rice production systems can be made more profitable and sustainable through the deployment of climate-resilient and resource-efficient technologies such as direct-seeded rice, alternate wetting and drying, laser land leveling, residue management, etc. These improved technologies not only help in addressing the climate risks, but also reduce the environmental footprints with enhanced rice productivity and reduced cost of cultivation, thereby ensuring sustainability of rice-based

The original version of the chapter has been revised. A correction to this chapter can be found at https://doi.org/10.1007/978-981-96-7699-6_21

D. Nath · M. K. Bhowmick · S. Kumar · P. Peramaiyan · S. Singh
International Rice Research Institute (IRRI)-South Asia Regional Centre (ISARC), Varanasi, Uttar Pradesh, India
e-mail: d.nath@cgiar.org

B. Pramanick
Department of Agronomy, Dr. Rajendra Prasad Central Agricultural University, Samastipur, Bihar, India

R. Laik
Department of Soil Science, Dr. Rajendra Prasad Central Agricultural University, Samastipur, Bihar, India

V. Kumari (✉)
Department of Soil Science, Advance Centre on Sericulture, Dr. Kalam Agricultural College, Bihar Agricultural University, Kishanganj, Bihar, India

B. Pramanick et al. (eds.), *Climate-Smart Agricultural Technologies*, Sustainability Sciences in Asia and Africa, https://doi.org/10.1007/978-981-96-7699-6_8

farming systems and food security as well. Developing and disseminating climate-smart resource-conserving technologies will contribute toward achieving the United Nations' Sustainable Development Goals of "Zero Hunger" and "Climate Action" by 2030.

Keywords Food security · Climate-smart rice production · Resilience · Sustainability

8.1 Introduction

Rice is grown on 47.8 million hectares of land in India, making it the world's second-largest rice producer after China during 2023–24 (PIB, 2025). To feed the growing population, the country will need at least 130 million tons of milled rice per year by 2030 (Gujja and Thiyagarajan 2009). Rapid urbanization on the one hand, and climate change on the other, limit the potential to expand the area under rice cultivation. It is also possible to increase rice production under existing cultivated areas by increasing productivity per unit area. The changing climatic conditions not only pose a challenge of increasing food production, but also of adapting to the water scarcity and mitigating greenhouse gas (GHG) emissions from the rice fields (Kar et al. 2021). Millions of resource-poor farmers depend on rice for food, nutrition, and livelihood in South Asia. In India, rice is an economically important commodity, and feeds more than 60% of the population. Inappropriate management practices along with injudicious use of agricultural inputs not only affect the quality of natural resources, but they also contribute to the climatic variability that adversely affects rice productivity, posing a significant threat to overall food security (Pramanick et al. 2018). Climate change will continue to have a significant impact on agricultural growth in the future, according to the Intergovernmental Panel on Climate Change (IPCC). Droughts, heavy rainfall, flooding, and high temperatures will become more frequent and intense as a result of climate change. Additionally, agriculture-related GHG emissions, such as those resulting from the burning of crop residues, are a primary cause of climate change. Farmers are struggling to cope with the situations due to lack of knowledge about adaptation to adverse climatic conditions, which ultimately lead to food security problems. As a result, it is crucial to transition from traditional agricultural practices toward climate-smart practices in order to tackle climate-related issues and establish sustainable agriculture. Climate-smart agriculture (CSA) is an integrated approach that reduces GHG emissions, increases productivity, improves resilience, and ensures food, nutrition, and livelihood security in a changing climate Water-saving technologies must be introduced for crop diversification, resource conservation, and sustainable crop management (Pramanick et al. 2022, 2023). Rice production systems can be made more profitable and sustainable through deployment of climate-resilient and resource-efficient technologies such as alternate wetting and drying (AWD), direct-seeded rice (DSR), laser land leveling, residue management, soil health management, community-based staggered rice nursery enterprise, mechanical transplanting of rice, site-specific nutrient and weed management, etc. For example, mechanized DSR saves water (20–25%), labor (40%), production cost (~USD 120/ha), increases net profit (~USD 100–125/ha), reduces drudgery and global warming potential (20–44%), and saves crop duration by 7–15 days, leading to an early harvest of rice which enables the timely sowing and higher yield of the succeeding crops such as wheat, chickpea, and mustard. These improved climate-smart technologies not only help in addressing the climate risks, but also reduce the environmental footprints with enhanced rice productivity and reduced cost of cultivation, thereby ensuring sustainability of rice-based farming systems and food security as well. Developing climate-smart production technologies in rice cultivating areas of India will contribute significantly to sustainable development, besides combating climate change challenges and enhancing farmers' income. The objective of this chapter is to illustrate and propagate the evidence-based climate-smart technologies for rice production that will be helpful to achieve the United Nations Sustainable Development Goals of "Zero Hunger", and "Climate Action" in India.

8.2 Prospective Crop Establishment Methods for Profitable Rice Production Under Changing Climate

8.2.1 Direct-Seeded Rice

Rice is traditionally grown with the resource-intensive method of puddled transplanting of rice (PTR) across the globe, especially in Asia. This conventional production system has significant drawbacks in terms of labor, water, and energy requirements, leading to reduced profitability due to the scarcity of these resources. Furthermore, it negatively impacts soil quality, hampers the growth of subsequent crops, and contributes to the emission of GHGs (Ladha et al. 2009). This calls for the promotion of DSR instead of PTR. Numerous studies have shown the advantages of DSR over PTR, including comparable yields, reduced irrigation water usage, labor and production cost savings, higher net economic returns, and lower GHG emissions (Kumar and Ladha 2011; Dharminder et al. 2021). In direct seeding, rice seeds are sown directly into the field, eliminating the need for nursery preparation and labor requirements for transplanting. Depending on the local conditions and farmers' preferences, different methods are in practice for sowing seeds, including manual broadcasting, mechanical seeding, or with the use of specialized seeding machines (Mahender Kumar et al. 2025).

There are mainly two different types of DSR, viz., dry-DSR and wet-DSR.

1. Dry-DSR involves broadcasting or drilling dry seeds onto unpuddled soil after dry tillage, zero tillage, or on raised beds, whereas
2. Wet-DSR involves broadcasting or sowing pre-germinated or sprouted seeds on wet or puddled soil.

Major Benefits of DSR (Kumar and Ladha 2011)

- Labor savings of up to 46 and 4–60% in wet- and dry-DSR, respectively.
- Reduced drudgery by eliminating the need for transplanting rice seedlings.
- Reduced water losses due to percolation, resulting in water savings of 12–35%.
- Lower methane (CH_4) emissions (by 6–92% reduction, depending on DSR type and water management).
- Production cost savings of 2–16 and 6–32% in wet- and dry-DSR, respectively.
- Early harvesting of kharif rice, allowing farmers to timely sow the subsequent crops.

Major Limitations Associated with DSR

- Risk of crop damage due to unexpected rainfall immediately after sowing
- Emergence of weedy rice in the DSR-based production system
- Increased population of nematodes

8.2.2 Mechanical Transplanting of Rice

Mechanization is the need of the hour to make rice farming profitable due to the labor shortage in farm operations. Self-propelled rice transplanters make the transplanting operation easier and more attractive. Because of small land holdings and poor economic status, farmers cannot purchase the machines individually. Mat nursery preparation requires about 40% of the total energy requirement in mechanical transplanting, while traditional nurseries under manual transplanting only account for about 11% (Baruah et al. 2001). In recent years, mat-type seedlings have become more popular for mechanical transplanting of rice (MTR) because of less labor requirements (Dixit et al. 2007). Mechanical transplanting significantly increases grain yields by 23%, 37%, and 63%, and straw yields by 17%, 14%, and 22% over manual transplanting, dry-DSR, and wet-DSR, respectively (Singh et al. 2006) (Fig. 8.1).

Major benefits of Mechanical Transplanting

- Reduced root damage during transplanting, resulting in better seedling vigor and uniform crop stand.
- Quicker and easier transplanting with reduced drudgery, and health risks for farm laborers.
- Customized mechanical transplanting services as a means of generating additional income for the rural youths and entrepreneurs.
- Less water requirements due to shallow tillage operations or even under unpuddled condition.
- Reduced labor requirements and costs, leading to increased net incomes.
- Timely planting of rice and succeeding crops in rotation.

Fig. 8.1 Mechanical transplanting of rice

8.2.3 *Aerobic Rice*

An aerobic rice production system requires well-drained soil without puddles, well-drained water, and good drainage. To reduce the use of water, rice is grown in this system similarly to an irrigated upland crop, such as wheat or maize, in order to reduce the amount of water use in the cultivation process. Water efficiency is one of the main drivers of aerobic rice. Aerobic method can save 37–45% water (Kadiyala et al. 2012) and provide 60.7% higher water productivity than the conventional flooded rice (Jinsy et al. 2015). Yield decline occurs more during the dry season than during the wet season when aerobic rice is continuously grown (Peng et al. 2006). There is no doubt that aerobic rice is a more attractive alternative to conventional rice, but it must be considered that yield penalty and yield stability must also be considered before this water-saving technology is promoted.

8.2.4 *System of Rice Intensification*

System of rice intensification (SRI) was developed in Madagascar during 1980–90 to ensure high yields even on infertile soils (Stoop et al. 2002). When seedlings are in their vegetative growth phase, they are transplanted singly in wide square patterns, using organic sources of nutrients, and keeping the soil moist. SRI exhibits a significant change in plant structure and function along with increased yields (50–100%) despite using half of the total amount of water required for conventional method (Uphoff et al. 2010).

8.2.5 *Furrow Irrigated Raised Bed*

Using beds for rice transplantation minimizes the effects of puddling and eliminates its harmful effects. In the furrow irrigated raised bed (FIRB), a furrow is created between each bed for irrigation, and rice is grown in raised beds. Raised beds are more effective at conserving water and increasing grain yield in comparison to the other methods of planting crops (Naresh et al. 2014). Compared to puddled plots, seedlings transplanted onto slopes of newly constructed beds can save 15% of the irrigation water without compromising grain yields (Sandhu et al. 2012). Dry-seeded rice crops grown on flat land or raised beds can reduce total water input by 35–42% when the soil is kept near saturation, and by 47% and 51% if the soil is dried out to a moisture tension of 20 and 40 kPa, respectively. As compared to the flat puddled plantings, irrigation water productivity in beds becomes 13.9% and 13.36% higher (Gathala et al. 2013).

8.3 Efficient Water Management

8.3.1 *Laser Land Leveling*

Almost 95% of the water used for agriculture in South Asian countries is withdrawn from the ground, compared to 70% globally (Babel and Wahid 2008). Overexploitation of groundwater and poor water management have been leading to a steady decline in groundwater depths in rice-wheat areas. Several factors threaten the sustainability and productivity of the system, including inefficient production practices, scarce resources, and socioeconomic changes (Ladha et al. 2003). Traditional land leveling practices in the Indo-Gangetic Plains (IGP) have resulted in poorly leveled agricultural land (Ladha et al. 2009). This issue is more prominent in the eastern IGP as compared to the western IGP. The average height difference between the highest and lowest portions of rice fields in South Asia is approximately 160 mm, which requires additional water application for complete field coverage. Despite farmers' rigorous efforts, unlevelled fields still have an elevation difference of 80–150 mm (Jat et al. 2006). Even the slope of the land varies from transect I to transect V in the IGP, with an average slope of 3–5° (Jat et al. 2006). In unlined irrigation channels and uneven fields, irrigation water is substantially lost (Gill 1994). Laser-assisted land leveling has emerged as an advanced technology for achieving accurate leveling. Appropriate irrigation scheduling with laser-assisted land leveling can save water and increase water productivity (Fig. 8.2).

Laser leveling improves water use efficiency and fertilizer use efficiency, by reducing nutrient losses, facilitating uniform seedling placement, and resulting in higher yields. Furthermore, it ensures uniform germination and enhances input use efficiency by distributing soil moisture uniformly. In contrast, undulated land

Fig. 8.2 Laser land leveler for precision land leveling

conditions increase energy consumption and production costs. The use of laser land leveling improves rice-wheat productivity by reducing slope variations and conserving moisture (Jat et al. 2006). Since the land surface is leveled within 2 cm of average elevation by laser-guided buckets, the slopes in the fields are always between 0% and 0.2%. Before leveling, the field is surveyed to determine the desired elevation, which is then averaged. Using this information, the scraper is guided at the designated level (Jat et al. 2006).

Laser Land Leveling Provides the Following Benefits

- Saving of irrigation water
- Increasing the cultivable area
- Better crop establishment
- Uniform crop maturity
- Improved efficiency of water application
- Increased cropping intensity
- Higher crop yields
- Easier management in saline environments
- Reduced weed problems and better weed control

8.3.2 Alternate Wetting and Drying

Alternate wetting and drying (AWD) is one of the most popular and farmer-friendly water-saving technologies (Fig. 8.3). There is a reduction in irrigation water input (by up to 30%) with the adoption of AWD practices, which increase water

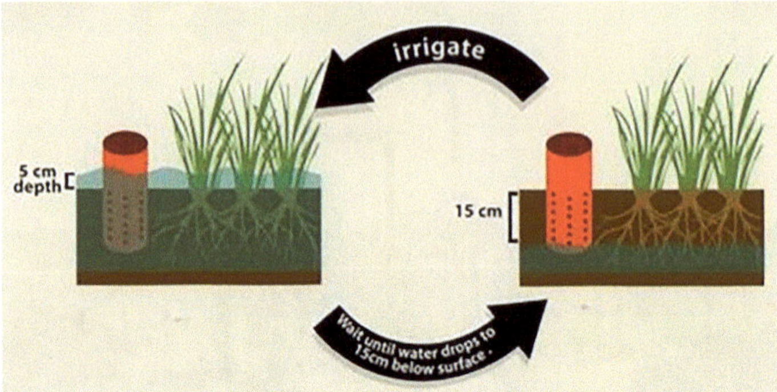

Fig. 8.3 Alternate wetting and drying technique in rice cultivation

productivity and nutrient uptake as well (Sarkar 2001; Akter et al. 2018; Biswas et al. 2021; Busari et al. 2019; Chidthaisong et al. 2018). AWD also increases grain productivity due to a higher number of productive tillers (Norton et al. 2017; Carrijo et al. 2018; Sriphirom et al. 2019).

8.3.3 Surface and Sub-surface Drip Irrigation

Rice can be irrigated using drip irrigation (Chen et al. 2015; Sharda et al. 2017). During the growing season, drip laterals need to be removed and placed multiple times during conventional tillage-based management systems. Compared to surface drip irrigation (SDI), subsurface drip irrigation (SSDI) eliminates the need to anchor laterals during the crop-growing season, thereby extending the economic life of laterals (Ayars et al. 1999). As a result of the SSDI system, soil evaporation is reduced, nutrients and water can be delivered directly to the root zone, which means fertilizer application is more efficient, weed growth is discouraged, labor costs are reduced, and cultural farming practices are easier than in SDI (Ayars et al. 1999). It can save irrigation water, reduce energy use, increase farmers' profits, and complement SSDI systems (Gathala et al. 2013; Jat et al. 2014) by allowing crop residues (which would otherwise be burned) to be recycled (Gathala et al. 2013; Jat et al. 2014).

Advantages of Surface and Sub-surface Irrigation

- Increased irrigation efficiency especially for areas with limited water availability for irrigation.
- High potentiality in water saving.
- Potential yield increments and under moisture deficit conditions SDI resulted in higher yields.
- Energy savings.

8.4 Soil Health and Nutrient Management for Climate-smart Rice Production

As soil microbes mineralize organic nitrogen in soil, they produce ammonia (NH_3) and ammonium ions (NH_4^+), which are then converted into nitrates. Clay particles and soil organic matter (SOM) particles are not attracted to nitrate ions due to their negative charge. Consequently, they can be leached out by soil water. When soil oxygen levels are low, the denitrifying organisms produce nitrous oxide (NO_2), a very powerful GHG. Ploughing and hoeing disturb the soil, and create new aggregates, which allow for the growth of bacteria, leading to the mineralization of SOM. Agronomic management can reduce nitrate leaching and GHG emissions. The provision of adequate structural carbohydrates (e.g., lignin) and nitrogen during cropping is important for reducing losses and contributing to crop growth (Huggins et al. 1998; Gregorich et al. 2001). The effectiveness of conservation agriculture (CA) and zero-tillage (ZT) practices in reducing nitrate leaching has been well demonstrated (Macdonald et al. 1989). CA and ZT reduce nitrate production by avoiding mechanical tilling. As a result of cover crops, nitrogen is absorbed by the soil and reduced. The soil does not lose unused mineralized nitrogen when it is left unused (Bergström 1995; Davies et al. 1996; Gors et al. 1993).

8.4.1 Carbon Management and GHG Emission Mitigations

The "enhanced greenhouse effect" is a result of the concentration of GHGs in the atmosphere increasing over time. Crop productivity is significantly affected by the increase in temperature in the atmosphere. The majority of GHGs emitted by agricultural systems are carbon dioxide (CO_2), methane (CH_4), and nitrous oxide (N_2O) (Bhattacharyya et al. 2019). Around 20% of the global GHG emissions are attributed to agriculture on a global scale. During crop growth stages, CO_2 is mostly emitted during the decomposition of soil organic carbon (SOC), CH_4 from submerged or flooded rice fields, and N_2O from nitrogenous fertilizer applications during the growth stage of the crop (IPCC 2014).

8.4.1.1 Altered Irrigation Pattern

Changing irrigation patterns control GHG emissions in the rice-based cropping system. Several water management options such as AWD, intermittent irrigation, regular mid-season drainage, and controlled irrigation through crop growth stages minimize the GHG emissions as compared to traditional flooded rice conditions (Pathak et al. 2010; Hussain et al. 2015). Crop yields can be maintained in varying climatic and soil conditions (Table 8.1). Compared to traditional flooded rice,

Table 8.1 Water management practices and their relative mitigation potentials (GHG emissions) in rice farming

Improved management practices	GHGs	Mitigation potential (%)	References
Intermittent irrigation	CH_4	38	Yagi et al. (1996)
	CH_4	15	Adhya et al. (2000)
	CH_4	26	Minamikawa and Sakai (2005)
	CH_4, N_2O	34	Hadi et al. (2010)
	CH_4, N_2O	54	Feng et al. (2013)
Mid-season drainage	CH_4	50	Cai et al. (1997)
	CH_4	43	Corton et al. (2000)
	CH_4	36	Zheng et al. (2000)
	CH_4	64	Minamikawa and Sakai (2005)
	CH_4, N_2O	27	Towprayoon et al. (2005)
	CH_4, N_2O	42	Zou et al. (2005)
	CH_4	37	Tyagi et al. (2010)
	CH_4, N_2O	72	Itoh et al. (2011)
	CH_4, N_2O, CO_2	33	Pathak et al. (2012)
Controlled irrigation	CH_4, N_2O	67	Yang et al. (2012)
	CH_4, N_2O	27	Hou et al. (2012)
Water saving irrigation	CH_4, N_2O	60	Win et al. (2015)

Source: Modified and adapted from Hussain et al. (2015)

intermittent irrigation reduces GHG emissions by 15–54% (Feng et al. 2013), whereas mid-season drainage reduces the emissions by 27–72% (Table 8.1).

8.4.1.2 Alternative Crop Establishment

Alternative crop establishment methods like DSR are effective in mitigating CH_4 emissions from rice systems. When the soil is continuously submerged or flooded for an extended period of time in PTR fields, CH_4 gets released from the soil. DSR does not require continuous submergence, which means that it is capable of significantly reducing CH_4 emissions (16–54%), and thereby reducing the GWP (even up to 75%) as compared to PTR systems (Pathak et al. 2010).

8.4.2 Conservation Agriculture

Carbon sequestration and GHG reduction can be achieved through conservation agriculture (CA) in rice-based crops. CH_4 emission is reduced by 56% with the switch over from PTR to ZT-DSR (Tirol-Padre et al. 2016), whereas no CH_4

emissions are detected in ZT-DSR, regardless of residue management in alluvial soils of northwest India (Sapkota et al. 2015). Additionally, Lal (2004) calculated that chisel ploughing (4.5–11.1 kg C ha^{-1}), moldboard plowing (13.4–20 kg C ha^{-1}), subsoilers (8.5–14.1 kg C ha^{-1}), and rotary hoeing (1.2–2.9 kg C ha^{-1}) emit significant amounts of CO_2. Consequently, CA over conventional tillage can reduce carbon emissions (30–35 kg ha^{-1}), besides improving soil fertility (Lal 2004). It has been demonstrated that CA-based resource-conserving technologies (RCTs) are energy- and input-efficient, as well as solution for emerging environmental and soil health hazards. Several CA technologies can improve soil health primarily by increasing SOC concentrations, including no-tillage and minimal-tillage methods, bed planting, and direct seeding, residue management (primarily residue retention), and crop diversification. The RCTs offer a number of potential benefits, including reduced use of water and energy use (fossil fuels and electricity), reduced GHG emissions, less soil erosion, and less labor shortages.

8.4.3 Residue Incorporation

As a result of high carbon content (36.7–42.3%), low nitrogen content (0.47–0.85%), and high lignin content (1.9–4.5%), rice residues are considered to be low-quality organic materials. In moist soils, they rapidly decompose, contribute to CO_2 evolution (Samahadthai et al. 2010; Puttaso et al. 2011), and serve as a substrate for CH_4 production (Samahadthai et al. 2010). In a study from China, straw incorporation boosted annual SOC sequestration in topsoil by 0.24–0.43 t C ha^{-1} $year^{-1}$. Despite this, Yuan et al. (2014) reported a 44–138 kg ha^{-1} $year^{-1}$ increase in CH_4-C and 1.49 kg ha^{-1} $year^{-1}$ increase in N_2O-N emissions as well. As compared to the rice residue (RR), wheat residue (WR) and wheat + rice residue (WR + RR) enhanced the annual emissions of CH_4 by 38% and 61%, respectively.

8.4.4 Site-Specific Nutrient Management

PTR-based cropping systems emit GHGs primarily due to injudicious fertilizer management, particularly nitrogen management practices, which primarily affect CH_4 and N_2O emissions. Higher doses of N can reduce total CH_4 emissions by 30–50% (Yao et al. 2012). When N fertilizer was applied more heavily, CH_4 emissions decreased, but N_2O emissions increased (Aulakh et al. 2001). A reduction of 75% of CH_4 emissions and an increase of 58% in N_2O emissions in rice was reported when N-fertilizer application rates were increased from 150 to 400 kg N ha^{-1} (Zou et al. 2005). For reducing N_2O emissions, site-specific nitrogen management (SSNM) and nitrification inhibitors are the most effective management practices. The SSNM system involves the use of principles and tools that enable plants to receive nutrients they need to grow more successfully. It involves synchronizing

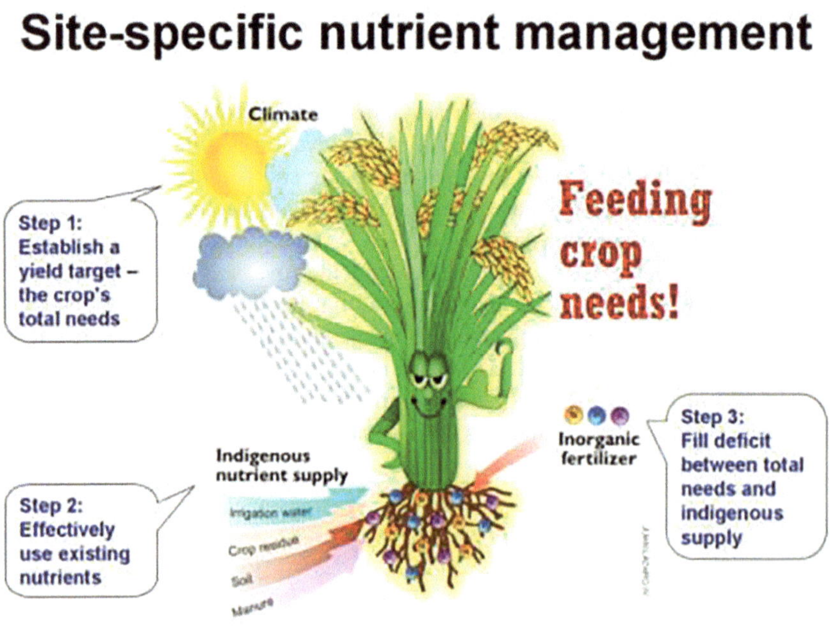

Fig. 8.4 Site-specific nutrient management (SSNM)

supply and demand based on climatic yield potential, yield targets, and the availability of nutrients from indigenous sources, which vary from place to place. Utilizing SSNM ensures that the right amount of fertilizer is applied to the plant at the right time to minimize losses (Fig. 8.4). SSNM-based N application increases N-use efficiency of irrigated rice by 30–40% in several parts of South Asia. SSNM involves using tools such as leaf color charts, chlorophyll meters, green seeker for real-time nitrogen management, and other tools such as soil plant analysis development (SPAD) chlorophyll meters.

8.4.4.1 Rice Crop Manager (RCM)

The International Rice Research Institute (IRRI) has developed a web-based SSNM tool, rice crop manager (RCM) (Fig. 8.5), which is based on the principles of SSNM. For example RCM gives region-specific NPK recommendations in Odisha, India, based on crop history to rice growers of Odisha through ICAR-Central Rice Research Institute (CRRI), Cuttack and Odisha University of Agriculture and Technology (OUAT), Bhubaneswar. Compared to farmer's practice, the RCM recommendations increase rice grain yield by 3–14%. Farmers can enhance rice productivity by 1.0 t ha^{-1} by using RCM instead of traditional fertilizers. The income also is increased by approximately Rs. 13,000/- per season. An online

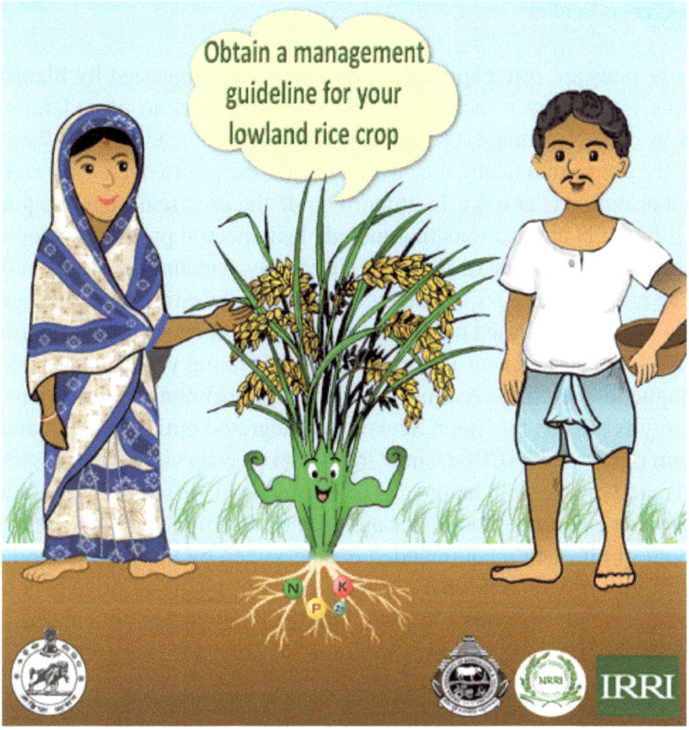

Fig. 8.5 Rice crop manager

decision-making tool based on SSNM principles is now available for different Indian regions for nutrient management. Taking appropriate decisions about fertilizer management can enhance productivity for farmers and extension workers.

8.4.4.2 Leaf Color Chart

Leaf nitrogen status can be determined by changes in rice leaf color within a growing season. Leaf color chart (LCC) is a real-time, low-cost, and easy-to-use tool to determine the change in leaf color when and how much N fertilizer to apply. There are at least four color panels, ranging from yellowish-green to dark green. Rice leaves exhibit a range of colors corresponding to a range of leaf N concentrations, from a leaf N deficiency to an excessive leaf N concentration. The ICAR-Central Rice Research Institute (CRRI), Cuttack has developed a customized leaf color chart (CLCC) for varieties grown in eastern India. Field trials have demonstrated that CLCC-based N recommendations can save 25% of fertilizer and reduce N_2O emissions by 13–21% in rice.

8.4.4.3 Green Seeker

Fertilizer N rates are often applied even higher than suggested by blanket recommendations by farmers to achieve high yields. There is no consideration for the variation in crop N demand over time in these recommendations. When fertilizer nutrients are used excessively, the cost increases, the return decreases, and the environmental pollution is at risks. In order to guide the precise nitrogen requirement in rice, need-based N applications that include leaf spectral properties should be used more rationally (Wells and Turner 1984). Recently, a technique has been developed to estimate crop nitrogen requirements based on early estimates of nitrogen uptake and yields in the field, based on measurements of reflectance in the red (defined by chlorophyll content) and near-infrared (defined by living vegetation) portions of the electromagnetic spectrum. A more efficient and accurate way to apply fertilizer immediately is offered by Green Seeker, an integrated optical sensing and application system (Billa et al. 2020). Using Green Seeker, crop vigor can be assessed in an affordable and innovative manner. The normalized difference vegetation index (NDVI) and an environmental factor are used to determine a crop's yield potential. The amount of nitrogen recommended is determined by the yield potential and the response of the crop to additional nitrogen. Combining prescriptive N doses during different stages with corrective N doses led by Green Seeker optical sensors shows promise for high yields. Green Seeker can improve fertilizer use efficiency through SSNM.

8.4.5 Soil Health Assessment Through Dry Chemistry-Based Spectroscopy

Soil properties are usually evaluated using routine conventional laboratory analyses that are time-consuming, labor-intensive, and use many chemicals that are not friendly to the environment. Spectroscopy in the mid-infrared region (approximately 4000–400 cm^{-1}) is an important analytical technique for obtaining qualitative and quantitative data in a short period of time (Ludwig et al. 2008).

A mid-infrared spectrum is fast, timely, inexpensive, non-destructive, straightforward, and often more accurate than a conventional analysis (Fig. 8.6). Additionally, a single spectrum is sufficient to characterize various soil properties simultaneously. A number of soil properties can be rapidly and cost-effectively measured with mid-infrared spectroscopy (MIRS). There are a few other added advantages associated with MIR analysis. It takes just 30 seconds for MIR spectra to acquire information, and only a few seconds to predict soil properties simultaneously. It is also possible to acquire spectra in batches and make bulk predictions for each property.

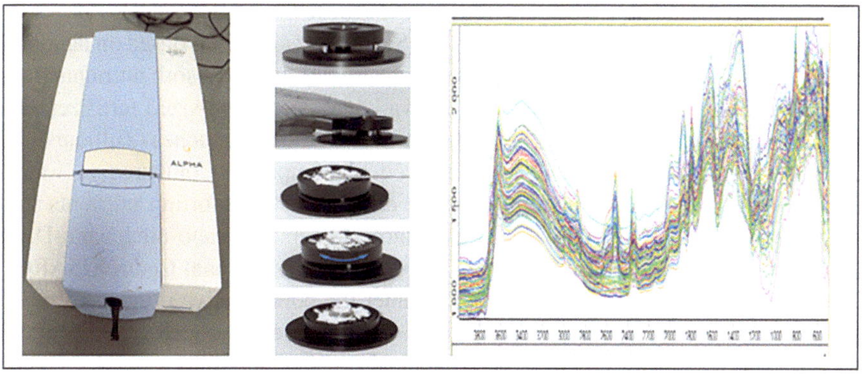

Fig. 8.6 Soil sample analysis through mid-infrared spectroscopy for rapid and low-cost soil health evaluation

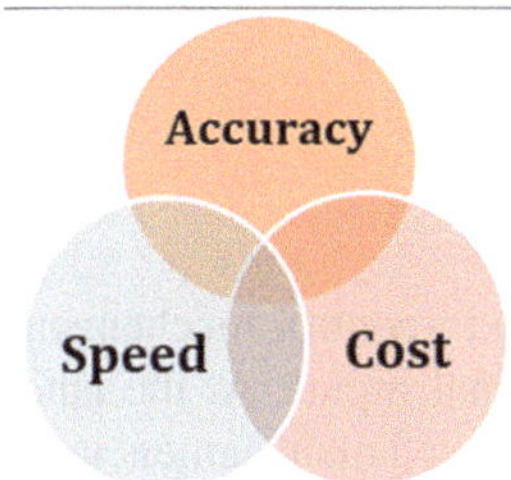

- Accuracy: Soil properties like soil organic carbon, sand, silt, clay etc. can be predicted very accurately
- Speed:A soil sample is scanned in 30 seconds.
- Cost: It is low cost. No reagents are required for sample preparation. Analyzing samples with this method is non-destructive.

Conventional lab analysis takes huge labor, time, and money for proper soil sample analysis. Therefore, MIRS can be an effective alternative (Nath et al. 2021, 2022). Government policy briefs can be effective in exploring the potential of MIRS to partially replace conventional lab testing. In India, there has been negligible research in using MIR spectroscopy to study soil. There is a need to develop and validate soil property prediction models for soils of India. For soil health evaluation, a library of soil mid-infrared spectra should be made from soils of India. MIRS can be an effective tool for the development of cost-effective hand-held sensors for nutrient recommendation to the farmers under the soil health card scheme.

8.4.6 Use of Nanofertilizers

Use of nanofertilizers is a sustainable way to meet the nutritional needs of plants without compromising yield (Janmohammadi et al. 2016). Plants grow and reproduce better when they receive one or more nutrients from nanofertilizers (Liu and Lal 2014). Nanofertilizers improve the efficiency of nutrient use by plants with high absorption rates and high utilization efficacy, facilitating plant uptake of nutrients with minimum losses (Fatima et al. 2021). Even nanofertilizers may be a better alternative to dissolvable inorganic fertilizers because they release discharging supplements slowly during crop production (El-Saadony et al. 2021). Based on nanoparticles' (NPs) morphology and synthesis characteristics, plants absorb nutrients from soil. Insoluble nutrients are more readily available when nano-sized fertilizers are used (Prasad et al. 2017). There are different engineering approaches (Zulfiqar et al. 2019) to encapsulate nutrients for increased plant uptake. Compared to water-soluble fertilizers, nanofertilizers are more efficient at transporting nutrients into plants (Al-Mamun et al. 2021). Different nanofertilizers like nano-urea, nano-DAP, and nano-zinc have been evaluated at IRRI South Asia Regional Centre (ISARC), and simultaneously in the farmers' fields under the collaborative project of Indian Farmers Fertiliser Cooperative Limited (IFFCO) to understand the extent of reduction in using chemical fertilizers both under direct-seeded and transplanted conditions in rainfed and irrigated ecologies. Preliminary results reveal an improved nutrient use efficiency and comparable crop yields, which would offer more environmental benefits with reduced GHG emissions.

8.5 Conclusion

Climate-smart agriculture (CSA) relies more on natural auto-control mechanisms rather than the unstable systems that mostly rely on external inputs. At the landscape level, this is mainly reflected in the use of ecosystem-based management techniques, as well as the use of integrated management techniques. A number of climate-smart strategies and practices can help farmers improve productivity and income by minimizing negative effects, creating an integrated farming system, managing nutrient levels and residues, intercropping with legumes, and using CA. In addition to reducing negative impacts, climate-smart agronomic interventions can increase rice productivity and farmers' income while strengthening their livelihood security. Agricultural practices can be designed using location-specific medium-range weather forecasts in the event of extreme climatic events like high rainfall, drought, frost, hail storms, and heat waves. The accuracy and precision of weather forecasting will help make continuous changes to crop varieties and also the cropping systems. Researchers need to develop technologies for precisely applying water, nutrients, and pesticides in accordance with the socio-economic conditions of smallholder farmers. There is a need for research on crop residue management, minimum tillage, and mulches. Majority of CSA options combine traditional practices, technologies, and services appropriate to the location with innovative ones. Sustainable, economically viable, and CSA practices in rice based systems are essential to ensure food security with better climatic resilience. Developing and disseminating climate-smart resource conserving technologies will contribute towards achieving the United Nations' Sustainable Development Goals of "Zero Hunger" and "Climate Action" by 2030.

References

Adhya TK, Bharati K, Mohanty SR, Ramakrishnan B, Rao VR, Sethunathan N, Wassmann R (2000) Methane emission from rice fields at Cuttack, India. Nutr Cycl Agroecosyst 58(1):95–105

Akter M, Deroo H, Kamal AM, Kader MA, Verhoeven E, Decock C, Boeckx P, Sleutel S (2018) Impact of irrigation management on paddy soil N supply and depth distribution of abiotic drivers. Agric Ecosyst Environ 261:12–24

Al-Mamun MR, Hasan MR, Ahommed MS, Bacchu MS, Ali MR, Khan MZH (2021) Nanofertilizers toward sustainable agriculture and environment. Environ Technol Innov 23:101658

Aulakh MS, Khera TS, Doran JW, Bronson KF (2001) Denitrification, N2 O and CO2 fluxes in rice-wheat cropping system as affected by crop residues, fertilizer N and legume green manure. Biol Fertil Soils 34(6):375–389

Ayars JE, Phene CJ, Hutmacher RB, Davis KR, Schoneman RA, Vail SS, Mead RM (1999) Subsurface drip irrigation of row crops: a review of 15 years of research at the Water Management Research Laboratory. Agric Water Manag 42(1):1–27

Babel MS, Wahid SW (2008) Freshwater under threat in South Asia. UNEP Report. United Nations Environment Programme (UNEP), Nairobi

Baruah DC, Goswami NG, Saikia R (2001) Manual transplanting vs. mechanical transplanting of paddy: a techno-economic analysis in Assam. J Agric Eng 38(3):66–72

Bergström L (1995) Leaching of dichlorprop and nitrate in structured soil. Environ Pollut 87:189–195

Bhattacharyya P, Dash PK, Swain CK, Padhy SR, Roy KS, Neogi S, Berliner J, Adak T, Pokhare SS, Baig MJ, Mohapatra T (2019) Mechanism of plant mediated methane emission in tropical lowland rice. Sci Total Environ 651:84–92

Billa SK, Murthy KR, Ramana A, Jagannadham J (2020) The potential of green seeker in nitrogen management of transplanted rice crop under north coastal environment of Andhra Pradesh, India. Int J Agricult Stat Sci 16(1):1925–1929

Biswas A, Mailapalli DR, Raghuwanshi NS (2021) APSIM-Oryza model for simulating paddy consumptive water footprints under alternate wetting and drying practice for Kharagpur, West Bengal, India. Paddy Water Environ 19(3):481–498

Busari TI, Senzanje A, Odindo AO, Buckley CA (2019) Irrigation management techniques with anaerobic baffled reactor effluent: effect on rice growth, yield and water productivity. Water Pract Technol 14(1):88–100

Cai Z, Xing G, Yan X, Xu H, Tsuruta H, Yagi K, Minami K (1997) Methane and nitrous oxide emissions from rice paddy fields as affected by nitrogen fertilisers and water management. Plant Soil 196(1):7–14

Carrijo DR, Akbar N, Reis AF, Li C, Gaudin AC, Parikh SJ, Linquist BA (2018) Impacts of variable soil drying in alternate wetting and drying rice systems on yields, grain arsenic concentration and soil moisture dynamics. Field Crop Res 222:101–110

Chen R, Cheng W, Cui J, Liao J, Fan H, Zheng Z, Ma F (2015) Lateral spacing in drip-irrigated wheat: the effects on soil moisture, yield, and water use efficiency. Field Crop Res 179:52–62

Chidthaisong A, Cha-un N, Rossopa B, Buddaboon C, Kunuthai C, Sriphirom P, Towprayoon S, Tokida T, Padre AT, Minamikawa K (2018) Evaluating the effects of alternate wetting and drying (AWD) on methane and nitrous oxide emissions from a paddy field in Thailand. Soil Sci Plant Nutr 64(1):31–38

Corton TM, Bajita JB, Grospe FS, Pamplona RR, Assis CA, Wassmann R, Lantin RS, Buendia LV (2000) Methane emission from irrigated and intensively managed rice fields in Central Luzon (Philippines). Nutr Cycl Agroecosyst 58(1):37–53

Davies DW, Garwood TWO, Rochford ADH (1996) Factors affecting nitrate leaching from a calcareous loam in east Anglia. J Agric Sci 126:75–86. Cambridge University Press

Dharminder, Singh RK, Kumar V, Pramanick B, Alsanie WF, Gaber A, Hossain A (2021) The use of municipal solid waste compost in combination with proper irrigation scheduling influences the productivity, microbial activity and water use efficiency of direct seeded rice. Agriculture 11:941. https://doi.org/10.3390/agriculture11100941

Dixit A, Khurana R, Singh J, Singh G (2007) Comparative performance of different paddy transplanters developed in India – a review. Agric Rev 28(4):262–269

El-Saadony MT, ALmoshadak AS, Shafi ME, Albaqami NM, Saad AM et al (2021) Vital roles of sustainable nano-fertilizers in improving plant quality and quantity an updated review. Saudi J Biol Sci 28(12):7349–7359

Fatima F, Hashim A, Anees S (2021) Efficacy of nanoparticles as nanofertilizer production: a review. Environ Sci Pollut Res 28(2):1292–1303

Feng J, Chen C, Zhang Y, Song Z, Deng A, Zheng C, Zhang W (2013) Impacts of cropping practices on yield-scaled greenhouse gas emissions from rice fields in China: a meta-analysis. Agric Ecosyst Environ 164:220–228

Gathala MK, Kumar V, Sharma PC, Saharawat YS, Jat HS, Singh M et al (2013) Optimizing intensive cerealbased systems addressing current and future drivers of agricultural change in the north-western Indo-Gangetic Plains of India. Agric Ecosyst Environ 177:85–97

Gill MA (1994) On-farm water management: a historical overview. In: Inayatullah C (ed) Water and community: an assessment of on-farm water management programme. SDPI, Islamabad, pp 24–39

Gors MJ, Howsc PW, Lane PW, Christian DG, Harris GL (1993) Losses of nitrate-nitrogen in water draining from under autumn-sown crops established by direct drilling or mouldboard ploughing. J Soil Sci 44(1):35–48

Gregorich EG, Drury CF, Baldock JA (2001) Changes in soil carbon under long-term maize in monoculture and legume-based rotation. Can J Soil Sci 81(1):21–31

Gujja B, Thiyagarajan TM (2009) New hope for Indian food security? The system of rice intensification, Gate keeper 143. International Institute for Environment and Development, London

Hadi A, Inubushi K, Yagi K (2010) Effect of water management on greenhouse gas emissions and microbial properties of paddy soils in Japan and Indonesia. Paddy Water Environ 8(4):319–324

Hou H, Peng S, Xu J, Yang S, Mao Z (2012) Seasonal variations of CH4 and N2 O emissions in response to water management of paddy fields located in Southeast China. Chemosphere 89(7):884–892

Huggins DR, Clapp CE, Allmaras RR, Lamb JA, Layese MF (1998) Carbon dynamics in corn-soybean sequences as estimated from natural carbon-13 abundance. Soil Sci Soc Am J 62:195–203

Hussain S, Peng S, Fahad S, Khaliq A, Huang J, Cui K, Nie L (2015) Rice management interventions to mitigate greenhouse gas emissions: a review. Environ Sci Pollut Res 22:3342–3360

IPCC (2014) In: Field CB, Barros VR, Dokken DJ, Mach KJ, Mastrandrea MD, Bilir TE, Chatterjee M, Ebi KL, Estrada YO, Genova RC, Girma B, Kissel ES, Levy AN, MacCracken S, Mastrandrea PR, White LL (eds) Climate change 2014: impacts, adaptation, and vulnerability. Part A: global and sectoral aspects. Contribution of Working Group II to the fifth assessment report of the Intergovernmental Panel on Climate Change. Cambridge University Press, Cambridge/New York

Itoh M, Sudo S, Mori S, Saito H, Yoshida T, Shiratori Y, Suga S, Yoshikawa N, Suzue Y, Mizukami H, Mochida T (2011) Mitigation of methane emissions from paddy fields by prolonging mid-season drainage. Agric Ecosyst Environ 141(3–4):359–372

Janmohammadi M, Amanzadeh T, Sabaghnia N, Dashti S (2016) Impact of foliar application of nano micronutrient fertilizers and titanium dioxide nanoparticles on the growth and yield components of barley under supplemental irrigation. Acta Agric Slov 107:265–276

Jat ML, Chandna P, Gupta RK, Sharma SK, Gill MA (2006) Laser land leveling: a precursor technology for resource conservation. Rice-Wheat Consortium Technical Bulletin Series 7, Rice-Wheat Consortium for the Indo-Gangetic Plains, New Delhi

Jat RK, Sapkota TB, Singh RG, Jat ML, Kumar M, Gupta RK (2014) Seven years of conservation agriculture in a rice–wheat rotation of Eastern Gangetic Plains of South Asia: yield trends and economic profitability. Field Crop Res 164:199–213

Jinsy VS, Pillai PS, Jacob J (2015) Productivity analysis of aerobic rice in the lowlands of Southern Kerala. J Trop Agric 53(1):1–7

Kadiyala MDM, Mylavarapu RS, Reddy GB, Reddy MD (2012) Impact of aerobic rice cultivation on growth, yield, and water productivity of rice–maize rotation in semiarid tropics. Agron J 104(6):1751–1765

Kar S, Pramanick B, Brahmachari K, Saha G, Mahapatra BS, Saha A, Kumar A (2021) Exploring the best tillage option in rice based diversified cropping systems in alluvial soil of eastern India. Soil Tillage Res 205:104761. https://doi.org/10.1016/j.still.2020.104761

Kumar V, Ladha JK (2011) Direct seeding of rice: recent developments and future research needs. Adv Agron 111:297–413

Ladha JK, Pathak H, Padre AT, Dawe D, Gupta RK (2003) Productivity trends in intensive rice–wheat cropping systems in Asia. In: Ladha JK et al (eds) Improving the productivity and sustainability of rice–wheat systems: issues and impacts. ASA Special Publication 65, ASA, CSSA, and SSSA, Madison, pp 45–76

Ladha JK, Kumar V, Alam MM, Sharma S, Gathala MK, Chandna P, Saharawat YS, Balasubramanian V (2009) Integrating crop and resource management technologies for enhanced productivity, profitability and sustainability of the rice-wheat system in South Asia. In: Ladha JK, Yadvinder-Singh, Erenstein O, Hardy B (eds) Integrated crop and resource management in the rice-wheat system in South Asia. International Rice Research Institute, Los Banos, pp 69–108

Lal R (2004) Carbon emission from farm operations. Environ Int 30:981–990. https://doi.org/10.1016/j.envint.2004.03.005

Liu R, Lal R (2014) Synthetic apatite nanoparticles as a phosphorus fertilizer for soybean (Glycine max). Sci Rep 4(1):1–6

Ludwig B, Nitschke R, Terhoeven-Urselmans T, Michel K, Flessa H (2008) Use of midinfrared spectroscopy in the diffuse-reflectance mode for the prediction of the composition of organic matter in soil and litter. J Plant Nutr Soil Sci 171:384–391

Macdonald AJ, Powlson DS, Poulton PR, Jenkinson DS (1989) Unused fertiliser nitrogen in arable soils-its contribution to nitrate leaching. J Sci Food Agric 46:407–419

Mahender Kumar R, Vijay Kumar S, Sreedevi B, Tuti MD, Prasad Babu MBB, Surekha K, Rathod S, Sai Prasad SV, Krishna Murthy SL, Arun Kumar S, Senguthuvel P, Brajendra Sailaja B, Padmavathi C, Muthu Raman P, Nirmala B, Sreedhar Y, Somashekar N, Chavan SN, Pundarikakshudu Sreenivasa Prasad M, Bandeppa B, Anantha MS, Saha S, Spandana Bhatt P, Dhillon BS, Singh DK, Bhowmick MK, Kavi Raju S, Srinivas D, Kumar V, Dixit S, Sundaram RM (2025) Status of Direct Seeded Rice (DSR) – Management for Sustainable Rice Production. ICAR-IIRR Book No. 132. ICAR-Indian Institute of Rice Research (IIRR), Rajendranagar, Hyderabad, Telangana State, India. 230 p.

Minamikawa K, Sakai N (2005) The effect of water management based on soil redox potential on methane emission from two kinds of paddy soils in Japan. Agric Ecosyst Environ 107(4):397–407

Naresh RK, Tomar SS, Samsher P, Singh SP, Kumar D, Dwivedi A et al (2014) Experiences with rice grown on permanent raised beds: effect of water regime and planting techniques on rice yield, water use, soil properties and water productivity. Rice Sci 21(3):170–180

Nath D, Laik R, Meena VS, Pramanick B, Singh SK (2021) Can mid-infrared (mid-IR) spectroscopy evaluate soil conditions by predicting soil biological properties? Soil Secur 4:100008

Nath D, Laik R, Meena VS, Kumari V, Singh SK, Pramanick B, Sattar A (2022) Strategies to admittance soil quality using mid-infrared (mid-IR) spectroscopy an alternate tool for conventional lab analysis: a global perspective. Environ Chall 7:100469

Norton GJ, Shafaei M, Travis AJ, Deacon CM, Danku J, Pond D, Cochrane N, Lockhart K, Salt D, Zhang H, Dodd IC, Hossain M, Islam MR, Price AH (2017) Impact of alternate wetting and drying on rice physiology grain production, and grain quality. Field Crop Res 205:1–13

Pathak H, Bhatia A, Jain N, Aggarwal PK (2010) Greenhouse gas emission and mitigation in Indian agriculture–a review. ING bulletins on regional assessment of reactive nitrogen, bulletin, vol 19, pp 1–34

Pathak H, Chakrabarti B, Bhatia A, Jain N, Aggarwal PK (2012) Potential and cost of low carbon technologies in rice and wheat systems: a case study for the Indo-Gangetic Plains. Low carbon technologies for agriculture: a study on rice and wheat systems in the Indo-Gangetic Plains, 12

Peng S, Bouman B, Visperas RM, Nie CA, Park HK (2006) Comparison between aerobic and flooded rice in the tropics: agronomic performance in an eight-season experiment. Field Crop Res 96:252–259

PIB (2025) Economic Survey 2024-25: Statistical Appendix, Government of India (GoI). 222 p.

Pramanick B, Brahmachari K, Ghosh D, Bera PS (2018) Influence of foliar application seaweed (*Kappaphycus* and Gracilaria) saps in rice (*Oryza sativa*)–potato (*Solanum tuberosum*)–blackgram (*Vigna mungo*) sequence. Indian J Agron 63(1):7–12

Pramanick B, Kumar M, Naik BM, Kumar M, Singh SK, Maitra S, Naik BSSS, Rajput VD, Minkina T (2022) Long-term conservation tillage and precision nutrient management in maize–wheat cropping system: effect on soil properties, crop production, and economics. Agronomy 12:2766. https://doi.org/10.3390/agronomy12112766

Pramanick B, Mahapatra BS, Datta D, Dey P, Singh SP, Kumar A, Paramanik B, Awasthi N (2023) An innovative approach to improve oil production and quality of mustard (*Brassica juncea L.*) with multi-nutrient-rich polyhalite. Heliyon 9(3):e13997. https://doi.org/10.1016/j.heliyon.2023.e13997

Prasad R, Bhattacharyya A, Nguyen QD (2017) Nanotechnology in sustainable agriculture: recent developments, challenges, and perspectives. Front Microbiol 8:1014

Puttaso A, Vityakon P, Saenjan P, Trelo-ges V, Cadisch G (2011) Relationship between residue quality, decompoistion patterns, and soil organic matter accumulation in a tropical sandy soil after 13 years. Nutr Cycling Agroecosyst 89:159–174. https://doi.org/10.1007/s10705-010-9385-1

Samahadthai P, Vityakon P, Saenjan P (2010) Effects of different quality plant residues on soil carbon accumulation and aggregate formation in a tropical sandy soil in Northeast Thailand as revealed by a 10-year field experiment. Land Degrad Dev 21:463–473. https://doi.org/10.1002/ldr.982

Sandhu SS, Mahal SS, Vashist KK, Buttar GS, Brar AS, Singh M (2012) Crop and water productivity of bed transplanted rice as influenced by various levels of nitrogen and irrigation in Northwest India. Agric Water Manag 104:32–39

Sapkota TB, Jat ML, Shankar V, Singh LK, Rai M, Grewal MS, Stirling CM (2015) Tillage, residue and nitrogen management effects on methane and nitrous oxide emission from ricewheat system of Indian Northwest Indo-Gangetic Plains. J Integr Environ Sci 12:31–46. https://doi.org/10.1080/1943815x.2015.1110181

Sarkar S (2001) Effect of water stress on growth, productivity and water expense efficiency of summer rice. Indian J Agric Sci 71(3):153–158

Sharda R, Mahajan G, Siag M, Singh A, Chauhan BS (2017) Performance of drip irrigated dry-seeded rice (Oryza sativa L.) in South Asia. Paddy Water Environ 15:93–100

Singh KK, Lohan SK, Jat AS, Rani T (2006) New technologies for planting rice for higher production. Res Crops 7(2):369–371

Sriphirom P, Chidthaisong A, Towprayoon S (2019) Effect of alternate wetting and drying water management on rice cultivation with low emissions and low water used during wet and dry season. J Clean Prod 223:980–988

Stoop WA, Uphoff N, Kassam A (2002) A review of agricultural research issues raised by the system of rice intensification (SRI) from Madagascar: opportunities for improving farming systems for resource-poor farmers. Agric Syst 71(3):249–274

Tirol-Padre A, Rai M, Kumar V, Gathala M, Sharma PC, Sharma S, Nagar RK, Deshwal S, Singh LK, Jat HS, Sharma DK, Wassmann R, Ladha J (2016) Quantifying changes to the global warming potential of rice wheat systems with the adoption of conservation agriculture in northwestern India. Agric Ecosyst Environ 219:125–137. https://doi.org/10.1016/j.agee.2015.12.020

Towprayoon S, Smakgahn K, Poonkaew S (2005) Mitigation of methane and nitrous oxide emissions from drained irrigated rice fields. Chemosphere 59(11):1547–1556

Tyagi L, Kumari B, Singh SN (2010) Water management-a tool for methane mitigation from irrigated paddy fields. Sci Total Environ 408(5):1085–1090
Uphoff N, Kassan A, Harwood R (2010) SRI as a methodology for raising crop and water productivity: productive adaptations in rice agronomy and irrigation water management. Paddy Water Environ 9(1):3–11
Wells BR, Turner FT (1984) Nitrogen use in flooded rice soils. In: Hauck RD (ed) Nitrogen use in crop production. American Society of Agronomy, Madison
Win KT, Nonaka R, Win AT, Sasada Y, Toyota K, Motobayashi T (2015) Effects of water saving irrigation and rice variety on greenhouse gas emissions and water use efficiency in a paddy field fertilized with anaerobically digested pig slurry. Paddy Water Environ 13(1):51–60
Yagi K, Tsuruta H, Kanda KI, Minami K (1996) Effect of water management on methane emission from a Japanese rice paddy field: automated methane monitoring. Glob Biogeochem Cycles 10(2):255–267
Yang S, Peng S, Xu J, Luo Y, Li D (2012) Methane and nitrous oxide emissions from paddy field as affected by water-saving irrigation. Phys Chem Earth Parts A/B/C 53:30–37
Yao Z, Zheng X, Dong H, Wang R, Mei B, Zhu J (2012) A 3-year record of N2 O and CH4 emissions from a sandy loam paddy during rice seasons as affected by different nitrogen application rates. Agric Ecosyst Environ 152:1–9
Yuan Q, Pump J, Conrad R (2014) Straw application in paddy soil enhances methane production also from other carbon sources. Biogeosciences 11:237–246. https://doi.org/10.5194/bg-11-237-2014
Zheng X, Mingxing W, Yuesi W, Renxing S, Jing L, Heyer J, Koegge M, Papen H, Jisheng J, Laotu L (2000) Mitigation options for methane, nitrous oxide and nitric oxide emissions from agricultural ecosystems. Adv Atmos Sci 17(1):83–92
Zou J, Huang Y, Jiang J, Zheng X, Sass RL (2005) A 3-year field measurement of methane and nitrous oxide emissions from rice paddies in China: effects of water regime, crop residue, and fertilizer application. Glob Biogeochem Cycles 19(2). https://doi.org/10.1029/2004GB002401
Zulfiqar F, Navarro M, Ashraf M, Akram NA, Munné-Bosch S (2019) Nanofertilizer use for sustainable agriculture: advantages and limitations. Plant Sci 289:110270

Chapter 9
Climate Change Impacts on Pulse Production and Mitigation Strategies

Pijush Das, Saumi Goswami, Ramyajit Mandal, Tanuj Kumar Mandal, Subhendu Bikash Goswami, and Milan Skalicky

Abstract Such changing climate is destroying the production of pulses worldwide from erratic rainfall patterns, rising temperature, and devastating pest outbreaks over these essential crop species that require protein. They are resilient. However, nowadays, pulses can easily be predisposed to increased environmental stresses most especially at such critical growth stage, which gives way to reductions in yields plus compromised nitrogen fixation. This review discusses the issues of climate changes on major pulse crops and proffers actual solutions to improve the challenges thereof. Strategies developed include drought-tolerant pulse varieties, water-conserving farming, and integrated pest management. The chapter is also filled with innovative breeding programs and adaptive agronomic techniques which can build greater resilience and productivity. It gives a clear track for the farmer to adapt to changing conditions in focusing on scalable solutions and actionable strategies. This chapter concludes by pointing out the requirement for proper cooperation between researchers, policymakers, and farming communities to ensure the sustainability of pulse production in the long term.

Keywords Pulse production · Environmental stress · Adaptation strategies · Crop resilience · Breeding programs

---

P. Das · S. Goswami · T. K. Mandal
School of Agriculture and Allied Sciences, The Neotia University, Kolkata, West Bengal, India

R. Mandal
School of Agriculture, Seacom Skills University, Santiniketan, West Bengal, India

S. B. Goswami
Department of Agronomy, Bidhan Chandra Krishi Viswavidyalaya, Mohanpur, West Bengal, India

M. Skalicky (✉)
Department of Botany and Plant Physiology, Faculty of Agrobiology, Food, and Natural Resources, Czech University of Life Sciences Prague, Prague, Czechia
e-mail: skalicky@af.czu.cz

B. Pramanick et al. (eds.), *Climate-Smart Agricultural Technologies*, Sustainability Sciences in Asia and Africa, https://doi.org/10.1007/978-981-96-7699-6_9

9.1 Introduction

Climate change has now become the most pressing environmental concern owing to its consequent effects on global food security and supply (Berg et al. 2013). In food-insecure regions, particularly in developing nations like India, Bangladesh, Pakistan, Sri Lanka, Myanmar, and Afghanistan, agricultural production is at risk due to climate change. India had made significant growth in agriculture in the past, but currently, it is dealing with a number of challenges, namely, stagnant net sown area, plateauing production levels, deteriorating soil quality, and decrease in per capita availability of land including the negative impact of climate change. According to INCCA (2010), India is the most susceptible country to climate change among all the developing nations. Although India has achieved self-sufficiency in food grain production through Green Revolution (Abrol and Sangar 2006; Pramanick et al. 2023), agriculture production is now becoming vulnerable due to climate change, particularly to drought, as two-thirds of India's agricultural land is rainfed and even the irrigated areas now depend on monsoon rain. In many regions of the nation, particularly in the east where flood events occur frequently, flooding is a major issue. Heat waves in the central and northern regions, frost in the northwest, and cyclones on the eastern coast all contribute to extreme hazards. Due to the rising temperature of the atmosphere, these climatic extremes have been occurring more frequently, increasing the danger of substantial losses in agricultural production. In India, a significant loss in crop yield is seen due to frequent drought occurrences since from 1966 (Auffhammer et al. 2012). Climate change will cut yields in the short run (2010–2039) by 4.5–9%, while in the long run (2070–2099), it will significantly reduce yields by at least 25% in the absence of required adaptation (Guiteras 2009). The average annual crop loss amounts to about 0.25% of India's GDP due to aberrant weather conditions and their later consequences (Singh et al. 2019).

Climate change has introduced new dimensions to current problems of reduced soil fertility, surface and groundwater pollution, water logging, and increased insect-disease pest incidence posing a serious threat to the agriculture production system in India (Rao et al. 2016). India observed significant rising temperatures in association with the incidence of numerous heat waves, droughts, instances of extremely heavy rain, and high cyclonic activity in recent years (Ray et al. 2019; Rohini et al. 2016; Sharma and Mujumdar 2017). Therefore, the adoption of climate-resilient agricultural practices and technologies is required to counteract these negative consequences of climate change and ensure sustained productivity (Pramanick et al. 2022). Under climate change scenarios, adaptation technology has the potential to increase the total productivity and profitability of any cropping system by changing crop management practices such as sowing time, planting density, fertilizer application, and irrigation water supply. Since global warming is already well underway, it is imperative that adaptation technology must be adjusted, especially for the poorest and most vulnerable nations. The impacts of climate change should be minimized at the farm, regional, and national levels by implementing an effective climate policy. However, more than 80% of Indian farmers are heterogeneous, unorganized, small, and marginal with inadequate capability to adapt against climate change, which is anticipated to exacerbate the issue of food security by placing stress on agriculture and undermining its sustainability.

Pulses are the major sources of protein for the vegetarian diet of the world population and supply almost 20–25% of protein as needed on a daily basis (Marwein and Ray 2019; Yadav et al. 2019; Pramanick et al. 2018). They also offer environmental services by maintaining the integrity of the soil health. Demand for pulses has been gradually rising over the past few years, resulting in a discrepancy between supply and demand. A growing population could result in increased consumption, which would widen this imbalance. India, the second largest populous nation in the world, will need about 39 Mt of total pulses by 2050 (Statista 2022). This will require a yearly growth rate of 2.2% to achieve this. India, which occupies an area of 30.5 million hectares under pulse production, is the largest producer among all nations accounting for around 25% of the total global production (Fig. 9.1).

Pulse crop production varies extensively under rainfed as well as irrigated agriculture systems (Pramanick et al. 2016). Since pulses make up a large portion of rainfed agriculture, yield reductions due to climate change are anticipated to be more pronounced for pulse crops. Rainfed agriculture in India will be primarily affected due to changing climatic conditions like increasing temperature, variable rainfall, and fewer numbers of rainy days. About 55% of the net sown area is under rainfed agriculture and exposed to various biotic as well as abiotic stresses arising from weather variability and climate change. Pulses are believed to be, in particular, susceptible to heat stress. A short time exposure to increased temperature during the flowering stage can result in significant yield reduction through falling down of anther, flower dropping, and pod damage. Major pulses grown under rainfed conditions, such as chickpea, pigeon pea, urd bean, mung bean, and lentil, are vulnerable to a number of challenges including extreme hot and cool temperatures, drought,

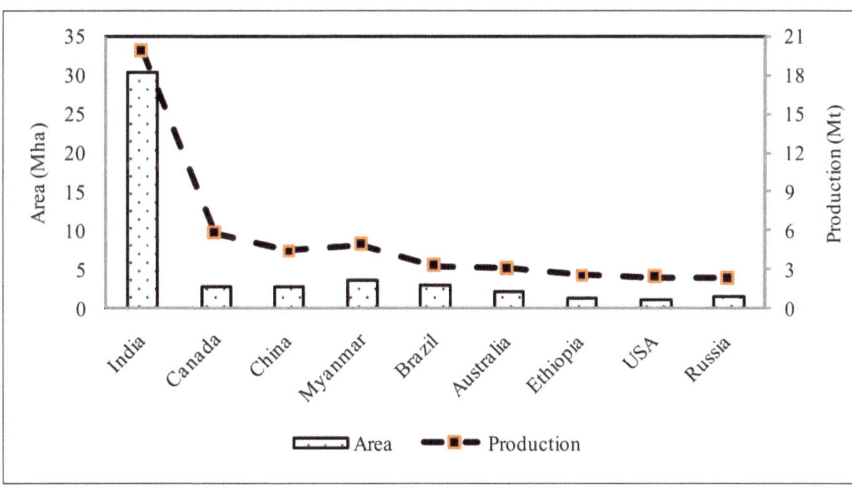

Fig. 9.1 Country-Wise Pulse Production. (Source: Indian Institute of Pulse Research (2014))

intense solar radiation, salinity, and waterlogging. Again, rising CO_2 levels pose a serious threat since it leads to increased average global temperature. The potential for pulse production is expected to be impacted by the predicted changes in temperature and their effects, water accessibility, pests, disease, and extreme weather events. The adverse situation will be more pronounced for pulses if the current trend of climate change continues as predicted by many weather prediction models (Pramanick et al. 2014).

9.2 Climatic Parameters in Relation to Pulse Crop Production

Pulse crops are mainly grown in India as kharif pulse (pigeon pea, green gram, black gram, etc.), rabi pulse (chickpea, field pea, lentil, etc.) and summer pulse (green gram, black gram, cowpea, etc.) throughout the agricultural year. Climatic parameters, namely, temperature requirement, amount of rainfall, relative humidity percentage, average sunshine hours, etc., vary widely for these pulse crops due to different agroclimatic zones where they grow. Kharif pulses generally prefer tropical and subtropical to humid climate whereas rabi pulses grow well in the regions under dry subhumid to semiarid climatic conditions. Climatic data which are suitable for pulse crop growth and production is discussed here.

9.2.1 Temperature

Kharif pulses require high temperature throughout the crop growing period. An average temperature of 30–35 °C for seed germination, 27–30 °C for higher vegetative growth, 15–18 °C for flowering, and 25–30 °C for good maturity is optimum for all pulse crops grown during kharif season. Temperature requirements of rabi pulses are 28–30 °C, 15–25 °C, 20–25 °C, and 25–30 °C for all the crop growth stages, respectively. Summer pulses require warm climate throughout the crop growing period. Extreme heat during the vegetative growth of pulses results in lower yield due to reduced number of fruit and lesser fruit setting, whereas severe cold and incidence of frost especially during the flowering stage are very much harmful for all rabi pulses except lentil which can tolerate it to some extent.

9.2.2 Rainfall

Pulses are generally grown as rainfed crop all over India and well adaptive when grown with irrigation facilities during dry season of the year. Water requirement for major pulse crops varies between 400–600 mm during warm season and 200–400 mm during cool season. Among the major pulses, pigeon pea uses maximum water of 540–600 mm throughout its crop growing period and lentil utilizes the least. It can also be noted that pulse crop having longer duration shows higher water use than short duration type apart from agroclimatic condition and location. Therefore, regions that have plenty of rainfall (on an average 600–800 mm) annually are suitable for pulse crop production. In contrast, water logging due to excessive rainfall during the flowering and later stage is detrimental which ultimately results in decreased yield of pulses.

9.2.3 Relative Humidity

Relative humidity (RH) is one of the important environmental factors for any crop production as it alters the rate of leaf transpiration and may have an impact on water balance in pulse crops. Low relative humidity in combination with high air velocity may result in water stress and closing of stomata thereby affecting photosynthesis. Best RH level for maximizing the photosynthetic rate is approximately 80% for any crop plant. The optimal RH for maximum plant growth lies within 60–70% RH (Zhang et al. 2018). Pulse crop grown with 85% of RH value shows faster growth rate in comparison to those grown with 50% RH. But pulse crops like pea show higher incidence of diseases with high RH. A combination of high RH and lack of air movement can cause serious crop damage.

9.2.4 Solar Radiation

Solar radiation is the most important factor influencing photosynthetic activity of all crops including pulses. It brings energy to the metabolic process of the plants. A significant quantity and high-intensity solar radiation is required throughout the day to maintain net assimilation rate above the compensation limit and positive carbon balance within plant body system. The intensity of solar radiation influences most of the growth stages of pulse crops, that is, germination, stem elongation, leaf growth, and time of flowering and fruiting. Major pulse crops like pigeon pea, green gram, and black gram require bright sunny days during flowering and ripening stage for better fruit setting.

9.2.5 Wind

Wind is a major factor in determining weather and climate. Many daily weather patterns depend on wind. Crop growth is heavily influenced by wind direction and speed. Wind promotes atmospheric turbulence, which improves the supply of carbon dioxide to the plants and boosts photosynthesis rates. Wind speed increases evapotranspiration, thereby increasing plant water use. A wind speed of more than 10 km/h results in reduced crop growth and accelerates the problem of flower drops in pulses. Wind alters the balance of hormones and enzymes in plant body. Wind also induces distant pollination by carrying pollen from one place to new areas.

9.2.6 Carbon Dioxide Concentration

The carbon dioxide (CO_2) content of the atmosphere is of major importance to plant growth. Photosynthesis is approximately proportional to the concentration of CO_2 in the air surrounding the foliage of the crop. A rising CO_2 concentration improves crop yield once the photosynthetic substrate and gradient of CO_2 concentration in between leaf and atmosphere increase. The efficiency of the photosynthetic process in pulse crops increases with a higher concentration of CO_2 in the atmosphere. On the contrary, increased CO_2 concentration accompanying with increase in air temperature may shorten growth cycles of major pulses thereby declining the crop yield.

9.2.7 Microclimate

Microclimate describes the climate that exists just above, inside, and beneath the crop canopy as well as in the soil root zone and can be influenced by day-to-day management activities on a range of time intervals. It is the condition at around plant, from the root zone depth to the highest top canopy. In microclimate, climate parameters are viewed as a continuous changing process in the atmosphere around the plant. The best crop microclimate is one that provides the most favorable environment for the desired plant response in order to increase the crop production and land productivity. The microclimate is influenced by the particular type of crop and may be strikingly different from the climate of the surrounding environment. Microclimate modification techniques are beneficial adaptive measures for managing harsh weather conditions and climatic risks in agriculture.

9.3 Climate Change Scenario in Indian Context and Its Effects in Agriculture

Nations along the Himalayas and South Asian coast must constantly be ready to resist the hazards of climate change (Stern 2007). It is predicted that temperature of South Asian countries may get increased by 2–6 °C in the twenty-first century (Ravindranath 2007). Many countries are becoming sensitive nowadays to all kinds of problems from current climate change particularly the Indian subcontinent.

As per World Bank data (2008), average temperature of 24.41 °C has been observed in India in between 1901 and 2021, with a maximum average value of 25.48 °C in 2009 and a lowest average value of 23.49 °C in 1917 (Fig. 9.2). An increase in temperature in the Indian subcontinent is expected to be 0.4–2.0 °C during kharif and 1.1–4.5 °C during rabi by the year 2070 (Khan et al. 2009b). According to the IPCC's fourth assessment report, India's temperature is projected to rise by 2.7–4.3 °C by the year 2080s (IPCC 2007). As per the prediction of coexistence of regions (B2) and heterogeneous world (A2) scenarios of the Special Report on Emission Scenario (SRES), India's temperature would rise by 2.9 °C and 4.1 °C, respectively, in the 2080s in comparison to the 1970s (Kolli et al. 2006). The balanced energy source (A1B) scenario of the Coupled Model Intercomparison Project Phase 3 (CMIP3)-based model predicts a rising temperature of 3.19 °C by the 2080s as compared with 1970s baseline (Chaturvedi et al. 2012).

Annual precipitation of any region as well as its intensity and frequency is greatly affected by climate change. World Bank data shows that annual precipitation decreased to 1209.84 mm in 2021 from 1231.95 mm in 2020 in India (Fig. 9.3). The SRES model predicts that the mean rainfall in India would rise by 10% between 2010 and 2070 for both the kharif and rabi seasons (Khan et al. 2009a). Beyond the 2040s, there will be significant increase in seasonal rainfall in India during twenty-first century (Kolli et al. 2006). In addition, it is also predicted that rainy days in

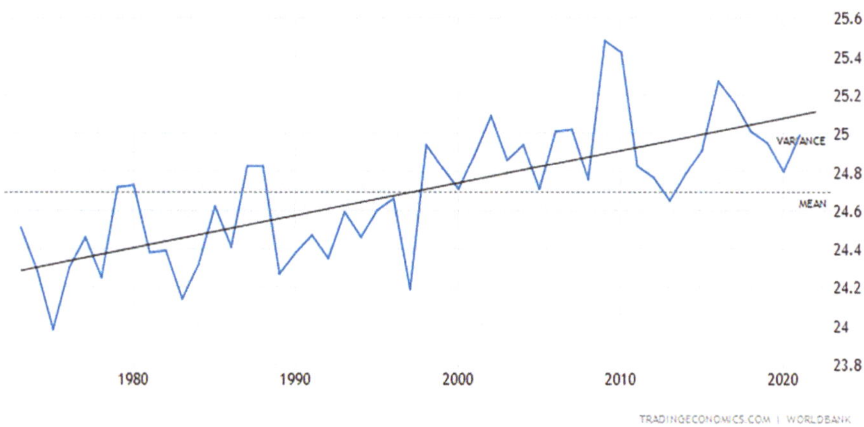

Fig. 9.2 Trend of average temperature (°C, last 50 years) of India. (Source: World Bank (2023))

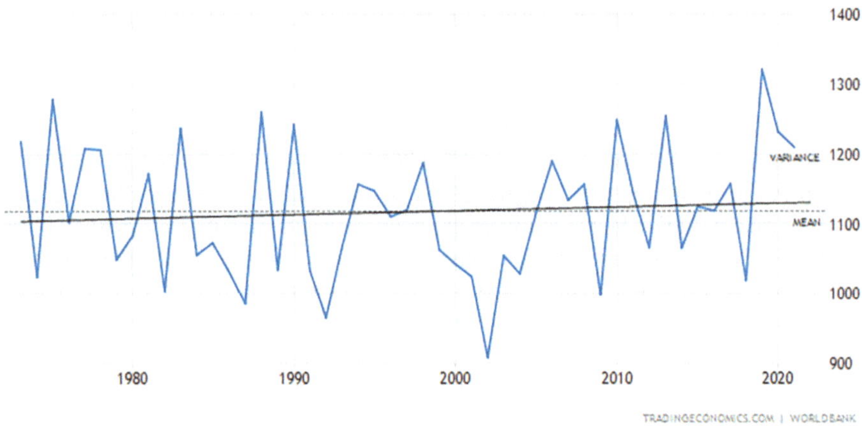

Fig. 9.3 Trend of average precipitation (mm, last 50 years) of India. (Source: World Bank (2023))

Northeast India and the foothills of the Himalaya may rise by 5–10 days. However, most areas of central and northern India have seen a propensity to witness a drop in rainfall during both the southwest and northeast monsoon seasons (Annamalai et al. 2010). Based on various Representative Concentration Pathway (RCP) scenarios, the total annual rainfall in India will rise by 1.2–2.4% in the year 2030s and 3.5–11.3% by the 2080s. The RCP8.5 scenario experiences highest increase in precipitation with projected precipitation ranging from 5% to 45%, while RCP2.6 sees the least increase in precipitation (Chaturvedi et al. 2012).

Climatic issues have direct impact on agriculture operations like choice of crops, date of sowing, space given to a particular crop, and the specific soil type, available resources, etc. Climatic hazards such as variations in temperature, precipitation, and average sunshine hours have an impact on all aspects of the agricultural ecosystem, including the livestock, arable, and hydrological sectors. Crop productivity is expected to decline by 10–40% by 2100, according to the global predictions report. Resource-poor like small and marginal farmers would be most affected by climate change issues in respect of agricultural crop yield. By the middle of the twenty-first century, it is expected that South Asian nations will see an average 30% decline in grain output. Due to unpredictable changes in rainfall and temperature, the states of North India are particularly vulnerable in terms of production efficiency (World Bank 2008). For instance, about 1.5 °C rise in temperature and 2 mm drop in precipitation result in 3–15% reduction in rice yield (Ahluwalia and Malhotra 2006). Climate changes brought on by rising greenhouse gas concentrations (GHGs) may have an impact on the productivity and output of agricultural crops, particularly pulses. The abrupt changes in climate disrupt the infections' progressive phases, which ultimately have a detrimental impact on crop development and yields. They also may increase the population of pests and insects, which would have a destructive effect on production as a whole.

9.4 Impacts of Climate Factors on Growth of Pulse

Due to the consequent effects on agriculture production and global food security, climate change has become one of the most important environmental challenges facing the entire world. The impact of extreme climate events like enhanced storm intensity and frequency, precipitation variance, subsequent drought and flooding, rising temperature, and CO_2 concentration has serious implications on global food production and food security. On rainfed crops, notably pulses, the changing climatic circumstances have a significant impact. As a major component of rainfed agriculture, pulses are expected to see a greater output decline due to climate change. The end outcome of global climate change will be a decline in the productivity and stability of dryland farming's field crops, notably pulses, which are more susceptible to weather variations and changes (Bahl 2015). The average world temperature is rising due to the concentration of greenhouse gases like CO_2, which is a severe hazard. In contrary, CO_2 has been favorably connected with improved biomass and yield, especially in C_3 plants. Temperature, rainfall, and humidity are all very sensitive factors that can have a big impact on the growth and yield of pulses. The growth, yield, and quality of pulse crops have been shown to be significantly impacted by variations in temperature and precipitation patterns (Table 9.1). Climate variables can interact with other environmental variables, like soil quality and nutrient availability, to influence the growth of pulse crops. This manuscript investigates how climatic variables affect the development of pulses, an important crop class that serves as a considerable source of protein for world population. The effects of extreme climate conditions on the growth of pulse crops must therefore be studied using a comprehensive strategy.

9.4.1 Effect of Excess Temperature

Major pulses such as chickpea, pigeon pea, lentil, and other pulse crops are generally grown in rainfed situations all over the world where high temperatures and dry conditions are prevalent as well as vulnerable to a number of challenges, including drought, extreme heat and cold, strong solar radiation, etc. Rising temperatures are expected to have a negative influence on the productivity and stability of the production of pulse crops by increasing the range and severity of a number of biotic and abiotic stressors (Bahl 2015). Interfering with plant growth and development, shrinking seed size, and reducing flower and pod production, high temperatures can impair the output of pulse crops. Rising temperature decreases rate of net photosynthesis during the period of the constraint which results in reduced plant growth and development. High temperatures and a lack of water consequently have an indirect impact on the quantity and weight of seeds. Eventually, changes in seed N content, one of the key factors affecting grain legume quality, can also be influenced by temperature changes. In case of chickpea which is subjected to high temperature for

Table 9.1 Consequences of changed weather parameters on the yield of major pulses

Year	Increase in temperature (°C)		Variation in temperature difference (°C) from the base year 2001–2003	Trends in actual and declined rainfall (mm) from the baseline year of 2001–2003 (860.07 mm)	Actual and decline yield (kg/ha) of major pulse from the base year 2004		
	Maximum temperature	Minimum temperature			Chickpea	Lentil	Pigeon pea
2004	+0.15	+2.45	−3.26	698 (162)	1550 (–)	1800 (–)	1400 (–)
2005	+0.16	+2.11	−2.93	680 (180)	1500 (−50)	1750 (−50)	1325 (−75)
2006	+1.15	+2.95	−2.55	534 (326)	1200 (−350)	1525 (−275)	1200 (−200)
2007	+0.56	+3.03	−3.37	556 (304)	1100 (−450)	1300 (−500)	1105 (−295)
2008	+1.81	+1.78	−0.93	600 (260)	1075 (−475)	1225 (−575)	1045 (−355)
Average	+0.86	+2.46	−2.60	614 (268)	1285 (−331)	1520 (−350)	1215 (−231)
Change in yield with every 0.1 °C increase in maximum temperature					−38.49 kg/ha	−40.7 kg/ha	−26.86 kg/ha
Change in yield with every 0.1 °C increase in minimum temperature					−13.46 kg/ha	−14.22 kg/ha	−9.39 kg/ha
Change in yield with every 0.1 °C increase in temperature differences					−12.73 kg/ha	−13.46 kg/ha	−2.90 kg/ha
Change in yield with every 10 mm drop in the rainfall					−12.35 kg/ha	−13.05 kg/ha	−8.60 kg/ha

Source: Dubey et al. (2011)

about ten days at pod formation stage, it resulted in production of fewer number of pods during post stress recovery than equal stress imposed at early flowering (Wang et al. 2006). Increased temperature above 21 °C (26/16 °C day/night) lowers the Bengal gram yield considerably under controlled greenhouse condition (Jumrani and Bhatia 2014). Excessive temperature decreases stigma receptivity and fertilization, both in Bengal gram (Kaushal et al. 2013) and lentil (Sehgal et al. 2017). During the reproductive phase, lentil is particularly sensitive to high temperatures (>30 °C), which results in the abortion of the pod and flower as well as a considerable drop in grain yield and quality (Gaur et al. 2015). Even during summer, green gram and black gram if subjected to high temperatures above 40–42 °C, they exhibit significant reduction in seed yield (Basu et al. 2016). Additionally, heat stress can have a substantial impact on the production of pulses. Extreme heat stress can result in blossom abortion, which directly reduces the amount of seeds. Pulse crop quality can also be impacted by heat stress, which can lower protein content, change starch and fiber content, and raise the risk of deterioration and discoloration. Pollen abnormalities, such as tiny, shrunken, and empty pollen grains, are caused by heat stress in the early blooming stages of the chickpea. At flowering stage of chickpea, common bean, cowpea, peanut, soybean, and lentil, heat stress reduces pollen viability and pollen output. The narrow temperature tolerance limit for optimal physiological performance applied to a C_3 crops particularly pulses is typically in the range of 30–35 °C. High temperatures can also make it more common for pests and diseases, like mites, aphids, and fungal infections, to impact pulse crop production.

9.4.2 Effect of Precipitation

The arrival of the southwest monsoon, as well as the amount and distribution of rainfall, has a significant impact on how well agriculture performs. Erratic rainfall has a negative effect on crop productivity. Frequency and quantity of precipitation have vital impact on pulse productivity as the reproductive phase is most sensitive to moisture stress. Shorter rainy season and high risk of water stress are two detrimental factors of climate change. Besides, sudden and aberrant precipitation during winter hampers the rabi pulse productivity (Rathore et al. 2022). Pulses are quite vulnerable to heavy rain, especially during the early vegetative stage and flowering, and a lot of rain can make crops infested with a greater number of diseases. Winter pulses are fond of dew occurrence which is nowadays not regular. Even if winter rains do not come, crops may grow significantly in December and January with normal dew water. Due to the warm night temperatures, the dew precipitation is actually slowly disappearing during the winter. An earlier maturity and a shorter crop life are predicted as effects of the rising average temperature. Cool-season legumes are more frequently affected by extreme atmospheric drought and may need additional watering if the current trend continues. Scarcity of irrigation water for pulses at peak crop growth stages arrest the plant growth system especially in the absence of supplemental irrigation (Basu et al. 2016). A deviation in precipitation can lead to droughts, floods, and severe climatic consequences like famines and rural poverty.

9.4.3 Effect of Frost

Frost is the ice crystal development on the ground or other surfaces brought on by the freezing of water vapor in the air when abrupt drop in temperature occurs. Global climate change is one of the major reasons for irregular frost occurrence which threatens the cool season pulse crop production. Frost can have a substantial negative effect on growth and productivity of pulse crops that include pea, beans, lentil, and chickpea. Pulse crops are particularly susceptible to frost damage since they are often grown in cooler season and are frequently planted early in the growing season. Frost can harm the delicate shoots and foliage of pulse crops, reducing their capacity to photosynthesize and yield food. The time and length of the frost occurrence, the temperature at which the frost occurs, and the stage of the crop growth all have severe impact on pulse crops. When frost strikes early in the growing season when the plants are still in the vegetative stage, for instance, the effects may be less severe than when frost strikes later in the season when the plants are in the reproductive stage. In legumes, frost occurrence has a detrimental effect on overall growth period of the plant especially during reproductive growth stage of plant. Alongside, nitrogen-fixing bacteria and their symbiotic relationship with host plants also get affected by the climatic aberrations (Stanley 2003). Due to frost sensitivity of pulses during flowering and pod development stages, the ambient temperature at reproductive stage put forth a significant impact on final crop yield.

9.4.4 Effect of Relative Humidity

Changes in climatic situation exhibit fluctuation of relative humidity, and this directly influences the water relations of the plant and indirectly affects leaf growth, photosynthesis, pollination, pest and disease outbreak, and finally the reduction of yield and economic loss. The relative humidity has a significant impact on evapotranspiration and, thus, on the amount of water that plants need to grow. By altering the gradient of vapor pressure from leaf to air at constant temperature, variations in atmospheric humidity have an impact on transpiration. Low relative humidity is typical in arid environments, which severely depletes water in pulses and lowers leaf water potential. Photosynthesis is also indirectly affected by fluctuations of relative humidity. Again, a higher percentage of relative humidity surrounding the crop canopy is conducive to disease infestation in pulse crop.

9.4.5 *Effect of Wind*

Wind speed and velocity have also directed as well as indirect effect on pulse production due to the climatic changes. Deforestation and its drastic resultant, that is, increased concentration of GHGs in the atmosphere frequently creating abnormal weather hazard like frequent formation of low pressure which in turn affects the atmosphere in the form of severe storms, cyclone, etc. Pulse crops that are grown in the summer and kharif season are more prone to these cyclonic hazards. High wind speed causes flower drop, pollen and anther desiccation, poor pod setting, etc. During maturity, high wind speed is a major threat for pulses as it induces shattering of pod which ultimately results in substantial yield loss.

9.4.6 *Effect of CO_2 Concentration*

CO_2, the prime component of global climate change, has versatile effect on the crop production. Besides the increased temperature and hazardous effect of heat wave, CO_2 reported compensatory effect on crop growth like the increase in atmospheric CO_2 has been shown to promote crop growth and productivity in C_3 plants, according to reports and predictions based on crop modeling and simulations (Rathore et al. 2022). Low CO_2 concentration in the substomatal chamber leads to high intrinsic water usage efficiency or the ratio of photosynthetic and transpiration rates at the leaf level. Increasing atmospheric CO_2 levels would maintain ideal CO_2 levels in the substomatal chamber at a lower level of stomata opening, leading to decreased rates of transpiration and preserving water. As a result, we can anticipate that the increased CO_2 conditions brought on by climate change will improve the overall water balance and productivity of plants including pulses, as has already been demonstrated (Serraj 2003).

9.5 Adaptation Measures to Combat the Hazards of Climatic Change

Food system is considerably influenced by a number of environmental concerns thus escalating the vulnerability of agricultural production system. Therefore, it is of utmost need to understand all the possible adaptation strategies by the farmers or producers in respect of global climatic change. Farmers' perception about the threat and severity of climate change is the most significant impetus factor in order to take required mitigation and adaption measures to combat the hazards of adverse climate. However, adaptation measures are very much dependent on the available information related to the particular issues. Traditional and agroecological management systems likely biodiversification, crop and soil management practices, water harvesting structures, etc. are useful to the farmers for adopting climate-resilient technologies (Altieri and Nicholls 2017). These management strategies ensure improved soil health and quality through increased carbon sequestration and reduced soil erosion that ultimately assure global food security in this critical juncture of climate change. These management strategies show huge potentials in mitigating the extreme climate-related issues and taking appropriate adaptation measures. Therefore, it is of immediate need to categorize the adaptation measures for agricultural sustainability under diverse alternative scenarios. Some possible mitigation options and adaptation measures are discussed here below.

9.5.1 New Variety Development

The extreme temperatures, drought, and flooding brought about by climate change are affecting the pulse crop production. New and changed pathogens are threatening to exacerbate these issues with new types of diseases in many crops. Therefore, it is necessary to develop new cultivars that are adapted to these threats by genetic transformation based on useful candidate genes. Development of new crop variety or cultivar can be an important option to cope with respect to resistance and tolerance to different biotic and abiotic stresses, adapted to climate change for sustainable agriculture. Selection of appropriate germplasm, reducing the breeding cycles, multilocational trials for suitability of new pulse crop variety, etc. are some of the ways to address the issue. Again the development of various stress-tolerant (biotic as well as abiotic) crop cultivars of pulses and its introduction to farmers' field should be exacerbated as one of the mitigation strategies to climate change (Table 9.2). Replacement of old susceptible crop varieties with these newly developed resistant varieties is better adapted to match the rapidly changing climate conditions. Removing older crop varieties from seed supply chain and accordingly promoting all these new varieties to the farmers' field need to be executed on immediate basis against presently occurring various climatic hazards.

9.5.2 Cultivation Practices

Cultivation practices, such as a shift in sowing time, population density, nutrient management and irrigation water supply, etc., are efficient management approaches in curbing the negative impact of climate change. Yield of chickpea is increased by 56% when the planting is done during 15th–30th of October instead of 15th–30th of November (Mansur et al. 2010). Long-duration pulse crop varieties perform better when spaced widely while short-duration varieties give higher yield with close spacing. Mung bean crop grown with a wider spacing of 45 × 15 cm gives higher crop yield as compared to narrow spacing of 30 × 15 cm (Singh et al. 2012). Mismanagement of nutrient sources in intensified conventional cultivation system is

Table 9.2 Pulse crop variety tolerant to biotic and abiotic stresses

Crop	Variety	Area of adaptation	Resistance/tolerance
Chickpea	DCP 92-3	North West Plain Zone (NWPZ)	Tolerant to crop lodging, resistant to wilt disease
	IPC 97-67 (SCS-3)	Jammu	Terminal moisture stress tolerant and resistant to wilt
	RSG-44	Rajasthan	Tolerant to drought and frost
	Phule G-81-1-1 (Vijay)	Central Zone (CZ)	Tolerant to terminal moisture stress
	IPCK 2002–29 (Shubhra)	CZ	Moderately resistant to wilt, tolerant to terminal heat and moisture stress
	IPCK 2004–29 (Ujjawal)	CZ	Moderately resistant to fusarium wilt, tolerant to terminal heat and moisture stress
	GPF 2	NWPZ	Resistant to crop lodging
Field pea	IPFD 9-2	Uttar Pradesh	Resistant to powdery mildew and tolerant to rust
	IPFD 12-2	CZ	Resistant to powdery mildew, pod borer, and moderately resistant to aphids and leaf miner
	IPFD 1-10 (Prakash)	Northern Hill Zone (NHZ) and CZ	Powdery mildew resistance
	IPF 5-19 (Aman)	NWPZ	Powdery mildew resistance, tolerant to rust
	IPF 4-9	Uttar Pradesh	Powdery mildew resistance, tolerant to rust

(continued)

Table 9.2 (continued)

Crop	Variety	Area of adaptation	Resistance/tolerance
Lentil	IPL 220	Eastern Uttar Pradesh, Bihar, West Bengal and Assam	Resistance major diseases including rust, fusarium wilt, and stemphylium blight
	IPL 321	Uttar Pradesh	Higher level of resistance to rust and wilt
	IPL 316	CZ	Resistant to rust and moderately tolerant to wilt
	DPL 62 (Sheri)	NWPZ	Resistant to rust and wilt
	DPL 15 (Priya)	NWPZ	Resistant to rust
Mung bean	IPM 302-2 (Kanika)	Uttar Pradesh	Highly resistant to yellow mosaic disease
	IPM 205-7 (Virat)	NWPZ, SZ, and CZ	Resistant to yellow mosaic disease
	IPM 410-3 (Shikha)	NWPZ and CZ	Highly resistant to mung bean yellow mosaic disease and powdery mildew
	IPM 02-03	NWPZ	Highly resistant to mung bean yellow mosaic disease
	PDM 139 (Samrat)	Uttar Pradesh	Resistant to mung bean yellow mosaic disease
Pigeon pea	Pusa 2002	CZ	Tolerant to moisture stress
	IPA 203	North East Plain Zone (NEPZ)	Resistance to sterility mosaic disease (SMD) and phytophthora blight
	IPA 206	Uttar Pradesh	Resistant to wilt
	IPH 09-5	NEPZ	Resistant to SMD and fusarium wilt
Urd bean	Vamban-2	Tamil Nadu	Tolerant to drought and yellow mosaic virus
	PDU 1 (Basant Bahar)	NWPZ and CZ	Tolerant to mung bean yellow mosaic virus
	IPU 94-1 (Uttara)	NWPZ and NEPZ	Resistant to mung bean yellow mosaic virus

Source: Indian Institute of Pulse Research (IIPR), Kanpur, Uttar Pradesh, India

responsible for huge economic losses in recent-day agriculture. Site-specific nutrient management (SSNM) can help in reducing the fertilizer application without affecting the productivity and profitability. In this regard, precision agriculture is thought to be more profitable compared to management of the entire crop field. Moreover, agriculture practices such as no-till farming, crop rotation, cultivation of cover crops, organic manuring, agroforestry, etc. can promote carbon sequestration and enhance soil organic carbon (SOC) status. Microirrigation techniques like sprinkler and drip irrigation also can help to mitigate climate change scenarios and deliver a sustained economic benefit. This has the potential to reduce the overdraft of groundwater, thereby minimizing the depletion of groundwater table and be resilient to the global climate change effects.

9.5.3 Risk Minimization Technology

9.5.3.1 Intercropping

Intercropping is the practice of growing two or more crop species simultaneously while they cohabit for a considerable portion of the crop cycle (Maitra et al. 2021). Intercropping practice focuses on yield enhancement per unit land area by making better use of soil resources that would otherwise be used by a single crop. To reduce agriculture's carbon footprint, increase food production, and make resilience to extreme weather, farmers must innovate through ecological intensification. Intercrops provide quality habitat to beneficial insects that augments the process of pollination and natural pest control, thereby stabilizing agroecosystems in the face of climate change (Huss et al. 2022). Combining deep-rooted and shallow-rooted plants together in the intercropping system opens up the possibility of bioirrigation. Additionally, the legumes foster an environment that is conducive to the habitation of several beneficial bacteria that promote increased ecosystem services. Legumes grown in intercropping systems improve the chemotaxis and rhizosphere behavior of good root-associated bacteria. Incorporation of legumes in intercropping systems enhances the soil fertility and physicochemico-biological properties and improves the resource use efficiency (Fig. 9.4). Comparing sole cropping to legume/nonlegume intercropping systems, the increase in water use efficiency is attributed to complementary interactions for water uptake due to differences in root distribution across soil layers and a reduction in evaporation loss by improved vegetative cover (Layek et al. 2018). The cereal + legume intercropping system is a popular practice as an insurance policy against crop failure for monocropping under rainfed conditions (Begam et al. 2020). In fact, intercropping system reflects the combined effects of all these aspects, and as a result, it may be regarded as a low-input agriculture practice with greater output in terms of increased farm output and agricultural sustainability.

9.5.3.2 Modified Planting Methods

The land and natural resources that India's farmers rely on not only to feed the country but also support their livelihoods are being put under increasing stress as a result of climate change. In India, novel methods are being developed to help farmers mitigate the effects of extreme weather conditions and adapt to the challenges posed by climate change. Therefore, to minimize this effect, alternate raised and sunken bed system, ridge and furrow method, broad bed planting, etc. are few of the options. Raised and sunken bed land configuration increases water productivity by two times

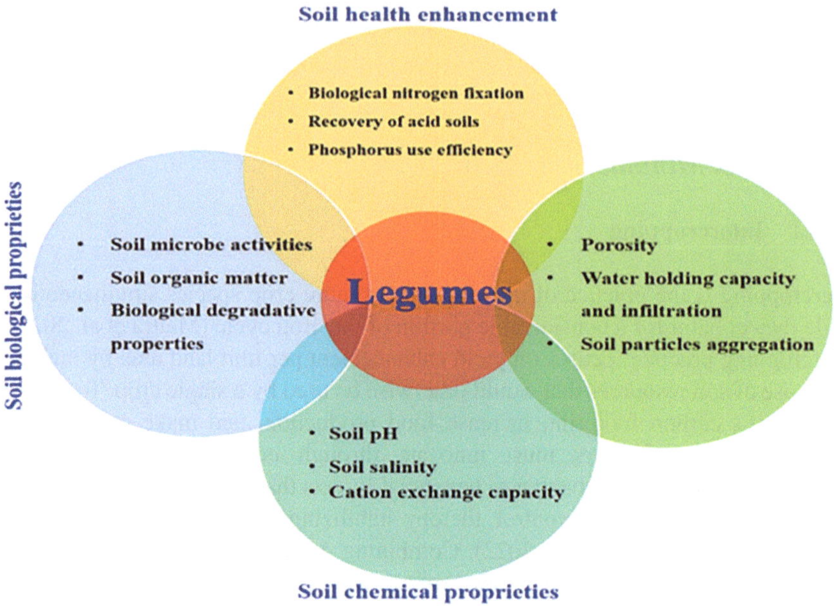

Fig. 9.4 Beneficial impacts of legumes on soil characteristics are shown schematically. (Source: Chamkhi et al. (2022))

and saves about 40–45% underground irrigation water in comparison to traditional cultivation method (Das et al. 2021). Planting methods like raised bed and ridge-furrow planting increases pulse crop yield by 25–30% compared to flatbed planting due to better drainage of excess moisture from the crop root zone and less incidence of insect-disease pest (Das et al. 2014). Ridge and furrow technology promotes the utilization of rainfall water that can be effectively controlled over time and space (Liu et al. 2022). The superposition of the water in the ditch, and the infiltration and transfer of water by the ridge, makes it possible to effectively accumulate natural precipitation and increase the effectiveness of rainfall utilization. The in situ conservation of rainwater in furrows improves drainage of excess water, and proper aeration of the ridge and root zone are just a few of the many benefits of this technology. Again, ridge or bed planting has been shown to reduce lodging, provide easy access to intercultural operations, facilitate aeration, and increase water productivity by 20–40% when compared to conventional planting or flatbed planting (Jani et al. 2008). The bed planting method alters the environment of the soil to provide moisture to pulses, which are susceptible to flooding or water stress. This type of agronomic strategy is very important to grow crops utilizing less water to improve crop water productivity, that is, "more crops per drop."

9.5.3.3 Paira and Utera Cultivation

Paira cropping is the practice of sowing a second crop after the flowering stage but before the main crop (like rice) is harvested. Paira or relay cropping is typically used when rice is harvested late due to high soil saturation resulting from late onset of monsoon. In low to medium land ecosystem, fields need another 15–20 days to bring soil moisture at optimum condition because of the high soil saturation experienced after rice harvesting. It compels majority of the farmers to keep their land fallow in the region. In the contrary, paira cropping practice of grass peas and lentil after rice harvesting facilitates to utilize those fallow lands effectively by using residual soil moisture. Cultivation of chickpea is also possible as a paira crop after rice in low to medium land ecosystems (Hedayetullah et al. 2018). Relay cropping can enhance soil water availability by increasing the profile contribution (PC) from the soil reserve in rainfed lowland ecosystem (Das and Goswami 2017). Under the paira cropping system, fertilization is not possible during the initial crop growth. At the vegetative stage or prior to flowering, nutrient solutions like urea and DAP are foliar sprayed at a rate of 2% under nutrient stress conditions. Again, with this minimal utilization of residual soil moisture in second crops and time-bound agronomic management, natural resources can be effectively utilized.

9.5.3.4 Conservation Tillage

In agricultural soils, excessive tillage may increase fertility in the short term but degrade soils over time. Erosion, loss of organic matter, structural degradation, and a decline in biodiversity are all to be expected. In many parts of the world, this led to the promotion of conservation tillage, particularly zero tillage. Zero tillage, direct drilling, minimum tillage, and/or ridge tillage are all commonly referred to as conservation tillage, which is a collective term used to indicate that the particular practice meets some resource conservation goals. Planting, growing, and harvesting crops with the least amount of surface disturbance are possible through conservation tillage. Conservation tillage practices are also recommended to enhance carbon sequestration and mitigate the issues of GHG emission (Kong et al. 2009). Several experiments with rabi crops such as lentil, field pea, fababean, lathyrus, and chickpea after harvesting of rice have resulted equally well with both conventional and conservation tillage practices. Conservation-effective tillage and pulses-based diversified cropping can increase productivity, profitability, and eco-efficiency besides improving soil fertility (Babu et al. 2023).

9.5.3.5 Mulching

Mulching is a practice of conserving natural resources in which organic and inorganic materials are applied to the surface of the soil to form a blanket. Mulching results in increased water productivity and yield enhancement in many pulses.

Agricultural crop yield including pulses is influenced by soil temperature and moisture content, both of which are affected by mulching. In situ soil moisture stress can be alleviated by using a variety of organic mulches (Ray et al. 2023). Organic mulching conserves soil moisture by minimizing surface disturbance to the soil. Mulching has numerous advantages, depending on the kind of mulch used, including weed growth inhibition, such as maintaining stable soil temperatures, checking the spread of soil-borne diseases, increasing soil fertility by mulch decomposition, boosting water infiltration, avoiding soil compaction, and improving the aesthetics of landscapes (Bahadur et al. 2018). Straw mulching is an effective agronomic practice in controlling weeds and for better conservation of soil moisture (Mandal et al. 2022). Besides, mulching also has different important uses, such as reducing soil water loss and erosion, enriching soil fauna, enhancing soil properties, and facilitating the nutrient cycle in the soil. Nowadays, biodegradable plastic mulches (BDMs) are used as a new type of mulch that performs all the purposes in contrast to regular polyethylene mulch. BDMs are incorporated into the soil where they are anticipated to biodegrade resulting in enhanced soil microbial population and enzymatic activities.

9.5.3.6 Seed Priming

Plants struggle for survival when they are subjected to heat stress, which significantly alters their growth and metabolic processes (Chakraborty and Dwivedi 2021). Crops' physiological and biochemical progressions through reproductive stages, such as grain/pod filling, its development, maturation, and ripening, are disrupted by terminal heat stress, which ultimately results in yield loss. Seed priming is considered as one of the most straightforward and affordable sophisticated strategies to reduce this loss. This method involves soaking seeds in various nutrient-rich solutions to increase their quality for quick and uniform seed germination, producing healthy and robust seedlings. The pregerminative metabolic responses that are elicited by this method of seed soaking prompt early germination and accelerate crop establishment (Jana et al. 2022). Some of the more recent priming techniques include hydropriming, halopriming, osmopriming, sand matric priming, and nutripriming (priming with water, salt, osmoticum, wet sand, and nutrition). Seed priming is a simple, useful, cost-effective, risk-averse, and practically viable technology for resource-poor farmers to ensure uniform germination of seed, quick seedling emergence, successful crop establishment, and increased yield of pulse crops in stress-prone conditions like adverse soil temperature, inconsistent soil moisture, and soil hardiness (Bhowmick 2018). Priming of pulse seeds prior to sowing increases speed and extent of emergence, seedling vigor, deeper roots with improved nutrient uptake, accelerated flowering and maturity, conferred insect resistance, and increased crop yield. Therefore, seed priming of target pulse crops is a useful and goal-oriented option for agronomists in climate-changing scenario.

9.5.3.7 Foliar Nutrition

Crop growth and yield are severely restricted by abiotic stresses like high temperature and moisture stress. Due to the intensification of adverse crop environments brought on by climate change, agricultural and horticultural crops suffer significant economic losses. Low water availability is frequently linked to high temperatures. The increased risk of combined heat stress and drought events in future highlights the need to research how crops respond to these combined stressors and cost-effective management solutions (IPCC 2014). When compared to applying nutrients to the soil, applying major and micronutrients to the leaves helps nutrients move quickly through the plant, which is important for reducing stress in plants (Sarkar et al. 2021). Foliar nutrients usually penetrate the leaf cuticle or stomata and enter the cells facilitating easy and rapid utilization of nutrients. Nutrients can be delivered directly to the site of food synthesis using this method, reducing the need for fertilizers and preventing nutrients from being wasted. In addition to addressing deficiencies in the soil, foliar feeding also helps overcome the soil's inability to transfer nutrients to the plant when moisture levels are low. Therefore, exogenous nutrition administration could be a useful strategy for reducing the adverse effects of environment.

9.5.3.8 Biofertilizer

Biofertilizers provide a way to replace chemically produced fertilizers and mitigate the effects of climate change on agricultural production. These substances are essential components of an integrated soil management strategy for sustainable crop production. These are preparation entities with living and latent microbial strains that may aid in the uptake of nutrients from the root rhizosphere zone and primarily fix nutrients from the atmosphere. Plant nutrient fixation is influenced in different ways by various biofertilizer strains. These preparations can be mixed with organic manures and applied directly in the field or used to treat seeds before sowing. When applied to plants, a wide variety of biofertilizers that contain one or more microorganisms may assist in providing numerous essential nutrients necessary for plant growth (Sana et al. 2020). They assist plants in nitrogen fixation, phosphorus solubilization and largely access, abiotic stress tolerance, biocontrol, and siderophore production. Phosphate-solubilizing microorganisms (PSMs), arbuscular mycorrhiza (AM), plant growth-promoting rhizobacteria (PGPR), blue green algae (BGA), and azolla are a few of the important biofertilizers commonly used in agricultural crop production. Biofertilizers are beneficial to the production of pulse crops as they play a crucial role in the fixation of nutrients from the atmosphere maintaining a good relation with the crop species.

9.5.4 *Breeding Strategies*

9.5.4.1 Trait-Based Breeding: Phenotyping to Genotyping

Identification of the desirable qualities, such as adaptation to a water-scarce environment, exporting carbohydrates to the reproductive organ under stress, staying green traits, and diverse root architecture are desirable for a stressful environment (Uga et al. 2013). Plant trait, which result in ultimate reproductive gain, is its ability of biomass partitioning that confers resilience in many crops. It should be enough flexible for maximum gain under stressful conditions by improved energy partition with source-sink economic balance. Flowering behavior under different stress conditions especially during heat and drought controls the yield in pulse that need some complex polygenic inheritance patterns of flowering character under both the conditions and plays a major role for any breeding program.

9.5.4.2 Indirect Selection

Direct selection of traits for yield improvement under any stress condition is a frequent practice in pulse breeding programs. Since grain yield is a quantitative trait, factors such as environmental influences, genotype-environment interactions, low heritability, linkage drag, and nonadditive gene actions make it ineffective for fruitful selection of desirable traits. These factors guided plant breeders to attempt on indirect selection of secondary traits which are easy to estimate, highly heritable, and comprehensively associated with yield for a meaningful selection of traits. Therefore, breaking down complicated qualities into their component traits, such as stress tolerance and yield, provides a way to breed good varieties with progressive genetic gain across numerous indirect selection cycles. Factors like ideotype, phenology, number of nodes and branches, pod and seed number, seed size, etc. affect productivity in pulses significantly. Therefore, it is essential to understand the nature, size, and relationships of the genes responsible for component qualities for choosing the best breeding technique and parental lines.

9.5.4.3 Speed Breeding and Rapid Generation Advance (RGA)

Maintaining high light intensity, continuous light phase with short dark period, relative humidity (60–70%), and optimum temperature range (17–22 °C), illuminating by various light sources in a greenhouse or controlled chambers, shortens generation time drastically. In crops like wheat, barley, pea, etc., six generations per year can be seen in a greenhouse with maintained light sources and other climatic factors required for plant growth, while four generations per year are observed in canola following the principle of speed breeding and rapid generation advance program. Generation time of 7, 6.2, and 6.0 per years has also been achieved in early-, medium-, and late-maturing genotypes of chickpea with controlled climatic conditions and pregerminated immature seeds from different agroecological conditions. Since a long time ago, it has been possible to grow crops in winter nurseries and reduce crop cycles more quickly by using the double haploid approach. Single seed descent (SSD) speed breeding minimizes the expense of the breeding program, shortens the breeding cycle, and improves the creation of favorable breeding lines. SSD speed breeding has been employed for a long time to isolate desirable transgressive segregants.

9.5.4.4 Marker-Assisted Selection (MAS)

Marker-assisted selection (MAS) is a promising breeding technique with respect to resistance and tolerance to various environmental stresses and adverse climatic conditions (Fig. 9.5). The drawbacks of phenotype-based selections can easily be eliminated through marker-assisted selection (MAS). The use of molecular markers has enabled breeders to create varieties with previously challenging features. Early generation marker-assisted selection refers to the early selection and eradication of undesirable plants from the breeding programs. This MAS enables plant breeders to

Fig. 9.5 Breeding strategies deployed for cool-season legume breeding. (Source: Kumari et al. (2021))

focus on a smaller number of promising plants, increasing selection efficiency and precision. In order to successfully breed cyst nematode-resistant soybean, anthracnose-resistant bean, and fusarium wilt-resistant chickpea, numerous MAS techniques such as marker-assisted recurrent selection (MARS), marker-assisted backcross (MABC), and genomic selection (GS) become imperative and have been utilized successfully for continuous development. Introgressive hybridization and pyramiding of the multiple genes which affect similar type of characters are quite difficult tasks for majority of the plant breeders. MAS allow to more precisely assembling target alleles in a same genotype with least linkage drags and minimum selection cycles. Commercial breeding programs have demonstrated twice as much genetic gain over phenotypic selection when using MAS techniques. Markers have become widely used in pulse breeding programs to speed up varietal development and improve the efficacy of selection. It results in a significant time savings when compared to traditional backcross breeding.

9.6 Government Initiatives for Climate Change Adaptation

Different parts of the world are anticipated to see varying future consequences of climate change and climate vulnerability on social systems. In various geographical and socioeconomic circumstances, various modeling studies, innovative technology systems, and adaptation techniques would therefore be needed. There are currently high degrees of uncertainty about the possible effects of climate change in different parts of country. The Indian government has taken a number of initiatives to create the most effective missions to combat global warming and get prepared for climate change. The National Action Plan on Climate Change (NAPCC) has been formulated in 2008 with the goal of raising awareness among the general public, government agencies, industry, scientists, and society about the threats posed by global climate changes. The National Adaptation Fund for Climate Change (NAFCC) has been established to provide funding for adaptation efforts in India's States and Union Territories (UTs) that are more susceptible to the negative impacts of climate change. Some major initiatives launched by the Government of India are mentioned below with its key objectives (Table 9.3).

9.7 Conclusion

Climate change has undoubtedly a direct impact on the livelihood of global population, particularly of the developing countries of Asia. Variability in climatic parameters has become clear due to indiscriminate deforestation, burning of fossil fuels, industrialization, and intensive agricultural practices. Climatic hazards like extreme temperatures, heat waves, drought, flood, cyclone, hailstorms, etc. show adverse effects on crop yields especially pulse crop production in India. It is now of a great

Table 9.3 Government initiatives/schemes and their key objectives to deal the climate change

Government initiatives	Year of commencement	Key objective
National Mission on Sustainable Agriculture (NMSA)	2014–2015	To conserve natural resource base primarily and soil fertility through adopting complete soil health management practices for making agriculture system more sustainable
National Adaptation Fund for Climate Change (NAFCC)	2015–2016	To mitigate the negative effects of adverse climatic conditions in various sectors like agriculture, animal husbandry, forestry, etc.
Pradhan Mantri Krishi Sinchayee Yojona (PMSKY)	2015	To provide support in conservation of water resources and their efficient utilization in agriculture with the vision of "Har Khet Ko Paani"
Pradhan Mantri Fasal Bima Yojna (PMFBY)	2016	To provide comprehensive insurance cover against crop failure due to various climatic hazards and/or natural calamities, thus stabilizing the income of distress farmer and encouraging them for new innovative technology adaptation
Soil Health Card (SHC)	2015	To make the farmers more aware about soil health of their land providing a complete information of soil nutrient content along with the recommendation of optimum fertilizer dose, judicious use of any inputs, and other soil management practices
Green India Mission (GIM)	2014	To fight against adverse effects of climate change with suitable mitigation options and adaptation techniques for the protection and restoration of forest areas in India
National Water Mission (NWM)	2011	To conserve water resources and minimize its wastage ensuring more equitable distribution of it through integrated water resource development and management
Paramparagat Krishi Vikas Yojona (PKVY)	2015	To encourage the conservation of natural resource and on farm recycling of nutrients and minimize farmer's reliance on off-farm inputs for climate resilient sustainable farming and enhancing natural resource base
National Innovations on Climate Resilient Agriculture (NICRA)	2011	To enhance resilience against the negative effects of adverse climatic conditions on Indian agriculture through strategic action plans and adoption of innovative technologies
Submission on Agro-forestry	2018–2017	To lessen the impact of climate change and enhance agricultural sustainability while easing demand on natural forests
National Action Programme to Combat Desertification	2001	To combat desertification through alleviating the impact of regular drought occurrence in arid regions with community-based drought management approach

concern to have a comprehensive assessment of the negative impacts of climate change in contrast with a better understanding of required mitigation options and adaptation technology to sustain human life. Adaptation technology that involves modification in crop management practices like sowing date, population density, fertilizer application, etc. and crop-specific breeding strategies like trait-based breeding, indirect selection, speed breeding, marker-assisted selection, etc. become more potential in augmenting the productivity and profitability of pulse crops under adverse climatic situations. So far these adaptation technologies have shown positive results in Indian agriculture although much more strategies will be needed in future for food security and sustainable production under climate change scenario.

References

Abrol IP, Sangar S (2006) Sustaining Indian agriculture–conservation agriculture the way forward. Curr Sci 91:1020–1025

Ahluwalia VK, Malhotra S (2006) Environmental science, a book. Pvt. Ltd. Goalie Enterprises, India

Altieri MA, Nicholls CI (2017) The adaptation and mitigation potential of traditional agriculture in a changing climate. Clim Chang 140:33–45

Annamalai H, Kida S, Hafner J (2010) Potential impact of the tropical Indian Ocean–Indonesian seas on El Niño characteristics. J Clim 23(14):3933–3952

Indian Institute of Pulse Research (2014) Annual report 2014, Kanpur, India

Auffhammer M, Ramanathan V, Vincent JR (2012) Climate change, the monsoon, and rice yield in India. Clim Chang 111:411–424

Babu S, Singh R, Avasthe R, Rathore SS, Kumar S, Das A, Layek J, Sharma V, Wani OA, Singh VK (2023) Conservation tillage and diversified cropping enhance system productivity and eco-efficiency and reduce greenhouse gas intensity in organic farming. Front Sustain Food Syst 7:1114617

Bahadur S, Pradhan S, Verma S, Maurya R, Verma SK (2018) Role of plastic mulch in soil health and crop productivity. In: Climate change and its implications on crop production and food security, Mahima Publications, Varanasi, India, pp 338–344

Bahl PN (2015) Climate change and pulses: approaches to combat its impact. Agric Res 4:103–108

Basu PS, Singh UMMED, Kumar ANIL, Praharaj CS, Shivran RK (2016) Climate change and its mitigation strategies in pulses production. Indian J Agron 61:S71–S82

Begam A, Mondal R, Dutta S, Banerjee H (2020) Impact of cereal+legume intercropping systems on productivity and soil health-a review. Int J Bioresour Stress Manag 11(3):274–286

Berg A, de Noblet-Ducoudré N, Sultan B, Lengaigne M, Guimberteau M (2013) Projections of climate change impacts on potential C4 crop productivity over tropical regions. Agric For Meteorol 170:89–102

Bhowmick MK (2018) Seed priming: a low-cost technology for resource-poor farmers in improving pulse productivity. In: Advances in seed priming. Springer, Singapore, pp 187–208

Chakraborty P, Dwivedi P (2021) Seed priming and its role in mitigating heat stress responses in crop plants. J Soil Sci Plant Nutr 21(2):1718–1734

Chamkhi I, Cheto S, Geistlinger J, Zeroual Y, Kouisni L, Bargaz A, Ghoulam C (2022) Legume-based intercropping systems promote beneficial rhizobacterial community and crop yield under stressing conditions. Ind Crop Prod 183:114958

Chaturvedi RK, Joshi J, Jayaraman M, Bala G, Ravindranath NH (2012) Multi-model climate change projections for India under representative concentration pathways. Curr Sci 103:791–802

Das P, Goswami SB (2017) Crop water productivity of rainfed toria (Brassica rapa L.) as influenced by residual effect of organic manures and direct fertilizer management. Environ Ecol 35(2B):1126–1130

Das TK, Choudhary AK, Sepat S, Vyas AK, Das A, Bana RS, Pooniya V (2014) Conservation agriculture: a sustainable alternative to enhance agricultural productivity and resources-use efficiency. Technical Extension Folder, IARI, New Delhi

Das P, Pramanick B, Goswami SB, Maitra S, Ibrahim SM, Laing AM, Hossain A (2021) Innovative land arrangement in combination with irrigation methods improves the crop and water productivity of rice (Oryza sativa L.) grown with okra (Abelmoschus esculentus L.) under raised and sunken bed systems. Agronomy 11(10):2087

Dubey SK, Sah U, Singh SK (2011) Impact of climate change on pulse productivity and adaptation strategies as practiced by the pulse growers of Bundelkhand region of Uttar Pradesh. J Food Legumes 24(3):230–234

Gaur PM, Samineni S, Krishnamurthy L, Kumar S, Ghanem ME, Beebe S, Rao I, Chaturvedi SK, Basu PS, Nayyar H, Jayalakshmi V, Babbar A, Varshney RK (2015) High temperature tolerance in grain legumes. Legume Perspectives 7:23–24

Guiteras R (2009) The impact of climate change on Indian agriculture manuscript. Department of Economics, University of Maryland, College Park

Hedayetullah MD, Chowdhury AR, Ghosh M, Hazra KK, Nath CP, Sadhukhan R, Zaman P (2018) Chickpea under rice fallow in lowland ecosystem of West Bengal, India. Agri Res Tech Open Access J 13(1):555870

Huss CP, Holmes KD, Blubaugh CK (2022) Benefits and risks of intercropping for crop resilience and pest management. J Econ Entomol 115(5):1350–1362

Indian Network for Climate Change Assessment (INCCA) (2010) India: greenhouse gas emissions 2007. Ministry of Environment & Forests, India

IPCC (2014) Climate change. Synthesis report. Contribution of Working Groups I. In: II and III to the fifth assessment report of the Intergovernmental Panel on Climate Change, Geneva, Switzerland, vol 151, no. 10.1017

Jana K, Sarkar A, Mondal R, Agrawal K (2022) Quality and green fodder productivity of forage maize (Zea mays L.) as influenced by different seed priming techniques under rainfed situation. J Crop Weed 18(3):71–77

Jani PP, Tanwar SPS, Singhi SM (2008) Performance of durum wheat (Triticum durum Desf.) varieties under varying sowing methods and input levels. Indian J Fertilisers 4(7):71

Jumrani K, Bhatia VS (2014) Impact of elevated temperatures on growth and yield of chickpea (Cicer arietinum L.). Field Crop Res 164:90–97

Kaushal N, Awasthi R, Gupta K, Gaur P, Siddique KHM, Nayyar H (2013) Heat-stress-induced reproductive failures in chickpea (Cicer arietinum) are associated with impaired sucrose metabolism in leaves and anthers. Funct Plant Biol 40(12):1334–1349

Khan S, Khan MA, Hanjra MA, Jianxin M (2009a) Pathways to reduce the environmental footprints of water and energy inputs in food production. Food Policy 34(2):141–149

Khan SA, Kumar S, Hussain MZ, Kalra N (2009b) Climate change, climate variability and Indian agriculture: impacts vulnerability and adaptation strategies. In: Climate change and crops. Springer Berlin Heidelberg, Berlin, Heidelberg, pp 19–38

Kolli R, Sahai AK, Kumar KK, Patwardhan S, Mishra PK, Revadekar JV, Kamala K, Pant GB (2006) High-resolution climate change scenarios for India for the 21st century. Curr Sci 90:334–345

Kong AYY, Fonte SJ, van Kessel C, Six J (2009) Transitioning from standard to minimum tillage: trade-offs between soil organic matter stabilization, nitrous oxide emissions, and N availability in irrigated cropping systems. Soil Tillage Res 104(2):256–262

Kumari VV, Roy A, Vijayan R, Banerjee P, Verma VC, Nalia A, Pramanik M et al (2021) Drought and heat stress in cool-season food legumes in sub-tropical regions: consequences, adaptation, and mitigation strategies. Plan Theory 10(6):1038

Layek J, Das A, Mitran T, Nath C, Meena RS, Yadav GS, Shivakumar BG, Kumar S, Lal R (2018) Cereal+legume intercropping: an option for improving productivity and sustaining soil health. In: Legumes for soil health and sustainable management. Springer, Singapore, pp 347–386

Liu P, Zhang T, Zhang F, Ren X, Chen X, Zhao X (2022) Ridge and furrow configuration improved grain yield by optimizing the soil hydrothermal environment and maize canopy traits in Northwest China. Plant Soil:1–14. https://doi.org/10.1007/s11104-022-05518-x

Maitra S, Hossain A, Brestic M, Skalicky M, Ondrisik P, Gitari H, Brahmachari K et al (2021) Intercropping—a low input agricultural strategy for food and environmental security. Agronomy 11(2):343

Mandal TK, Das P, Mondal M (2022) Effect of irrigation and mulching on growth, yield attributes and yield of safflower under safflower-chickpea intercropping system in Indian Sub-continent. IJBS 9(01):59–63

Mansur CP, Palled YB, Halikatti SI, Chetti MB, Salimath PM (2010) Effect of dates of sowing and irrigation levels on growth, yield parameters, yield and economics of kabuli chickpea. Karnataka J Agric Sci 23(3):461–463

Marwein Y, Ray LIP (2019) Performance of rajma (Phaseolus vulgaris) cultivars under organic mulches in Meghalayan Plateau of North Eastern India. Legume Res Int J 42(1):114–118

Pramanick B, Brahmachari K, Ghosh A, Zodape ST (2014) Foliar nutrient management through *Kappaphycus* and *Gracilaria* saps in rice-potato-greengram crop sequence. J Sci Ind Res 73:613–617

Pramanick B, Brahmachari K, Ghosh A, Zodape ST (2016) Effect of seaweed saps derived from two marine algae *Kappaphycus* and *Gracilaria* on growth and yield improvement of blackgram. Indian J Geo Mar Sci 45(6):789–794

Pramanick B, Brahmachari K, Ghosh D, Bera PS (2018) Influence of foliar application seaweed (*Kappaphycus* and *Gracilaria*) saps in rice (*Oryza sativa*)–potato (*Solanum tuberosum*)–blackgram (*Vigna mungo*) sequence. Indian J Agron 63(1):7–12

Pramanick B, Kumar M, Naik BM, Kumar M, Singh SK, Maitra S, Naik BSSS, Rajput VD, Minkina T (2022) Long-term conservation tillage and precision nutrient management in maize–wheat cropping system: effect on soil properties, crop production, and economics. Agronomy 12:2766. https://doi.org/10.3390/agronomy12112766

Pramanick B, Mahapatra BS, Datta D, Dey P, Singh SP, Kumar A, Paramanik B, Awasthi N (2023) An innovative approach to improve oil production and quality of mustard (*Brassica juncea* L.) with multi-nutrient-rich polyhalite. Heliyon 9(3):e13997. https://doi.org/10.1016/j.heliyon.2023.e13997

Rao CAR, Raju BMK, Subba Rao AVM, Rao KV, Rao VUM, Ramachandran K, Venkateswarlu B et al (2016) A district level assessment of vulnerability of Indian agriculture to climate change. Curr Sci 110:1939–1946

Rathore SS, Babu S, Shekhawat K, Singh R, Yadav SK, Singh VK, Singh C (2022) Designing energy cum carbon-efficient environmentally clean production system for achieving green economy in agriculture. Sustain Energy Technol Assess 52:102190

Ravindranath NH (2007) Forests in India-Take action now. The Hindu Survey of the Environment, The Hindu, Special Issue, New Delhi, India

Ray LK, Goel NK, Arora M (2019) Trend analysis and change point detection of temperature over parts of India. Theor Appl Climatol 138:153–167

Ray LIP, Swetha K, Singh AK, Singh NJ (2023) Water productivity of major pulses–a review. Agric Water Manag 281:108249

Rohini P, Rajeevan M, Srivastava AK (2016) On the variability and increasing trends of heat waves over India. Sci Rep 6(1):1–9

Sana M, Jana K, Mondal R, Mondal K, Banerjee H (2020) Integration of chemical fertilizer with biological products on growth and production of rice during wet season. ORYZA Int J Rice 57(3):211–218

Sarkar A, Jana K, Mondal R (2021) Growth and yield of hybrid mustard (Brassica juncea L.) as influenced by foliar nutrition in Gangetic plains of West Bengal. J Crop Weed 17(3):35–40

Sehgal A, Sita K, Kumar J, Kumar S, Singh S, Siddique KHM, Nayyar H (2017) Effects of drought, heat and their interaction on the growth, yield and photosynthetic function of lentil (Lens culinaris Medikus) genotypes varying in heat and drought sensitivity. Front Plant Sci 8:1776

Serraj R (2003) Atmospheric CO2 increase benefits symbiotic N2 fixation by legumes under drought. Curr Sci 85:1341–1343

Sharma S, Mujumdar P (2017) Increasing frequency and spatial extent of concurrent meteorological droughts and heatwaves in India. Sci Rep 7(1):1–9

Singh SP, Sandhu SK, Dhaliwal LK, Singh I (2012) Effect of planting geometry on microclimate, growth and yield of mung-bean (Vigna radiata L.). J Agric Phys 12(1):70–73

Singh C, Rio CRD, Soundarajan V, Nath N, Shivaranjani V (2019) Assessing India's mounting climate losses to Financial Institutions. Accessed on 28 Feb 2020

Stanley M (2003) Frost damage in crops and pastures. Primary Industries and Resources SA Fact Sheet

Statista 2020 (2022) Volume of pulses produced worldwide 2010–2020. Published by Shahbandeh

Stern NH (2007) The economics of climate change: the Stern review. Cambridge University Press, Cambridge

Team, Core Writing (2007) Contribution of working groups I, II and III to the fourth assessment report of the intergovernmental panel on climate change. In: IPCC 2007: climate change 2007: synthesis report 104, Geneva, Switzerland

Uga Y, Sugimoto K, Ogawa S, Rane J, Ishitani M, Hara N, Kitomi Y et al (2013) Control of root system architecture by DEEPER ROOTING 1 increases rice yield under drought conditions. Nat Genet 45(9):1097–1102

Wang J, Gan YT, Clarke F, McDonald CL (2006) Response of chickpea yield to high temperature stress during reproductive development. Crop Sci 46(5):2171–2178

World Bank (2008) World Bank's approach to climate change in South Asia: an overview. Bank Information Center, Washington DC 20433, USA

World Bank (2023) World Bank Annual Report 2023: A New Era in Development. Washington D.C World Bank Group

Yadav AS, Kumar S, Kumar N, Ram H (2019) Pulses production and productivity: status, potential and way forward for enhancing farmers income. Int J Curr Microbiol App Sci 8(4):2315–2322

Zhang X, He D, Niu G, Yan Z, Song J (2018) Effects of environment lighting on the growth, photosynthesis, and quality of hydroponic lettuce in a plant factory. Int J Agric Biol Eng 11(2):33–40

Chapter 10
Climate-Smart Technologies for Maize Cultivation

Sanju Choudhary, Sudhir Kumar Rajpoot, Devilal Birla, Manju Choudhary, Ambikesh Tripathi, Mahipal Singh Choudhary, and Ahmed Gaber

Abstract Maize is being severely challenged by climate change. Weather patterns have become unpredictable, and the rising temperatures are posing threats to crop yields and the livelihoods of farmers. This chapter provides practical, climate-smart solutions to help farmers adapt and thrive in such changing conditions. Water-saving techniques, such as rainwater harvesting and precision irrigation, and nutrient management practices ensure crops receive the right nutrients at the right time. Conservation tillage and other energy-saving techniques to enhance soil health are also discussed as well as methods for reducing farming costs. It further focuses on the role of stress-tolerant maize varieties and the role of crop insurance and weather-based advisories that can be helpful for farmers to prepare and recover from extreme weather conditions. This chapter provides an action plan, clear and precise, on how to develop resilient maize farming systems that would be able to withstand the forces of a changing climate by marrying local knowledge with modern innovations.

Keywords Maize production · Conservation tillage · Stress-tolerant varieties · Nutrient management · Crop insurance · Climate adaptation

---

S. Choudhary · S. K. Rajpoot · A. Tripathi
Department of Agronomy, Institute of Agricultural Sciences, BHU, Varanasi, Uttar Pradesh, India

D. Birla · M. Choudhary
Department of Agronomy, Anand Agriculture University, Anand, Gujarat, India

M. S. Choudhary
Department of Agronomy, Maharana Pratap University of Agriculture and Technology, Udaipur, Rajasthan, India

A. Gaber (✉)
Department of Biology, College of Science, Taif University, Taif, Saudi Arabia

B. Pramanick et al. (eds.), *Climate-Smart Agricultural Technologies*, Sustainability Sciences in Asia and Africa, https://doi.org/10.1007/978-981-96-7699-6_10

10.1 Introduction

The Earth surface temperature has increased by 0.74 °C over the past century (Solomon et al. 2007) and the temperature will rise by 1–1.5 °C by 2100 (Ju et al. 2013). By the end of this century, regions with proportionally warm and exceptionally dry weather may increase due to global warming (Kogan et al. 2013). People are unable to control climate change through farmland management because it is so unstable (Schmidt et al. 2022). Worldwide, climate change affects supply of water and threatens food production (Mitchell et al. 2016). Climate change will significantly impact agroecological food systems in the coming decades or centuries or even centuries (Schattman et al. 2018). The combination of climate change and population growth presents a major challenge to food security on limited arable land in the twenty-first century (Gong et al. 2015; Pramanick et al. 2022). Climate vulnerabilities, such as heat waves, droughts, or rainstorms, which occur during crucial growth stages, like anthesis and filling, can severely affect yield (Lesk et al. 2016). Elevated temperatures during the grain filling stage can hinder the rate at which grains are filled (Hunt et al. 1991), shorten grain filling time (Sofield et al. 1977), and finally cause a substantial grain yield reduction. There is a negative correlation between global warming and the productivity of certain crops like wheat, barley, and maize (Lobell and Field 2007).

In the face of climate change, achieving a balance between global food demand and supply necessitates the pursuit of three main goals: adapting agriculture to adverse weather conditions, promoting carbon sequestration, and enhancing crop yields (Fischer 2018). Maize (*Zea mays* L.) is one of the important source nutrition (Jimoh et al. 2021). Around one million smallholder farmers rely on maize cultivation for their livelihoods and local consumption (Harvey et al. 2018). The production of maize is negatively affected by extreme climate changes, since it is responsive to increasing temperatures and irregular rainfall patterns. The production, flowering, and yields of maize are negatively affected by high temperatures and drought conditions (Díaz-Ambrona et al. 2013). It has been suggested that enhancing resilience in the essential for food production systems, such as maize production, can have a greater effect on ensuring the resilience of the farming system (Liu et al. 2015). Maize is most commonly grown crop, which is susceptible to extreme weather conditions.

To support farmers in facing the challenges of climate change, it is crucial to adopt climate-smart agriculture (CSA), which enables them to manage and mitigate its impacts. The Food and Agriculture Organization (FAO 2022) emphasises that agriculture must be "climate smart" to ensure global food security and promote sustainable rural development. In the context of climate change, CSA serves as a strategy to establish favourable and sustainable conditions for agricultural development and food security in the long term (FAO 2022). It aims to achieve national food security and development goals through three key objectives: enhancing agricultural productivity and incomes, increasing adaptability and resilience to climate change, and mitigating or reducing greenhouse gas emissions (FAO 2022). CSA is a versatile strategy that plays a crucial role in enabling agricultural systems to sustain rural development and ensure food security amidst a changing climate. Its significance lies in its ability to help achieve the three policy objectives mentioned earlier: enhancing agricultural productivity and incomes, increasing adaptability and resilience to climate change, and mitigating or reducing greenhouse gas emissions (FAO 2018; Lipper et al. 2014). Multiple studies (Antwi-Agyei et al. 2021) have found that implementing various CSA practices can enhance food security by mitigating the impacts of climate change and improving nutrient availability. These practices focus on enhancing agricultural productivity in a changing climate by increasing adaptation and resilience. Additionally, CSA practices show promise in strengthening agriculture under climate change by promoting soil moisture retention, preserving soil biodiversity, and facilitating carbon storage (FAO 2013).

CSA is essential to ongoing efforts to promote the Sustainable Development Goals (SDGs) of the United Nations (UN), including eradicating hunger and combating climate change, especially in low- and medium-income countries (FAO 2018; IPCC 2018). The yield and yield components of crops are influenced by various factors, including genotype, climatic conditions, agronomic management, cropping systems, and interactions between plants (Sun et al. 2018). To sustain agricultural output, it is crucial to employ appropriate climate-smart crop management practices.

10.2 Climate-Smart Agronomy of Maize

In broad terms, CSA focuses on developing resilient agricultural production systems that ensure both food and financial securities in the face of ongoing climate change and its associated uncertainties (Lipper et al. 2014). Agriculture can adapt to climate hazards in many places by changing crop and animal management techniques, using new farm technology, and changing land use patterns. These approaches encompass various strategies such as utilising improved climate-resilient seeds, adopting changes in tillage practices (e.g. transitioning from conventional to minimum tillage), implementing crop nutrient and water management practices, and adjusting sowing time based on climate information (Dunnett et al. 2018). Adoption of drought-resistant cultivars is one strategy used to offset the impacts of climate change on yield of maize (Shiferaw et al. 2011). Practices like mulching have been tested effective in improving soil moisture storage, water use efficiency, and maize production (Ngangom et al. 2020; Pramanick et al. 2022). The use of halfmoon pits led to a remarkable increase in maize production, boosting yields by 63–74% as a result of enhanced soil moisture storage (Sawadogo 2011). Additionally, the adoption of permanent planting basins enhanced soil water availability, conservation for crop use, and rainfed maize yield (Nyamangara et al. 2014), consequently increasing the resilience of cropping systems to progressive climate change consequences.

10.2.1 Weather-Smart Practices

Weather-Based Agroadvisory

Agroadvisory services may assist farmers in choosing the right technology, marketing approach, and inseason crop management decisions (Gangopadhyay et al. 2019). Customised agroadvisories, which are often provided by service providers, frequently give extremely specific evidence for crop management to reduce the risks related to agriculture caused by climate change on an individual basis.

Information and Communication Technologies (ICT)

The utilisation of information and communication technologies (ICTs), including radio, television, Internet services, and cell phones, is an emerging field that offers opportunities for delivering climate-smart agriculture (CSA) technologies, services, and practices to rural farmers. Additionally, according to some studies, ICT-driven climatic information and agroadvisory services can provide valuable assistance to farmers in choosing the right technology, marketing approach, and crop management practices during the growing season (Mittal et al. 2010). However, the distribution of CSA information to the rural agricultural communities depends on cost-effective and dependable ICT services to shape future initiatives in managing climate risks in agriculture; it is essential to identify the areas where climate information services are needed and explore suitable ICT-based tools for disseminating climate information-based agroadvisories.

Contingent Crop Planning

It reduces the detrimental effects of climate hazards through crop adjustment. The production of food grains, the supply of raw materials to agroindustries, the availability of fodder for livestock, the livelihoods of rural labour, and the amount of food grain supplies available to the central pool for public supply are all negatively impacted by any variation from the regular rainfall pattern. It causes price increases and inflation, which ultimately have an impact on the share of agriculture in GDP. The governments have implemented contingency measures, notably for droughts, to meet delays in monsoon or mid-season interruptions (Prasad et al. 2012; Pramanick et al. 2023). A plan for managing climate risk is needed to deal with significant weather-related risks including drought, flooding, and heat/cold stresses during the growing season (Khatri-Chhetri et al. 2017).

10.2.2 Crop Establishment Practices

Conservation Tillage

Conservation agriculture is supported widely to boost the environmental sustainability and productivity of maize production systems and is generally recorded as a CSA application. Though the effects of conservation agriculture (CA) on food security, adaptive extent, and mitigating climate change vary depending on site, it is unclear if CA can concurrently fulfil the many CSA objectives (Kimaro et al. 2016).

Conservation tillage, like no-tillage or minimum tillage integrated with mulching along with crop rotation, can minimise evaporation, soil erosion, runoff, and soil degradation (Bossio et al. 2010). According to a study, straw mulching and minimum tillage together can effectively lower soil temperature while improving soil moisture and maize production (Bhatt and Khera 2006). Furthermore, the use of minimal or no tillage in conjunction with straw return may effectively improve soil macrocohesiveness and prevent the improper use of conventional tillage techniques to boost soil fertility and crop yield (Pittelkow et al. 2015; Kar et al. 2021). Straw mulching acts by soaking the kinetic energy of raindrops to minimise compaction of the soil and erosion induced by rainfall (Nyssen et al. 2008). In well-drained soils, conservation tillage may improve maize production, but minimum tillage does not appear to have a direct effect on yield stability (Rusinamhodzi et al. 2011; Pramanick et al. 2024). The temperature and rainfall have notable effects on crop production, and long-term appropriate tillage under warm-dry or cool-wet regions can improve the yield and yield components of maize (Liu et al. 2022). There are 80% of instances in Southern Africa where CA-produced maize yields are greater than non-CA-produced maize yields, with yield benefits ranging from 482 to 1173 kg ha^{-1} depending on the CA system (Thierfelder et al. 2015).

Change in Planting Time

Planting date has significant effects on the elongation, development, and production of maize. The most suitable planting time has become vital for greater crop yield. Planting dates directly affect growth, pest and disease incidence, and plant establishment. Different crop varieties respond to planting times in different ways. A variety of morphoanatomical, biochemical, and physiological variations in plants are caused by early or late planting dates to maize, and these changes can have a significant adverse impact on elongation and yield of crop. Maize growth and development consist of a variety of biochemical processes that are susceptible to variations in weather conditions as influenced by planting times. Due to a decrease in the cumulative interception of photosynthetically active radiation (PAR), later planting times impact aspects such as anthesis silking period, dry matter production photosynthesis, and physiological maturity. Late in planting dates causes greater non-structural carbohydrate concentration in stem at mid-grain filling stage because of the low temperature exposure of crop limiting growth of kernel and photosynthesis. By predicting ideal planting dates using crop modelling studies, the negative impact of late planting times can be reduced (Shrestha et al. 2018; Dey et al. 2022).

Warming temperatures not only hasten plant development but also reduce the length of growth period required for optimum plant and seed size. The time of the peak water consumption of plants is also affected by quicker phenological development. The temperature rises during maize pollination in hot growing environments pose a threat to turning midseason crop failure into the norm. Declines in maize maturity time indicate that warming itself is limiting yield potential more and more, irrespective of influences on water availability. Simple adjustments to planting dates and maize variety types might help farmers to avoid some losses (Harrison et al. 2011).

Climate-Resilient Varieties

Production of maize is severely constrained by drought, particularly in rainfed agriculture. A DT variety of maize may define as a variety that may yield around 30% of its possible production (1–3 t ha^{-1}) after facing water stress for 45 days before and at the time of flowering and grain-filling (Magorokosho et al. 2008). In controlled experimental studies, DT maize cultivars beat the majority of current commercial hybrids in terms of yield and drought resistance (CIMMYT 2013). As a result, DT maize provides some protection against mid-season droughts and dry period and has the ability to produce an adequate maize crop even under moderate drought situations (Setimela et al. 2014). Aside from that, ex ante economic assessments indicate that, if mostly used by smallholder farmers, DT maize can offer significant financial advantages through improved yield and decreased threat. This has significant effects on reducing food insecurity as well as saving a significant amount of resources for the household. DT maize varieties have been introduced in response to this threat with the goal of ensuring maize production in mild drought situations at the national level, which is used to supplement food purchases during the lean season (Lunduka et al. 2019).

Intercropping

Intercropping is the agricultural activity of growing of two or more than two crops in the same place at the same time (Andrews and Kassam 1976). Better resource management, weed control, pest and disease control, soil loss control, yield advantage, and economic benefit are all advantages of intercropping (Seran and Brintha 2010). Compared to maize monoculture, systems that include maize and legumes can use less nitrogen from the soil. Intercropped legumes will fix nitrogen from the environment in the absence of nitrogen fertiliser and will not contend with maize for nitrogen sources (Adu-Gyamfi et al. 2007). The maize-French bean combination produced greater maize equivalent yields than sole maize crop (Hugar and Palled 2008), and kernel yield of maize was not affected in maize-French bean intercropping. Maize vegetable intercropping is also beneficial for changing climatic condition (Adeniyan et al. 2007).

Use of Pneumatic Planter

Traditional sowing of maize is labour intensive, expensive, and time consuming. The production cost is increased when a conventional drill is used for planting because higher seed rate and thinning operations are needed. Commercially accessible maize planters are available for sowing operations, but their acceptance rate is lower because the majority of farmers sow it manually. The efficiency of seed viability improved, and the seed rate decreased with the use of planters. Additionally, thinning operations are not required with precision planters such as inclined plate and pneumatic planters (Dixit et al. 2011). A precise and exact rate of seed placement is provided by pneumatic planters. In comparison to manual ridge planting, the overall operating cost of planting with pneumatic planters was significantly lower (70%) (Singh et al. 2018) (Fig. 10.1).

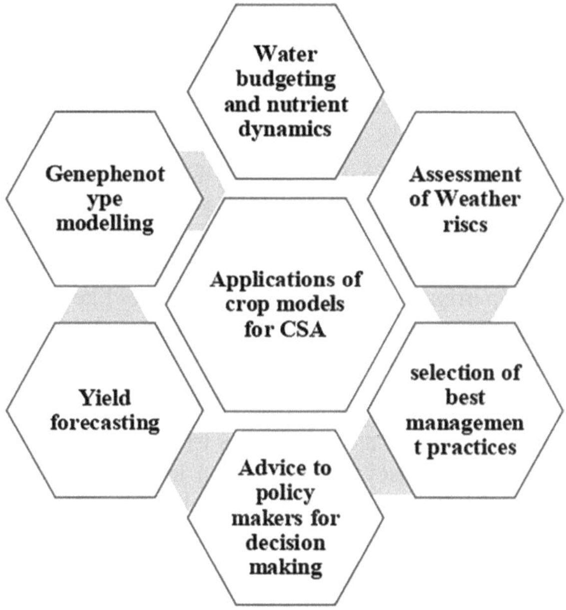

Fig. 10.1 Applications of crop models for CSA

Crop Rotation

Crop rotational diversity is understood to affect soil microorganisms and crop productivity; so far, the specific contributions of microorganisms to rotational benefits are not known (Benitez et al. 2021). In all conditions, a monoculture produced less maize than a crop rotation. In monocultures, reductions in yields of maize were mostly noted following a dry winter, especially if the summer had been similarly dry. In a monoculture, pathogenic agents that are promoted by the weather might be primarily responsible for the decreased yield of maize (Berzsenyi et al. 2000).

10.2.3 Water-Smart Practices

To adopt the climate change and unpredictability, water-smart practices incorporate both conventional and cutting-edge practices, technology, and services (CIAT 2014). With careful design and execution, location-specific water-smart practices, either alone or in mixture, offer a significant potential way of reducing the things of climate alteration on water supplies. Accordant to a meta-analysis of agricultural simulations below various weather situations, farm level adjustments might boost yields by 7–15% and save 25–50% more water than they would have otherwise (Jain et al. 2014).

Drip and Sprinkler Irrigation

Trickle irrigation, an ingenious approach, involves the precise distribution of water straight to the root region of the soil, showcasing its exceptional sophistication. Sprinkler irrigation, centre pivot and low energy precision application (LEPA), microjet irrigation, drip/trickle irrigation, and surface or subsurface watering are only a few examples of pressure irrigation techniques. In particulars of output, water consumption efficiency, and water savings, drip irrigation systems outperform alternative surface watering techniques. Using this technique of irrigation results in higher crop yields and lower costs for fertilisers, insecticides, and irrigation electricity. Drip irrigation reduces traditional losses like deep percolation, runoff, and evaporation as a result (FICCI 2016). Singh et al. (2022) reported that drip irrigation increases yield by 20.06% in maize as compared to surface irrigation.

Laser Land Levelling

Embracing laser land levelling, a tried-and-tested agronomic on-farm innovation, non-only decreases the reliance on irrigation nonetheless also revolutionises the way we optimise water usage. This cutting-edge technology saves valuable time by streamlining irrigation processes while significantly enhancing water efficiency. Furthermore, it minimises the ingesting of seeds, fertilisers, substances, and firewood through the creation of a uniform and flawless soil surface.

In the realm of agricultural advancement, laser land levelling has firmly established itself as a pioneering technique. It not only addresses the pressing need to reduce farm irrigation but also revolutionises irrigation practices by significantly reducing time and enhancing water utilisation efficiency. By implementing this innovative approach, farmers can optimise their water resources while ensuring a sustainable and efficient use of irrigation, thus fostering a more productive and environmentally conscious agricultural landscape, ingesting of seeds, fertilisers, substances, and firewood decreased by a more even and smooth soil surface. According to Naresh et al. (2011), laser levelling may allow for a 20–25% water savings in irrigation. Precision levelling, according to Aggarwal et al. (2009), greatly enhanced the soil and canopy microenvironment and supported irrigation water conservation (27–33%) in maize fields.

Raised Bed Planting

For the cultivation of vegetables, medicinal, aromatic, and cereal crops, raised beds with alternating furrows of 1–5 m in width are frequently laid out. Typically, a bed or ridge will have two rows of plants planted on either side (Govaerts et al. 2007). Unveiling an ingenious approach, raised bed planting stands out as the optimal technique for corn cultivation throughout both the rainy season and winter time, be it in environments characterised by excessive moisture or limited water availability in rainfed conditions. This innovative technique empowers farmers to harness the full potential of their maize crops, enabling successful cultivation while enhancing water efficiency and promoting robust germination practices. The method employed guarantees substantial conservation of water by allowing the water in the furrow to flow horizontally in soil beneath the bed or edge, meeting the crop water requirements. This approach effectively prevents salt build-up at the plant's base and minimises salt-related damage to crops, particularly in regions where raised bed and furrow salt accumulation pose challenges. To ensure optimal germination, sowing or planting on the southern side of the east-west ridges/beds is highly recommended. Moreover, this technique provides enhanced control over irrigation and drainage, along with efficient management of rainwater during the monsoon season. By adopting the cutting-edge raised bed planting technology, farmers can revolutionise their irrigation practices and achieve remarkable results. This innovative approach enables a remarkable water saving of 20–30% while boosting productivity to unprecedented levels. The implementation of permanent beds proves particularly advantageous in areas with excessive soil moisture, as the higher infiltration rate effectively safeguards crops against temporary waterlogging injuries. With this forward-thinking method, farmers can optimise water usage, mitigate risks, and secure higher yields, all while embracing sustainable and efficient agronomic practices (Hassan et al. 2005).

IOT Sensor

The corn plant exhibits heightened sensitivity to variations in water levels, whether they are too low or too high. Notably, its water demands escalate during the growing season due to the substantial production of organic matter. Furthermore, corn plants display a remarkable sensitivity to shade, with a reduction in light intensity by 30–40% resulting in a delayed ripening period of approximately 5–6 days (Cicioğlu and Çalhan 2021). Wind play a pivotal act in the development of corn as it facilitates the crucial process of wind pollination. However, strong winds can pose a threat to the crop, leading to lodging. Successful corn cultivation relies on various ecological factors that necessitate careful consideration in the context of precision agriculture applications. The goal of smart agriculture is to streamline farmers' workload and enhance farm productivity through the integration of cutting-edge technologies like Internet of Things (IoT), sensor and actuator networks, drones, and more. AI-enabled sensors function as intelligent tools, while the emergence of IoT has revolutionised agriculture by providing diverse sensor-based apparatus and devices that prove highly advantageous in the field (Ullo and Sinha 2021). The IoT concept combines different hardware and software peripherals to achieve interoperability. This interconnected system enables effective communication and data exchange, fostering innovative solutions across various domains (Thakur et al. 2020). IoT technologies hold boundless potential for creating new solutions and fostering smarter application development, revolutionising every facet of the agricultural industry (Glaroudis et al. 2020). At present, technological advancements are revolutionising conventional farming practices. Innovations such as drones, satellite imagery, IoT, and sensor/actuator technologies are making a profound impact on the agricultural industry, bringing about significant changes (Boursianis et al. 2022). Smart farming is a crop management thought that allows farmers to efficiently address spatial and temporal erraticism. It involves decreasing resource usage; managing irrigation, production, and fertiliser; safeguarding against intrusion attacks; and real-time data monitoring. This approach harnesses advanced technologies for optimised agricultural practices (Thakur et al. 2019). These capabilities can be facilitated through the deployment of specific sensors strategically positioned within the fields (Cui et al. 2018).

Subsurface Irrigation

Subsurface irrigation, also known as subirrigation, is an irrigation technique that involves applying water directly beneath the soil surface to nourish plant roots. This method enhances water use efficiency on a field scale. Unlike surface irrigation systems, where large areas are wetted to irrigate less cropped area, subsurface irrigation minimises loss of water through evapotranspiration, a mutual issue in dry and semidry environments. By decreasing the saturated zone and deep percolation, the adoption of subsurface irrigation techniques conserves water resources effectively (Megersa and Abdulahi 2015). Therefore, the implementation of subsurface irrigation offers the potential to decrease expenses associated with energy, water, and agronomic chemicals (Sinobas et al. 2012). Pitcher irrigation, an age-old technique, has been employed in diverse variations throughout Iran, India, and African countries (Siyal et al. 2009). Utilisation of clay pots for irrigation demonstrated a remarkable water saving of 70% when compared to the usage of buckets and spray irrigation in maize farming (Daka 1991). The technique of pitcher irrigation is characterised by its simplicity, affordability, substantial water-saving potential, and inherent autoregulative capabilities (Vasudevan et al. 2011).

10.2.4 Nutrient-Smart Practices

New-Generation Fertiliser

The use of coated urea, a controlled-release fertiliser, ensures that 9.3% rise in grain produce of corn compared to conventional urea. The coated urea also enhanced efficiency of nitrogen by 17.8% and reduced soil nitrate accumulation, suggesting potential benefits for both productivity and environmental sustainability (Chen and Wei 2018). The use of polymer-coated urea, combined with cover crops, increased maize corn yield by 20% likened to conventional urea. Improved yield was attributed to better synchronisation of nitrogen release with crop demand and reduced nitrogen losses from the soil. Next-generation fertilisers can enhance maize grain yield under various conditions, likely due to their ability to improve nutrient use efficiency and match nutrient availability with crop requirements. However, the magnitude of the yield response to next-generation fertilisers can be prejudiced by several aspects, for example, soil properties, environment, and practices of management (Sachan et al. 2023).

Customised Fertiliser

Customised fertilisers are several-nutrient carrier molecules developed to hold major and micronutrients. In another words, customised fertiliser is a several-nutrient transporter that specially made to fulfil the particular basic nutrient requirements of a plant. Farmer can be offered customised fertilisers based on their agroclimatic conditions to remedy nutrient deficiencies, particularly secondary and micronutrient deficiencies. Customised fertilisers might help in this situation by promoting a balanced nutrient supply to obtain the greatest possible usage efficacy of the supplied nutrients in an efficient way.

The grain yield of maize enhanced over grades by progressively increasing of customised fertiliser. Maximum corn yield in alkaline and neutral soil (7.69 and 9.11 kg ha^{-1}, respectively) was found by the application of 450 kg ha^{-1} of customised fertiliser which supplied 150 N: 99 P_2O_5: 77 K_2O kg ha^{-1}. A considerable increase in corn yield over RDF by 7% and 38% in alkaline and neutral soil, respectively, by the application of 450 kg ha^{-1} of customised fertiliser comparative to the complete controlled and RDF (Nagabovanalli et al. 2020).

Residue Recycling

The organic matter or plant residue promote to reuse the nutrients to precise their deficiencies. Using organic bases can keep a better N:P ratio and surge production (Singh et al. 2023). Farmer's concern about long-term, continuous use of mineral fertilisers, especially in decreasing soil productivity and causing environmental pollution, leads them to seek out organic manures for recycling (Kiani et al. 2005). Incorporating organic matter or plant residue into soil recovers soil productiveness and reduces fertiliser nitrogen requirements, resultant in higher crop yield.

In no-tillage systems, the usage of cover crops is a crucial agricultural practise. The dynamics of plant residue breakdown has an impact on soil coverage and nutrient recycling. In the savannah, Central Brazil, research was done to determine how the synthetic fertiliser and decay rates of cover plants affected the production of maize in no-tillage settings. The following plant species' hemicellulose, cellulose, and lignin concentrations and the breakdown rates were measured during blooming and at harvest (De Carvalho et al. 2013).

In two successive cropping seasons, 5 t ha^{-1} trashes of sugarcane and mustard straw were added to an alfisol soil (acid clay loam) along with—0, 60, and 120 kg of fertiliser nitrogen. To test the residual soil fertility brought about by the addition of organic matter, wheat was grown as the crop that would benefit right away and maize as the one that would follow. Wheat output was greatly raised when sugarcane waste was added and the crop's ability to recover and absorb nitrogen fertiliser more effectively. Only the integration of sugarcane trash, as opposed to the other two crop wastes, was shown to have a substantial lasting impact on the subsequent corn plant, as per seen through the rise in corn production and regaining of applied nitrogen, despite the typical impact of agricultural leftovers (Dev and Bhardwaj 1991).

Biofertiliser

Climate-smart agriculture was developed after increasing agricultural resilience to change climate in the current environment of rising temperatures, shifting rainfall patterns, and a rise in the incidence of thrilling climatic actions with cyclones, droughts, and floods. Numerous ecological procedures in agroecosystems, for example, the breakdown of biological matter, cycling of nutrient, nitrogen fixation, solubilisation of phosphate, nutrient gaining, and the newly found involvement of probiotic bacterium, depend on the presence of microorganisms (Mohanty and Swain 2018).

In sustainable agriculture, biofertiliser is a natural input that may be used in place of or in addition to inorganic fertiliser (Ebrahimpour et al. 2011). Using biofertilisers in combination offers a less expensive, less capital-intensive, and environmentally friendly way to increase agricultural output (Thavaprakaash et al. 2005). About 33% of the trials using *Azotobacter* inoculation on crops were carried out in India and showed inoculation responses. The use of biofertiliser and inorganic fertilisers together considerably boosted maize production. Utilising organic-manure and biofertiliser together with nitrogen application at each level significantly enhanced the protein content of maize grain compared to applying each nitrogen level alone (Meena et al. 2013).

The application of microbial inoculants has been achieved the greatest biomass and plantlet height. According to this protected cultivation study, half the microbial inoculants application had the same effect as organic fertiliser or inorganic fertiliser. Microbial inoculants enhanced both soil qualities, such as the amount of carbon-based matter and total nitrogen in the soil, along with plant nutritional absorption (total nitrogen, phosphorus, and potash). The arbuscular mycorrhizal fungi (AMF) had a greater root contagion amount in the existence of bacterial inoculum. In contrast, the P-solubilising bacteria appeared to be inhibited by the AMF. The lack of nutrients in the soil led to an increase in the inhabitants of nitrogen-fixing microorganisms and the annexation of AMF (Wu et al. 2005).

10.2.5 Weed Management Practices

Weeds inflict significant economic losses, reaching billions of US dollars, as reported by WSSA in 2019. These losses arise from escalated production expenses, diminished quality and quantity of crops, compromised aesthetics of landscapes, adverse health effects on humans and animals, and unwanted consequences like acting as fuel for forest fires. Throughout history, various methods including mechanical, biological, and cultural approaches have been employed to control weeds. Chemical control of weed by using inorganic molecule was widely employed from the late-nineteenth to the mid-twentieth century, and evidence suggests its earliest use dates back to the Roman era (Timmons 2005). Herbicides quickly became the primary method for weed management, surpassing alternative approaches due to their exceptional effectiveness, cost efficiency, selectivity, and targeted control. They offered benefits like improved productivity, improved crop quality, decreased labour in hand weeding, and diminished topsoil loss through limited cultivation and tillage, facilitated by lower fossil fuel consumption. However, excessive reliance on herbicides led to the evolution of herbicide-resistant weed species. Hence, timely and diverse weed management practices are crucial for achieving higher yields. Atrazine, known for its selectivity and broad-spectrum efficacy, effectively controls the emergence of a large range of weeds in maize crops or Alachlor (Lasso) or Metolachlor (Dual) proves to be an effective method for controlling various annual and BLWs. In cases of heavy weed infestation, protected spray using hoods with postemergence use of Paraquat can also be utilised.

Herbicide-Tolerant Maize

Herbicide-tolerant (HT) crops provide farmers with a valuable tool for combating weeds while being compatible with no-till farming practices, which aid in preserving topsoil. These crops offer farmers the flexibility to apply herbicides selectively when necessary, allowing for better control over the overall herbicide input and the ability to choose herbicides with desirable environmental characteristics. This combination of herbicide tolerance and no-till methods enhances weed management strategies while promoting more sustainable agricultural practices (Tables 10.1 and 10.2).

Smothering Crops

A smother crop refers to a dense and fast-growing crop used to effectively suppress or halt the growth of weeds. These smother crops possess robust root systems that outcompete weeds for water, weakening the weed's root systems in the process. This competition for resources helps to suppress weed growth and enhance weed

Table 10.1 Recent transgenic herbicide-resistant maize and associated trait genes

Selection method	Herbicide family	Year of disclosure
Tissue culture	Imidazolinone	1991
Pollen mutagenesis	Imidazolinone	1992
Tissue culture	ACCase inhibitor	1992

Adapted from Duke (2005), Green and Castle (2010)
ACCase acetyl-CoA carboxylase

Table 10.2 Recent trait genes of transgenic herbicide-resistant maize

Resistance trait	Trait gene	Trait designation	First sales	Trade name
Glufosinate	*Pat*	T14 and T25	1996	LibertyLink System®
Glyphosate	Three modified maize *epsps*	GA21	1998	Roundup Ready®
Glyphosate	Two *cp4 epsps*	NK603	2001	Roundup Ready® 2
2,4-D	*tfdA*	DHT1	2019	Enlist™ Weed Control System
AOPP	*Aad*	DHT1	2019	Enlist™ Weed Control System

Partly adapted from Green and Castle (2010) and Green (2009)
AOPP aryloxyphenoxypropionate, *2,4-D* 2,4-dichorophenoxyacetic acid

control in agricultural fields (Kumar and Shivay 2008). Smother crops effectively compete with weeds for vital resources like space, water, light, and nutrients, thereby inhibiting weed germination and growth (Wedryk and Cardina 2012). Smother crops, once ploughed under with weakened weeds, act as green fertiliser. They effectively compete with weeds while minimising competition with the main crop. This reduces reliance on chemical weed control. Spring-planted smother plants have potential for weed management through interplant competition and surface residues. These specialised cover crops suppress weeds without herbicides.

Various annual medic species, berseem clover, and yellow mustard demonstrated diverse effectiveness in weed control when planted in a strip alongside corn. Smother crop establishment and weed control were notably improved, ranging from 20% to 90%. In addition to their potential as an alternative weed management approach, these crops contribute to soil coverage, organic matter enrichment, decrease soil erosion, increase soil moisture-holding capacity, enhanced soil quality, and increased yield by 20–30% (Buhler et al. 1998) (Table 10.3).

10.3 Conclusion

CSA approaches are related to the local climate, socioeconomic factors, and desire to pay for accessible technology. A farmer's priorities for CSA technology may change depending on the climate and threats that are thought to exist. Farmers, for instance, choose risk mitigation methods like crop assurance, climate-based agro-advisories, and precipitation collection in places with low annual precipitation and large CV in annual precipitation. In areas with heavy rainfall, their choice for CSA technology substantially varies. Therefore, policies and initiatives designed to advance CSA technology should concentrate on particular place that are useful for the local growers. The administration and other organisations may influence farmers' liking for CSA technologies. Likewise, their choices may vary depending on the annexation of CSA technology and their possible benefit for adaptations to climate change.

References

Adeniyan ON, Akande SR, Balogun MO, Saka JO (2007) Evaluation of crop yield of African yam bean, maize and kenaf under intercropping systems. Am Eurasian J Agric Environ Sci 2(1):99–102

Table 10.3 Selected climate-smart options to analyse farmers' preferences

Climate-smart options	Adaptation/mitigation potential
1. *Water-smart*	Improve water use efficiency
Drip irrigation	Water is applied directly to the root zone of plants to reduce water loss.
Drainage management	Removal excess water using control structure
Laser land levelling	By levelling the field, evenly water distribution, less water loss and improves nutrient use efficiency
Rainwater harvesting	Runoff is used on-site for agricultural usage in rainfed or dry areas.
Furrow irrigated bed planting	Effective irrigation and drainage management, rainwater management, and improves nutrient use efficiency
Cover crops	Reduces evaporation loss and adds nutrients
2. *Nutrient-smart*	Approaches that improve nutrient use efficiency
Intercropping with legumes	Legumes and cereal crops are grown in alternate rows or mixed. This practise increases soil quality and nitrogen availability.
Leaf colour chart	Provide best time of application of nitrogen based on greenness of crops. Mostly used in rice but also suitable for maize and wheat crops
Green manuring	Legumes are grown for manuring; increases soil quality and nitrogen availability
Site-specific/integrated nutrient management	The best possible supply source of nutrients at right time and space, according to the needs of the crops, at the proper product, rate, and location
3. *Energy-smart*	To improve energy use efficiency
Conservation tillage (zero/minimum tillage)	Reduces the energy required and enhances the soil organic matter and water retention
4. *Carbon-smart*	Approaches that reduce GHG emissions
Integrated Pest Management	Reduces dependency on agrochemicals
Agro-Forestry	Promoting sustainable land use management and carbon sequestration
5. *Knowledge-smart*	Combined use of local knowledge and science
Contingent Crop Planning	Plan for managing climate risk to prepare for severe weather-related events such as drought, flooding, and heat/cold stresses during the growing season
Improved Crop Varieties	Tolerant varieties to drought, flood, and heat/cold stresses reduce climate risk.
6. *Weather-smart*	Approaches that provide services related to income security and weather advisories to farmers
Crop Insurance	To cover income losses caused by weather events
Weather based Crop Agro-advisory	Value-added agroadvisories for farmers based on climate information

Source: Khatri-Chhetri et al. (2017)

Adu-Gyamfi JJ, Myaka FA, Sakala WD, Odgaard R, Vesterager JM, Høgh-Jensen H (2007) Biological nitrogen fixation and nitrogen and phosphorus budgets in farmer-managed intercrops of maize–pigeonpea in semi-arid southern and eastern Africa. Plant Soil 295:127–136

Aggarwal PK, Singh AK, Samra JS, Singh G, Gogoi AK, Rao GGSN, Ramakrishna YS (2009) Global climate change and Indian agriculture: case studies from ICAR network project. ICAR, New Delhi

Andrews DJ, Kassam AH (1976) The importance of multiple cropping in increasing world food supplies. Mult Cropp 27:1–10

Antwi-Agyei P, Abalo EM, Dougill AJ, Baffour-Ata F (2021) Motivations, enablers and barriers to the adoption of climate-smart agricultural practices by smallholder farmers: evidence from the transitional and savannah agroecological zones of Ghana. Reg Sustain 2(4):375–386

Benitez M-S, Ewing PM, Osborne SL, Michael Lehman R (2021) Rhizosphere microbial communities explain positive effects of diverse crop rotations on maize and soybean performance. Soil Biol Biochem 159:108309

Berzsenyi Z, Győrffy B, Lap DQ (2000) Effect of crop rotation and fertilisation on maize and wheat yields and yield stability in a long-term experiment. Eur J Agron 13(2–3):225–244

Bhatt R, Khera KL (2006) Effect of tillage and mode of straw mulch application on soil erosion in the submontaneous tract of Punjab, India. Soil Tillage Res 88(1–2):107–115

Bossio D, Geheb K, Critchley W (2010) Managing water by managing land: addressing land degradation to improve water productivity and rural livelihoods. Agric Water Manag 97(4):536–542

Boursianis AD, Papadopoulou MS, Diamantoulakis P, Liopa-Tsakalidi A, Barouchas P, Salahas G, Karagiannidis G, Wan S, Goudos SK (2022) Internet of things (IoT) and agricultural unmanned aerial vehicles (UAVs) in smart farming: a comprehensive review. Internet Things 18:100187

Buhler DD, Kohler KA, Foster MS (1998) Spring-seeded smother plants for weed control in corn and soybean. J Soil Water Conserv 53(3):272–275

CIAT (2014) Climate-smart agriculture investment prioritization framework. International Center for Tropical Agriculture (CIAT), Cali

Cicioğlu M, Çalhan A (2021) Smart agriculture with internet of things in cornfields. Comput Electr Eng 90:106982

CIMMYT (2013) The drought tolerant maize for Africa project. DTMA Brief. Retrieved from http://dtma.cimmyt.org/index.php/about/background

Chen J, Wei X (2018) Controlled-Release Fertilizers as a Means to Reduce Nitrogen Leaching and Runoff in Container-Grown Plant Production. InTech. doi:https://doi.org/10.5772/intechopen.73055

Cui S, Ling P, Zhu H, Keener HM (2018) Plant pest detection using an artificial nose system: a review. Sensors 18(2):378

Daka AE (1991) Conservative irrigation using ceramic pitchers as ancillary media for water conservation. Greece Belkema Press, p 5

De Carvalho AM, Coelho MC, de Dantas RA, Fonseca OP, Guimarães Júnior R, Figueiredo CC (2013) Chemical composition of cover plants and its effect on maize yield in no-tillage systems in the Brazilian savanna. Crop Pasture Sci 63(12):1075–1081

Dev SP, Bhardwaj KKR (1991) Recycling of crop wastes for improving crop yields and nitrogen-use efficiency in a wheat-maize system. Bioresour Technol 37(2):135–139

Dey P, Mahapatra BS, Pramanick B, Pyne S, Pandit P (2022) Optimization of seed rate and nutrient management levels can reduce lodging damage and improve yield, quality and energetics of sub-tropical flax. Biomass Bioenergy 157:106355. https://doi.org/10.1016/j.biombioe.2022.106355

Díaz-Ambrona CGH, Gigena R, Carlos H, Mendoza O (2013) Climate change impacts on maize and dry bean yields of smallholder farmers in Honduras. Iberoam J Dev Stud 2(1):4–22

Dixit A, Mahal JS, Manes GS, Khurana R, Nare B (2011) Comparative performance of tractor-operated inclined plate and pneumatic planters. Agric Eng Today 35(1):33–37

Duke SO (2005) Taking stock of herbicide-resistant crops ten years after introduction. Pest Manag Sci 61(3):211–218

Dunnett A, Shirsath PB, Aggarwal PK, Thornton P, Joshi PK, Pal BD, Khatri-Chhetri A, Ghosh J (2018) Multi-objective land use allocation modelling for prioritizing climate-smart agricultural interventions. Ecol Model 381:23–35

Ebrahimpour F, Eidizadeh KH, Damghani AM (2011) Sustainable nutrient management in maize with integrated application of biological and chemical fertilizers. Int J Agrisci 1(7):423–426

FAO (2013) Climate smart agriculture sourcebook. FAO, Rome. Available at: www.fao.org/docrep/018/i3325e/i3325e.pdf

FAO, FAOSTAT (2018) Food and agriculture organization of the United Nations. FAO, Rome. Available at: http://faostat.fao.org

FICCI (2016) Knowledge paper series 3/2016. FICCI, New Delhi, India

Fischer G (2018) Transforming the global food system. Nature 562:501–502

Food and Agriculture Organization (FAO) (2022) Climate smart agriculture: managing ecosystem for sustainable livelihoods. Undated. Available online: www.fao.org/climatechange/climatesmart

Gangopadhyay PK, Khatri-Chhetri A, Shirsath PB, Aggarwal PK (2019) Spatial targeting of ICT-based weather and agro-advisory services for climate risk management in agriculture. Clim Chang 154(1–2):241–256

Glaroudis D, Iossifides A, Chatzimisios P (2020) Survey, comparison and research challenges of IoT application protocols for smart farming. Comput Netw 168:107037

Gong F, Xiaolin W, Zhang H, Chen Y, Wang W (2015) Making better maize plants for sustainable grain production in a changing climate. Front Plant Sci 6:835

Govaerts B, Sayre KD, Lichter K, Dendooven L, Deckers J (2007) Influence of permanent raised bed planting and residue management on physical and chemical soil quality in rain fed maize/wheat systems. Plant Soil 291:39–54

Green JM (2009) Evolution of glyphosate-resistant crop technology. Weed Sci 57(1):108–117

Green JM, Castle LA (2010) Transitioning from single to multiple herbicide-resistant crops. In: Glyphosate resistance in crops and weeds: history, development, and management. Wiley, Hoboken, pp 67–91

Harrison L, Michaelsen J, Funk C, Husak G (2011) Effects of temperature changes on maize production in Mozambique. Clim Res 46(3):211–222

Harvey CA, Milagro Saborio-Rodríguez M, Martinez-Rodríguez R, Viguera B, Chain-Guadarrama A, Vignola R, Alpizar F (2018) Climate change impacts and adaptation among smallholder farmers in Central America. Agric Food Sec 7(1):1–20

Hassan I, Hussain Z, Akbar G (2005) Effect of permanent raised beds on water productivity for irrigated maize–wheat cropping system. In: Evaluation and performance of permanent raised bed cropping systems in Asia, Australia and Mexico. ACIAR Proceedings No. 121. Australian Centre for International Agricultural Research, Canberra

Hugar HY, Palled YB (2008) Effect of intercropped vegetables on maize and associated weeds in maize-vegetable intercropping systems. Karnataka J Agric Sci 21(2):159–161

Hunt LA, van der Poorten G, Pararajasingham S (1991) Postanthesis temperature effects on duration and rate of grain filling in some winter and spring wheats. Can J Plant Sci 71(3):609–617

IPCC (Intergovernmental Panel on Climate Change) (2018) Global warming of 1.5 °C: an IPCC special report on the impacts of global warming of 1.5 °C above pre-industrial levels and related global greenhouse gas emission pathways, in the context of strengthening the global response to the threat of climate change, sustainable development, and efforts to eradicate poverty. Cambridge University Press, Cambridge

Jain N, Dubey R, Dubey DS, Singh J, Khanna M, Pathak H, Bhatia A (2014) Mitigation of greenhouse gas emission with system of rice intensification in the Indo-Gangetic Plains. Paddy Water Environ 12:355–363

Jimoh LO, Akintunde OK, Kehinde AL, Agboola TO, Alabi AA (2021) Structure, conduct and performance of maize marketing in Irewole Local Government area, Osun State. Niger J Basic Appl Sci 29(2):59–66

Ju H, van der Velde M, Lin E, Xiong W, Li Y (2013) The impacts of climate change on agricultural production systems in China. Clim Chang 120:313–324

Kar S, Pramanick B, Brahmachari K, Saha G, Mahapatra BS, Saha A, Kumar A (2021) Exploring the best tillage option in rice based diversified cropping systems in alluvial soil of eastern India. Soil Tillage Res 205:104761. https://doi.org/10.1016/j.still.2020.104761

Khatri-Chhetri A, Aggarwal PK, Joshi PK, Vyas S (2017) Farmers' prioritization of climate-smart agriculture (CSA) technologies. Agric Syst 151:184–191

Kiani MJ, Kaleem Abbasi M, Rahim N (2005) Use of organic manure with mineral N fertilizer increases wheat yield at Rawalakot Azad Jammu and Kashmir. Arch Agron Soil Sci 51(3):299–309

Kimaro AA, Mpanda M, Rioux J, Aynekulu E, Shaba S, Thiong'o M, Mutuo P et al (2016) Is conservation agriculture "climate-smart" for maize farmers in the highlands of Tanzania? Nutr Cycl Agroecosyst 105:217–228

Kogan F, Adamenko T, Guo W (2013) Global and regional drought dynamics in the climate warming era. Remote Sens Lett 4(4):364–372

Kumar D, Shivay YS (2008) Definitional glossary of agricultural terms, vol 1. K. International Pub House

Lesk C, Rowhani P, Ramankutty N (2016) Influence of extreme weather disasters on global crop production. Nature 529(7584):84–87

Lipper L, Thornton P, Campbell BM, Baedeker T, Braimoh A, Bwalya M, Caron P et al (2014) Climate-smart agriculture for food security. Nat Clim Chang 4(12):1068–1072

Liu RJ, Sheng P, Hui HB, Lin Q, Chen YL (2015) Integrating irrigation management for improved grain yield of winter wheat and rhizosphere AM fungal diversity in a semi-arid cropping system. Agric Syst 132:167–173

Liu Z, Wang N, Lü J, Wang L, Li G, Ning T (2022) Climate-smart tillage practices with straw return to sustain crop productivity. Agronomy 12(10):2452

Lobell DB, Field CB (2007) Global scale climate–crop yield relationships and the impacts of recent warming. Environ Res Lett 2(1):014002

Lunduka RW, Mateva KI, Magorokosho C, Manjeru P (2019) Impact of adoption of drought-tolerant maize varieties on total maize production in south Eastern Zimbabwe. Clim Dev 11(1):35–46

Magorokosho C, Vivek BS, MacRobert J (2008) Characterization of maize germplasm grown in eastern and southern Africa: results of the 2007 regional trials coordinated by CIMMYT. CIMMYT

Meena MD, Tiwari DD, Chaudhari SK, Biswas DR, Narjary B, Meena AL, Meena RB, Meena RB (2013) Effect of biofertilizer and nutrient levels on yield and nutrient uptake by maize (*Zea mays* L.). Ann Agribio Res 18(2):176–181

Megersa G, Abdulahi J (2015) Irrigation system in Israel: a review. Int J Water Resour Environ Eng 7(3):29–37

Mitchell D, Heaviside C, Vardoulakis S, Huntingford C, Masato G, Guillod BP, Frumhoff P, Bowery A, Wallom D, Allen M (2016) Attributing human mortality during extreme heat waves to anthropogenic climate change. Environ Res Lett 11(7):074006

Mittal S, Gandhi S, Tripathi G (2010) Socio-economic impact of mobile phones on Indian agriculture. ICRIER Working Paper No. 246

Mohanty S, Swain CK (2018) Role of microbes in climate smart agriculture. In: Microorganisms for green revolution: Volume 2: Microbes for sustainable agro-ecosystem. Springer, Singapore, pp 129–140

Nagabovanalli PB, Majumdar S, Kollalu S (2020) Status of silicon in ecosystem, silicon solubilization by rhizospheric microorganisms and their impact on crop productivity. In: Sharma SK, Singh UB, Sahu PK, Singh HV, Sharma PK (eds) Rhizosphere Microbes. Microorganisms for Sustainability, vol 23. Springer, Singapore. https://doi.org/10.1007/978-981-15-9154-9_16

Naresh RK, Gupta RK, Kumar A, Prakesh S, Tomar SS, Singh A, Rathi RC, Misra AK, Singh M (2011) Impact of laser land leveler for enhancing water productivity in Western Uttar Pradesh. Int J Agric Eng 4(2):133–147

Ngangom B, Das A, Lal R, Idapuganti RG, Layek J, Basavaraj S, Babu S, Yadav GS, Ghosh PK (2020) Double mulching improves soil properties and productivity of maize-based cropping system in eastern Indian Himalayas. Int Soil Water Conserv Res 8(3):308–320

Nyamangara J, Nyengerai K, Masvaya EN, Tirivavi R, Mashingaidze N, Mupangwa W, Dimes J, Hove L, Twomlow S (2014) Effect of conservation agriculture on maize yield in the semi-arid areas of Zimbabwe. Exp Agric 50(2):159–177

Nyssen J, Poesen J, Moeyersons J, Haile M, Deckers J (2008) Dynamics of soil erosion rates and controlling factors in the Northern Ethiopian Highlands – towards a sediment budget. Earth Surf Process Landf 33(5):695–711

Pittelkow CM, Liang X, Linquist BA, Van Groenigen KJ, Lee J, Lundy ME, Van Gestel N, Six J, Venterea RT, Van Kessel C (2015) Productivity limits and potentials of the principles of conservation agriculture. Nature 517(7534):365–368

Pramanick B, Kumar M, Naik BM, Kumar M, Singh SK, Maitra S, Naik BSSS, Rajput VD, Minkina T (2022) Long-term conservation tillage and precision nutrient management in maize–wheat cropping system: effect on soil properties, crop production, and economics. Agronomy 12:2766. https://doi.org/10.3390/agronomy12112766

Pramanick B, Mahapatra BS, Datta D, Dey P, Singh SP, Kumar A, Paramanik B, Awasthi N (2023) An innovative approach to improve oil production and quality of mustard (*Brassica juncea* L.) with multi-nutrient-rich polyhalite. Heliyon 9(3):e13997. https://doi.org/10.1016/j.heliyon.2023.e13997

Pramanick B, Kumar M, Naik BM, Singh SK, Singh SV (2024) Soil carbon-nutrient cycling, energetics, and carbon footprint in calcareous soils with adoption of long-term conservation tillage practices and cropping systems diversification. Sci Total Environ 912:169421. https://doi.org/10.1016/j.scitotenv.2023.169421

Prasad YG, Venkateswarlu B, Ravindra Chary G, Srinivasarao C, Rao KV, Ramana DBV, Subba Reddy G, Rao VUM, Singh AK (2012) Contingency crop planning for 100 districts in peninsular India. Central Research Institute for Dryland Agriculture, Hyderabad, p 302

Rusinamhodzi L, Corbeels M, Van Wijk MT, Rufino MC, Nyamangara J, Giller KE (2011) A meta-analysis of long-term effects of conservation agriculture on maize grain yield under rainfed conditions. Agron Sustain Dev 31:657–673

Sachan DS, Jaisimha Reddy K, Saini Y, Rai AK, Singh O, Laxman T (2023) Assessing grain yield and achieving enhanced quality in maize by next generation fertilizer: a review. Int J Environ Clim Chang 13(8):626–637

Sawadogo H (2011) Using soil and water conservation techniques to rehabilitate degraded lands in northwestern Burkina Faso. Int J Agric Sustain 9(1):120–128

Schattman RE, Ernesto Méndez V, Merrill SC, Zia A (2018) Mixed methods approach to understanding farmer and agricultural advisor perceptions of climate change and adaptation in Vermont, United States. Agroecol Sustain Food Syst 42(2):121–148

Schmidt L, Odening M, Schlanstein J, Ritter M (2022) Exploring the weather-yield nexus with artificial neural networks. Agric Syst 196:103345

Seran TH, Brintha I (2010) Review on maize based intercropping. J Agron 9(3):135–145

Setimela PS, Kassie GT, Erenstein O, Lunduka R, MacRoberts J, Magorokosho C, Ndoro O (2014) Maize regional on-farm variety trials in Eastern and Southern Africa 2011 and 2012. CIMMYT

Shiferaw B, Prasanna B, Hellin J, Banziger M (2011) Feeding a hungry world: past successes and future challenges to global food security in maize. Food Secur 3:307–327

Shrestha J, Kandel M, Chaudhary A (2018) Effects of planting time on growth, development and productivity of maize (*Zea mays* L.). J Agric Nat Resour 1(1):43–50

Singh M, Kumar M, Prakash A, Sharma K, Mishra PK (2018) Comparative field performance of pneumatic planters for planting of maize crop. Agric Eng Today 42(3):12–18

Singh D, Kumar M, Birla D, Gajanand, Shrivastava S (2022) Effect of irrigation methods, nutrient management and intercropping system on grain yield, maize equivalent yield, protein content and economics of maize (*Zea mays* L.). Int J Plant Soil Sci 34(23):1248–1253

Singh SV, Chaturvedi S, Nayak P, Dhyani VC, Nadipamu TMK, Singh DK, Gudapaty P, Mathyam P, Srinivasarao K, Gavindaraju K (2023) Carbon offset potential of biochar based straw management under rice – wheat system along Indo-Gangetic Plains of India. Sci Total Environ 897:165176. https://doi.org/10.1016/j.scitotenv.2023.165176

Sinobas LR, Rodríguez MG, Lee TS (2012) A review of subsurface drip irrigation and its management. In: Water quality, soil and managing irrigation of crops. IntechOpen, pp 171–194

Siyal AA, van Genuchten MT, Skaggs TH (2009) Performance of pitcher irrigation system. Soil Sci 174(6):312–320

Sofield I, Evans LT, Cook MG, Wardlaw IF (1977) Factors influencing the rate and duration of grain filling in wheat. Funct Plant Biol 4(5):785–797

Solomon S, Qin D, Manning M, Averyt K, Marquis M (eds) (2007) Climate change 2007: the physical science basis: Working Group I contribution to the fourth assessment report of the IPCC, vol 4. Cambridge University Press, Cambridge

Sun S, Yang X, Lin X, Sassenrath GF, Li K (2018) Climate-smart management can further improve winter wheat yield in China. Agric Syst 162:10–18

Thakur D, Kumar Y, Kumar A, Singh PK (2019) Applicability of wireless sensor networks in precision agriculture: a review. Wirel Pers Commun 107:471–512

Thakur D, Kumar Y, Vijendra S (2020) Smart irrigation and intrusions detection in agricultural fields using IoT. Procedia Comput Sci 167:154–162

Thavaprakaash N, Velayudham K, Muthukumar VB (2005) Effect of crop geometry, intercropping systems and integrated nutrient management practices on productivity of baby corn (*Zea mays* L.) based intercropping systems. Res J Agric Biol Sci 1(4):295–302

Thierfelder C, Matemba-Mutasa R, Rusinamhodzi L (2015) Yield response of maize (*Zea mays* L.) to conservation agriculture cropping system in Southern Africa. Soil Tillage Res 146:230–242

Timmons FL (2005) A history of weed control in the United States and Canada. Weed Sci 53(6):748–761

Ullo SL, Sinha GR (2021) Advances in IoT and smart sensors for remote sensing and agriculture applications. Remote Sens 13(13):2585

Vasudevan P, Thapliyal A, Sen PK, Dastidar MG, Davies P (2011) Buried clay pot irrigation for efficient and controlled water delivery. J Sci Ind Res 70:645–652

Wedryk S, Cardina J (2012) Smother crop mixtures for Canada thistle (*Cirsium arvense*) suppression in organic transition. Weed Sci 60(4):618–623

Wu SC, Cao ZH, Li ZG, Cheung KC, Wong MH (2005) Effects of biofertilizer containing N-fixer, P and K solubilizers and AM fungi on maize growth: a greenhouse trial. Geoderma 125(1–2):155–166

Chapter 11
Climate-Smart Millet Cultivation: A Promising Option for Next-Generation Agriculture

Satyabrata Mangaraj, Shriram Ratan Pradhan, Biswa Ranjan Pattnayak, Prasannajit Mishra, Debadatta Sethi, and Sagar Maitra

Abstract Climate-smart millet cultivation offers a sustainable solution for next-generation agriculture, addressing food security and resilience in the face of climate change. Millets, known for their resilience to extreme weather conditions, require minimal water and chemical inputs compared to major cereals. This makes them ideal for arid and semiarid regions, where climate variability poses significant agricultural challenges. By integrating climate-smart practices—such as crop diversification, organic farming, and efficient water management—millet cultivation can improve soil health, reduce greenhouse gas emissions, and enhance biodiversity. Additionally, millets' high nutritional value supports dietary diversity, which is crucial for addressing malnutrition. This approach aligns with global goals for sust-Bainable agriculture, making millet cultivation an essential component of climate-resilient farming systems. With increased research, policy support, and awareness, climate-smart millet cultivation can offer a viable pathway for future agricultural sustainability and adaptation to climate change.

Keywords Climate-smart millets · Nutritional diversity · Water-efficient crops · Climate-adapted crops · Next-generation agriculture

S. Mangaraj (✉) · B. R. Pattnayak
Krishi Vigyan Kendra, Sonepur, Odisha University of Agriculture and Technology, Bhubaneswar, Odisha, India
e-mail: smangaraj@ouat.ac.in

S. R. Pradhan · P. Mishra · D. Sethi
Odisha University of Agriculture and Technology, Bhubaneswar, Odisha, India

S. Maitra
Centurion University of Technology and Management, Paralakhemundi, Odisha, India

B. Pramanick et al. (eds.), *Climate-Smart Agricultural Technologies*, Sustainability Sciences in Asia and Africa, https://doi.org/10.1007/978-981-96-7699-6_11

11.1 Introduction

Millets, one of the ancient grown underutilized coarse grain cereals belonging to family *Poaceae*, have been cultivated in Southeast Asia and Africa from hundreds of years. A wide range of small-seeded grasses that are referred to as millets are widely grown as fodder and cereal crops around the world (Amadou et al. 2013; Maitra et al. 2022a). The French term "Mile" is where the word "millet" first appeared which implies that only a few millets are enclosed with thousands of grains. Several millets like foxtail millet, barnyard millet, finger millet, etc. have been referenced in ancient Indian text books. In contrary to rice and wheat, which require considerable inputs including water as well as fertilizer, millets thrive in arid climates as rainfed crops. Millets serve as a means of survival for farmers in the present day of climate change, which has disrupted our weather pattern (Saha et al. 2016; Maitra et al. 2022b). The capacity to adjust to circumstances, overcome obstacles, minimize disruptions, and exploit opportunities is referred to as climate resilience, and millet farmers can be seen as proponents of this skill in the current agricultural environment. Millet is recognized as a climate smart crop owing to its numerous advantages, notably minimal upkeep, pathogen and pest resistance, energy content, consumer demand, feed value, and sustainability advantages. Millets, additionally referred to as C_4 plants, are far more successful at taking in and utilizing carbon dioxide. A wide range of millet types have become recognized because of their resilience as well as their capacity to cope with extended droughts as well as elevated warmth while continuing to generate grains and fodder. Considering that it is smoother to grow, resilient, and environment-proof than rice or wheat, millet might be the perfect choice for "climate-smart agriculture."

Millets, being abundant in in vitamins like beta-carotene and niacin, can be regarded as low glycemic index crop owing to its substantial quantity of complex carbohydrate as well as its low content of simple carbohydrates (Pramanick et al. 2024). Additionally helping the digestive system by boosting the beneficial gut flora is millet. It is abundant in both soluble and insoluble dietary fibers. By promoting beneficial bacteria in stomach, millet may assist to improve eubiosis. Additionally, it may increase the volume of stools, lowering the risk of colon cancer. Additionally, millet is a good source of vitamin A, vitamin B, phosphorus, calcium, iron, and antioxidants. Potassium promotes healthy kidney and heart function (Table 11.1).

Two major groups of millets can be distinguished in general. (1) major millets, such as sorghum [*Sorghum bicolor* (L.)] and pearl millet [*Pannisetum glaucum* (L.)], and (2) minor or small millets, such as finger millet [*Eleusine coracana* (L.) Gaertn.], proso millet [*Panicum miliaceum* (L.)], foxtail millet [*Setaria italica* (L.) Beauv.], barnyard millet (*Echinochloa* spp.), and little millet [*Panicum sumatrense* Roth ex. Roem. and Schult.] (Maitra 2020). More than 93 countries throughout the world cultivate millets (including pearl millet and minor millets). The most extensively grown millet crop is sorghum, which is farmed on 42.1 million hectares of land across 105 nations. Production data for pearl millet and other minor millets are known from 93 countries. Developing nations, particularly those in Africa and Asia,

Table 11.1 Millets with their scientific name

Millet	Common name	Scientific name
Sorghum	*Jowar*	*Sorghum bicolor*
Pearl millet	*Bajra*	*Pennisetum glaucum*
Finger millet	*Ragi*	*Eleusine coracana*
Proso millet	*Cheena*/common millet	*Panicum miliaceum*
Foxtail millet	Indian paspalum	*Setaria italica*
Kodo millet	*Kodo*	*Paspalum scrobiculatum*
Barnyard millet	*Suan*	*Echiocloa frumentacea*
Little millet	*Kutki*	*Panicum sumatrense*
Brown top millet	*Korale*	*Brachiaria ramosa*

produce and consume 97% of the world's millets. With 26.6% of the global and 83% of the Asian millet producing areas, India is the world's largest millet producer. Millets have long been a staple of tribal cuisine in India, particularly in the states of Odisha, Madhya Pradesh, Jharkhand, Rajasthan, Karnataka, and Uttarakhand.

11.2 Nutritive Value of Millets

Millets range from about 307–347 Kcal in energy value. The percentages of carbs, protein, total fat, and ash in millets are roughly 60–68%, 6–12%, 1.7–5.4%, and 1.3–2.7%, respectively. In total, millet grains contain 6–12% dietary fiber. The micronutrients in millets, such as minerals and B vitamins, as well as phytochemicals with health advantages, are good sources (Muthamilarasan et al. 2013; Banerjee and Maitra 2020). The main nutritional component of millet grains, like other cereals, is carbohydrates. The primary carbohydrate in millets is starch, which accounts for around 57–78% of the grain's weight. Millets contain just a trace amount of soluble sugars, such as sucrose, glucose, fructose, raffinose, and stachyose. The grain's pericarp and endosperm contain the majority of the dietary fiber components. Dietary fiber is abundant in millets. Both large and small millets have dietary fiber levels that are higher than those of rice, while pearl, finger, and barnyard millet have levels that are on par with or higher than those of wheat (Saleh et al. 2013). The grains of millet are a good source of protein.

They are comparable to wheat in terms of protein content and have a greater level than rice and maize. Foxtail and proso millet are the millet grains with the greatest protein content (12.3% and 11.5%, respectively, surpassing wheat's (10.6%) level) (Gupta et al. 2012). Like most cereals, millets are lacking in lysine when it comes to the makeup of amino acids. Among millet grains, pearl millet has a balanced composition of essential amino acids (Shan et al. 2014). These grains offer a source of protein for those with celiac disease because they do not contain gluten. On the germ layer of millet grains, fat or lipids are primarily concentrated. These grains range from 1.7% in sorghum to 5.4% in pearl millet in terms of fat content (Sun et al. 2020). The lower shelf-life of pearl millet flour can be partially attributed to the grain's significantly higher fat content. In terms of the fatty acid makeup, millets include unsaturated fatty acids, including the two important fatty acids, linoleic acid and linolenic acid. Millets are a great source of vitamins, minerals, and micronutrients, particularly B-vitamins (Maitra et al. 2023). A decent source of phosphorus (P), magnesium (Mg), potassium (K), and other minerals may be found in most millets, although finger millet is particularly high in calcium (Ca). More than one-fourth of an adult person's RDA for calcium can be satisfied by just 100 g of finger millet (Shobana et al. 2009). Additionally, millets, particularly pearl millet, are a good source of zinc (Zn) and iron (Fe). The B-vitamins B1 (thiamine), B2 (riboflavin), B3 (niacin), and folic acid are all present in millets in good amounts, just like they are in other grains (Pudake et al. 2017). The B-vitamins are primarily found in the bran and germ layers; as a result, decortication, a widely used processing method to turn small millets into edible form, is known to diminish these micronutrients. Millets have antinutrients such as phytic acid, which prevent minerals from being absorbed. It is well known that sorghum and finger millet contain tannins, polyphenolic substances with some antinutritional qualities. Due to the location of antinutrients in the grain's outer layers, common household processing techniques like decortication, milling, soaking, roasting, germination, fermentation, and cooking are known to lessen their detrimental effects. Millets contain a large number of phytochemicals, including phenolic compounds, phytosterols, and other molecules that are thought to have many positive health effects. The phenolic compounds found in millets, such as phenolic acids, flavonoids, and others, have bioactivities that include antioxidant, antimicrobial, antidiabetic, and antihypertensive effects. The 3-deoxy anthocyanins present in sorghum are known to exhibit a number of bioactivities, including the capacity to guard against specific malignancies.

11.3 Health Benefits of Millets

There are several ways to ingest sorghum grain. African nations favor beverages made from sorghum. Traditionally, sorghum is consumed in India as cooked rice and roti/chapati, an unleavened flatbread. You may make bread and other delicacies with composite flour based on sorghum. The slowest starch digestion is a well-known characteristic of sorghum. The grains have adequate amounts of iron and zinc as well as good amounts of potassium, phosphorus, and calcium. Sorghum consumption is linked to a multitude of health advantages, including the reduction of chronic lifestyle problems. Sorghum's abundance in resistant starch aids in colon health maintenance and also helps to prevent obesity and its associated consequences.

Sorghum phytochemicals are recognized to have possible health benefits, such as lowering the risk of developing certain cancers, promoting cardiovascular health, and preventing diabetic problems (Babitha et al. 2015). After the seed coat is removed, pearl millet is eaten as cooked grains or as flour-based, unleavened bread.

In many regions of the world, it is used to produce weaning dishes along with beans. It is primarily eaten as bajra roti in India, while fermented or unfermented pearl millet porridges are popular in Africa. Pearl millet is a highly fatty food that is high in calories. Compared to other millets, it has a comparatively high concentration of the amino acid lysine, making it a rich source of protein. Consuming pearl millet would be beneficial for those with diabetes because it contains slow-releasing carbohydrates. Pearl millet grains are ideally suited for pregnant and nursing women as well as anemic patients due to their high iron and zinc content.

Since the phytonutrients have anticancer and antioxidant properties, they lower the risk of cardiovascular illnesses. Particularly in India, finger millet is a significant weaning meal. Ragi rotis, dosas, and porridges are more ways to eat it.

For creating bread, its composite flour is frequently used. Methionine and amino acids are found in high concentrations in finger millet. For pregnant and nursing women, it is especially advantageous because it is high in dietary fiber, as well as minerals including calcium, iron, and magnesium (Maharajan et al. 2022). In reality, finger millet has the highest calcium concentration (350 mg/100 g) when compared to the majority of cereal grains. Consuming finger millet improves grow-ups and elderly people's bones by strengthening them because of its high calcium content. Consuming finger millet is believed to help prevent fractures, particularly those brought on by osteoporosis. Additionally, finger millet's phenolics, slow-digesting carbs, and fiber help lower the risk of diabetes and enhance intestinal health. Before eating, one must remove the thin coating of husk from the foxtail millet grains. Cooked rice, porridge, and roti are made from foxtail millet. The manufacture of puddings, breads, cakes, and noodles all employ its composite flour. When compared to rice, foxtail millet is higher in protein. It contains a lot of mineral stuff as well. It is very advantageous for preventing type 2 diabetes and cardiovascular illnesses due to its low glycemic index and high fiber content. Due to their capacity to neutralize free radicals, the phytochemicals in foxtail millet have a number of beneficial effects on health (Mohapatra et al. 2019). Cooked rice and other fermented foods like idly and dosa are the two main ways that little millet is consumed. Multigrain flours have recently been used in creating bread and bakery goods, such as biscuits plus cakes. Little millet is an excellent source of fat and protein and is a top-notch supplier of dietary fiber. The little millet is also abundant in zinc, magnesium, and iron. The little millet known to have low cholesterol and blood sugar. Hence, frequent consumption is advantageous for management of cardiovascular illnesses and diabetes. Dehulled Proso millet is sometimes ground into roti and eaten in various regions of India, where it is frequently eaten as cooked rice (Hegde and Chandra 2005). Proso millet grains are well known for having a high level of protein (up to 12.5%). Leucine, isoleucine, and sulfur-containing amino acids like methionine and cysteine are reported to be present in high concentrations in proso millet protein. Niacin and folic acid, among other B-vitamins, are abundant in its grains. Additionally, it is an excellent supplier of the minerals, manganese, iron, and potassium. Proso millet is just like other millets when it comes to dietary fiber and polyphenol content. According to studies, eating proso millet may have a number of health advantages, such as controlling cholesterol metabolism by raising

HDL levels, protecting the liver from damage, and reducing the risk of type 2 diabetes and obesity. Due to its high lecithin concentration, proso millet eating is also recognized to enhance neurological health. Kodo millet is typically eaten like rice and is simple to digest. Additionally, it is used to make soup, idly, dosa, and roti. Total dietary fiber and B-vitamins like niacin, pantothenic acid, biotin, and folate are abundant in kodo millet. Kodo millet is suitable for inclusion in the diet of those with anemia due to its high folate level (39.5 mg/100 g). Additionally, kodo millet is a good source of the mineral's calcium, iron, potassium, magnesium, and zinc. Due to its high lecithin content, it is utilized as a supplement to support the neurological system. It also contains a large number of phenolic phytochemicals. Kodo millet's phenolic phytochemicals have antioxidant properties that have positive impacts on human health, such as lowering the risk of certain diseases. Dehulled, cooked, and eaten as rice is barnyard millet. Its porridge is well known in Himalayan regions, but its grain is primarily utilized in traditional dishes like idly and dosa in the southern portions of India. Nowadays, bakery products including biscuits, noodles, and bread are frequently made with barnyard millet composite flour. Barnyard millet grains are a great source of protein, dietary fiber, and mineral elements like magnesium, iron, and zinc. Its grains are a good source of iron and zinc for pregnant and nursing mothers. It is also advantageous for those with type 2 diabetes and obesity due to its high fiber content.

The main environmental and soil factors that limit the growth of millet include moisture stress, nutritional stress, salinity, alkalinity, acidity, and heat stress. Given that millet is typically farmed by resource-constrained farmers in drylands, moisture stress is thought to be the main barrier to millet production. Any physiological stage of millets' growth is susceptible to drought. Drought is one of the most significant stresses for millet production in African nations, affecting Odisha, Madhya Pradesh, Jharkhand, Rajasthan, Karnataka, and Uttarakhand. However, due to their incredible nutraceutical potential, they have recently started to gain popularity in urban areas as well. Sorghum and finger millet are millets that are grown in India throughout the rainy and postrainy seasons.

The main cause of postrainy season sorghum's low productivity is last stages of drought stress. Due to the fact that millets are produced on marginal low soil organic carbon, salinity, and alkalinity, as well as low soil fertility, results in low production of millets. During the seedling emergence stage, the crop is adversely impacted by the soil's salinity and poor drainage. Because of evolving dietary patterns and customer preferences, land is now being used to grow other high-value cereal grains, which reduces the output of millets.

11.4 Traits Responsible for Climate Resilience of Millets

Millets have various morphophysiological, molecular, and biochemical characteristics that improve their ability to withstand climate change. Millets' shorter life cycle (seed to seed)—12–14 weeks compared to 20–24 weeks for grains like rice—helps people relax. Small leaf area, thickening cell wall, and capacity to develop dense root system are characteristics that help people avoid stressful situations. In the C4 system, RuBP oxygenation and photorespiration are suppressed by the concentration of CO_2 near (ribulose bisphosphate carboxylase/oxygenase) RuBISCO. Accordingly, the C4 mechanism increases the amount of CO_2 in the bundle sheath, which decreases photorespiration (by about 80% depending on temperature) and raises the catalytic activity of RuBisCO in plants (Serna-Saldivar and Espinosa-Ramírez 2019). Millets have increased photosynthetic rates in hotter climates because RuBISCO operates at greater CO_2 levels, and they have superior water and nitrogen usage efficiency than C3 plants. Millets receive additional benefits from C4 photosynthesis in addition to WUE and NUE, such as better flexibility in biomass allocation patterns and decreased hydraulic conductivity per unit leaf area in warm temperatures.

Compared to other grains, millets are more morphophysiological, molecular, and biochemically adapted to withstand environmental challenges. Millets have a short life cycle (12–14 weeks from seed to seed), compared to rice and wheat, which can take up to 20–24 weeks. This helps people deal with stress better. However, some characteristics, such as being low in stature, having a limited leaf area, having thickened cell walls, and having the potential to create a dense root system, might mitigate the occurrence of stress situations and their effects. Furthermore, millets greatly benefit from the C4 photosynthetic characteristic. Ribulose-1,5-bisphosphate carboxylase/oxygenase (RuBisCO) is surrounded by a concentration of carbon dioxide (CO_2) in the C4 system, which inhibits photorespiration and oxygenation of ribulose 1,5-bisphosphate (RuBP). C4 mechanism increases the concentration of CO_2 in the bundle sheath, which reduces photorespiration (by about 80% depending on temperature), and raises the catalytic activity of RuBisCO in planta. Millets have enhanced photosynthetic rates in warm climates and confer instantaneous water use efficiency (WUE) and nitrogen use efficiency (NUE), which are 1.5–4 times higher than C3 photosynthesis. This is because RuBisCO of C4 plants operates at high CO_2 levels (Liang et al. 2010). Millets receive additional benefits from C4 photosynthesis in addition to WUE and NUE, such as better flexibility in biomass allocation patterns and decreased hydraulic conductivity per unit leaf area in warm temperatures. These characteristics of millets make them next-generation crops with the potential for study into climatically resilient traits, and they can be used to gain knowledge for the improvement of major grains. Millets are a tractable system to research stress-responsive features at the cellular, molecular, and physiological levels because of their excellent tolerance to a variety of abiotic stimuli, including drought, salinity, light, and heat. Millets have demonstrated their mechanisms for coping with stress in a number of morphophysiological and biochemical studies (Kumar et al. 2015). Abiotic stresses have also been shown to increase biochemical activities in foxtail millet, including elevated levels of antioxidants, reactive oxygen species and their scavenging enzymes, enzyme activity of catalase and superoxide, and synthesis of osmolytes and other stress-related proteins. Despite this development, research that shed light on the molecular mechanisms underpinning millets' stress tolerance are still mostly missing. Additionally, little is known about the genetic components of stress tolerance discovered through association mapping and biparental mapping. To fully utilize the underlying genetic potential using conventional/molecular breeding procedures and transgenic technologies, thorough phenotypic screening to observe the natural genetic variability in stress tolerance across distinct millet germplasms is highly necessary in this context.

11.5 Millets and Drought: Effects and Responses

The production of crops is impacted by a variety of biotic and abiotic stressors. Abiotic stressors were imposed at crucial plant growth and development stages, resulting in yield losses, as a result of the rapid climate changes that resulted in significant losses of arable area used for crop production. Abiotic stressors include drought, severe temperature (cold, frost, and heat), flooding, salinity, etc. that are the main yield-limiting stressors in semiarid and arid environments. Major grains are less tolerant of environmental challenges than millets because millets have a variety of morphophysiological, molecular, and biochemical characteristics.

Numerous processes, including sensing, signaling, transcription, processing of transcripts, translation, and posttranslational protein modifications, which are controlled by both genetic and epigenetic factors, are among the molecular mechanisms behind the plants' responses to abiotic stressors. Increased heat and drought brought on by climate change have the greatest negative effects on crop output and result in the highest yearly losses of all major abiotic stresses. Drought stress is expected to remain a significant abiotic constraint on food production, according to climate change models. Numerous studies link drought to rising rates of malnutrition, ill health, hunger, starvation, and food and water insecurity in a nonlinear fashion. Millets are the greatest alternative in semiarid and dry regions since they are naturally drought tolerant. Drought has been shown to have an impact on millets' growth, yield, membrane integrity, pigment, osmotic adjustment, water relations, photosynthetic activity, and root-associated microbiome structure and activity, according to several research. Compared to major grains, millets have more morphophysiological, biochemical, and molecular characteristics that improve their ability to withstand drought. Millets are especially helpful in reducing stress because of their short life cycle (10–12 weeks) compared to other important crops (20–24 weeks). A number of characteristics, including short stature, limited leaf area, thickened cell walls, and a dense root system, help avoid stress and its long-term effects. They also contain a C4 (Hatch and Slack pathway) photosynthetic system, which is very favorable for surviving in high temperatures and poor moisture. The Calvin cycle and light-dependent processes (which take place in bundle-sheath cells in the C4 system) are physically different from one another. As a result, plants can keep their stomata closed during the day to prevent water loss thanks to the C4 mechanism's enhancement of the concentration of CO_2 in the bundle sheath, which outperforms photorespiration. Millets' improved photosynthetic rates in warm weather provide rapid water and nitrogen consumption efficiency that is

1.5–4 times greater than C3 photosynthesis. Millets also gain additional advantages from C4 photosynthesis, such as higher biomass allocation flexibility, lower hydraulic conductivity per unit leaf area, and improved growth and ecological performance in hotter climates. In the context of a changing environment, these examples present prospects for raising the production level of various crops (Shrestha et al. 2023). Millets have been found to have a number of mechanisms for coping with and adapting to stress. Millets have been shown to accumulate osmolytes and antioxidants in response to abiotic stressors. These characteristics of millets make them a viable model system with the potential for research into the characteristics of stress-responsiveness and the identification of the physiological, biochemical, and molecular mechanisms of stress. Information is necessary for millet improvement.

11.6 Millets for Post-COVID-19 Nutrition and Wellness Security

In any crisis or epidemic event, the poorer segments of society are always the hardest hit, especially in developing nations. For instance, small and marginal farmers make up around 90% of India's agricultural industry, making them especially susceptible to economic shocks like COVID-19 hardship (Muthamilarasan and Prasad 2021). The best defense is to strengthen the immune system in order to boost the body's inherent defenses against infections. An increase in metabolism due to this hyperactivity necessitates energy sources, biosynthetic substrates, and regulatory molecules. As a result, it is crucial to supply the immune system with the proper amount and kind of nutrients, which are all ultimately obtained from the diet we typically consume. It is critical to restore the immune response to normal by consuming healthy foods after exposure to or infection with the fatal coronavirus, which compromises the immune system. After recovering from COVID-19, patients must maintain a healthy lifestyle and follow a rigorous diet to strengthen their immunity (Mishra et al. 2021). Compared to other major grains, millets are more nutrient-dense due to their calcium, iron, potassium, magnesium, and zinc content as well as other vital compounds including vitamins, amino acids, and fatty acids. Numerous vitamins (A, B6, B12, C, D, E, and folate) and trace minerals (zinc, copper, selenium, and iron) have been shown to play important roles in boosting the human immune system. Additionally essential to maintaining and controlling the immune system is the gut flora. A characteristic of many disorders, including many infectious diseases, including COVID-19, is gut dysbiosis. The immune system can be helped by eating habits that result in a healthy human microbiome. Millet, which is high in magnesium and potassium, lowers blood pressure by acting as a vasodilator, lowering the risk of heart attacks or strokes—particularly in people with atherosclerosis—and improving circulation is one of the best strategies to safeguard cardiovascular health. Millets are high in lignans, which the gut bacteria can convert into animal lignans after digestion. Animal lignans have been found to offer protection from some chronic diseases, including cancer and heart disease. Proteins, dietary fiber, and phenols are just a few of the component's millets offer that are beneficial to your health. They are advantageous for diabetics because of their high dietary fiber and phenolic content as well as their low glycemic index (Feng et al. 2016). The body's ability to control diabetes is thought to be aided by magnesium through enhancing the function of insulin and glucose receptors. Populations divided into diets with or without magnesium have shown a 30% reduction in diabetes. By removing harmful "bad cholesterol" (LDL) from the body and enhancing the benefits of "good cholesterol" (HDL), foods with high dietary fiber content are also excellent for decreasing cholesterol (Bandyopadhyay et al. 2017). Having issues with constipation, too much gas, bloating, and cramps can be resolved with millet's high fiber and mineral content. Improved nutritional absorption and liver and kidney function, which are both crucial for the body's metabolic processes and eventually the immune system, are made possible by gastrointestinal regulation.

11.7 Approaches for Incorporating Millets into Conventional Agriculture

Combating pandemics and climate change provide a number of similar difficulties. In terms of their beginning, scope, and degree of severity, both catastrophes exhibit significant ambiguity (Beta et al. 2000). These ambiguities and disruptions compound each other when they occur, which could result in a serious economic imbalance and food insecurity. Important lessons can be learned from the present pandemic and natural disasters about climate change policy and agricultural sustainability. Scientists now have the ability to more accurately forecast natural disasters and pandemics and to launch early warning systems. Governments that depend on scientific advice are likely to manage natural and human disasters more effectively, even when science does not have all the answers. In order to address crises like this, science must contribute to the development of risk governance solutions, specifically institutions, regulations, and capabilities that strike a balance between national and regional responsibilities in order to produce socially beneficial results (Priyadarshini and Abhilash 2021). Hopefully, these crises will increase the amount of time that scientists devote to studying issues like climate change.

A multistakeholder structure was necessary to adopt coordinated measures on the local, regional, and worldwide levels in order to fully realize the promise of orphan crops. The global sustainable development goals will govern the implementation of these initiatives. Large-scale production, however, cannot be accomplished without the coordinated activities of stakeholders along the entire value chain, from research to production to marketing to end use. In order to promote increased awareness, production, marketing, and integration of orphan crops and wild edible species in public initiatives to mainstream them in contemporary agriculture, policymaking measures are essential for bridging knowledge gaps and building links between stakeholders in many sectors. Growing consumer and producer interest in millets' climatically robust qualities and their potential for health benefits highlights the need for increased funding for research and development. The large staple crops (maize, rice, and wheat) must no longer be tended to; policies must now support the orphan or underutilized crops. Nutritional security needs to encourage the underutilized crops for a change in the food system in the post-COVID-19 future.

11.8 Value Addition of Millets

Millet crops are primarily processed and used on a small scale at home, which limits its production to catmint and prevents a vast population from accessing its beneficial nutrients. Various millets are consumed in India in the form of different foods and various regional specialties (Balasubramanian 2013). Millets have drawn attention recently since they have an unmatched nutritional profile and are simple to take in any manner that is appropriate. Millets may be made into several products with added value thanks to their great nutritional and functional properties. Apart from the high nutritional value, only traditional consumers use millets as a staple food, and this is mostly due to the lack of consumer-friendly, ready-to-eat food products. Poor quality characteristics of millets and their traditional first method of ingestion, such as flours, have limited their use. These characteristics include dark and dull color, coarse and gritty texture, astringent flavor, high fiber content, longer cooking time, antinutritional ingredients, and low shelf life (Platel 2013). Millet is a potential and successful ingredient in many traditional and ready-to-eat health foods when processed using the proper methods.

Millets have been transformed into a variety of processed foods, including weaning foods, flours that have been popped, flaked, puffed, extruded, roller dried, fermented, malted, and composite. The nutritional value of some processing technologies, such as the extrusion of pearl millet in weaning foods, which improves protein digestion, is promising, similar to how probiotic fermentation and germination improved the protein profile and mineral availability (Malik 2015).

11.9 Conclusion

The road to a climate-smart agriculture finds another robust ally through the cultivation of millets, with these grain-less crops supplying not only means for survival through tough times but also serve as a guide for thriving as a sustainable culture of farming while major cereals were being accorded preference on their own rather than millets. Millets, however, have lower input use and are rather more naturally adapted to a changing climatic scenario yet much more promises from the side of millets themselves. The real challenge, of course, is to make those nutritious grains into foods that modern consumers are attracted to. Current home-scale processing and traditional preparation methods, valuable as they are, restrict millets' outreach and impact. We must bridge this gap with focused investment in research, particularly the development of convenient, appealing millet-based products. In a postpandemic world, we cannot look forward to food security without being beyond conventional crops. Combining ancient wisdom with modern innovation and supporting such with favorable policies, millet farming can change the face of traditional agriculture into an anchor of new-age agriculture. The road to these calls for collaboration from a farmer to the food processor and from a researcher to a policymaker working toward unlocking millets' potential for building resilient food systems.

References

Amadou I, Gounga ME, Le GW (2013) Millets: nutritional composition, some health benefits and processing-a review. Emir J Food Agric 25:501–508

Babitha KC, Vemanna RS, Nataraja KN, Udayakumar M (2015) Overexpression of EcbHLH57 transcription factor from Eleusine coracana L. in tobacco confers tolerance to salt, oxidative and drought stress. PLoS One 10(9):e0137098

Balasubramanian S (2013) Processing of millets. Paper presented National Seminar on recent

Bandyopadhyay T, Muthamilarasan M, Prasad M (2017) Millets for next generation climate-smart agriculture. Front Plant Sci 8:1266

Banerjee P, Maitra S (2020) The role of small millets as functional food to combat malnutrition in developing countries. Indian J Nat Sci 10(60):20412–20417

Beta T, Rooney LW, Marovatsanga LT, Taylor JRN (2000) Effect of chemical treatments on polyphenols and malt quality in sorghum. J Cereal Sci 31:295–302

Feng ZJ, Xu ZS, Sun J, Li LC, Chen M, Yang GX et al (2016) Investigation of the ASR family in foxtail millet and the role of ASR1 in drought/oxidative stress tolerance. Plant Cell Rep 35:115–128. https://doi.org/10.1007/s00299-015-1873-y

Gupta N, Srivastava AK, Pandey VN (2012) Biodiversity and nutraceutical quality of some Indian millets. Proc Natl Acad Sci India B Biol Sci 82:265. https://doi.org/10.1007/s40011-012-0035-z. http://www.springerlink.com

Hegde PS, Chandra TS (2005) ESR spectroscopic study reveals higher free radical quenching potential in kodo millet (Paspalum scrobiculatum) compared to other millets. Food Chem 92(1):177–182

Kumar A, Gaur VS, Goel A, Gupta AK (2015) De novo assembly and characterization of developing spikes transcriptome of finger millet (Eleusine coracana): a minor crop having nutraceutical properties. Plant Mol Biol Report 33:905–922. https://doi.org/10.1007/s11105-014-0802-5

Liang S, Yang G, Ma Y (2010) Chemical characteristics and fatty acid profile of foxtail millet bran oil. J Am Oil Chem Soc 87(1):63–67

Maharajan T, Ceasar SA, Ajeesh Krishna TP (2022) Finger millet (Eleusine coracana (L.) Gaertn): nutritional importance and nutrient transporters. Crit Rev Plant Sci 41:1–31. https://doi.org/10.1080/07352689.2022.2037834

Maitra S (2020) Potential horizon of brown-top millet cultivation in drylands: a review. Crop Res 55(1–2):57–63

Maitra S, Pine S, Banerjee P, Pramanick B, Shankar T (2022a) Millets: robust entrants to functional food sector. In: Pirzadah TB, Malik B, Bhat RA, Hakeem KR (eds) Bioresource technology: concept, tools and experiences. Wiley, Hoboken, pp 1–27

Maitra S, Praharaj S, Hossain A, Patro TSSK, Pramanick B, Shankar T, Pudake RN, Gitari HI, Palai JB, Sairam M, Sagar L, Sahoo U (2022b) Small millets: the next-generation smart crops in the modern era of climate change. In: Pukade RN, Solanke AU, Sevanthi AM, Rajendrakumar P (eds) Omics of climate resilient small millets. Springer Nature, pp 1–25

Maitra S, Patro TSSK, Reddy A, Hossain A, Pramanick B, Brahmachari K, Krishna Prasad K, Santosh DT, Mandal M, Shankar T, Sagar L, Banerjee M, Palai JB, Subhashisa Praharaj S, Sairam M (2023) Brown top millet (*Brachiaria ramosa* L. Stapf; *Panicum ramosum* L.) – the neglected and smart crop in fighting against hunger and malnutrition. In: Farooq M, Siddique KHM (eds) Neglected and underutilized crops, pp 221–245. https://doi.org/10.1016/B978-0-323-90537-4.00012-0

Malik S (2015) Pearl millet-nutritional value and medicinal uses. Int J Adv Res Innov Ideas Educ 1(3):414–418

Mishra A, Bruno E, Zilberman D (2021) Compound natural and human disasters: managing drought and COVID-19 to sustain global agriculture and food sectors. Sci Total Environ 754:142210. https://doi.org/10.1016/j.scitotenv.2020.142210

Mohapatra D, Patel AS, Kar A, Deshpande SS, Tripathi MK (2019) Effect of different processing conditions on proximate composition, anti-oxidants, anti-nutrients and amino acid profile of grain sorghum. Food Chem 271:129–135

Muthamilarasan M, Prasad M (2021) Small millets for enduring food security amidst pandemics. Trends Plant Sci 26:33–40. https://doi.org/10.1016/j.tplants.2020.08.008

Muthamilarasan M, Theriappan P, Prasad M (2013) Recent advances in crop genomics for ensuring food security. Curr Sci 105(2):155–158

Platel K (2013) Millet flours as a vehicle for fortification with iron and zinc. In: Preedy VR, Srirajaskanthan R, Patel VB (eds) Handbook of food fortification and health. Springer, New York, pp 115–123

Pramanick B, Choudhary S, Kumar M, Singh SK, Jha RK, Singh SK, Salmen SH, Ansari MJ, Hossain A (2024) Can site-specific nutrient management improve the productivity and resource use efficiency of climate-resilient finger millet in calcareous soils in India? Heliyon 10(12):e32774. https://doi.org/10.1016/j.heliyon.2024.e32774

Priyadarshini P, Abhilash PC (2021) Agri-food systems in India: concerns and policy recommendations for building resilience in post COVID-19 pandemic times. Glob Food Sec 29:100537. https://doi.org/10.1016/j.gfs.2021.100537

Pudake RN, Mehta CM, Mohanta TK, Sharma S, Varma A, Sharma AK (2017) Expression of four phosphate transporter genes from Finger millet (Eleusine coracana L.) in response to mycorrhizal colonization and Pi stress. 3 Biotech 7(1):17

Saha D, Gowda MC, Arya L, Verma M, Bansal KC (2016) Genetic and genomic resources of small millets. Crit Rev Plant Sci 35(1):56–79

Saleh ASM, Zhang Q, Chen J, Shen Q (2013) Millet grains: nutritional quality, processing, and potential health benefits. Compr Rev Food Sci Food Saf 12:281–295

Serna-Saldivar SO, Espinosa-Ramírez J (2019) Grain structure and grain chemical composition. In: Sorghum and millets: chemistry, technology, and nutritional attributes. Woodhead Publishing, Duxford

Shan S, Li Z, Newton IP, Zhao C, Li Z, Guo M (2014) A novel protein extracted from foxtail millet bran displays anti-carcinogenic effects in human colon cancer cells. Toxicol Lett 227(2):129–138

Shobana S, Sreerama YN, Malleshi NG (2009) Composition and enzyme inhibitory properties of finger millet (Eleusine coracana L.) seed coat phenolics: mode of inhibition of a-glucosidase and pancreatic amylase. Food Chem 115(4):1268–1273

Shrestha N, Hu H, Shrestha K and Doust AN (2023) Pearl millet response to drought: A review. Front Plant Sci, 14, 1059574

Sun M, Huang D, Zhang A, Khan I, Yan H, Wang X et al (2020) Transcriptome analysis of heat stress and drought stress in pearl millet based on Pacbio full-length transcriptome sequencing. BMC Plant Biol 20:1–15. https://doi.org/10.1186/s12870-020-02530-0

Chapter 12
Soil-Centric Cropping Systems Building Climate Resilience

Sk Asraful Ali, K. M. Krishna, Kadagonda Nithinkumar, Megha Kumari, Twinkle Jena, Unti Miiri Ezing, A. R. Devika, Nilutpal Saikia, Rohit Bapurao Borate, Rashmi Jha, Tony Manoj Kumar Nandipamu, Rashmi Sharma, and Shiv Vendra Singh

Abstract Cropping system plays a crucial role in ensuring global food security and agricultural sustainability, particularly in the face of climate change. It emphasizes the limitations of conventional monoculture systems, advocating for diverse cropping approaches. The importance of climate resilient practices and soil management is highlighted, accentuating the need for sustainable agricultural strategies to address the challenges posed by a growing population and changing climate. Cropping system encompasses crop sequences, spatial distribution, and management practices on a field over time. Various cropping systems globally, such as multiple and monoculture, play a crucial role in climate-resilient agriculture.

The original version of the chapter has been revised. A correction to this chapter can be found at https://doi.org/10.1007/978-981-96-7699-6_21

S. A. Ali (✉) · K. M. Krishna
Division of Agronomy, ICAR-Indian Agricultural Research Institute, New Delhi, India

Department of Agronomy, G.B. Pant University of Agriculture and Technology, Pantnagar, Uttarakhand, India

K. Nithinkumar · M. Kumari · T. Jena · A. R. Devika · N. Saikia · R. B. Borate
Division of Agronomy, ICAR-Indian Agricultural Research Institute, New Delhi, India

U. M. Ezing
Division of Plant Physiology, ICAR-Indian Agricultural Research Institute, New Delhi, India

R. Jha
Department of Seed Science and Technology, Tamil Nadu Agricultural University, Coimbatore, Tamil Nadu, India

T. M. K. Nandipamu · R. Sharma
Department of Agronomy, G.B. Pant University of Agriculture and Technology, Pantnagar, Uttarakhand, India

S. V. Singh
Department of Agronomy, Rani Lakshmi Bai Central Agricultural University, Jhansi, Uttar Pradesh, India

B. Pramanick et al. (eds.), *Climate-Smart Agricultural Technologies*, Sustainability Sciences in Asia and Africa, https://doi.org/10.1007/978-981-96-7699-6_12

Climate change impacts cropping systems through temperature shifts, affecting crop phenology and pest dynamics. Sustainable soil practices like conservation tillage, cover cropping, etc. contribute to resilient agriculture by enhancing soil health, water retention, and nutrient cycling. The imperative role of cropping system-based soil management in ensuring sustainable agriculture under climate change scenario also explores technological innovations such as remote sensing, data-driven decision-making tools, robotics, and advances in soil monitoring. Integration of traditional and modern practices, including conservation agriculture and agroforestry, is highlighted. Addressing climate-induced soil challenges, emphasizing tailored nutrient management, understanding soil microbial communities, and comprehensive investigations into organic matter dynamics are crucial. Emerging technologies like precision agriculture, IoT devices, and climate modeling are key contributors to climate resilient agriculture, stressing the importance of research gaps for further exploration.

Keywords Cropping system · Climate change · Climate-resilient agriculture · Conservation agriculture

12.1 Introduction

Cropping system forms a critical component of farming system and is the backbone of food security and agricultural sustainability. By definition, cropping system is the temporal and spatial arrangement of various crops and fallow in the farm, along with the different management practices, technologies, and other enterprises adopted in the farm (Gangwar et al. 2012). The sustainability of agricultural systems is closely linked to the cropping systems adopted. With the global population expected to reach ten billion by 2050 (The Royal Society 2021), it is essential for food, feed, and fuel production from agriculture to increase by 50% compared to production in 2012 (FAO 2017). However, IPCC predicts an average 2% per decade reduction in food grain production as an impact of climate change by 2050 (IPCC 2014). Worsening the scenario, soil degradation and desertification spread to 23 ha arable land every single minute (The Royal Society 2021), leading to further reduction in agricultural productivity. This situation necessitates wise adoption of cropping systems to feed the world. Conventionally followed cropping systems such as monoculture are mainly designed to maximize the economic yield. They excessively deplete natural resources, including groundwater and the inherent fertility of the soil, which ultimately undermines their long-term sustainability. However, in this scenario of increasing population, decreasing per capita land availability and declining crop productivity worsened by the vagaries of climate, there is an immediate need to restructure the crop production practices. Therefore, the recent approaches aim at designing more sustainable and climate resilient cropping systems that ensure world food security and nutrition, without imperiling soil health, ecosystem, and the environment.

Among the various cropping systems, monoculture (single season cultivation) is the most prevalent cropping system (88%) followed globally, producing large quantities especially of cereals due to market demand and grain price (Yang et al. 2020). Multiple cropping systems account for only 12% of the total cropped area that too with maximum coverage under cereal-cereal rotations such as irrigated rice-rice system (32.66 Mha), irrigated double cropping of wheat with rice and with maize (14.38 and 12.16 Mha, respectively), etc. (Waha et al. 2020). These cereal-based conventional systems lack diversity, degrade the soil fertility, are less climate change resilient, and have higher carbon footprint, thus making themselves less sustainable. Monoculture and sequence cropping with crops belonging to the same group decrease soil microbial diversity, enhance population of host-specific microorganism, and also imbalance the soil nutrient content, all resulting in reduced soil health and crop yield (Yang et al. 2020). Extensive tillage operations, which is a prominent feature of these systems, reduce the soil organic carbon reserve (Singh et al. 2023a; Pramanick et al. 2024) and disturb soil microbial food webs, resulting in proportional increase in bacterial population and decrease in fungal (including arbuscular mycorrhiza) and earthworm populations (Montgomery and Biklé 2021). After harvest, soil is left bare in monocropped area, accelerating erosion and consequent soil degradation. Moreover, excessive trafficking of the soil with heavy machinery leads to soil compaction, which further hampers crop growth and reduces yields (Shaheb et al. 2021). Monoculture and cereal-based sequence cropping are sustainable neither from the viewpoint of food and nutritional security. These cropping systems have less "insurance effect" against climate change (Wright et al. 2020) and also cannot assure nutritional security of the world as they lack diversity. Therefore, further intensification of these systems, to ensure food security, and their diversification to ensure nutritional security, climate change resilience, and environmental safety, along with adoption of more scientific and advanced technologies and management practices, which build sustainability without compromising with profitability, are the needs of the day (Fig. 12.1).

Adopting more diverse cropping systems is a viable option to minimize the adverse effects of the conventional monoculture systems. It improves the nutrient cycling and resource use efficiency thereby making the system more sustainable, profitable, and resilient to climate and biological vagaries (Acharya et al. 2011).

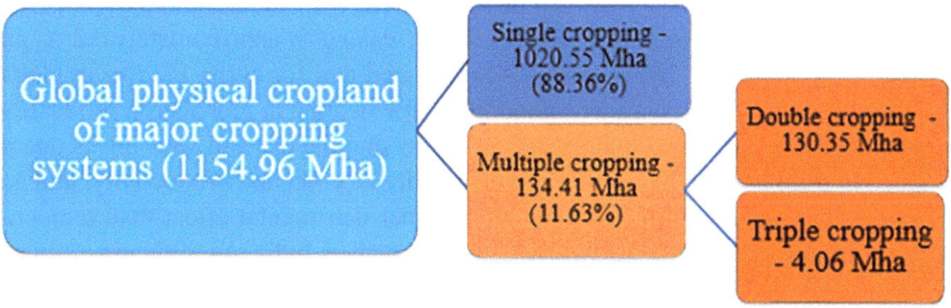

Fig. 12.1 Global physical cropland of major cropping systems (Waha et al. 2020)

Adoption of more diverse crop rotation, mixed and intercropping, altering the seeding time and cropping pattern, incorporation of legumes and pulses in otherwise cereal dominated systems, adoption of a second crop in single cropped area, inclusion of perennial alleys in the farm, etc. increases the farm diversity in space, time, and genetics (Hufnagel et al. 2020). Intercropping is a feasible option for diversifying monoculture systems without decreasing the land under main crop. It has the potential to rejuvenate the soils deteriorated by maize monoculture (for 30 years) by improving the soil organic carbon, bacterial diversity, chemical properties, and enzymatic activities (Wolinska et al. 2022). Inclusion of legumes as third crop in cereal-cereal rotation has the potential to increase the main crop yield improving the soil health parameters including nutrient availability. Lowland rice yield and upland maize yield were increased by 10–14% and 5–11%, respectively, when mung bean was introduced as third crop in rice-wheat and maize-wheat rotations (Ghosh et al. 2020). Inclusion of cover crops in the cropping system has numerous benefits of organic matter addition, soil erosion control, improving soil water, nutrient availability, etc. apart from helping in more carbon sequestration. Significant reduction in the runoff discharge and leaching of nitrate and phosphorus, along with improvement in the main crop yield and soil physical properties, as a result of cover crop incorporation in corn-soyabean rotations, has been reported by Adler et al. (2020). Crop diversification through various methods has the potential to reduce yield loss due to drought as a result of increase in organic matter (Renwick et al. 2021). Following maize-potato rotation instead of maize monoculture can avoid ground water depletion and also increase the water use efficiency of grain/tubers by 15.52–23.39% (Wang et al. 2021a). Adoption of crop rotation has the potential to increase around 20% yield compared monoculture, with maximum effect when the rotation cycles are longer and when legume crops are incorporated in it (Zhao et al. 2020). Wise selection of cropping systems plays a crucial role in addressing climate change. In China, the global warming potential of the ryegrass-sweet sorghum-wheat-maize rotation and the spring maize-wheat-maize rotation was found to be 22% and 19% lower, respectively, compared to the wheat-maize rotation. Additionally, incorporating sweet potatoes or legumes like groundnut and soybean into these systems could lead to reductions of 51%, 43%, and 32%, respectively (Yang et al. 2024).

Success of any cropping system depends on its adaptability to environmental conditions and the management practices followed. Over the last three decades, the global average surface temperature has been increasing by approximately 0.2 °C per decade (Valipour et al. 2021), and it is anticipated that for every 1% rise in global temperature, there will be a corresponding 0.19% decline in global crop production (Cakan and Tipi 2024). Damania (2020) reports that the dry and wet shocks due to change in rainfall pattern can result in variability in yield and productivity, and the intraseasonal rainfall variability is having higher detrimental effect than temperature variations, on crop production in countries such as India. As mitigating climate change is a slow process and manipulating climate is not a feasible option on a large scale, the possible way out is making the system climate resilient through better management options. The system has to be designed with such crop and resource management practices that promise higher and sustainable production under changing climate, while still preserving and rejuvenating the finite natural resources such as innate soil fertility, soil water, etc.

Healthy soil is a prerequisite for stable yield under changing climate (Wolf et al. 2023), with effective soil management serving as a cornerstone of climate-smart agriculture. This approach aims to adapt farming systems to climate variability, reduce greenhouse gas emissions, and enhance productivity (Bayu 2020). Soil properties such as nutrient availability, moisture retention, biological activity, root penetration ease, porosity, thermal conductivity, etc. directly influence crop resilience to abiotic stresses like extreme heat, drought, and waterlogging. These factors collectively determine both crop yields and its stability under changing climate (Chen et al. 2021; Koresh 2021; Wolf et al. 2023).

As agricultural soils are both a cause and solution for climate change, soil management practices that increase soil organic carbon and decrease carbon emission are the key to climate-smart agriculture. Increased soil organic carbon can also have beneficial effect on the soil biota, which enhance crop productivity through the synthesis of nutrient-solubilizing enzymes (Wang et al. 2022). Reduction of soil disturbance through reduced tillage or no-till cultivation cuts down on the oxidative loss of soil organic carbon and also enhances nutrient (carbon, nitrogen, and phosphorus) acquiring enzymes and antioxidative enzymes present in the soil (Li et al. 2024). Addition of organic amendments, crop diversification, green manuring, cover cropping, legume cropping, and stubble incorporation are all considered efficient ways of enhancing soil organic carbon and crop productivity, making the agricultural system a net CO_2 sequester than an emitter (Bayu 2020). Following 4Rs stewardship to ensure balanced crop nutrition plays a very important role in climate-smart agriculture (Fixen 2020). The right dose of nitrogen is important as its excess application results in denitrification and volatilization, releasing N_2O and NH_3, respectively, both of which are detrimental to the atmosphere (Menegat et al. 2022). Nitrous oxide is a highly potent greenhouse gas while ammonia gets redeposited on the soil surface causing acidity. Phosphorus availability being highly dependent on soil moisture content, soil temperature, pH, and CO_2 concentration, care needs to be taken to ensure its optimum availability under changing climatic conditions (Maharajan et al. 2021). Potassium plays an important role in stress tolerance and turgidity maintenance (Sardans and Peñuelas 2021). Application of sulfur mitigates methane emission from paddy fields by enhancing the growth of nonmethanogenic bacteria that compete for same substrate (Linquist et al. 2018) and also helps in alleviating the adverse effect of terminal heat stress in wheat (Tao et al. 2018). Boron deficiency or toxicity aggravates abiotic stresses such as salinity, drought, etc. (García-Sánchez et al. 2020). The use of enhanced efficiency nitrogen fertilizers (EENF) such as coated urea, slow-release fertilizers, biochar-based fertilizers, etc. along with nitrification inhibitors and urease inhibitors helps in achieving higher yields and nitrogen use efficiency (Ma et al. 2021; Puga et al. 2020). Incorporating crops having the potential to release biological nitrification inhibitors, such as barley, sorghum, and grasses such as *Brachiaria humidicola*, also helps in decreasing the nitrous oxide emissions from the agricultural fields (Wang et al. 2021a).

Adoption of management options such as subsurface fertigation, remote sensing, variable rate application, robotics and automation, etc. for more precise nutrient application can further mitigate greenhouse gas emission, increase nutrient use efficiency, and enhance yield, thus making the farming practice more sustainable and profitable.

12.2 Understanding Cropping Systems

12.2.1 Predominant Cropping Systems Throughout Different Climatic Zones

A cropping system refers to the crop sequences and crops along with the management practices applied to a specific field over several years. It depicts the patterns of cropping that are employed on farms and how they interact with resources and other farming enterprises. The cropping system mainly comprises three components, namely, crop succession, cropping pattern, and crop management techniques (Begue et al. 2018). Crop succession refers to the practice of growing different crops on the same piece of land in sequential seasons or years and is mainly practiced for more than 1 year. Instead of planting the same crop repeatedly, farmers rotate or sequence different crops in a planned manner. This agricultural technique is designed to achieve several benefits, including improved soil fertility, pest and disease control, increased crop yield, and overall sustainable land management (Bowles et al. 2020). It includes practices such as monoculture in which the same crop is grown year after year on the same piece of land. Crop rotation is a systematic practice of growing different types of crops in sequential seasons or years on the same piece of land, and fallow cropping is a form of land management where a field is intentionally left without crops for a designated period. The fallow period allows the soil to replenish nutrients, conserve moisture, and improve overall soil health. In some cases, cover crops may be planted during the fallow period to provide additional benefits such as erosion control and nitrogen fixation.

Cropping pattern refers to the yearly sequence and spatial arrangement of crops. It constitutes single cropping in which a single crop is grown on a piece of land and multiple cropping in which more than one crop is grown on a piece of land. It can either be sequential cropping constituting double crop, triple crop, and relay crop or intercropping which can either be additive or replacement in nature. All the crop management techniques undertaken on a given farm like irrigation, tillage, fertilizer, pesticide application, weed control, etc. are also included in the cropping system.

An estimation by Waha et al. (2020) showed that multiple cropping is done in 134.41 Mha, that is, in 12% of the global cropland area, of which double and triple cropping accounts for 130.35 and 4.06 Mha land, respectively. Single cropping extents to 1020.55 Mha land. The total global cropland land area accounts for

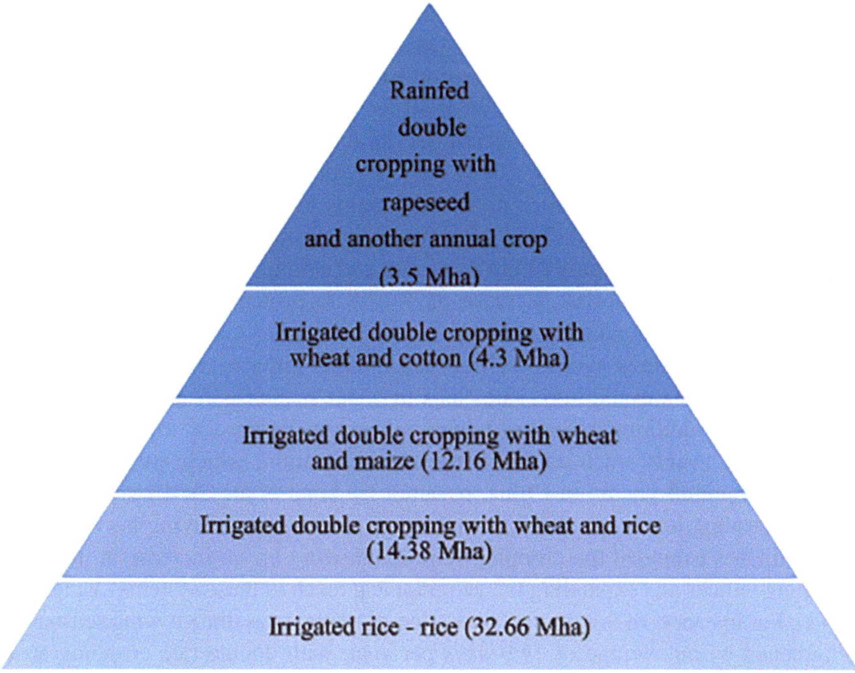

Fig. 12.2 Top five multiple cropping systems with largest physical area globally (Waha et al. 2020)

1.16 billion hectares without fallow; together with fallow (0.44 billion hectares), it constitutes 1.60 billion hectares. Globally, there are approximately 200 different multiple cropping systems that utilize a combination of 25 annual crops, with 29 of these systems accounting for 80% of the cultivated area. The areas dedicated to rice, wheat, and maize under multiple cropping represent 44% (49.63 Mha), 13% (24.12 Mha), and 10% (13.49 Mha) of their respective total areas. The five predominant multiple cropping systems worldwide include irrigated rice-rice (32.66 Mha), irrigated double cropping with wheat and rice (14.38 Mha), irrigated double cropping with wheat and maize (12.16 Mha), irrigated double cropping with wheat and cotton (4.3 Mha), and rainfed double cropping with rapeseed alongside another annual crop (3.56 Mha). Additionally, monocultures of wheat, maize, soybean, rice, and pulses cover a combined area of 468.8 Mha, representing over 50% of the global cropland area (Waha et al. 2020) (Fig. 12.2).

12.2.2 *Impact of Climate Change on System Productivity*

Anthropogenic increases in fossil fuel emissions have significantly raised atmospheric carbon dioxide (CO_2) levels and other greenhouse gases, leading to warmer temperatures, alterations in precipitation patterns, and a rise in extreme weather events worldwide. These emissions have been definitively linked to global warming, with global surface temperatures reaching 1.1 °C above the levels recorded between 1850 and 1900 during the period from 2011 to 2020 (IPCC 2023). Such climatic changes threaten crop yields, food security, and the sustainability of agricultural practices. While the predicted rise in atmospheric CO_2 concentrations to 550 ppm by 2050 (Ciais et al. 2014) may provide a “fertilization effect” for C_3 crops, it could also diminish their nutritional quality by lowering mineral concentrations in seeds (Myers et al. 2014). Recent studies indicate an average decline of about 8% in several minerals within the leaves and edible parts of C_3 crops (Loladze 2014), alongside a reported decrease of 3–17% in protein, zinc, and iron levels under elevated CO_2 conditions (Medek et al. 2017).

Rising temperatures have both direct and indirect impacts on cropping systems. Directly, increased evaporation rates lead to water stress for crops, particularly in arid regions. Additionally, higher temperatures can disrupt the timing of critical phenological events, such as flowering and maturation, which are essential for achieving optimal yields. In China, research by Song et al. (2010) indicated that global warming has resulted in earlier sowing dates while delaying harvest dates. This shift has extended the cropping season, allowing for an increase in multiple-cropping indices and expanding the geographical reach of these systems (Yang et al. 2011). For instance, in Southern China’s rice-growing areas, the growing season has lengthened by an average of 17.9 days per year, with double rice cropping areas moving northward by 300 km from 1961 to 2009 (Yan-Ling et al. 2011). Similarly, in Punjab, Pakistan, spring maize sowing was advanced by 4.6 days per decade from 1980 to 2014, while autumn maize emergence was delayed by 1.9–3.0 days per decade (Abbas et al. 2017b).

Shifts in temperature patterns affect the distribution, population dynamics, and overwintering survival rates of pests and diseases, which may introduce new threats to crops or worsen existing issues. Additionally, variations in precipitation can lead to more severe and extended droughts in certain areas, while others may encounter heightened rainfall and an increased risk of flooding. Without adaptation strategies (alteration in sowing date, that is, spring maize early sowing by 15 days and autumn maize delayed sowing by 20 days, fertigation, crop varieties) spring maize grain yield was reduced to 17–29% while autumn maize reduced to 19–33% but with the incorporation of these strategies grain yield was increased in both of the crops (Abbas et al. 2023). Thus, there is a need for management of the cropping system so that it can produce profitable yields in changing climatic scenarios.

12.2.3 *Impact of Climate Change on Soils*

Climate change impacts soil functions both directly and indirectly. Direct effects include changes in soil processes such as organic carbon transformations and nutrient cycling, driven by altered moisture and temperature conditions, as well as increased soil erosion from more frequent high-intensity rainfall events. Indirect

Fig. 12.3 Impact of climate change on soils

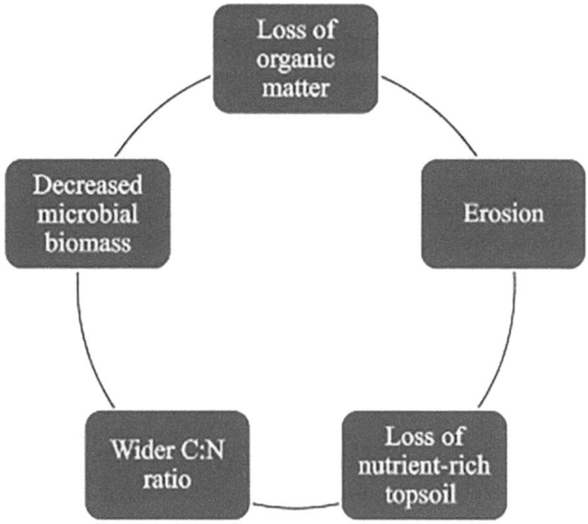

effects arise from adaptations like irrigation, changes in crop rotation, and tillage practices (Hamidov et al. 2018). The quantity, intensity, timing, and type of precipitation significantly influence soil formation, while seasonal and daily temperature fluctuations affect moisture availability, biological activity, chemical reaction rates, and vegetation types (Sivakumar and Stefanski 2007) (Fig. 12.3).

Increased temperatures are expected to negatively impact carbon allocation to the soil, leading to declines in soil organic carbon (SOC) and creating a feedback loop in the global carbon cycle, where higher temperatures result in greater CO_2 emissions from soils, further exacerbating warming (Wan et al. 2011). A study conducted in a semiarid steppe by Link et al. (2003) found that soil warming and drying caused a 32% reduction in soil carbon over 5 years, a more rapid decline than what has been observed due to increased tillage practices. Additionally, modeling studies predict a decrease in SOC of 2.0–11.5% in the North Central United States by 2100 compared to 1990 levels (Grace et al. 2006).

Increased evaporation from soil and heightened transpiration from plants are expected to lead to soil moisture stress (Bhattacharya and Bhattacharya 2021). A changing climate may also affect the workability of mineral soils, increasing their susceptibility to erosion, compaction, and reduced water retention. Drier soil conditions can hinder root growth and organic matter decomposition, making soils more vulnerable to erosion. In regions experiencing heavier winter rainfall, some soils may become more prone to erosion as well. For instance, water erosion models in the United Kingdom have predicted that a 10% rise in winter rainfall could potentially increase annual soil erosion by up to 150% during particularly wet years, although long-term averages of soil erosion may only show a modest increase compared to current levels (Favis-Mortlock and Boardman 1995).

Rising atmospheric temperatures and increased rainfall enhance mineralization and decomposition processes in soil, leading to greater microbial activity and biomass accumulation while reducing the soil carbon-to-nitrogen (C:N) ratio (Lal 2004). However, intense rainfall events can result in the loss of organic matter and nutrient leaching. For instance, Mo et al. (2022) found that as rainfall intensity increased, urea nitrogen leaching rose significantly from 0.04% to 18.09%. Climate not only affects soil properties but also regulates the climate itself through the uptake and release of greenhouse gases such as carbon dioxide, methane, and nitrous oxide. Soil can function as both a carbon source and sink, influenced by land use and climatic conditions (Eglin et al. 2010). Higher temperatures may accelerate microbial decomposition of organic matter, negatively impacting soil fertility over time. Additionally, increased rainfall in already moist regions can lead to heightened mineral leaching, particularly of nitrates.

12.2.4 Role of Cropping System-Based Soil Management in Climate-Resilient Agriculture

Cropping systems are essential for developing and implementing climate-resilient agriculture, providing a comprehensive approach to tackle the challenges posed by climate change. As temperatures rise, precipitation patterns shift, and extreme weather events become more frequent, the importance of resilient agricultural practices has become increasingly critical. To create a climate-smart cropping system, it is vital to utilize high-quality seeds or cultivars suited to specific agroclimatic regions, implement soil and water conservation techniques, adopt sustainable mechanization, and diversify crops either spatially (through crop substitution) or temporally (via crop rotation) (Panda et al. 2019). These strategies enhance the resilience of agricultural systems against climate variability.

Drought causes losses in yield for crops such as wheat and maize (Kim et al. 2019). The use of drought-tolerant varieties such as Pusa Wheat 8802 and CAH 201, CAH 202, and CAH 203 in maize released by CIMMYT can be used in cropping systems to combat climate change. Similarly, in the case of terminal heat stress in wheat, the newly released variety HD 3385 serves the purpose. Maize hybrids have also been released to address the issue of heat, cold, or frosts such as HQPM-1 and HHM-1 which is frost and cold-tolerant. Therefore, the use of heat-, drought-, flood-, and pest (insect and disease)-tolerant varieties needs to be incorporated into the cropping system. Introduction of a new crop, replacing the existing crops with the more efficient ones, water-saving technologies in rice via the system of rice intensification (SRI), aerobic rice, and custom hiring centers (CHCs) for farm machinery ensuring timely sowing can be used. Alternation in the sowing like early sowing of wheat not only enhances its productivity (236–241 kg/ha) and profitability but also saves the crop from terminal heat stress experienced by the crop during the grain filling stage (Paudel et al. 2023).

Diversification and crop rotation are essential practices in climate-resilient agriculture, allowing farmers to mitigate risks associated with climate-related stressors by cultivating a variety of crops in a planned sequence. Different crops have varying tolerances to temperature extremes, water availability, and soil conditions, which helps hedge against potential losses. Crop rotation disrupts pest and disease life cycles, reducing the likelihood of outbreaks that may be intensified by climate change. Additionally, conservation agriculture practices such as no-till farming, cover cropping, and agroforestry significantly enhance the resilience of cropping systems. Effective water management is crucial for adapting to climate variability, and cropping systems play a vital role in optimizing water use through techniques like drip irrigation, rainwater harvesting, and precision agriculture, enabling farmers to better cope with changing water availability.

12.3 Cropping System-Based Soil Management Strategies

It has long been realized by agricultural scientists that soil management practices are crucial not only for maximizing agricultural production but also for mitigating environmental pollution (Powlson et al. 2011). Therefore, it is essential to focus on protecting soil from erosion, which can lead to land shortages, as well as adopting practices that prevent soil contamination and degradation. Soil management practices can be categorized based on their impact on tillage, nutrient placement, and residue incorporation. Understanding how changes in these practices influence soil water balance, temperature, biological activity, and gas exchange between soil and the atmosphere is vital. Follett (2001) highlighted that the effects of soil management on soil organic carbon (SOC) are linked to tillage and management systems, increased crop cover, and enhanced efficiency in nutrient and water use within cropping systems. To effectively adapt soil management practices in response to climate change, it is important to comprehend how these factors influence greenhouse gas emissions.

12.3.1 Conservation Tillage

Conservation tillage can enhance soil physical structure, improve water retention, protect moisture, and boost crop yields. However, consistently using a single tillage method over the long term may lead to negative impacts on soil and the surrounding ecological environment, despite increases in crop production (Li et al. 2020a). The dynamics of carbon and nitrogen cycling in the soil are influenced by the distribution of soil water over time. Generally, reduced tillage intensity results in higher soil water content due to decreased evaporation rates from the soil surface, which is protected by crop residues and less disturbed soil (Hatfield et al. 2012). In two field studies comparing no-tillage with conventional tillage following wheat silage harvest and before corn planting, Mitchell et al. (2012) found that no-tillage retained an additional 0.89 and 0.97 inches of water compared to tilled soils.

Higher levels of available N, P, K, and soil moisture (ranging from 18.6% to 31.4%) have been observed in no-tillage systems. In the eastern Himalayas, no-till flatbed planting demonstrated the highest soil organic carbon (SOC) stock at a depth of 0–30 cm, while the no-till raised bed system exhibited the greatest SOC sequestration rate of 0.85 Mg/ha/year (Yadav et al. 2021). An 11-year study conducted in China's Loess Plateau found that both no-tillage (NT) and no-tillage with straw mulching (NTS) resulted in higher SOC and total nitrogen content in the top 0–10 cm of soil (Li et al. 2018). Qin et al. (2006) noted that no-tillage increased root length density in summer corn at a depth of 0–5 cm compared to traditional rotary tillage, contributing to greater carbon pool inputs. Liu et al. (2004) reported that conservation tillage could enhance corn and wheat yields by 11.8% and 9.7%, respectively, within a corn-wheat cropping system. The impact of no-tillage on greenhouse gas emissions has been widely debated, with studies showing varied outcomes: some indicate significant increases (Zhang et al. 2016), while others report decreases (Lu et al. 2015) or no effect (Bayer et al. 2015) on CO_2, CH_4, and

Table 12.1 Effects of different tillage systems on CO_2, N_2O, and CH_4 emissions

Study area	Cropping system	Findings	References
Beijing (China)	Soybean-wheat-maize, soybean-wheat, maize-wheat	↑ 28.89% (2019) and 37.94% (2020) cumulative N_2O emissions under Rotary tillage than no tillage ↑ 20.93% (2019) and 19.6% (2020) cumulative CH_4 emissions under Rotary tillage than no tillage	Virk et al. (2022)
Changsha (China)	Double-cropping rice	↑ N_2O emissions from paddy fields with no tillage ↓ CH_4 emissions from paddy fields with no tillage ↓ Per yield GWP CO_2 with no tillage	Tang et al. (2022)
Shandong Province (China)	Wheat maize	↓ CO_2 (35.43%), CH_4 (67.33%), and N_2O (339.07%) emissions in no tillage than conventional tillage ↓ GWP (37.25%) in no tillage than conventional tillage	Li et al. (2023)
North China Plain (China)	Wheat-maize	↓ N_2O emission in winter wheat under no tillage ↑ CH_4 uptake in summer maize under no tillage	Pu et al. (2022)
West Bengal (India)	Rice	↓ N_2O emission under conventional tillage than no tillage ↑ CH_4 and CO_2 fluxes under the conventional tillage than no tillage	Gangopadhyay et al. (2023)
Gazipur (Bangladesh)	Rice-mustard-rice	↓ CH_4 emission factor (24–47%), CH_4 flux (20–32%), GHG intensity (32–46%), and GWP (31–47%) under strip tillage than conventional tillage	Haque et al. (2023)
Kentucky (USA)	Corn-soybean	↓ Soil CO_2 (1.6% for corn and 4.53% for soybean) and N_2O emissions (10.5% for corn and 19.6% for soybean) under no tillage than conventional tillage	Huang et al. (2022)

N_2O emissions from agricultural soils. Overall, conservation tillage is vital for successful cropping systems, promoting sustainable soil management practices that enhance soil health, reduce environmental impacts, and improve the resilience of agricultural ecosystems (Table 12.1).

12.3.2 Cover Cropping

Cover cropping is a conservation practice widely adopted by farmers to mitigate soil erosion while enhancing soil organic matter, nitrogen (N) cycling, soil structure, and reducing pest infestations and nitrogen leaching compared to systems without cover crops (Daryanto et al. 2018). Generally, legume cover crops are more effective at supplying nitrogen to subsequent crops through biological nitrogen fixation than nonlegume or no cover crops (Gabriel et al. 2012). Conversely, nonlegume cover crops can be more beneficial for increasing soil organic matter and minimizing nitrogen leaching due to their higher biomass production (Sainju et al. 2003).

Cover crops enhance soil health by reducing soil penetration resistance by an average of 5%, improving wet aggregate stability by an average of 16%, and increasing cumulative infiltration by an average of 43% (Blanco-Canqui and Ruis 2020). Studies show that cover crops like hairy vetch, crimson clover, pea, turnip brassica, oats, and rye significantly increase soil organic carbon (SOC) concentrations compared to plots without cover crops (Olson et al. 2014). This increase in SOC is attributed to the addition of both above and belowground biomass to the soil (Poeplau and Don 2015). For example, in the Eastern Cape, South Africa, cover crop mixtures such as 70% oats + 30% vetch and sole vetch contributed higher total carbon to deep alluvial soils compared to sole oats and control groups (Mukumbareza et al. 2016). Cover crops not only increase nutrient availability and store carbon and nitrogen in the soil but also mitigate the effects of global warming (Dabney et al. 2010). Soils treated with hairy vetch and crimson clover have shown a marked increase in nitrogen concentration compared to fallow soils (Sainju et al. 2002). Hairy vetch can provide the equivalent of 50–155 kg/ha (Seo et al. 2000) and 150–250 kg/ha of fertilizer N, respectively, to subsequent corn crops. Similarly, in wheat rotations, soil nitrogen levels increased by 258 kg/ha and 279 kg/ha under soybean and sun hemp, respectively, compared to treatments without cover crops (Blanco-Canqui et al. 2011). Legume cover crops supply significant quantities of plant-available nitrogen to subsequent crops (Thilakarathna et al. 2015), enhancing soil fertility and reducing the need for nitrogen fertilizers (Fortuna et al. 2008). Besides improving soil fertility, cover crops contribute to nutrient recycling, altering nutrient cycles and the timing of nutrient availability (Fageria et al. 2005). Legume species, in particular, contribute to net nitrogen inputs through biological fixation, potentially adding over 100 kg N/ha, even when cultivated for a short period (Buchi et al. 2015).

Soils treated with cover crops such as hairy vetch, common vetch, and white mustard have been shown to enhance plant production, resulting in crop yield increases of up to 24% in reduced tillage organic systems, compared to only a 2% increase in conventional tillage systems (Wittwer et al. 2017). Additionally, crops following leguminous cover crops yielded 11% more than those without cover crops (Chahal and Van Eerd 2023). Overall, cover crops significantly improved total phospholipid-derived fatty acids (PLFA), microbial biomass carbon (MBC), and microbial biomass nitrogen (MBN) by 24%, 40%, and 51%, respectively. The presence of cover crops also led to a 7–31% increase in total bacteria and fungi populations compared to no cover crop scenarios, with fungi showing a greater response than bacteria, as indicated by a higher fungi-to-bacteria ratio (Muhammad et al. 2021). In addition, bacterial phospholipid fatty acid levels and enzyme activities rose by 6.8% and 20%, respectively, in soils amended with mixtures of oats and radish or oats, radish, and vetch (Chavarria et al. 2016) (Table 12.2).

Table 12.2 Crop cover impacts on compaction/penetration resistance and aggregate stability

Study area	Cover crops	Findings	References
Northeastern Mississippi (USA)	Elbon rye, Daikon radish, Austrian winter field peas	↓ 4.9% and 6.29% soil BD in peas (B and G), 7.69% and 6.29% soil BD in rye (D and F) than no cover crop treatment ↑ The proportions of 1–0.5 mm and 0.25–0.5 mm aggregates 86.37% in rye (F) and 88.78% in peas (B) than no cover crop treatment at 0–5 cm depth ↓ The proportion of <0.25 mm aggregates 28.16% in peas (B) than no cover crop treatment at 0–5 cm depth	Dai et al. (2024)
Parana State (Brazil)	Black oat, turnip, white oat, rye, white lupine, and common vetch	↓ Soil BD up to 5% ↑ Soil air-filled porosity up to 27% ↑ Saturated hydraulic conductivity up to 300%	Pott et al. (2023)
Middle Black Sea Region (Turkey)	*Vicia villosa*	↑ 10.9% total porosity, 161.2% saturated hydraulic conductivity, 9.9% ↓ Soil BD 11.7%	Demir and Isik (2020)
Southwestern United States (USA)	Hairy vetch, rye, and mustard	↑ 18% wet aggregate stability in mustard and 27% in mustard-vetch plots ↑ Mean weight diameter of soil aggregates	Antosh et al. (2020)
Missouri (USA)	Hairy vetch, red clover, turnips, buckwheat, cereal rye, winter peas, barley, triticale, radish, and cowpeas	↑ Total pores and macropores under cover crop than no-cover crop ↑ 104.7% Saturated hydraulic conductivity under cover crop than no-cover crop	Çerçioğlu (2020)

12.3.3 Organic Amendments

Organic soil amendments, including animal manures, composts, green manures, crop residues, straws, and biochar, are utilized to enhance soil and plant health, contributing to sustainable agricultural practices (Ansari et al. 2017; Singh et al. 2023a, b). Incorporating these organic inputs into the soil not only boosts organic matter content but also improves various physical properties of the soil (Williams et al. 2017). The addition of organic matter enhances soil characteristics such as water-holding capacity, aggregate stability, infiltration rates, drainage, and porosity. For instance, the application of biochar has been shown to increase the formation of water-stable aggregates (>0.25 mm) in the top 15 cm of soil (Liu et al. 2014). Long-term studies indicate that organic amendments can raise soil organic carbon (SOC) levels by an average of 49% compared to unfertilized controls or by 29% compared to fertilized controls (Chen et al. 2018). In an 18-year study on paddy rice production, annual applications of manure and straw resulted in SOC increases of 72% and 103%, respectively (Pan et al. 2009). Organic amendments also provide direct supplies of macro- and micronutrients, with nitrogen availability particularly influenced by the mineralization rates of these organic materials. The nutrient content varies significantly depending on the source and quality of the amendments (Quilty and Cattle 2011). Meta-analyses reveal that using organic amendments can lead to yield increases up to 60% greater than nonamended controls, while combining organic amendments with nitrogen fertilizers can enhance yields by as much as 114% (Chivenge et al. 2011). Liu et al. (2018) demonstrated that co-applying horse manure with chemical fertilizers significantly improved corn yields compared to using nitrogen, phosphorus, and potassium fertilizers alone. A combination of 50% recommended dose of fertilizers (RDF), full crop residues, farmyard manure at 5 t/ha, and biofertilizers resulted in reduced dry bulk density, higher specific volume, minimized water-filled pore space by 4–6%, increased available nitrogen, and enhanced enzyme activities compared to RDF alone (Nath et al. 2023). Overall, organic amendments are vital for cropping system-based soil management as they promote fertility, improve soil structure, foster microbial diversity, and support sustainable agricultural practices.

12.3.4 Water Conservation

Agriculture plays a dual role as both a contributor to and a victim of climate change, particularly in the context of water scarcity. As global warming intensifies, it disrupts the hydrological cycle, leading to reduced water availability in various regions. In India, for instance, the per capita annual water availability is projected to decline to 1465 m^3 by 2025 and further to 1235 m^3 by 2050 (Kumar et al. 2013). To effectively address these challenges, the adoption of smart water technologies and modern irrigation techniques is essential for enhancing water productivity while ensuring sustainable and climate-resilient farming practices. A comprehensive approach known as climate-smart agriculture (CSA) aims to increase soil organic carbon (SOC) while reducing greenhouse gas (GHG) emissions and the water footprint (WF) to secure food resources. Research by Zizinga et al. (2022) demonstrated that CSA practices such as mulching, permanent planting basins (PPB), and half-moon pits (HM) significantly improved the resilience of maize crops in rainfed systems by enhancing total soil water storage by 1–12% compared to control treatments. In cooler climates, Ground Cover Rice Production Systems (GCRPS) employing various types of mulches have been shown to decrease water usage by 60–85% relative to conventional flooded systems without compromising rice productivity (Huang et al. 1999). Furthermore, combining stone bunds with compost application has led to a remarkable increase of approximately 142% in sorghum grain yield, improving the cropping system's adaptability to climate change due to its resilience against both floods and droughts (Zougmoré et al. 2014). The alternate wetting and drying (AWD) technique is another effective method for conserving water in rice cultivation, allowing irrigation only after ponded water has disappeared for several days; studies indicate that combining AWD with biochar application results in irrigation savings and reduced methane emissions without negatively impacting yields or farmer incomes (Sriphirom et al. 2020). Aerobic rice cultivation represents an innovative approach that eliminates continuous flooding, reducing total water use by 60% compared to traditional methods while achieving 1.5–2 times higher water productivity (Surendran et al. 2021). The System of Rice Intensification (SRI), designed for resource-poor farmers, reduces water requirements by half while doubling yields compared to conventional rice farming methods (Uphoff and Randriamiharisoa 2002). As climate change increases the frequency of extreme weather events, no-till management practices will be crucial for enhancing infiltration rates and minimizing surface runoff losses (Liu and Basso 2020). Advanced techniques such as laser land leveling create uniformly flat fields, which reduce irrigation time and minimize evaporation losses (Humphreys et al. 1996). Microirrigation technologies, particularly drip irrigation systems, deliver water directly to the root zone more efficiently than traditional surface methods. The adoption of site-specific technologies tailored to agroecological conditions is essential for optimizing both water productivity and crop yield potential. These advancements not only contribute positively to reducing GHG emissions but also play a vital role in achieving sustainable development goals (SDGs).

12.3.5 Carbon Sequestration

Carbon sequestration in soil involves capturing carbon from the atmosphere and storing it in the soil, ensuring a high mean residence time and low reemission rates (Lal et al. 2015). This process enhances soil carbon content, which is beneficial for both mitigating and adapting to climate change (Ussiri and Lal 2017). As global temperatures rise due to climate change, the rate of soil organic carbon mineralization increases, necessitating management strategies that either reduce this mineralization or enhance carbon storage in soils (Tiefenbacher et al. 2021). Intensive tillage practices accelerate the oxidation of soil organic carbon, making no-till or reduced tillage methods preferable for increasing carbon sequestration in agricultural soils (Haddaway et al. 2017). By minimizing tillage, the rearrangement of organic matter and disintegration of soil aggregates, key components for storing soil organic carbon, are reduced, thereby protecting them from microbial degradation (Pinheiro et al. 2015). Additionally, practices such as mulching and retaining crop residues on the soil surface without burning them contribute to carbon sequestration in the soil (Chen et al. 2019). These crop residues add carbon to the soil and reduce mineralization; if provided, their C:N ratio is sufficiently high. Adopting suitable changes in the cropping system plays an important role in carbon sequestration. Crops with C_4 pathway of carbon fixation are found to be better sequesters of carbon than the ones having C_3 pathway (Wang et al. 2012). Following crop rotation including crops from various groups that can exploit various depths of soils and whose nutrient requirements are different helps in maintaining soil productivity and sequester more carbon than monoculture systems (Triberti et al. 2016). Including green manure incorporation and cover cropping in the cropping systems is beneficial. Cover cropping decreases soil erosion, adds organic matter to the soil, reduces nitrous oxide emission by using up residual nitrogen from the previous crop, and increases infiltration, water-holding capacity, and soil biodiversity, thus making the system climate smart (Kaye and Quemada 2017). Incorporation of deep-rooted crops and crops with higher root biomass in the system is found to sequester larger amounts of carbon in deeper layers of soil, thus reducing the chances of carbon loss (Gherardi and Sala 2020). Conservation agriculture with all the above said principles of reduced soil disturbance, crop rotation, and soil cover throughout the year has high potential to sequester carbon and thus mitigate and adapt to climate change (Gonzalez-Sanchez et al. 2019). Adoption of efficient nutrient management with integration of both chemical and organic manure can increase carbon sequestration by increasing the crop productivity and biomass addition to soil (Lal et al. 2015). Nitrogen application increases both biomass accumulation and mineralization rate, and optimization of its dose, along with balanced supply of other nutrients, is required to ensure maximum carbon sequestration (Van Groenigen et al. 2017). Owing to the high stability and large-scale production potential, biochar is a highly potent carbon sequester. Biochar is a carbon-rich soil ameliorant made from pyrolysis of organic matter and has long residence time in soil and low decay rate, with potential to increase nutrient and water availability, yield, microbial biomass, and microbial biodiversity and ultimately reduce the carbon footprint of the crops (Gross et al. 2021). Organic matter converted into biochar will not be easily mineralized by microorganisms to release carbon dioxide, resulting in their longer residence time in soil. Organic farming that solely depends on manures (rich in carbon) as a source of nutrient follows crop diversification, includes cover cropping and incorporation of crop residue, and sequesters high quantities of carbon than the conventional systems that are mostly dependent only on chemical fertilizers for fertilization (Mandal et al. 2022). Incorporating forestry component in agriculture (agroforestry) increases both above- and below-ground carbon sequestration potential of the system, because of high shoot and root biomass of trees that are lignin rich (Abbas et al. 2017a, b). Apart from this, agroforestry system also provides food, fuel, and fodder, helps in increasing biodiversity, moderates microclimate, reduces greenhouse gas emission, controls erosion, and increases farm productivity and sustainability (Sollen-Norrlin et al. 2020). Raising bioenergy crops rich in lignin, such as miscanthus, is also good measure of sequestering carbon in agricultural system (Chimento et al. 2016). Apart from these measures, adoption of suitable package of practices, such as timely sowing, good crop stand, efficient nutrient and water management, pest and disease management, etc. that ensure good crop growth, results in better carbon sequestration (Table 12.3).

12.3.6 Nutrient Management

Nutrient management is crucial for enhancing crop productivity and sustainability in the context of climate change, as it helps mitigate the factors contributing to climate change while also making cropping systems more resilient to its effects. As

Table 12.3 Carbon sequestration rates in different production systems

Study area	Methods	Cropping system	Findings	References
Hunan Province (China)	Biochar feedstock: 70% maize stover and 30% saw dust at 450–500 °C	Rice-rice, maize-rice	↑ 22% and 19% SOC in R–R system, 30% and 31% SOC in M–R system relative to control and straw, respectively ↑ 4% SOC stocks in M–R system than R–R system with biochar addition	Mehmood et al. (2020)
Tripura (India)	Tillage and land configurations	Maize-field pea	Highest SOC stock (0–30 cm) under no-tillage flatbed planting Maximum SOC sequestration rate (0.85 Mg ha^{-1} year^{-1}) under no-tillage-raised bed planting	Yadav et al. (2021)
Faisalabad district (Pakistan)	Multi-nutrient enriched compost (NPKEC)	Maize-cowpea	Highest SOC sequestration under multinutrient-enriched compost (0.30 Mg C ha^{-1} year^{-1}) than other treatments	Roohi et al. (2022)
Jharkhand (India)	Gamhar tree-based agroforestry	Gamhar + Arhar	↑ 138.1% carbon stock, 136.64% carbon sequestration, 136.64% carbon credit, and 136.8% carbon price than sole Gamhar in 2016–2017 in pooled data	Kumar et al. (2024)
Northwest China (China)	Green manure: *Vicia villosa*	Maize-wheat	↑ 72.1% C sequestration rates in the 0–60 cm layer than mineral nitrogen fertilizer	Ma et al. (2024)

enhanced CO_2 concentration can increase the biomass production and resultant exhaustion of soil innate fertility, balanced nutrition following the 4Rs stewardship is most important in enhancing the nutrient use efficiency and yield and fighting climate change (Norton 2014). Employment of site-specific nutrient management practices such as STCR-, Nutrient Expert-, LCC-, SPAD-, QUEFTS-, NDVI-, remote sensing-, etc.-based fertilization with integration of both chemical and organic source of nutrients can help in making the goal of "sustainable agriculture under changing climate" achievable (Singh et al. 2021). Right dose of nitrogen can improve the yield under adverse climate conditions and also reduce gaseous emissions responsible for the same. While lower dose of nitrogen restricts the productivity of the more vigorously growing crop under increased atmospheric CO_2, higher doses can increase the nitrous oxide emission to the atmosphere, which itself is a greenhouse gas and also promote the formation tropospheric ozone, which is detrimental to the plant (Pathak et al. 2016). Excess nitrogen also results in its volatilization in the form of ammonia and reacts with oxides of nitrogen and sulfur present in the atmosphere to form sulphate nitrate ammonium (SNA) particles, which form PM 2.5 particles, causing air pollution (Fung et al. 2019). Optimization of the nitrogen release rate from the nitrogen sources using smart fertilizers which employ various microorganisms, nanomaterials, nitrification inhibitors of biological and chemical origin, sensors of biological and chemical nature that respond to external conditions, semipermeable coatings, etc. helps in increasing the nitrogen use efficiency (Calabi-Floody et al. 2018). Phosphorus availability is also influenced by climate change. Fluctuation in the temperature and rainfall patter can affect the phosphorus release from the sources (as release rate is dependent on temperature, pH, soil moisture, and microbial population), its diffusion rate in the soil, and transport with runoff water (Forber et al. 2018). Elevated CO_2 levels and saltwater intrusion as a result of climate change also increase phosphorus loss from soil into water bodies, increasing eutrophication effect (Lucas et al. 2023). Optimum phosphorus supply to ensure good crop growth and to avoid phosphorus leaching, adoption of cover crop instead of fallowing in seasons with high rainfall, and establishment of riparian strips to absorb phosphorus from runoff water can increase phosphorus use efficiency and avoid eutrophication (Jeppesen et al. 2009). Thought potassium and calcium fertilization are not given much importance in the nutrient management schedule; both nutrients play an important role in crop production under abiotic and biotic stress conditions. While potassium maintains osmotic pressure and helps in pest tolerance, calcium acts as sensors for moisture stress and potassium deficiency in the soil, both together helping in achieving better water use efficiency under drought conditions (Ramírez-Builes and Küsters 2021). Sulphur fertilization also plays an important role in mitigating and adapting to climate change. Sulphur is an integral component of compounds which help in scavenging reactive oxygen species formed in plant as a result of abiotic stress, chelate the heavy metals entering the plant, involve in abscisic acid signalling during salinity stress, and fight against diseases and pests (Procházková et al. 2016). Sulphur, being a component of nitrate reductase enzyme, plays an important role in nitrogen assimilation and thus increases nitrogen use efficiency (Zenda et al. 2021). Sulphur fertilization under terminal heat stress is reported to decrease yield loss and increase protein content, elasticity, and gluten content of wheat (Tao et al. 2018). Moreover, sulfur fertilization also helps in mitigating methane emission from the wetland paddy soils by promoting the sulphate-reducing bacteria which compete with methanogens for the same substrate materials (Denier van der Gon et al. 2001). Application of ammonium sulphate as nitrogen source in paddy resulted in 21.6% less CH_4 emission than urea application (Malyan et al. 2016). Zinc and iron application is also recommended under changing climate as accumulation of the element in the grain under elevated CO_2 condition is found to be suboptimal and much below the human dietary recommendations (Nakandalage et al. 2016; Wang et al. 2016). Changes in temperature and CO_2 concentration also alter availability of micronutrients in soil as well as their uptake and translocation by plant (Wang et al. 2016) denoting their significance in climate-smart agriculture. Integration of organic fertilizers along with chemical supplement is also important. Their combined application meets the immediate need of higher quantity of primary nutrients while simultaneously increasing soil organic carbon, improving physical properties of soil, and gradually releasing micronutrients (Shi et al. 2023). Such integrated nutrient management can not only increase nutrient use efficiency and yield but also contribute to a reduction in greenhouse gas and ammonia emissions (Sharma et al. 2021).

12.4 Technological Innovation in Cropping System-Based Soil Management

12.4.1 Remote Sensing

Remote sensing data, obtained from various technologies such as satellites, aerial photography, GPS, LiDAR, ground-based sensors, and radar systems, presents significant opportunities for soil classification, mapping soil resources, assessing soil moisture, and identifying soil degradation like salinity. Hyperspectral remote sensing (HRS) is particularly advantageous due to its capacity to measure reflectance across numerous narrow wavelength bands. Agricultural soils typically exhibit simpler spectral signatures than plants, with a consistent increase in reflectance observed in the visible and near-infrared (NIR) regions (Price 1990). The spectral signature of soil is influenced by several factors including mineral composition, organic matter content, soil moisture levels, texture, and crop residue (Nagler et al. 2000). Wet and organic-rich soils tend to reflect less light compared to dry, smooth-surfaced soils which appear brighter (Daughtry 2001). Specific minerals such as iron oxides can be identified through their unique spectral characteristics, notably increased red reflectance. Spectral reflectance analysis is a common practice in remote sensing; the visible range (400–700 nm) indicates organic matter and iron oxide minerals, while NIR reflectance (700–1300 nm) provides insights into soil moisture and clay minerals. The shortwave infrared reflectance (1300–2500 nm) is particularly useful for analyzing soil organic carbon content and identifying minerals like gypsum or calcite (Balaram and Sawant 2022). In the shortwave infrared range, soil spectra display more features compared to shorter wavelengths (Daughtry 2001). Additionally, thermal infrared imaging measures thermal radiation in the longwave infrared region (8–14 μm), aiding in the detection of variability in soil compaction, fertility, moisture content, texture, and organic matter levels (Farella et al. 2022). Radar remote sensing employs microwave signals to discern subsurface characteristics such as soil moisture content, surface roughness, and texture.

The normalized difference vegetation index (NDVI) and soil-adjusted vegetation index (SAVI), derived from optical imagery, serve as effective tools for evaluating water stress and soil moisture levels in various crops. The crop water stress index (CWSI), based on thermal remote sensing, is particularly useful for irrigation scheduling in orchard crops (Khanal et al. 2017). Optical and thermal remote sensing data have been employed for estimating soil moisture and evapotranspiration (ET) using methods like the land surface temperature-vegetation index (LST-VI) approach and the optical trapezoid model (OPTRAM). High-resolution land surface temperature and vegetation data, up to 1 km, obtained from the Moderate Resolution Imaging Spectroradiometer (MODIS), can facilitate soil moisture estimation. Various vegetation indices such as NDVI and SAVI are correlated with chlorophyll content and photosynthetic activity. Handheld and tractor-mounted remote sensors like Green Seeker, Yara N-sensor, and Crop Circle utilize crop reflectance to assess spatial variability of fertility in real time (Ali et al. 2017). Blasch et al. (2015) used multitemporal satellite imagery from RapidEye and principal component analysis to create field-scale maps of soil organic matter for precision management practices. Di Gennaro et al. (2016) found a strong correlation between NDVI and grapevine leaf stripe disease. Abdulridha et al. (2020) applied a machine learning approach with vegetation indices from hyperspectral UAV images, achieving 96% accuracy in early citrus canker detection. The unique spectral signatures related to phonological or morphological traits enable differentiation between weeds and crop plants, with recent advancements in machine learning methodologies enhancing weed mapping efforts (Huang et al. 2020). Remote sensing data regarding crop growth metrics such as Leaf Area Index (LAI) and biomass provide insights into site-specific characteristics, management practices, and various biotic and abiotic stressors affecting crops (Campos et al. 2019).

12.4.2 Data-Driven Decision-Making Tools

The application of remote sensing in agriculture has facilitated the systematic collection of extensive data across large geographical areas. These datasets are heterogeneous, arising from multiple sources with varying spatial, temporal, and spectral attributes. Machine learning techniques, especially those utilizing non- and semiparametric structures, are commonly employed for their adaptability in managing large input datasets and addressing nonlinear problems.

Soil sensors, coupled with advanced artificial intelligence (AI) and machine learning (ML) algorithms, provide valuable decision-making support, particularly in customizing irrigation strategies and monitoring hydrological risks. Research has focused on developing and evaluating models such as support vector regression (SVR), artificial neural networks (ANN), and random forests (RF) to predict soil organic carbon (SOC) stocks, utilizing auxiliary data including soil characteristics, climate information, topography, and remote sensing data (Were et al. 2015). The study's findings included the creation of maps that forecast SOC stocks in the area. Accurate assessments of soil properties are essential for optimizing soil management, nutrient planning, and land-use decisions (Lahoche et al. 2002). In agriculture, land management practices aim to prevent floods and landslides to reduce negative environmental impacts. Estimating soil moisture provides detailed water content maps, which, when combined with temperature and weather forecasts, enhance irrigation systems and help maintain climatic balance (Hassan-Esfahani et al. 2017).

Various deep learning models have been applied in agriculture for performance analysis, including hybrids of principal component analysis, autoencoders, and logistic regression for land cover classification (Chen et al. 2014). Additionally, the use of convolutional neural networks (CNN) combined with support vector machines (SVM) has been explored for segmenting soil and roots (Douarre et al. 2016). A Deep Belief Network-based Macroscopic Cellular Automata (DBN-MCA) approach was utilized for predicting soil moisture content by Song et al. (2016), while deep learning models have also been employed for crop yield estimation. Furthermore, Monís et al. (2020) developed a Genetic Algorithm (GA) to optimize the design of irrigation pumping systems powered by photovoltaic (PV) energy.

12.4.3 Robotics and Automation

The implementation of automated operations in agriculture plays a crucial role in enhancing crop management efficiency, leading to cost and energy savings. Autonomous robotic platforms, equipped with various sensors, contribute to decision support by optimizing soil variability (Fentanes et al. 2018). Robotics can be leveraged to detect variations in soil fertility, microbial activities, soil moisture content, and other physical properties. Mapping the agricultural environment using robotics involves advanced technologies such as laser-induced breakdown spectroscopy (LIBS) sensors for soil fertility measurement (Yan et al. 2018), cosmic neutron detectors for moisture assessment (Badiee et al. 2021), and infrared gas sensors for monitoring CO_2 emissions across fields. The autonomous sampling and measurement of soil by robotic systems not only save time but also reduce costs, offering the potential to significantly improve precision and decision support for farmers.

In the realm of arable cropping, Controlled Traffic Farming (CTF) systems utilize Real-Time Kinematic Global Positioning System (RTK-GPS) data to guide field vehicles precisely along established traffic lanes. This innovative approach aims to minimize soil compaction by confining it to designated lanes, thereby aiding in the reduction of N_2O gas emissions, a potent greenhouse gas. Robotic platforms have the capability to revolutionize the current level of mechanization by exploring the deployment of extensive fleets of ultralight weight machines (weighing >5 Mg) (Polvara et al. 2021). With challenges such as limited availability and high labor costs in agricultural crop production (Marinoudi et al. 2019), the evolution of robotic systems becomes essential to increase autonomy and overall capability (Bechar and Vigneault 2016).

A sensing system utilizing infrared technology was developed to quantify the sowing rate of seeds for chickpea, wheat, and alfalfa, utilizing infrared light emitting diodes (IR LEDs) and photodiodes (Besharati et al. 2019). John Deere introduced the Sesam 2, an electric robot tractor with a power output of 300 kW (400 HP), which plays a significant role in both tilling and harvesting and can collaborate with multiple other robots (Heidrich et al. 2022). Additionally, Kim et al. (2020) created an autonomous spraying system based on the SegNet model for pesticide application in orchards, which integrates hardware configuration, semantic segmentation, and the fusion of depth and RGB data for improved performance. Li et al. (2020a, b) developed an enhanced detection algorithm that achieved 94.6% accuracy in harvesting robots.

12.4.4 Advances in Soil Monitoring Technology

Delineating soil conditions is essential for optimizing resources to enhance agricultural yields and minimize environmental impacts, as well as for creating site-specific databases that link soil conditions to plant growth within sustainable agricultural systems. There is a pressing need for innovative technologies to efficiently gather soil information, addressing the requirements of smart and precision agriculture (Kayad et al. 2020). Fiber optic sensor-based (FOS) technology can be employed to measure and monitor various physical and chemical processes in the soil. These sensors are particularly effective for assessing small-scale processes like water dynamics due to their compact size and minimal intrusion. Additionally, FOS technology enables continuous real-time monitoring, and its multiplexing capabilities allow for extensive measurement coverage across large areas.

The estimation of soil water content can be effectively performed using the near-infrared (NIR) reflectance technique at wavelengths of 970, 1200, 1450, and 1970 nm (Stenberg et al. 2010). This method allows for rapid and nondestructive measurement of various soil properties, including organic matter, minerals, pH, and heavy metal content. Several electromagnetic sensors provide noninvasive, on-site measurements of soil moisture levels, including dielectric sensors, time domain reflectometry (TDR) sensors, time domain transmission (TDT) sensors, and capacitance sensors. The advent of wireless sensor networks (WSNs) has significantly improved our understanding of hydrological systems at a catchment scale by offering continuous three-dimensional soil moisture data with high temporal resolution and extensive spatial and vertical coverage beyond the root zone (Leone 2022).

12.5 Challenges and Opportunities

12.5.1 Addressing Climate-Induced Soil Challenges

The impact of climate change on agricultural productivity and food supply is significantly influenced by climate-induced soil degradation. Global soil and water resources are limited and vulnerable to misuse and mismanagement, closely linked to processes that also affect climate change. As the Earth's climate experiences unprecedented alterations, agriculture faces substantial challenges. Rising temperatures, shifting precipitation patterns, and more frequent extreme weather events render soils increasingly susceptible to degradation. This chapter examines the effects of climate change on soils and investigates innovative cropping systems designed to mitigate these challenges.

Global soil resources are limited and unevenly distributed across biomes and geographical regions, significantly impacted by climate change and variability, and are susceptible to degradation—whether physical, chemical, or biological due to land misuse and mismanagement. However, these soils can be restored through careful land use and appropriate management practices. Climate-induced soil challenges include erosion, nutrient depletion, and alterations in microbial activity. Title cropping systems present a promising solution by optimizing plant-soil interactions and enhancing the resilience of agricultural ecosystems. By understanding the specific soil challenges posed by climate change, farmers can implement targeted practices to improve soil health and productivity.

One-third of global soils are degraded, releasing 78 gigatons (Gt) of carbon dioxide into the atmosphere, which results in economic losses exceeding 10% of global GDP due to diminished biodiversity and ecosystem services (Poeplau 2019). Healthy soil is essential for achieving higher crop yields and enhancing resilience against drought, making soil health management a central focus in discussions about global food security and climate action (Lal et al. 2018). To meet the rising food demand and fulfill sustainable development goals (SDGs), efficient management of land and water resources is crucial. Rain-fed agriculture faces increasing challenges from climate change, with rising temperatures being a key factor contributing to reduced crop yields through accelerated decomposition of soil organic matter (SOM).

12.5.2 Integration of Traditional and Modern Practices

To effectively manage soils in the face of climate change, a harmonious integration of traditional and modern agricultural practices is essential. Indigenous knowledge, passed down through generations, provides valuable insights into sustainable land management. Combining this wisdom with cutting-edge technologies and modern scientific approaches creates a holistic strategy for climate-resilient agriculture. Exploring case studies where farmers have successfully integrated traditional crop rotation, cover cropping, and agroforestry practices with precision agriculture, sensor technologies, and data analytics would help to address the challenges in adoption of technology-driven soil management. This synthesis not only enhances soil fertility and structure but also optimizes resource use, making agriculture more sustainable in the long run.

Bottlenecks in adopting existing technologies and a lack of interest among farmers must be addressed through effective training and extension activities that explain their impact on soil dynamics. By integrating technologies such as conservation agriculture, precision agriculture, water harvesting, and agroforestry, along with efficient nutrient and water management practices, it is possible to create research priorities focused on soil and site-specific identification of suitable system-based agronomic packages. Promoting precision and conservation agriculture will significantly aid in tackling the various challenges posed by climate change.

12.6 Future Directions

In the era of rapid technological advancement, innovative solutions are emerging to tackle climate challenges in agriculture by providing real-time data on soil conditions, which empowers farmers to make informed decisions. These technologies can be integrated into cropping systems, allowing for customized soil management practices that cater to the specific requirements of each crop. With these tools, farmers can swiftly adapt to changing environmental conditions, thereby ensuring the resilience and sustainability of their agricultural operations. To maximize agriculture's potential for mitigation and adaptation, investments in new technologies are essential. Enhancing input efficiency and establishing incentives and monitoring systems that include small and marginal farmers will be crucial for the success of both mitigation and adaptation efforts, particularly those focused on adaptation-led mitigation. It is vital to prioritize adaptation strategies in key areas such as developing stress-tolerant crops, implementing storm warning systems, improving water storage and diversion methods, and strengthening contingency planning and infrastructure. The emerging technologies in the developing countries, namely, conservation agriculture, precision agriculture, deploying remotes sensing technologies, IoT devices, sensors, and models to assess the real-time soil dynamic conditions to support and deploy efficient management practices. The integration of IoT, sensors, drones, CNNs, LLMs, and big data holds immense potential to revolutionize soil management in cropping systems. Future applications encompass real-time monitoring of soil health through sensors and IoT, providing a mechanism for promptly detecting changes caused by anthropogenic activities. Drones and CNNs play a crucial role in identifying crop stress and nutrient deficiencies, facilitating timely corrective measures for enhanced crop productivity. LLMs contribute to predicting crop yields and optimizing fertilizer application. Precision agriculture techniques, driven by IoT and big data, offer a pathway to minimizing environmental impact by reducing the use of agrochemicals. Additionally, the synergy of big data and machine learning can foster the development of soil improving cropping systems, enhancing soil structure, promoting deeper rooting, and building carbon stocks for sustainable agriculture. These technological advancements empower farmers to make informed, data-driven decisions, optimizing resource utilization while promoting soil health and sustainable crop productivity. Besides, the sound policy framework and strong political will help to address the issues to redesign Agri-Scape and adopt the future-driven technologies to conserve the soil health toward good agricultural practices.

Cropping system-based soil management is vital for sustainable agriculture, demanding targeted research to address complexities. Exploring adaptive measures, especially tailored nutrient management, soil microbial communities, organic matter dynamics, and water management implications, is essential. Research gaps also include assessing crop resilience to climate change, addressing socioeconomic aspects, and integrating advanced technologies for optimized decision-making. Location-specific scientific results are crucial for effective adaptation strategies against climate variability and environmental constraints. Bridging these research gaps is crucial for formulating sophisticated, science-based strategies that improve soil health, productivity, and environmental sustainability within agricultural systems. The emphasis should be on the dynamics, diversity, and adaptability of solutions that can capitalize on the opportunities presented by evolving economic, technological, and institutional contexts (Jodha et al. 2012).

12.7 Conclusion

To ensure global food security and sustainability among challenges like climate change, monoculture cropping systems must be replaced by diverse and climate-resilient approaches. Incorporating practices like crop rotation, intercropping, and soil management, including reduced tillage and organic amendments, can enhance resilience, mitigate climate impact, and promote sustainable agriculture for a growing world population. Sustainable soil practices, including conservation tillage and cover cropping, optimize soil health, fostering resilience in agricultural ecosystems. Adapting cropping systems is essential for addressing climate challenges and ensuring long-term agricultural productivity. Organic amendments enhance soil quality, providing gradual nutrient release and fostering a diverse microbial ecosystem.

Climate-smart practices, such as water conservation and carbon sequestration, contribute to resilience. Nutrient management, guided by technological innovations like remote sensing, AI-driven decision tools, robotics, and advanced soil monitoring, optimizes crop production while mitigating environmental impact. Embracing these integrated approaches ensures soil health, water efficiency, and climate resilience, laying the foundation for sustainable agriculture in the face of evolving climatic challenges. To address the climate-induced soil challenges requires an integrated approach that combines traditional wisdom with modern technologies. Conservation agriculture, precision farming, agroforestry, and carbon sequestration techniques play pivotal roles. Future directions involve leveraging emerging technologies like IoT devices, drones, and machine learning for climate-resilient agriculture. Research must focus on tailored nutrient management, soil microbial communities, water implications, and socioeconomic aspects. Bridging these gaps will foster science-driven strategies, ensuring sustainable soil health and productivity in the face of climate change.

References

Abbas F, Hammad HM, Fahad S, Cerdà A, Rizwan M, Farhad W et al (2017a) Agroforestry: a sustainable environmental practice for carbon sequestration under the climate change scenarios—a review. Environ Sci Pollut Res 24:11177–11191

Abbas G, Ahmad S, Ahmad A, Nasim W, Fatima Z, Hussain S et al (2017b) Quantification the impacts of climate change and crop management on phenology of maize-based cropping system in Punjab, Pakistan. Agric For Meteorol 247:42–55

Abbas G, Ahmed M, Fatima Z, Hussain S, Kheir AM, Ercişli S, Ahmad S (2023) Modeling the potential impact of climate change on maize-maize cropping system in semi-arid environment and designing of adaptation options. Agric For Meteorol 341:109674

Abdulridha J, Ampatzidis Y, Kakarla SC, Roberts P (2020) Detection of target spot and bacterial spot diseases in tomato using UAV-based and benchtop-based hyperspectral imaging techniques. Precis Agric 21:955–978

Acharya SP, Basavaraja H, Kunnal LB, Mahajanashetti SB, Bhat AR (2011) Crop diversification in Karnataka: an economic analysis §. Agric Econ Res Rev 24(2):351–357

Adler RL, Singh G, Nelson KA, Weirich J, Motavalli PP, Miles RJ (2020) Cover crop impact on crop production and nutrient loss in a no-till terrace topography. J Soil Water Conserv 75(2):153–165

Ali MM, Al-Ani A, Eamus D, Tan DK (2017) Leaf nitrogen determination using non-destructive techniques–a review. J Plant Nutr 40(7):928–953

Ansari RA, Mahmood I, Rizvi R, Sumbul A, Safiuddin (2017) Siderophores: augmentation of soil health and crop productivity. In: Probiotics in agroecosystem. Springer, Singapore, pp 291–312

Antosh E, Idowu J, Schutte B, Lehnhoff E (2020) Winter cover crops effects on soil properties and sweet corn yield in semi-arid irrigated systems. Agron J 112(1):92–106

Badiee A, Wallbank JR, Fentanes JP, Trill E, Scarlet P, Zhu Y et al (2021) Using additional moderator to control the footprint of a COSMOS Rover for soil moisture measurement. Water Resour Res 57(6):e2020WR028478

Balaram V, Sawant SS (2022) Indicator minerals, pathfinder elements, and portable analytical instruments in mineral exploration studies. Fortschr Mineral 12(4):394

Bayer C, Zschornack T, Pedroso GM, da Rosa CM, Camargo ES, Boeni M et al (2015) A seven-year study on the effects of fall soil tillage on yield-scaled greenhouse gas emission from flood irrigated rice in a humid subtropical climate. Soil Tillage Res 145:118–125

Bayu T (2020) Review on contribution of integrated soil fertility management for climate change mitigation and agricultural sustainability. Cogent Environ Sci 6(1):1823631

Bechar A, Vigneault C (2016) Agricultural robots for field operations: concepts and components. Biosyst Eng 149:94–111

Begue A, Arvor D, Bellon B, Betbeder J, De Abelleyra D, Ferraz RPD et al (2018) Remote sensing and cropping practices: a review. Remote Sens 10(1):99

Besharati B, Navid H, Karimi H, Behfar H, Eskandari I (2019) Development of an infrared seed-sensing system to estimate flow rates based on physical properties of seeds. Comput Electron Agric 162:874–881

Bhattacharya A, Bhattacharya A (2021) Effect of soil water deficit on growth and development of plants: a review. In: Soil water deficit and physiological issues in plants. Springer, Singapore, pp 393–488

Blanco-Canqui H, Ruis SJ (2020) Cover crop impacts on soil physical properties: a review. Soil Sci Soc Am J 84(5):1527–1576

Blanco-Canqui H, Mikha MM, Presley DR, Claassen MM (2011) Addition of cover crops enhances no-till potential for improving soil physical properties. Soil Sci Soc Am J 75(4):1471–1482

Blasch G, Spengler D, Hohmann C, Neumann C, Itzerott S, Kaufmann H (2015) Multitemporal soil pattern analysis with multispectral remote sensing data at the field-scale. Comput Electron Agric 113:1–13

Bowles TM, Mooshammer M, Socolar Y, Calderón F, Cavigelli MA, Culman SW et al (2020) Long-term evidence shows that crop-rotation diversification increases agricultural resilience to adverse growing conditions in North America. One Earth 2(3):284–293

Buchi L, Gebhard CA, Liebisch F, Sinaj S, Ramseier H, Charles R (2015) Accumulation of biologically fixed nitrogen by legumes cultivated as cover crops in Switzerland. Plant Soil 393:163–175

Cakan VA, Tipi T (2024) What are the implications of climatic and non-climatic factors on crop production? Evidence from Turkey. Int J Environ Res 18(1):10

Calabi-Floody M, Medina J, Rumpel C, Condron LM, Hernandez M, Dumont M, de La Luz Mora M (2018) Smart fertilizers as a strategy for sustainable agriculture. Adv Agron 147:119–157

Campos I, González-Gómez L, Villodre J, Calera M, Campoy J, Jiménez N et al (2019) Mapping within-field variability in wheat yield and biomass using remote sensing vegetation indices. Precis Agric 20:214–236

Çerçioğlu M (2020) Changes in soil hydro-physical properties by cover crops relative to tillage management. Eurasian Soil Sci 53:1446–1454

Chahal I, Van Eerd LL (2023) Do cover crops increase subsequent crop yield in temperate climates? A meta-analysis. Sustainability 15(8):6517

Chavarria DN, Verdenelli RA, Serri DL, Restovich SB, Andriulo AE, Meriles JM, Vargas-Gil S (2016) Effect of cover crops on microbial community structure and related enzyme activities and macronutrient availability. Eur J Soil Biol 76:74–82

Chen Y, Lin Z, Zhao X, Wang G, Gu Y (2014) Deep learning-based classification of hyperspectral data. IEEE J Sel Top Appl Earth Obs Remote Sens 7(6):2094–2107

Chen Y, Camps-Arbestain M, Shen Q, Singh B, Cayuela ML (2018) The long-term role of organic amendments in building soil nutrient fertility: a meta-analysis and review. Nutr Cycl Agroecosyst 111:103–125

Chen J, Gong Y, Wang S, Guan B, Balkovic J, Kraxner F (2019) To burn or retain crop residues on croplands? An integrated analysis of crop residue management in China. Sci Total Environ 662:141–150

Chen M, Zhang S, Liu L, Wu L, Ding X (2021) Combined organic amendments and mineral fertilizer application increase rice yield by improving soil structure, P availability and root growth in saline-alkaline soil. Soil Tillage Res 212:105060

Chimento C, Almagro M, Amaducci S (2016) Carbon sequestration potential in perennial bioenergy crops: the importance of organic matter inputs and its physical protection. GCB Bioenergy 8(1):111–121
Chivenge P, Vanlauwe B, Six J (2011) Does the combined application of organic and mineral nutrient sources influence maize productivity? A meta-analysis. Plant Soil 342:1–30
Ciais P, Sabine C, Bala G, Bopp L, Brovkin V, Canadell J et al (2014) Carbon and other biogeochemical cycles. In: Climate change 2013: the physical science basis. Contribution of Working Group I to the fifth assessment report of the Intergovernmental Panel on Climate Change. Cambridge University Press, Cambridge, pp 465–570
Dabney SM, Delgado JA, Meisinger JJ, Schomberg HH, Liebig MA, Kaspar T et al (2010) Using cover crops and cropping systems for nitrogen management. In: Advances in nitrogen management for water quality, vol 66. Soil and Water Conservation Society, Ankeny, pp 231–282
Dai W, Feng G, Huang Y, Adeli A, Jenkins JN (2024) Influence of cover crops on soil aggregate stability, size distribution and related factors in a no-till field. Soil Tillage Res 244:106197
Damania R (2020) The economics of water scarcity and variability. Oxf Rev Econ Policy 36(1):24–44
Daryanto S, Fu B, Wang L, Jacinthe PA, Zhao W (2018) Quantitative synthesis on the ecosystem services of cover crops. Earth Sci Rev 185:357–373
Daughtry CS (2001) Discriminating crop residues from soil by shortwave infrared reflectance. Agron J 93(1):125–131
Demir Z, Isik D (2020) Using cover crops to improve soil quality and hazelnut yield. Fresenius Environ Bull 29(4):1974–1987
Denier van der Gon HA, van Bodegom PM, Wassmann R, Lantin RS, Metra-Corton TM (2001) Sulfate-containing amendments to reduce methane emissions from rice fields: mechanisms, effectiveness and costs. Mitig Adapt Strateg Glob Chang 6:71–89
Di Gennaro SF, Battiston E, Di Marco S, Facini O, Matese A, Nocentini M et al (2016) Unmanned Aerial Vehicle (UAV)-based remote sensing to monitor grapevine leaf stripe disease within a vineyard affected by esca complex. Phytopathol Mediterr 55:262–275
Douarre C, Schielein R, Frindel C, Gerth S, Rousseau D (2016) Deep learning based root-soil segmentation from X-ray tomography images. bioRxiv, 071662
Eglin T, Ciais P, Piao SL, Barré P, Bellassen V, Cadule P et al (2010) Historical and future perspectives of global soil carbon response to climate and land-use changes. Tellus B 62(5):700–718
Fageria NK, Baligar VC, Bailey BA (2005) Role of cover crops in improving soil and row crop productivity. Commun Soil Sci Plant Anal 36(19–20):2733–2757
FAO (2017) The future of food and agriculture—trends and challenges. FAO, Rome
Farella MM, Fisher JB, Jiao W, Key KB, Barnes ML (2022) Thermal remote sensing for plant ecology from leaf to globe. J Ecol 110(9):1996–2014
Favis-Mortlock D, Boardman J (1995) Nonlinear responses of soil erosion to climate change: a modelling study on the UK South Downs. Catena 25(1–4):365–387
Fentanes JP, Gould I, Duckett T, Pearson S, Cielniak G (2018) 3-D soil compaction mapping through kriging-based exploration with a mobile robot. IEEE Rob Autom Lett 3(4):3066–3072
Fixen PE (2020) A brief account of the genesis of 4R nutrient stewardship. Agron J 112(5):4511–4518
Follett RF (2001) Soil management concepts and carbon sequestration in cropland soils. Soil Tillage Res 61(1–2):77–92
Forber KJ, Withers PJA, Ockenden MC, Haygarth PM (2018) The phosphorus transfer continuum: a framework for exploring effects of climate change. Agric Environ Lett 3(1):180036
Fortuna A, Blevins R, Frye W, Grove J, Cornelius P (2008) Sustaining soil quality with legumes in no-tillage systems. Commun Soil Sci Plant Anal 39:1680–1699
Fung KM, Tai AP, Yong T, Liu X, Lam HM (2019) Co-benefits of intercropping as a sustainable farming method for safeguarding both food security and air quality. Environ Res Lett 14(4):044011

Gabriel JL, Muñoz-Carpena R, Quemada M (2012) The role of cover crops in irrigated systems: water balance, nitrate leaching and soil mineral nitrogen accumulation. Agric Ecosyst Environ 155:50–61

Gangopadhyay S, Chowdhuri I, Das N, Pal SC, Mandal S (2023) The effects of no-tillage and conventional tillage on greenhouse gas emissions from paddy fields with various rice varieties. Soil Tillage Res 232:105772

Gangwar B, Ravisankar N, Prasad K (2012) Agronomic research on cropping systems in India. Indian J Agron 57(3s):105–115

García-Sánchez F, Simón-Grao S, Martínez-Nicolás JJ, Alfosea-Simón M, Liu C, Chatzissavvidis C et al (2020) Multiple stresses occurring with boron toxicity and deficiency in plants. J Hazard Mater 397:122713

Gherardi LA, Sala OE (2020) Global patterns and climatic controls of belowground net carbon fixation. Proc Natl Acad Sci 117(33):20038–20043

Ghosh PK, Hazra KK, Venkatesh MS, Praharaj CS, Kumar N, Nath CP et al (2020) Grain legume inclusion in cereal–cereal rotation increased base crop productivity in the long run. Exp Agric 56(1):142–158

Gonzalez-Sanchez EJ, Veroz-Gonzalez O, Conway G, Moreno-Garcia M, Kassam A, Mkomwa S et al (2019) Meta-analysis on carbon sequestration through Conservation Agriculture in Africa. Soil Tillage Res 190:22–30

Grace PR, Colunga-Garcia M, Gage SH, Robertson GP, Safir GR (2006) The potential impact of agricultural management and climate change on soil organic carbon of the north central region of the United States. Ecosystems 9:816–827

Gross A, Bromm T, Glaser B (2021) Soil organic carbon sequestration after biochar application: a global meta-analysis. Agronomy 11(12):2474

Haddaway NR, Hedlund K, Jackson LE, Kätterer T, Lugato E, Thomsen IK et al (2017) How does tillage intensity affect soil organic carbon? A systematic review. Environ Evid 6(1):1–48

Hamidov A, Helming K, Bellocchi G, Bojar W, Dalgaard T, Ghaley BB et al (2018) Impacts of climate change adaptation options on soil functions: a review of European case-studies. Land Degrad Dev 29(8):2378–2389

Haque MM, Biswas JC, Salahin N, Alam MK, Akhter S, Akhtar S et al (2023) Tillage systems influence on greenhouse gas emission factor and global warming potential under rice-mustard-rice cropping system. Arch Agron Soil Sci 69(4):599–614

Hassan-Esfahani L, Torres-Rua A, Jensen A, Mckee M (2017) Spatial root zone soil water content estimation in agricultural lands using Bayesian-based artificial neural networks and high-resolution visual, NIR, and thermal imagery. Irrig Drain 66(2):273–288

Hatfield JL, Parkin TB, Sauer TJ, Prueger JH (2012) Mitigation opportunities from land management practices in a warming world: increasing potential sinks. In: Managing agricultural greenhouse gases: coordinated agricultural research through GRACEnet to address our changing climate. Academic Press, London, pp 487–504

Heidrich J, Gaulke M, Golling M, Alaydin BO, Barh A, Keller U (2022) 324-fs Pulses from a SESAM modelocked backside-cooled 2-μm VECSEL. IEEE Photon Technol Lett 34(6):337–340

Huang Y, Zhang Z, Wei F, Li J (1999) Ecophysiological effect of dry cultivated and plastic film mulched rice planting. Chin J Appl Ecol 10(3):305

Huang H, Lan Y, Yang A, Zhang Y, Wen S, Deng J (2020) Deep learning versus object-based image analysis (OBIA) in weed mapping of UAV imagery. Int J Remote Sens 41(9):3446–3479

Huang Y, Tao B, Yang Y, Zhu X, Yang X, Grove JH, Ren W (2022) Simulating no-tillage effects on crop yield and greenhouse gas emissions in Kentucky corn and soybean cropping systems: 1980–2018. Agric Syst 197:103355

Hufnagel J, Reckling M, Ewert F (2020) Diverse approaches to crop diversification in agricultural research. A review. Agron Sustain Dev 40:1–17

Humphreys E, Muirhead WA, Fawcett BJ, Townsend JT, Murray EA (1996) Puddling in mechanised rice culture: impacts on water use and the productivity of rice and post-rice crops. In: ACIAR proceedings. ACIAR, Canberra, pp 213–218
IPCC (2014) Climate change 2014: synthesis report. Contribution of Working Groups I, II and III to the fifth assessment report of the Intergovernmental Panel on Climate Change [Core Writing Team, R.K. Pachauri and L.A. Meyer (eds.)]. IPCC, Geneva, 151 pp
IPCC (2023) Climate change 2023: synthesis report, summary for policymakers. Contribution of Working Groups I, II and III to the sixth assessment report of the Intergovernmental Panel on Climate Change [Core Writing Team, H. Lee and J. Romero (eds.)]. IPCC, Geneva
Jeppesen E, Kronvang B, Meerhoff M, Søndergaard M, Hansen KM, Andersen HE et al (2009) Climate change effects on runoff, catchment phosphorus loading and lake ecological state, and potential adaptations. J Environ Qual 38(5):1930–1941
Jodha NS, Singh NP, Bantilan MCS (2012) Enhancing farmers' adaptation to climate change in arid and semi-arid agriculture of India: evidences from indigenous practices: developing international public goods from development-oriented projects. Working Paper Series No. 32
Kayad A, Paraforos DS, Marinello F, Fountas S (2020) Latest advances in sensor applications in agriculture. Agriculture 10(8):362
Kaye JP, Quemada M (2017) Using cover crops to mitigate and adapt to climate change. A review. Agron Sustain Dev 37:1–17
Khanal S, Fulton J, Shearer S (2017) An overview of current and potential applications of thermal remote sensing in precision agriculture. Comput Electron Agric 139:22–32
Kim W, Iizumi T, Nishimori M (2019) Global patterns of crop production losses associated with droughts from 1983 to 2009. J Appl Meteorol Climatol 58(6):1233–1244
Kim J, Seol J, Lee S, Hong SW, Son HI (2020) An intelligent spraying system with deep learning-based semantic segmentation of fruit trees in orchards. In: 2020 IEEE international conference on robotics and automation (ICRA). IEEE, pp 3923–3929
Koresh H (2021) Analysis of soil nutrients based on potential productivity tests with balanced minerals for maize-chickpea crop. J Electron Inform 3(1):23–35
Kumar M, Patle GT, Khanna M, Dagar JC (2013) Climate smart water conservation management technologies. Int J Water Resour Environ Eng 5(10):556–572
Kumar A, Malik MS, Shabnam S, Kumar R, Karmakar S, Das SS et al (2024) Carbon sequestration and credit potential of gamhar (Gmelina arborea Roxb.) based agroforestry system for zero carbon emission of India. Sci Rep 14(1):4828
Lahoche F, Godard C, Fourty T, Lelandais V, Lepoutre D (2002) An innovative approach based on neural networks for predicting soil component variability. In: Proceedings of the 6th international conference on precision agriculture and other precision resources management, Minneapolis, MN, USA. American Society of Agronomy, pp 14–17
Lal R (2004) Soil carbon sequestration to mitigate climate change. Geoderma 123(1–2):1–22
Lal R, Negassa W, Lorenz K (2015) Carbon sequestration in soil. Curr Opin Environ Sustain 15:79–86
Lal R, Smith P, Jungkunst HF, Mitsch WJ, Lehmann J, Nair PR et al (2018) The carbon sequestration potential of terrestrial ecosystems. J Soil Water Conserv 73(6):145A–152A
Leone M (2022) Advances in fiber optic sensors for soil moisture monitoring: a review. Results Opt 7:100213
Li Z, Yang X, Cui S, Yang Q, Yang X, Li J, Shen Y (2018) Developing sustainable cropping systems by integrating crop rotation with conservation tillage practices on the Loess Plateau, a long-term imperative. Field Crop Res 222:164–179
Li B, Yang Y, Qin C, Bai X, Wang L (2020a) Improved random sampling consensus algorithm for vision navigation of intelligent harvester robot. Ind Robot 47(6):881–887
Li Y, Li Z, Chang SX, Cui S, Jagadamma S, Zhang Q, Cai Y (2020b) Residue retention promotes soil carbon accumulation in minimum tillage systems: implications for conservation agriculture. Sci Total Environ 740:140147

Li Z, Zhang Q, Li Z, Qiao Y, Du K, Yue Z et al (2023) Different responses of agroecosystem greenhouse gas emissions to tillage practices in a Chinese wheat–maize cropping system. Carbon Res 2(1):7

Li T, Li G, Lu Z, Zhao D, Li Y, Wang Z et al (2024) Crop diversification increases soil extracellular enzyme activities under no tillage: a global meta-analysis. Soil Tillage Res 235:105870

Link SO, Smith JL, Halvorson JJ, Bolton JH (2003) Effect of climate change on a perennial bunchgrass and soil carbon and nitrogen pools in a semi-arid shrub-steppe ecosystem. Glob Chang Biol 9:1097–1105

Linquist BA, Marcos M, Adviento-Borbe MA, Anders M, Harrell D, Linscombe S et al (2018) Greenhouse gas emissions and management practices that affect emissions in US rice systems. J Environ Qual 47(3):395–409

Liu L, Basso B (2020) Impacts of climate variability and adaptation strategies on crop yields and soil organic carbon in the US Midwest. PLoS One 15(1):e0225433

Liu LJ, Gao HW, Li HW (2004) Conservation tillage for corn-wheat two crops a year region. Trans CSAE 20(3):70–73

Liu Z, Chen X, Jing Y, Li Q, Zhang J, Huang Q (2014) Effects of biochar amendment on rapeseed and sweet potato yields and water stable aggregate in upland red soil. Catena 123:45–51

Liu P, Zhao H, Li Y, Liu Z, Gao X, Zhang Y et al (2018) Corn yields with organic and inorganic amendments under changing climate. Nutr Cycl Agroecosyst 111:141–153

Loladze I (2014) Hidden shift of the ionome of plants exposed to elevated CO2 depletes minerals at the base of human nutrition. elife 3:e02245

Lu X, Lu X, Tanveer SK, Wen X, Liao Y (2015) Effects of tillage management on soil CO2 emission and wheat yield under rain-fed conditions. Soil Res 54(1):38–48

Lucas E, Kennedy B, Roswall T, Burgis C, Toor GS (2023) Climate change effects on phosphorus loss from agricultural land to water: a review. Curr Pollut Rep 9:1–23

Ma R, Zou J, Han Z, Yu K, Wu S, Li Z et al (2021) Global soil-derived ammonia emissions from agricultural nitrogen fertilizer application: a refinement based on regional and crop-specific emission factors. Glob Chang Biol 27(4):855–867

Ma Z, Liang T, Fu H, Ma Q, Chang D, Zhang J et al (2024) Long-term green manuring increases soil carbon sequestration via decreasing qCO2 caused by lower microbial phosphorus limitation in a dry land field. Agric Ecosyst Environ 374:109142

Maharajan T, Ceasar SA, Krishna TPA, Ignacimuthu S (2021) Management of phosphorus nutrient amid climate change for sustainable agriculture. J Environ Qual 50(6):1303–1324

Malyan SK, Bhatia A, Tomer R, Kumar O (2016) Impact of nitrogen fertilizers on methane emissions from flooded rice. Int J Agric Invent 1(2):124–128

Mandal A, Majumder A, Dhaliwal SS, Toor AS, Mani PK, Naresh RK et al (2022) Impact of agricultural management practices on soil carbon sequestration and its monitoring through simulation models and remote sensing techniques: a review. Crit Rev Environ Sci Technol 52(1):1–49

Marinoudi V, Sørensen CG, Pearson S, Bochtis D (2019) Robotics and labour in agriculture. A context consideration. Biosyst Eng 184:111–121

Medek DE, Schwartz J, Myers SS (2017) Estimated effects of future atmospheric CO2 concentrations on protein intake and the risk of protein deficiency by country and region. Environ Health Perspect 125(8):087002

Mehmood I, Qiao L, Chen H, Tang Q, Woolf D, Fan M (2020) Biochar addition leads to more soil organic carbon sequestration under a maize-rice cropping system than continuous flooded rice. Agric Ecosyst Environ 298:106965

Menegat S, Ledo A, Tirado R (2022) Greenhouse gas emissions from global production and use of nitrogen synthetic fertilizers in agriculture. Sci Rep 12(1):14490

Mitchell JP, Singh PN, Wallender WW, Munk DS, Wroble JF, Horwath WR et al (2012) No-tillage and high-residue practices reduce soil water evaporation. Calif Agric 66(2):55

Mo X, Peng H, Xin J, Wang S (2022) Analysis of urea nitrogen leaching under high-intensity rainfall using HYDRUS-1D. J Environ Manag 312:114900

Monís JI, López-Luque R, Reca J, Martínez J (2020) Multistage bounded evolutionary algorithm to optimize the design of sustainable photovoltaic (PV) pumping irrigation systems with storage. Sustainability 12(3):1026

Montgomery DR, Biklé A (2021) Soil health and nutrient density: beyond organic vs. conventional farming. Front Sustain Food Syst 5:417

Muhammad I, Wang J, Sainju UM, Zhang S, Zhao F, Khan A (2021) Cover cropping enhances soil microbial biomass and affects microbial community structure: a meta-analysis. Geoderma 381:114696

Mukumbareza C, Muchaonyerwa P, Chiduza C (2016) Bicultures of oat (Avena sativa L.) and grazing vetch (Vicia dasycarpa L.) cover crops increase contents of carbon pools and activities of selected enzymes in a loam soil under warm temperate conditions. Soil Sci Plant Nutr 62(5–6):447–455

Myers SS, Zanobetti A, Kloog I, Huybers P, Leakey AD, Bloom AJ et al (2014) Increasing CO2 threatens human nutrition. Nature 510(7503):139–142

Nagler PL, Daughtry CST, Goward SN (2000) Plant litter and soil reflectance. Remote Sens Environ 71(2):207–215

Nakandalage N, Nicolas M, Norton RM, Hirotsu N, Milham PJ, Seneweera S (2016) Improving rice zinc biofortification success rates through genetic and crop management approaches in a changing environment. Front Plant Sci 7:764

Nath CP, Kumar N, Dutta A, Hazra KK, Praharaj CS, Singh SS, Das K (2023) Pulse crop and organic amendments in cropping system improve soil quality in rice ecology: evidence from a long–term experiment of 16 years. Geoderma 430:116334

Norton R (2014) Combating climate change through improved agronomic practices and input-use efficiency. J Crop Improv 28(5):575–618

Olson K, Ebelhar SA, Lang JM (2014) Long-term effects of cover crops on crop yields, soil organic carbon stocks and sequestration. Open J Soil Sci 4:284–292

Pan G, Zhou P, Li Z, Smith P, Li L, Qiu D et al (2009) Combined inorganic/organic fertilization enhances N efficiency and increases rice productivity through organic carbon accumulation in a rice paddy from the Tai Lake region, China. Agric Ecosyst Environ 131(3–4):274–280

Panda BB, Nayak AK, Das P, Giri KS, Tripathi R, Nayak PK (2019) Cropping system approach to cope up with climate change. In: Climate resilient agricultural technologies for future. Training manual, model training course on climate resilient agricultural technologies for future, vol 3. ICAR-National Rice Research Institute, Cuttack, p 29

Pathak H, Jain N, Bhatia A, Kumar A, Chatterjee D (2016) Improved nitrogen management: a key to climate change adaptation and mitigation. Indian J Fertil 12(11):151–162

Paudel GP, Chamberlin J, Maharjan S, Nguyen TT, Craufurd P, McDonald AJ (2023) Insights for climate change adaptation from early sowing of wheat in the Northern Indo-Gangetic Basin. Int J Disaster Risk Reduct 92:103714

Pinheiro ÉFM, de Campos DVB, de Carvalho Balieiro F, dos Anjos LHC, Pereira MG (2015) Tillage systems effects on soil carbon stock and physical fractions of soil organic matter. Agric Syst 132:35–39

Poeplau C (2019) Measuring and modelling soil carbon stocks and stock changes in livestock production systems: guidelines for assessment; Version 1-Advanced copy. FAO, Rome

Poeplau C, Don A (2015) Carbon sequestration in agricultural soils via cultivation of cover crops–a meta-analysis. Agric Ecosyst Environ 200:33–41

Polvara R, Del Duchetto F, Neumann G, Hanheide M (2021) Navigate-and-seek: a robotics framework for people localization in agricultural environments. IEEE Rob Autom Lett 6(4):6577–6584

Pott CA, Conrado PM, Rampim L, Umburanas RC, Conrado AMC, Outeiro VH, Müller MML (2023) Mixture of winter cover crops improves soil physical properties under no-tillage system in a subtropical environment. Soil Tillage Res 234:105854

Powlson DS, Gregory PJ, Whalley WR, Quinton JN, Hopkins DW, Whitmore AP et al (2011) Soil management in relation to sustainable agriculture and ecosystem services. Food Policy 36:S72–S87

Pramanick B, Kumar M, Naik BM, Singh SK, Singh SV (2024) Soil carbon-nutrient cycling, energetics, and carbon footprint in calcareous soils with adoption of long-term conservation tillage practices and cropping systems diversification. Sci Total Environ 912:169421. https://doi.org/10.1016/j.scitotenv.2023.169421

Price JC (1990) On the information content of soil reflectance spectra. Remote Sens Environ 33(2):113–121

Procházková D, Pavlíková D, Pavlík M (2016) Sulphur: role in alleviation of environmental stress in crop plants. In: Plant-environment interaction: responses and approaches to mitigate stress. Wiley, Hoboken, pp 84–96

Pu C, Chen JS, Wang HD, Virk AL, Zhao X, Zhang HL (2022) Greenhouse gas emissions from the wheat-maize cropping system under different tillage and crop residue management practices in the North China Plain. Sci Total Environ 819:153089

Puga AP, Queiroz MCDA, Ligo MAV, Carvalho CS, Pires AMM, Marcatto JDOS, Andrade CAD (2020) Nitrogen availability and ammonia volatilization in biochar-based fertilizers. Arch Agron Soil Sci 66(7):992–1004

Qin R, Stamp P, Richner W (2006) Impact of tillage on maize rooting in a Cambisol and Luvisol in Switzerland. Soil Tillage Res 85(1–2):50–61

Quilty JR, Cattle SR (2011) Use and understanding of organic amendments in Australian agriculture: a review. Soil Res 49(1):1–26

Ramírez-Builes VH, Küsters J (2021) Calcium and potassium nutrition increases the water use efficiency in coffee: a promising strategy to adapt to climate change. Hydrology 8(2):75

Renwick LL, Deen W, Silva L, Gilbert ME, Maxwell T, Bowles TM, Gaudin AC (2021) Long-term crop rotation diversification enhances maize drought resistance through soil organic matter. Environ Res Lett 16(8):084067

Roohi M, Arif MS, Guillaume T, Yasmeen T, Riaz M, Shakoor A et al (2022) Role of fertilization regime on soil carbon sequestration and crop yield in a maize-cowpea intercropping system on low fertility soils. Geoderma 428:116152

Sainju UM, Singh BP, Whitehead WF (2002) Long-term effects of tillage, cover crops, and nitrogen fertilization on organic carbon and nitrogen concentrations in sandy loam soils in Georgia, USA. Soil Tillage Res 63(3–4):167–179

Sainju UM, Whitehead WF, Singh BP (2003) Cover crops and nitrogen fertilization effects on soil aggregation and carbon and nitrogen pools. Can J Soil Sci 83(2):155–165

Sardans J, Peñuelas J (2021) Potassium control of plant functions: ecological and agricultural implications. Plants 10(2):419

Seo JH, Lee HJ, Hur IB, Kim SJ, Kim CK, Jo HS (2000) Use of hairy vetch green manure as nitrogen fertilizer for corn production. Korean J Crop Sci 45(5):294–299

Shaheb MR, Venkatesh R, Shearer SA (2021) A review on the effect of soil compaction and its management for sustainable crop production. J Biosyst Eng 46:1–23

Sharma S, Singh P, Choudhary OP (2021) Nitrogen and rice straw incorporation impact nitrogen use efficiency, soil nitrogen pools and enzyme activity in rice-wheat system in north-western India. Field Crop Res 266:108131

Shi X, Chai N, Wei Y, Qin R, Yang J, Zhang M et al (2023) Harmonizing manure and mineral fertilizers can mitigate the impact of climate change on crop yields. Agric Ecosyst Environ 352:108526

Singh VK, Gautam P, Nanda G, Dhaliwal SS, Pramanick B, Meena SS et al (2021) Soil test based fertilizer application improves productivity, profitability and nutrient use efficiency of rice (Oryza sativa L.) under direct seeded condition. Agronomy 11(9):1756

Singh S, Luthra N, Mandal S, Kushwaha DK, Pathak SO, Datta D, Sharma R, Pramanick B (2023a) Distinct behavior of biochar modulating biogeochemistry of salt-affected and acidic soil: a review. J Soil Sci Plant Nutr 23(3):2981–2997. https://doi.org/10.1007/s42729-023-01370-9

Singh SV, Chaturvedi S, Nayak P, Dhyani VC, Nadipamu TMK, Singh DK, Gudapaty P, Mathyam P, Srinivasarao K, Gavindaraju K (2023b) Carbon offset potential of biochar based straw management under rice – wheat system along Indo-Gangetic Plains of India. Sci Total Environ 897:165176. https://doi.org/10.1016/j.scitotenv.2023.165176

Sivakumar MV, Stefanski R (2007) Climate and land degradation—an overview. In: Climate and land degradation. Springer, Berlin, Heidelberg, pp 105–135

Sollen-Norrlin M, Ghaley BB, Rintoul NLJ (2020) Agroforestry benefits and challenges for adoption in Europe and beyond. Sustainability 12(17):7001

Song Y, Linderholm HW, Chen D, Walther A (2010) Trends of the thermal growing season in China, 1951–2007. Int J Climatol 30(1):33–43

Song X, Zhang G, Liu F, Li D, Zhao Y, Yang J (2016) Modeling spatio-temporal distribution of soil moisture by deep learning-based cellular automata model. J Arid Land 8:734–748

Sriphirom P, Chidthaisong A, Yagi K, Tripetchkul S, Towprayoon S (2020) Evaluation of biochar applications combined with alternate wetting and drying (AWD) water management in rice field as a methane mitigation option for farmers' adoption. Soil Sci Plant Nutr 66(1):235–246

Stenberg B, Rossel RAV, Mouazen AM, Wetterlind J (2010) Visible and near infrared spectroscopy in soil science. Adv Agron 107:163–215

Surendran U, Raja P, Jayakumar M, Subramoniam SR (2021) Use of efficient water saving techniques for production of rice in India under climate change scenario: a critical review. J Clean Prod 309:127272

Tang H, Li C, Shi L, Cheng K, Wen L, Li W, Xiao X (2022) Effects of short-term tillage managements on CH4 and N2O emissions from a double-cropping rice field in southern of China. Agronomy 12(2):517

Tao Z, Chang X, Wang D, Wang Y, Ma S, Yang Y, Zhao G (2018) Effects of sulfur fertilization and short-term high temperature on wheat grain production and wheat flour proteins. Crop J 6(4):413–425

The Royal Society (2021) Nourishing ten billion sustainably: resilient food production in a time of climate change. Royal Society, London. Retrieved from https://policycommons.net/artifacts/1628409/nourishing-ten-billion-sustainably/2318330/. Accessed on 31 Jan 2024. CID: 20.500.12592/61kkz6

Thilakarathna MS, Serran S, Lauzon J, Janovicek K, Deen B (2015) Management of manure nitrogen using cover crops. Agron J 107(4):1595–1607

Tiefenbacher A, Sandén T, Haslmayr HP, Miloczki J, Wenzel W, Spiegel H (2021) Optimizing carbon sequestration in croplands: a synthesis. Agronomy 11(5):882

Triberti L, Nastri A, Baldoni G (2016) Long-term effects of crop rotation, manure and mineral fertilisation on carbon sequestration and soil fertility. Eur J Agron 74:47–55

Uphoff N, Randriamiharisoa R (2002) Reducing water use in irrigated rice production with the Madagascar System of Rice Intensification (SRI). In: Water-wise rice production. International Rice Research Institute/Plant Research International, Los Baños/Wageningen, pp 71–87

Ussiri DA, Lal R (2017) Carbon sequestration for climate change mitigation and adaptation. Springer, Cham, pp 287–325

Valipour M, Bateni SM, Jun C (2021) Global surface temperature: a new insight. Climate 9(5):81

Van Groenigen JW, Van Kessel C, Hungate BA, Oenema O, Powlson DS, Van Groenigen KJ (2017) Sequestering soil organic carbon: a nitrogen dilemma. Environ Sci Technol 51:4738

Virk AL, Liu WS, Chen Z, Yves N, Bohoussou D, Cheema MA et al (2022) Effects of different tillage systems and cropping sequences on soil physicochemical properties and greenhouse gas emissions. Agric Ecosyst Environ 335:108010

Waha K, Dietrich JP, Portmann FT, Siebert S, Thornton PK, Bondeau A, Herrero M (2020) Multiple cropping systems of the world and the potential for increasing cropping intensity. Glob Environ Chang 64:102131

Wan Y, Lin E, Xiong W, Guo L (2011) Modeling the impact of climate change on soil organic carbon stock in upland soils in the 21st century in China. Agric Ecosyst Environ 141(1–2):23–31

Wang C, Guo L, Li Y, Wang Z (2012) Systematic comparison of C3 and C4 plants based on metabolic network analysis. In: BMC systems biology, vol 6. BioMed Central, London, pp 1–14

Wang J, Zhang X, Li L, Cheng K, Zheng J, Zheng J et al (2016) Changes in micronutrient availability and plant uptake under simulated climate change in winter wheat field. J Soils Sediments 16:2666–2675

Wang H, Zhang X, Yu X, Hou H, Fang Y, Ma Y, Zhang G (2021a) Maize-potato rotation maintains soil water balance and improves productivity. Agron J 113(1):645–656

Wang X, Bai J, Xie T, Wang W, Zhang G, Yin S, Wang D (2021b) Effects of biological nitrification inhibitors on nitrogen use efficiency and greenhouse gas emissions in agricultural soils: a review. Ecotoxicol Environ Saf 220:112338

Wang X, Li J, Xing G, Mai S, Liu W, Jiang Y et al (2022) Soil organic carbon distribution, enzyme activities, and the temperature sensitivity of a tropical rainforest in Wuzhishan, Hainan Island. Forests 13(11):1943

Were K, Bui DT, Dick ØB, Singh BR (2015) A comparative assessment of support vector regression, artificial neural networks, and random forests for predicting and mapping soil organic carbon stocks across an Afromontane landscape. Ecol Indic 52:394–403

Williams DM, Blanco-Canqui H, Francis CA, Galusha TD (2017) Organic farming and soil physical properties: an assessment after 40 years. Agron J 109(2):600–609

Wittwer RA, Dorn B, Jossi W, van der Heijden MG (2017) Cover crops support ecological intensification of arable cropping systems. Sci Rep 7(1):41911

Wolf MK, Wiesmeier M, Macholdt J (2023) Importance of soil fertility for climate-resilient cropping systems: the farmer's perspective. Soil Secur 13:100119

Wolinska A, Kruczyńska A, Podlewski J, Słomczewski A, Grządziel J, Gałązka A, Kuźniar A (2022) Does the use of an intercropping mixture really improve the biology of monocultural soils?—a search for bacterial indicators of sensitivity and resistance to long-term maize monoculture. Agronomy 12(3):613

Wright AJ, Mommer L, Barry K, Van Ruijven J (2020) Stress gradients and biodiversity: monoculture vulnerability drives stronger biodiversity effects during drought years. Ecology 102(1):e03193

Yadav GS, Das A, Babu S, Mohapatra KP, Lal R, Rajkhowa D (2021) Potential of conservation tillage and altered land configuration to improve soil properties, carbon sequestration and productivity of maize based cropping system in eastern Himalayas, India. Int Soil Water Conserv Res 9(2):279–290

Yan XT, Donaldson KM, Davidson CM, Gao Y, Wu H, Houston AM, Kisdi A (2018) Effects of sample pretreatment and particle size on the determination of nitrogen in soil by portable LIBS and potential use on robotic-borne remote Martian and agricultural soil analysis systems. RSC Adv 8(64):36886–36894

Yang XG, Liu ZJ, Fu C (2011) The possible effect of climate warming on northern limits of cropping system and crop yield in China. Agric Sci China 10(4):585–594

Yang T, Siddique KH, Liu K (2020) Cropping systems in agriculture and their impact on soil health-a review. Global Ecol Conserv 23:e01118

Yang X, Xiong J, Du T, Ju X, Gan Y, Li S et al (2024) Diversifying crop rotation increases food production, reduces net greenhouse gas emissions and improves soil health. Nat Commun 15(1):198

Yan-Ling S, Bo L, Hai-Ling Z (2011) Impact of global warming on the rice cultivable area in Southern China in 1961-2009. Adv Clim Chang Res 7(4):259

Zenda T, Liu S, Dong A, Duan H (2021) Revisiting sulphur—the once neglected nutrient: it's roles in plant growth, metabolism, stress tolerance and crop production. Agriculture 11(7):626

Zhang ZS, Cao CG, Guo LJ, Li CF (2016) Emissions of CH 4 and CO 2 from paddy fields as affected by tillage practices and crop residues in central China. Paddy Water Environ 14:85–92

Zhao J, Yang Y, Zhang K, Jeong J, Zeng Z, Zang H (2020) Does crop rotation yield more in China? A meta-analysis. Field Crops Res 245:107659

Zizinga A, Mwanjalolo JGM, Tietjen B, Bedadi B, Gabiri G, Luswata KC (2022) Impacts of climate smart agriculture practices on soil water conservation and maize productivity in rainfed cropping systems of Uganda. Front Sustain Food Syst 6:889830

Zougmoré R, Jalloh A, Tioro A (2014) Climate-smart soil water and nutrient management options in semiarid West Africa: a review of evidence and analysis of stone bunds and zaï techniques. Agric Food Secur 3:1–8

Chapter 13
Adaptation of Conservation Agriculture as Climate-Smart Technology for Food Crops

Gopal Dutta, Subhadeep Mandal, Bappa Paramanik, Priyanka Ghatak, Partha Sarathi Patra, Ashok Choudhury, Jyotirmay Karforma, and Achyuta Basak

Abstract Climate change is causing increasing challenges to global agriculture, such as more frequent and intense weather events, shifting rainfall patterns, and rising temperatures. In this situation, it is very important to adopt climate-smart technologies to make the agricultural system more sustainable and resilient. An approach to farming known as conservation agriculture (CA), which emphasizes the use of

G. Dutta
Department of Agricultural Meteorology and Physics, Bidhan Chandra Krishi Viswavidyalaya, Haringhata, West Bengal, India

S. Mandal
Department of Soil Science and Agricultural Chemistry, Uttar Banga Krishi Viswavidyalaya, Cooch Behar, West Bengal, India

B. Paramanik (✉)
Subject Matter Specialist, Department of Soil Science, Dakshin Dinajpur KVK, Uttar Banga Krishi Viswavidyalaya, Dakshin Dinajpur, West Bengal, India

P. Ghatak
Soil Science and Agricultural Chemistry, Sister Nivedita University, Kolkata, West Bengal, India

P. S. Patra
Regional Research Station (Terai Zone), Uttar Banga Krishi Viswavidyalaya, Cooch Behar, West Bengal, India

A. Choudhury
Soil Science and Agricultural Chemistry, Uttar Banga Krishi Viswavidyalaya, Cooch Behar, West Bengal, India

J. Karforma
Regional Research Station (Old Alluvial Zone), Uttar Banga Krishi Viswavidyalaya, Dakshin Dinajpur, West Bengal, India

A. Basak
Department of Genetics and Plant Breeding, Uttar Banga Krishi Viswavidyalaya, Cooch Behar, West Bengal, India

B. Pramanick et al. (eds.), *Climate-Smart Agricultural Technologies*, Sustainability Sciences in Asia and Africa, https://doi.org/10.1007/978-981-96-7699-6_13

environmentally friendly techniques to improve soil quality, water management, and biodiversity, is gaining traction as a potentially useful strategy for mitigating the negative effects of climate change on agricultural production. Conservation agricultural practices like permanent soil cover, minimum soil disturbance, crop rotation, and diversification help to enhance the resilience of crops against biotic and abiotic stresses. Conservation of water resources, reduction of greenhouse gas emissions, and preservation of biodiversity integrate conservation agriculture with the principles of climate-smart agriculture. Conservation agriculture, which promotes soil health, water conservation, and biodiversity, is a promising climate-smart technology for food crops.

Keywords Conservation agriculture (CA) · climate-smart agriculture (CSA) · food crops

13.1 Introduction

According to the World Meteorological Organisation, 2020 was one of the top three warmest years on record (WMO 2021). Extreme weather events are occurring more frequently and are more severe as a result of rising temperatures. The International Panel on Climate Change (IPCC) also reported that there is a clear correlation between climate change and the occurrence of extreme weather events, and due to climate change, the frequency and intensity of heatwaves, cyclones, flood, and drought are increasing, and they have a serious impact on communities around the world (IPCC 2012). Moreover, the distribution and behavior of species are affected by changes in temperature and precipitation patterns, which in turn disrupts ecosystems and reduces biodiversity. IPCC reported that global surface temperatures are projected to exceed the 1.5 °C and 2 °C thresholds relative to preindustrial levels, which will increase the risk of climate-related consequences (IPCC 2021).

Agricultural system also affected due to the climate change. Growth and productivity of the field crops were affected by the increasing temperature, alteration of precipitation pattern, and frequently occurrence of extreme weather events. The food supply, the viability of agriculture, and the standard of living for farmers around the world are all threatened by these consequences. Climate change affects agricultural field crops in several ways like temperature rises that can have a direct impact on crop growth and development by affecting critical physiological functions including as photosynthesis, respiration, and reproductive stages. Heat stress during critical growth stages can cause crops to have lower yield and lower quality and nutritional values (Lobell and Gourdji 2012). Changes in the frequency, duration, and intensity of precipitation can lead to water shortages and floods, both of which threaten crop yield. Drought can reduce soil moisture, which in turn reduces the availability of nutrients and makes plants more susceptible to pests and diseases.

Intense precipitation and flood make the soil waterlogged and increases erosion and leaching of nutrients, all of which are detrimental to the growth and productivity of crops (Hatfield et al. 2011). Climate change has a considerable impact on pest and disease dynamics, which has significant effect on field crops. Changing climatic condition influenced the life cycle, abundance, and distribution of pest and pathogen like Pest's growth and reproduction can be sped up by higher temperature, leading to more frequent infestation and more extensive crop damages. Moreover, changes in climatic condition also aid to the spread of pest and diseases to new regions, which can have an effect on crops that were previously resistant to these problems (Gregory et al. 2009).

To address the challenges posed by climate change, farmers, researchers, and policymakers need to develop and implement adaptation and mitigation strategies. Given the growing global population, shifting climate patterns, and the imperative of sustainable agricultural practices, the adoption of conservation agriculture has emerged as a promising approach to augment food crop production while mitigating the effects of climate change (Basso et al. 2020). This chapter specifically concentrates on the techniques involved in adapting conservation agriculture as a climate-smart technology for food crops.

13.2 Climate-Smart Agriculture

The sustainability of the food system is a global challenge that needs to be tackled to maintain food security. Increasing population, poverty, malnutrition coupled with the stress from climate change, abrupt weather (Lipper et al. 2014), and faulty agricultural practice altogether are causing suppressed agricultural productivity with the emission of greenhouse gases (GHGs). According to the report, FAO (2017), greenhouse gas emissions are expected to rise by 30% by 2050. Consequently, the threat of climate change on agriculture and its allied sector needs to be addressed with smart approaches to transform the agricultural sector into a more climate-resilient system. Agriculture and allied sectors provide livelihoods to the majority of the population worldwide. Therefore, United Nations has formulated 17 Sustainable Development Goals (SDGs), where 13th SDG is titled as "Climate action" (Das and Ansari 2021) to provide sustainability of food grain production in the era of climate change.

In this context, the "Climate-Smart Agriculture" approach (CSA) seems to be an innovative path of reshaping agricultural development especially, for small and marginal farmers in developing countries (Branca et al. 2021). CSA is defined as an "agricultural approach that increases sustainable production with higher resilience (adaptation) enhances resilience (adaptation), reduces/removes emission of greenhouse gases (mitigation) whenever possible with enhancing the achievement of national food security and development goals."

To achieve environmental sustainability, the Conservation Agriculture (CA) system is an avenue that offers a variety of ecosystem services such as soil conservation, erosion control, restoration of the health of the soil, and improvement of crop productivity (Kassam et al. 2022; Pramanick et al. 2024a, b). Furthermore, the principles of sustainability and the CSA are somehow interrelated. The adoption of the CSA is based on elements such as strengthening resilience, the effective use of natural resources, and their conservation and protection. The goal of CSA is sustainable agricultural development and assuring food security in the climate change scenario.

13.3 Principles of Climate-Smart Agriculture

The concept of climate-smart agriculture was introduced in 2009, and broadly, the CSA's approach is centered on three pillars (i) to ensure food and nutritional security, (ii) to adapt to climatic stress, and (iii) to reduce greenhouse gas emissions from agriculture and allied sectors. The CSA approach seeks to merge traditional agricultural practices such as crop diversification, planting of cover crops, conservation agriculture, agroforestry, and rainwater harvesting (Altieri and Nicholls 2017) along with innovative practices, technologies, and services like information and communication technologies (ICTs), crop/livestock insurance, climate risk insurance, planting of heat- and drought-tolerant crops, and use of improved seeds. All of these CSA activities assist in minimizing the impact of climate change on rural livelihoods that are largely site-specific (Gangopadhyay et al. 2019; CIAT 2014). A variety of factors, including the socioeconomic characteristics of farmers, the physical environment of a specific region, and the characteristics of new technologies, determine the extent to which CSA technologies are to be adopted in a particular region/location. As a result, CSA uses techniques that strive for sustainability in production with the provision of food and livelihood security to support rural livelihood (Fig. 13.1).

13.4 Climate-Smart Agriculture Techniques

Climate-smart technology involves interventions ranging from field management to information related to weather awareness-related issues. All these information is required for the successful achievement of the sustainability goal (Nyasimi et al. 2017). The CSA strategies (Table 13.1):

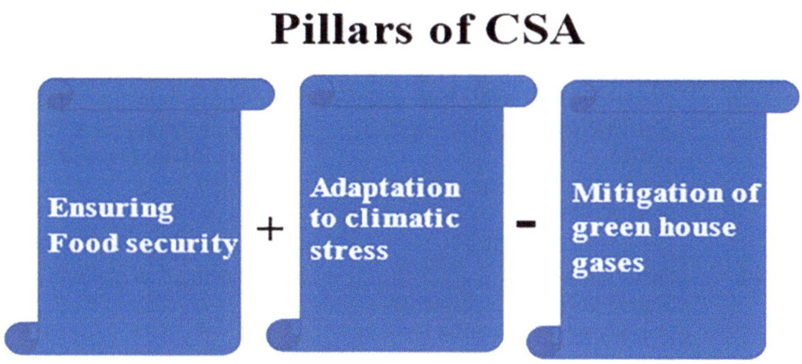

Fig. 13.1 Three pillars of climate-smart agriculture (CSA)

13.4.1 Conservation Agriculture

Conservation agriculture is an innovative and sustainable approach to farming that aims to protect and enhance the health of our planet's precious natural resources while ensuring food security for present and future generations. With the ever-increasing global population and the mounting challenges posed by climate change and environmental degradation, the need for sustainable agricultural practices has become more urgent than ever.

13.4.1.1 Ways in Which Conservation Agriculture Can Contribute to Climate-Smart Food Crop Production

Improve Soil Health

Conservation agriculture (CA) is a potential way-out to address sustainable agricultural practices, which is primarily aimed at minimizing soil degradation by restoring nutrient cycling, carbon (C) transformation, and soil structure as a whole (Pheap et al. 2019). One of the basics of CA is to improve soil health, which is referred as the soil system's ability not only to produce an ample quantity and quality of food and fiber but also provide various ecosystem services (Kibblewhite et al. 2008). Soil health refers to the capacity of soil to support the living system, that is, plant, animal, and human life. This concept synchronizes with three basic principles of CA which are minimal soil disturbances, crop cover/residue retention, and crop diversification through intercropping (Page et al. 2020).

Numerous onfarm trials conducted worldwide have documented soil health benefits that facilitate the regeneration of degraded soils while concurrently reducing farming inputs and expenses, including fertilizer and pesticide usage (Montgomery 2021). Through global studies, it has been revealed that the implementation of CA practices enhances soil health by positively impacting its physical, chemical, and biological properties (Pheap et al. 2019).

Table 13.1 Technological interventions to attain climate-smart agriculture

Technology	Focus area	Adaptation/mitigation strategy
Conservation agriculture	Prevents loss of arable land with regeneration of degraded land system. Key roles in attaining sustainability of an ecosystem	No tillage or minimum tillage, cover cropping, mulching, crop rotation including legumes to optimize to C/N ratio of soil
Crop diversification	Enhances soil fertility Boost productivity Supply of ecological diversity Enhances carbon sequestration in soil Less disease pest incidence	Includes practices like crop rotation, intercropping, and multiple cropping
Improved crop varieties/seed bank	Includes improved crop varieties for specific purpose	Tolerance to temperature, drought, and flood, weed, and salinity stress along with seed banks for ex situ conservation of seeds
Soil management	Soil fertility is one of the main factors in plant productivity, sustainability, and nutrient availability. Improve the fertilizer use efficiency.	Site-specific nutrient management (SSNM), integrated nutrient management (INM) integration of both organics and inorganics through green manuring (GM), farm yard manure (FYM), and slow-release nitrogenous fertilizers
Water management	Boosts productivity through proper water conservation, infiltration, and supplementation to increase water use efficiency by reducing runoff and seepage loss	Harvesting rainwater and using drip irrigation and appropriate drainage channel to lessen excessive runoff and seepage loss from cropland. Cover crops to reduce the effects of rainfall, soil erosion, and evapotranspiration
Agroforestry	Includes tress with crops and/or livestock Increasingly acknowledged to reduce climate-related risks. Source of revenue for rural livelihood	Increases soil fertility by sequestering carbon, used for fuel, food, and fodder purposes, used as shelterbelts and windbreaks and tree leaves used as mulch.
Post-harvest management	Required to reduce the food spoilage and enhance the food security	A better harvesting method, as well as processing, handling, transportation, storage, and packing improvements
Energy management	Renewable source of energy could be a more efficient way of reducing climatic stress	Biofuel usage, instead of fossil fuels emphasis on the use of solar energy, reduced mechanization, and precision agriculture
Seasonal and climatic stress management	For an assured production, forecast of all the climate-related information should reach to farmers, so they could plan for an alternate strategy to escape any crop failure and also protect livestock from unnatural climatic events.	Meteorological observatories and information and communication technologies (ICTs) aid in compiling the necessary data.

No Till and Soil Health

One of the objectives of CA is to mitigate soil disturbance through the implementation of practices that decrease or eliminate tillage operations. Tillage speeds up the decomposition process through rapid oxidation and microbial breakdown, consequently diminishing the quantity of soil organic matter (SOM) present. Excessive tillage damage causes the loss of SOM and disruption of soil structure and led to accelerated soil loss (Hou et al. 2012). Soil crusting is a common occurrence in alluvial soils, typically caused by poor soil structure, hindered infiltration and excessive tillage operations (Thierfelder et al. 2005). Tillage also results in immediate increase in microbial activity and results in early availability of the nitrogen (N) prior to critical growth stage of the crop (Francis and Knight 1993). This essentially reduces the fertilizer use efficiency and promote indiscriminate use of fertilizers. One of the principles of CA is reduced soil disturbance and aims to foster soil health by nurturing beneficial soil organisms. However, conventional tillage practices affect the diversity of microbial species in soil, including bacteria, worms, and fungal populations. The operation also reported to disrupt the mycorrhizal fungal roots and impedes the uptake of nutrients by plants (Montgomery 2021).

No-tillage (NT) plays a significant role in CA by minimizing disturbance to surface residues. It eliminates all preplanting operations except for creating holes in the ground for seed placement. Global data suggests that decrease in tillage intensity is associated with significant reductions in runoff (30–65%) and erosion (63–80%), resulting in reduced nutrient losses caused by erosion (Preiti et al. 2017). Naab et al. (2017) conducted a study in North Western Ghana and found that implementing CA practices, such as NT and residue retention, led to higher soil organic carbon (SOC) and total soil N levels. It also resulted in increased maize and soybean yields ranging from 23% to 39% and 20% to 29% lower production costs compared to conventional tillage.

Sequestration of Soil Organic Matter

Agricultural practices such as crop monoculture and deep tillage involving layer inversion have reported to gradual degradation of soil structure, compaction, and a decline in SOM content (Francaviglia et al. 2023; Pramanick et al. 2024a, b). The SOM enhances soil structure, water-holding capacity, nutrient retention, and microbial activity (Blanco-Canqui et al. 2013). CA practice modifies the distribution of SOM through vertical stratification achieved by adopting no-till practices and through the addition of crop residues and cover crops, thereby influencing SOM pools (Wacker et al. 2022). CA promotes crop diversification, organic inputs, and rhizodeposition, which in turn enhance microbial activity and contribute to the accumulation of stable SOM. The alteration of the microbial community has implications for its capacity to decompose C substrates of different structural complexity, potentially influencing the turnover rate of SOM (Mangalassery et al. 2015). In agricultural systems, rate of C sequestration varies depending on factors such as soil properties, climate, crop type, and initial SOC levels. Meta-analyses indicated a range of 0.27–1.1 t ha^{-1} $year^{-1}$ in Mediterranean woody cropping systems and 0.12–1.56 t SOC ha^{-1} $year^{-1}$ in tropical conditions, highlighting the variability of C sequestration rates (Francaviglia et al. 2023).

Implementing practices such as incorporating cover crops, utilizing crop residues, crop diversification, and applying organic amendments such as compost or manure all contribute to the augmentation of organic matter levels, thereby promoting improved soil health. Crop cover plays a pivotal role in improving soil structure and aggregate stability, leading to increased water infiltration, valuable protection against soil erosion, and enhanced microbial activity (McDaniel et al. 2014). A global meta-analysis suggests that CA increased soil microbial C (C_{mic}) to an extent of 33% in no-tillage condition with residue retention, while no-tillage with residue retention increased C_{mic} by 25% (Li et al. 2018). In a recent study by Zani et al. (2023), it was found that implementing diversified legume-intensive organic systems with a three-year grass clover ley periods and compost fertilization resulted in notably higher soil carbon stocks when compared to cereal-intensive conventional rotations. The increase in soil carbon stocks was observed at both soil depths (0–0.30 m and 0.30–0.60 m), with the organic rotation showing 28% higher carbon stocks in the topsoil and 46% higher carbon stocks in the subsoil compared to the conventional rotation.

All the discussed CA practices contribute to the long-term sustainability and productivity of agricultural systems, fostering a healthier and more resilient soil ecosystem. Thierfelder et al. (2013) conducted a long-term study in Zambia to assess the impact of CA on soil quality. The findings showed substantial yield improvements, with maize yields increasing by up to 78% after four seasons and 75–91% higher after six seasons compared to conventional practices. The on-station long-term trial also revealed significant carbon increases in CA plots, achieving a 12% increase in the top 30 cm of soil, while the conventional control resulted in a 15% decrease. Their study compared the CA system with conventionally tilled control and reported the improvement in key soil quality parameters with health improvement.

Improve Water Management

Soil moisture is essential for maintaining soil health, but insufficient water supply and frequent dry spells can adversely affect crop growth (Barron et al. 2003). Recent climate change models predict an increased depletion of soil water reserves and higher reference evapotranspiration, exacerbating the existing water availability challenge. Consequently, crop water extraction is projected to rise by approximately 30% due to various crop management practices, potentially leading to an imbalance between crop water demand and available resources (García-Tejero et al. 2020). This issue is particularly critical in arid and semiarid environments with limited water availability. In rainfed ecosystems, maintaining optimal soil moisture regime also poses challenges, highlighting the importance of improving soil physical properties in such circumstances.

Response in Relation to Hydraulic Conductivity and Soil Moisture Characteristic (SMC) Curve

The soil moisture characteristic (SMC) curve can be a valuable tool for assessing such functional relationship between soil water content and its energy state, specifically the matric potential, when the soil is in a state of equilibrium (Mohammadi and Meskini-Vishkaee 2013). Hysteresis, caused by factors such as the ink-bottle effect, nonuniform pore distribution, variations in contact angles and soil swelling, and shrinking during wetting and drying, can lead to variations in soil moisture at specific suction levels. Changes in hysteresis, particularly at the inflection point, are closely associated with significant alterations in soil physical characteristics. The implementation of CA management practices has the potential to influence the behavior of the SMC curve, particularly by altering the discharge point's slope (García-Tejero et al. 2020). This is an essential aspect of improving water management, as CA aims to address soil physical properties (i.e., pore distribution, structure, compaction, hydraulic conductivity, etc.) which can modify the SMC curve and enhance water retention in the soil.

Improved Soil Physical Condition: Reduction in Runoff Loss, Soil Erosion, and Pollution

CA has been extensively demonstrated to have a profound impact on crucial soil properties like bulk density, porosity, aggregate stability, sorptivity, and hydraulic conductivity, which play a vital role in regulating water retention and energy dynamics within the soil (Pheap et al. 2019). Increasing the stability of aggregates and organic matter content in the soil is considered one of the most reliable approaches to enhance soil health, leading to improved water dynamics. However, it is important to note that assessing soil quality improvements in a short-duration study requires careful quantification. Nonetheless, techniques such as no tillage, minimal tillage, cover cropping, crop rotation, and intercropping have shown their effectiveness in reducing evaporation, enhancing water use efficiency, minimizing surface runoff, improving micropore porosity, and increasing infiltration and water-holding capacity (Turmel et al. 2015). These approaches have been found to effectively mitigate soil erosion caused by water by promoting vegetation cover, leading to significant reductions in runoff (30–65%) and erosion (63–80%) (Martínez-Mena et al. 2020). In another study, Clausen et al. (1996) have shown that implementing CA practices can significantly reduce runoff by 15–89% and effectively decrease the presence of dissolved pesticides, nutrients, and sediments in runoff.

A study was conducted in Malawi of Southeastern Africa to compare the continuous no-till maize with conventional tillage rotation in two agroecological zones

(TerAvest et al. 2015). Over a period of four years (2011–2014), the impact of different cropping systems on soil moisture content, infiltration, sediment runoff, and soil organism abundance (termites and earthworms) was investigated. The results showed that in soils with low water-holding capacity, residue retention had a positive correlation with infiltration and CA exhibited an average increase in soil water content (~ 20 mm for 0–60 cm) compared to conventional tillage practices. Furthermore, both NT and CA practices significantly reduced sediment runoff compared to conventional practices, with reductions ranging from 209 kg ha^{-1} to 1537 kg ha^{-1}. However, the extent of these benefits is contingent upon the level of residue retention in the field, which plays a crucial role in increasing soil moisture content and reducing surface crusting across different agroecological conditions (Mupangwa et al. 2012). Additionally, increased residue retention promotes the proliferation of soil organisms such as earthworms and termites, facilitating improved infiltration through the creation of flow channels and the reduction of surface crusting (Stagnari et al. 2010). Earthworm also helps to produce higher amount of organic matter, restricting the movement of agrochemicals and reducing the potential risk of leaching and water pollution.

Maintaining Biodiversity

The term biodiversity was derived from the term "biological diversity." It encompasses the multitude of life forms present on Earth across various levels, ranging from genetic diversity to entire ecosystems (American Museum of Natural History 2023). The preservation of biodiversity is a comprehensive concept that encompasses the interconnected processes of evolution, ecology, and culture, which collectively contribute to the sustenance and maintenance of life. However, the importance of soil biodiversity in agriculture has not received adequate attention, potentially due to the predominant focus on increasing crop productivity through the application of various inputs such as fertilizers, pesticides, irrigation, and tillage operations (Stagnari et al. 2010).

Ecosystem Services and Conservation Agriculture

To address the issue of declining biodiversity, the European Union (EU) has implemented the Biodiversity Strategy for 2030, with the objective of reintegrating nature into agricultural land and safeguarding global ecosystems by ensuring their restoration, resilience, and effective protection by the year 2050 (European Commission 2011; Francaviglia et al. 2023). One of the approaches is the integration of the UK National Ecosystem Assessment (UKNEA) framework that have potential in conceptualizing ES and understanding their impact on human well-being through their inclusion in policy and planning (Watson et al. 2011). While these efforts show promise, further refinement is needed to determine the optimal number of variables required for comprehensive ES assessment. Additionally, anthropogenic activities, such as specific crop management practices, can significantly influence ecosystem functioning, thereby introducing complexity to the assessment of ecosystem multifunctionality (EMF) (Wittwer et al. 2021).

The Food and Agriculture Organization (FAO) put forth a recommendation endorsing CA as a sustainable approach for managing agroecosystems in a manner that ensures sustainable crop production while safeguarding natural resources and the environment (Jayaraman et al. 2021). In CA, incorporating grasses or legumes enhances soil biodiversity, with legumes playing a role in biological nitrogen fixation (BNF) and further supporting soil biodiversity (Srinivasarao et al. 2013). Crop rotation and diversification are vital for promoting local biodiversity, particularly when indigenous crop varieties are cultivated by farmers (Makate et al. 2016). Implementing crop rotations, cover crops, intercropping, and agroforestry is anticipated to amplify ecosystem services provided by microbial populations, pollinators, and pest controllers (Di Falco and Zoupanidou 2017). Additionally, the retention of crop residue in crop rotation contributes to nutrient cycling, facilitating the redistribution of carbon (C), nitrogen (N), phosphorus (P), sulfur (S), and other micronutrients, thereby promoting biodiversity (Al-Kaisi and Lal 2017). These practices not only enhance resilience and foster greater spatial and temporal biodiversity on the farm (Holling 1973) but also contribute to the improvement of soil fertility, making them valuable for sustainable crop production. Increased soil fertility also improves soil biodiversity with increase in number of predators and parasitoids promoting biological control of pathogens. When considering microbial diversity, CA has been found to result in an increase in microbial biomass of up to 83%, a 23% increase in the microbial biomass carbon (MBC): total carbon ratio and a 32% rise in the metabolic quotient (Choudhary et al. 2016). In a long-term farming experiment conducted by Wittwer et al. (2021) in Europe, they examined the characteristics of 43 agroecosystems to evaluate not only the productivity and economic performance of the agricultural system but also its multifunctionality. The results showed that conventional farming practices led to the highest yields but at the expense of reduced multifunctionality. On the other hand, CA demonstrated positive effects on various aspects, such as enhancing soil physical quality, contributing to climate mitigation and preserving biodiversity. The implementation of RT practices had a beneficial impact on aggregate stability, significantly reduced sediment delivery (~93% compared to conventional no-tillage), and increased the abundance of soil organisms like earthworms and arbuscular mycorrhizal fungi.

Nonetheless, despite the significance of these approaches, limited research has been conducted on the indirect effects of CA on various ES, highlighting the need for further investigation. To comprehend such intricate dynamics in CA, it is necessary to adopt a multidisciplinary approach that can address whether CA can provide significant ES to maintain biodiversity (Jayaraman et al. 2021). One potential indirect approach involves improving vegetation diversity and connectivity through the establishment of agroforestry plots, which can mitigate the requirement for encroaching upon natural habitats.

13.5 Climate Mitigation

Climate change presents a significant threat and poses a major challenge to sustainable crop production (Mandal and Banik 2023). Projections from the Intergovernmental Panel on Climate Change (IPCC) indicate that India is expected to witness a temperature increase of 3–4 °C by the end of the twenty-first century, leading to a potential decline of 3–26% in net agricultural revenues (Kumari et al. 2020). The climatic fluctuations contribute in elevated carbon dioxide (CO_2) levels, which have a twofold effect on food crops: on one hand, they stimulate plant growth, but on the other hand, they diminish the nutritional content of most crops by lowering protein and essential mineral concentrations, potentially posing a risk to human health (EPA 2017). Furthermore, heightened levels of CO_2 contribute to an escalation in weed growth, the prevalence of pest infestations, and the occurrence of fungal attacks. This, in turn, leads to escalated pesticide usage, further jeopardizing human health due to increased pest pressures and reduced pesticide efficacy. The repercussions are particularly severe for rural livelihoods, especially for smallholder farmers who heavily rely on the natural environment (Wheeler and Von Braun 2013).

13.5.1 Conservation Agriculture and Climate-Smart Approach

CA has the potential to function as a "climate-smart" agricultural technique that enhances agricultural productivity, mitigates the adverse effects on crop production, and enhances resilience against weather-induced stresses (FAO 2011). Through the improvement of soil quality and nutrient cycling, the adoption of CA can enhance crop resilience and facilitate better adaptation to local climate change, potentially increasing drought tolerance in specific regions (Hobbs and Govaerts 2010). However, it is important to note that the process of developing climate-resilient soils through CA requires a significant amount of time and does not yield immediate results. Although there are some evidences supporting the positive impact of CA, the availability of rigorous and comprehensive studies in this domain is limited and may be influenced by potential selection bias (Michler et al. 2019). Additionally, the wide variation in agroecosystems contributes to challenges in replicating and transferring results. This hampers the assessment of CA's effectiveness in terms of production outcomes and climate resilience. Michler et al. (2019) conducted an experiment utilizing panel data from Zimbabwe to assess the performance of CA across different rainfall conditions, including both shortfalls and surpluses. Through a rigorous analysis that accounted for potential sources of bias, they found that CA demonstrated effectiveness in mitigating the adverse effects of weather risks arising from deviations in rainfall. Over a four-year period, their study revealed that yields were generally more resilient to abnormal rainfall patterns when CA practices were employed compared to conventional farming methods.

13.5.2 Carbon Sequestration in the Context of Climate Change

Dimassi et al. (2013) proposed that there are limitations to the capacity for SOC accumulation, as reaching a new equilibrium value implies no further increase in carbon content. However, various studies indicate that the accumulated SOC is primarily in labile forms, such as particulate organic carbon, with limited evidence of significant increases in stable pools. Since the labile pool is reversible, it becomes essential to store SOC in more recalcitrant forms to effectively address climate change challenges (O'Rourke et al. 2015). This realization underscores the necessity for implementing new management practices. Furthermore, it is crucial that the management practices are specific and represent a net additional transfer of SOC from the atmosphere to the land to meet the criteria for climate change mitigation, rather than a mere spatial redistribution of SOC within the soil (Sommer and Bossio 2014). Powlson et al. (2014) reported that NT practices benefit soil quality and contribute to agricultural adaptation to climate change. However, NT has limited potential, leading to altered carbon distribution, and its role in climate change mitigation is frequently overstated.

In the given context, implementing all the three principles of CA (i.e., minimal soil disturbance, residue retention, and crop diversification) is tricky and complex. The combined effect of these components is essential but can be challenging to maintain them all simultaneously. This is particularly pronounced in arid regions where there is high competition for crop residues, making it difficult for farmers to retain a sufficient amount of residue on the soil, as highlighted by Govaerts et al. (2009). Insufficient residue retention can also result in soil degradation, even with RT practices. Thus, in order to achieve an accumulation of more stable fraction of SOM, it is essential not only to have C input from NT or crop residues but also from other practices such as crop diversification, which is often neglected. Makate et al. (2016) investigated the impact of crop diversification on enhanced climate resilience by conducting a study in rural Zimbabwe, utilizing data from 500 smallholder farmers. They employed a conditional mixed process framework to address the selection bias associated with crop diversification. Their findings revealed that diversified cropping systems, in the context of climate change, can lead to increased SOC accumulation, crop yields, income, and food security in smallholder farming systems. Consequently, they concluded that wider adoption of diversified cropping systems is crucial for promoting production stability in the face of a changing climate.

13.5.3 Improve Crop Yield

The impact of CA on crop yield is variable and dependent on different factors such as soil quality, climate, and other ecosystem variables that influence crop productivity. Positive effects on crop yield have been observed (Kumar et al. 2018; Page et al.

2019), whereas there are also reports of no change or decreases in yield associated with CA, as highlighted by Baudron et al. (2012) and Camarotto et al. (2018). Some researchers have found that specific CA practices and rotation interactions can increase crop yield, leading to improved profitability for farmers with reduced input requirements (Zheng et al. 2014). In dry climates, adopting CA has been linked to yield increases through residue retention, improved soil physical conditions, hydraulic properties, and enhanced water storage (Steward et al. 2018; Page et al. 2019). On the other hand, negative effects on yield may be attributed to altered soil physiochemical conditions (such as compaction, surface crusting, waterlogging, low infiltration, and low soil temperature), biological conditions, absence of cover crops, and poor management practices (e.g., improper fertilizer placement, inadequate residue management), which can lead to weed proliferation, pest outbreaks, and disease incidences (Steward et al. 2018).

The beneficial effects of CA practices have been consistently demonstrated, with CA being adopted on over 180 million hectares of cropland worldwide, equivalent to approximately 12.5% of the global cropland (Francaviglia et al. 2023). To comprehensively evaluate the effects of CA on crop yield, it is necessary to consider specific practices and take regional climatic characteristics into account. However, most of the studies are process based and based on the trial results, thus not sufficient to claim a wholesome soil health assessment (Pheap et al. 2019). This may be one of the reasons that is slowing down the adoption of CA, despite of its widespread promotion. Furthermore, the global applicability of CA is somewhat uncertain; barriers to its adoption may arise from cultural and as well as socioeconomic factors. There have been reports suggesting that the adoption of CA has been limited, potentially due to insufficient local participation and the failure to integrate local knowledge and practices into the scientific framework, as emphasized by Oppenheimer et al. (2015). To assess the long-term advantages of CA resulting from improved soil health, it is essential to prioritize the creation of comprehensive, long-term observational datasets on CA practices. However, a significant challenge arises in persuading an adequate number of farmers to consistently adopt these potentially costly agricultural practices, particularly when immediate benefits may not be evident in the short term (Michler et al. 2019). To bridge this void, it could be advantageous to embrace an integrated approach that takes into account the interdependencies among various soil components. Additionally, it is essential to prioritize the requirements of smallholder farmers over extensive protocols and huge data analytics. This can be linked with the customization of CA practices based on the specific location and farmers' need to address the constraints (Page et al. 2020). Such approaches have the potential to enhance decision-making in land management and facilitate the implementation of scaled-down technical interventions for easier adoption within the farming communities (Fig. 13.2).

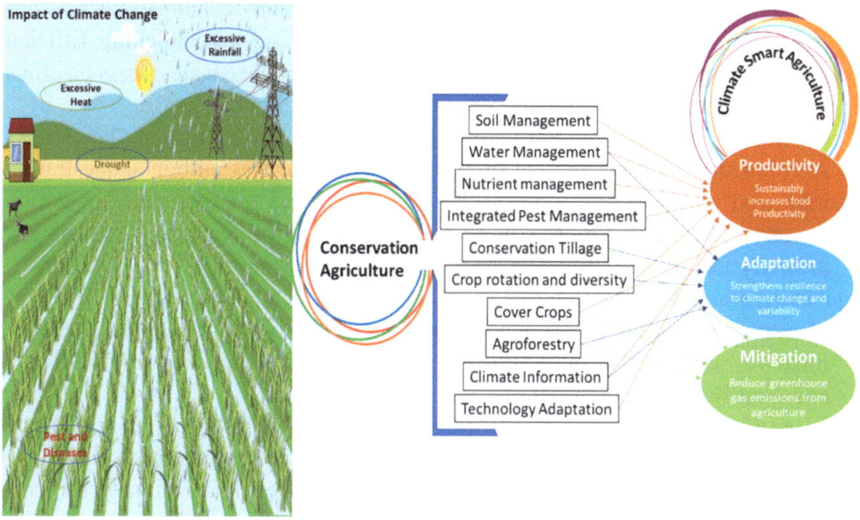

Fig. 13.2 Mechanism for mitigating climate change through conservation agriculture

13.6 How Can Conservation Agriculture Be More Climate Smart?

Conservation agriculture (CA) is an approach to farming that aims to improve sustainable agricultural practices while minimizing environmental impacts. Conservation agriculture (CA) can be made more climate smart by incorporating various practices and principles that promote climate change mitigation and adaptation. Here are some ways to enhance the climate smartness of conservation agriculture (Dutta et al. 2023):

Soil management: Emphasize the use of practices that improve soil health and fertility, such as minimal tillage or no-till farming. Adopt minimum or zero tillage techniques such as no-till or reduced tillage systems to preserve soil structure and organic matter, which helps sequester carbon and retain soil moisture thereby reducing greenhouse gas emissions and promoting climate resilience.

Crop rotation and diversity: Encourage the use of diverse crop rotations and intercropping systems. Crop rotation helps break pest and disease cycles, reduces the need for chemical inputs, improves soil structure, and enhances biodiversity. Diversification of crops can also increase resilience to climate change by spreading risks associated with extreme weather events.

Cover crops: Promote the use of cover crops during fallow periods or between cash crops. Cover crops protect the soil from erosion, enhance organic matter content, suppress weeds, improve water infiltration, and retain soil moisture. They also sequester carbon from the atmosphere, mitigating climate change.

Water management: Optimize water use efficiency through techniques such as precision irrigation (drip or sprinkler), rainwater harvesting, and mulching. Efficient water management helps conserve water resources, reduce water stress on crops, and adapt to changing precipitation patterns due to climate change.

Agroforestry: Incorporate trees and shrubs within agricultural landscapes. Agroforestry systems provide multiple benefits, such as carbon sequestration, improved soil structure, enhanced biodiversity, and microclimate regulation, thereby reducing the impacts of climate change. Agroforestry systems can also diversify farm income through the production of timber, fruits, nuts, or fodder.

Nutrient management: Optimum level of nutrient application through precision fertilization, application of organic fertilizer, and utilization of crop residues improves soil fertility and nutrient use efficiency while reduces nutrient losses and greenhouse gas emission.

Conservation of genetic resources: Utilizing local genetic resources and traditional crop varieties which were adapted to the specific climatic condition, as a result of this diversity, climate change impacts can be buffered and climate resilience can be enhanced.

Integrated pest management: Integrated pest management (IPM) promotes ecological techniques, biological control, and various cultural practices to reduced environmental pollution and improves ecosystem health. So adoption of integrated pest management practices instead of synthetic pesticides to control and manage pest and diseases, which increases resilience and better ecosystem.

Climate information and weather advisory services: The provision of timely and accurate climatic information, weather forecast, and agroadvisory services to the farmers is very essential. With this information, they can make informed decisions about crop planning, water management, and pest control according to changing weather conditions.

Knowledge sharing and capacity building: To adopt climate-smart technologies, it is very important to sharing of information among farmers, researchers, and extension services. Through capacity building initiatives, people gain a deeper understanding of climate change impacts and solutions, which helps foster resilience and adaptation.

Policy support: Better policy system including financial incentives for adoption of conservation agriculture, funding for research and development, and a better market structure which will helps to promote and adopt climate-smart technologies in a sustainable way.

Incorporation of above all strategies helps to make conservation agriculture more climate smart through reduction of greenhouse gas emission, conservation of natural resources, and enhanced climate resilience in agricultural systems.

13.7 Conclusion

In conclusion, the implementation of conservation agriculture as a climate-smart technology for food crops shows enormous promise in resolving the issues brought about by climate change to global agriculture sector. Conservation agriculture provides an effective solution to enhance the resilience and productivity of food crops by implementing sustainable agricultural practices which prioritize water conservation, soil health, and biodiversity. Agricultural systems become more resilient to drought, flood, and pest-diseases through addition of practices like minimum soil disturbance, permanent soil cover, crop rotation, and crop diversification. In addition, conservation agriculture adheres to the tenets of climate-smart agriculture by optimizing water use, reducing greenhouse gas emission and sustaining biodiversity. A reduction in tillage and the optimal use of synthetic inputs contributes to climate change mitigation efforts by both lowering GHG emissions and enhancing soil carbon sequestration.

However, barriers to the widespread adoption of conservation agriculture, such as a lack of knowledge and awareness, availability to adequate technology, and supportive policies, need to be addressed. At the same time, to overcome these challenges and to promote conservation agriculture worldwide requires extensive research, capacity building, and policy support. Therefore, by addressing all the challenges, conservation agriculture can be adopted as a climate-smart technology for food crops which reflects a promising opportunity to ensure food security in the context of climate change.

References

Al-Kaisi MM, Lal R (2017) Conservation agriculture systems to mitigate climate variability effects on soil health. In: Soil health and intensification of agroecosytems. Academic Press, pp 79–107

Altieri MA, Nicholls CI (2017) The adaptation and mitigation potential of traditional agriculture in a changing climate. Clim Chang 140:33–45

American Museum of Natural History (2023). Available at: https://www.amnh.org/research/center-for-biodiversity-conservation/what-is-biodiversity. Retrieved on 25 May 2023

Barron J, Rockström J, Gichuki F, Hatibu N (2003) Dry spell analysis and maize yields for two semi-arid locations in East Africa. Agric For Meteorol 117(1–2):23–37

Basso B, Giola P, Dumont B, Balzarini M (2020) Conservation agriculture systems for climate-change mitigation: an integrated assessment approach. Front Plant Sci 11:1129

Baudron F, Tittonell P, Corbeels M, Letourmy P, Giller KE (2012) Comparative performance of conservation agriculture and current smallholder farming practices in semi-arid Zimbabwe. Field Crop Res 132:117–128

Blanco-Canqui H, Shapiro CA, Wortmann CS, Drijber RA, Mamo M, Shaver TM, Ferguson RB (2013) Soil organic carbon: the value to soil properties. J Soil Water Conserv 68(5):129A–134A

Branca G, Arslan A, Paolantonio A, Grewer U, Cattaneo A, Cavatassi R, Lipper L, Hillier J, Vetter S (2021) Assessing the economic and mitigation benefits of climate-smart agriculture and its implications for political economy: a case study in Southern Africa. J Clean Prod 285:125161

Camarotto C, Dal Ferro N, Piccoli I, Polese R, Furlan L, Chiarini F, Morari F (2018) Conservation agriculture and cover crop practices to regulate water, carbon and nitrogen cycles in the low-lying venetian plain. Catena 167:236–249

Choudhary M, Ghasal PC, Kumar S, Yadav RP, Singh S, Meena VS, Bisht JK (2016) Conservation agriculture and climate change: an overview. In: Conservation agriculture: an approach to combat climate change in Indian Himalaya. Springer, pp 1–37

CIAT "Climate-smart Agriculture Investment Prioritization Framework" (2014) International Center for Tropical Agriculture (CIAT), Cali, Colombia

Clausen JC, Jokela WE, Potter Iii FI, Williams JW (1996) Paired watershed comparison of tillage effects on runoff, sediment, and pesticide losses. Vol. 25, no. 5. American Society of Agronomy, Crop Science Society of America, and Soil Science Society of America

Das U, Ansari MA (2021) The nexus of climate change, sustainable agriculture and farm livelihood: contextualizing climate smart agriculture. Clim Res 84:23–40

Di Falco S, Zoupanidou E (2017) Soil fertility, crop biodiversity, and farmers' revenues: evidence from Italy. Ambio 46:162–172

Dimassi B, Cohan J-P, Labreuche J, Mary B (2013) Changes in soil carbon and nitrogen following tillage conversion in a long-term experiment in northern France. Agric Ecosyst Environ 169:12–20

Dutta G, Paramanik B, Mandal A (2023) Climate smart agriculture for better production system. In: Technological advancement and use of artificial intelligence in climate smart agriculture. International Books and Periodical Supply Service, pp 41–58

EPA, U (2017) Climate impacts on agriculture and food supply. United States Environmental Protection Agency, Washington, DC

European Commission (2011) Communication from the commission to the European Parliament, the Council, the European Economic and Social Committee and the Committee of the Regions Youth Opportunities Initiative. European Commission, Brussels, Belgium

FAO (2011) State of food and agriculture 2010–11. "Women and Agriculture: Closing the Gender Gap for Development"

FAO (2017) Sustainable development goals. Food and Agriculture Organization of the United Nations. Retrieved from. http://www.fao.org/sustainable-development-goals/goals/goal-13/en/

Francaviglia R, Almagro M, Vicente-Vicente JL (2023) Conservation agriculture and soil organic carbon: principles, processes, practices and policy options. Soil Syst 7(1):17

Francis GS, Knight TL (1993) Long-term effects of conventional and no-tillage on selected soil properties and crop yields in Canterbury, New Zealand. Soil Tillage Res 26(3):193–210

Gangopadhyay PK, Khatri-Chhetri A, Shirsath PB, Aggarwal PK (2019) Spatial targeting of ICT-based weather and agro-advisory services for climate risk management in agriculture. Clim Chang 154(1–2):241–256

García-Tejero IF, Carbonell R, Ordoñez R, Torres FP, Zuazo VHD (2020) Conservation agriculture practices to improve the soil water management and soil carbon storage in Mediterranean rainfed agro-ecosystems. In: Soil health restoration and management. Springer, pp 203–230

Govaerts B, Verhulst N, Castellanos-Navarrete A, Sayre KD, Dixon J, Dendooven L (2009) Conservation agriculture and soil carbon sequestration: between myth and farmer reality. Crit Rev Plant Sci 28(3):97–122

Gregory PJ, Johnson SN, Newton AC, Ingram JSI (2009) Integrating pests and pathogens into the climate change/food security debate. J Exp Bot 60(10):2827–2838

Hatfield JL, Boote KJ, Kimball BA, Ziska LH, Izaurralde RC, Ort D, Thomson AM, Wolfe D (2011) Climate impacts on agriculture: implications for crop production. Agron J 103(2):351–370

Hobbs PR, Govaerts B (2010) How conservation agriculture can contribute to buffering climate change. In: Climate change and crop production. CABI, Wallingford, UK, pp 177–199

Holling CS (1973) Resilience and stability of ecological systems. Annu Rev Ecol Syst 4(1):1–23

Hou XQ, Li R, Jia ZK, Han QF, Yang BP, Nie JF (2012) Effects of rotational tillage practices on soil structure, organic carbon concentration and crop yields in semi-arid areas of Northwest China. Soil Use Manag 28(4):551–558

IPCC (2012) Managing the Risks of Extreme Events and Disasters to Advance Climate Change Adaptation (SREX). Retrieved from: https://www.ipcc.ch/site/assets/uploads/2018/03/SREX-Chap1_FINAL-1.pdf

IPCC (2021) Climate Change 2021: The Physical Science Basis (AR6). Retrieved from: https://www.ipcc.ch/report/ar6/wg1/

Jayaraman S, Dang YP, Naorem A, Page KL, Dalal RC (2021) Conservation agriculture as a system to enhance ecosystem services. Agriculture 11(8):718

Kassam A, Friedrich T, Derpsch R (2022) State of the global adoption and spread of conservation agriculture. In: Advances in conservation agriculture, adoption and spread. Burleigh Dodds, Cambridge, UK

Kibblewhite MG, Ritz K, Swift MJ (2008) Soil health in agricultural systems

Kumar V, Jat HS, Sharma PC, Gathala MK, Malik RK, Kamboj BR, Yadav AK et al (2018) Can productivity and profitability be enhanced in intensively managed cereal systems while reducing the environmental footprint of production? Assessing sustainable intensification options in the breadbasket of India. Agric Ecosyst Environ 252:132–147

Kumari S, George SG, Meshram MR, Esther D, Kumar P (2020) A review on climate change and its impact on agriculture in India. Curr J Appl Sci Technol 39(44):58–74

Li Y, Chang SX, Tian L, Zhang Q (2018) Conservation agriculture practices increase soil microbial biomass carbon and nitrogen in agricultural soils: a global meta-analysis. Soil Biol Biochem 121:50–58

Lipper L, Thornton P, Campbell BM, Baedeker T, Braimoh A, Bwalya M, Caron P et al (2014) Climate-smart agriculture for food security. Nat Clim Chang 4(12):1068–1072

Lobell DB, Gourdji SM (2012) The influence of climate change on global crop productivity. Plant Physiol 160(4):1686–1697

Makate C, Wang R, Makate M, Mango N (2016) Crop diversification and livelihoods of smallholder farmers in Zimbabwe: adaptive management for environmental change. Springerplus 5:1–18

Mandal S, Banik GC (2023) Biosensors for precision agriculture. In: Narayan R (ed) *Encyclopedia of sensors and biosensors*, vol 4. Elsevier, pp 709–727

Mangalassery S, Mooney SJ, Sparkes DL, Fraser WT, Sjögersten S (2015) Impacts of zero tillage on soil enzyme activities, microbial characteristics and organic matter functional chemistry in temperate soils. Eur J Soil Biol 68:9–17

Martínez-Mena M, Carrillo-López E, Boix-Fayos C, Almagro M, García Franco N, Díaz-Pereira E, Montoya I, De Vente J (2020) Long-term effectiveness of sustainable land management practices to control runoff, soil erosion, and nutrient loss and the role of rainfall intensity in Mediterranean rainfed agroecosystems. Catena 187:104352

McDaniel MD, Tiemann LK, Grandy AS (2014) Does agricultural crop diversity enhance soil microbial biomass and organic matter dynamics? A meta-analysis. Ecol Appl 24(3):560–570

Michler JD, Baylis K, Arends-Kuenning M, Mazvimavi K (2019) Conservation agriculture and climate resilience. J Environ Econ Manag 93:148–169

Mohammadi MH, Meskini-Vishkaee F (2013) Predicting soil moisture characteristic curves from continuous particle-size distribution data. Pedosphere 23(1):70–80

Montgomery DR (2021) Soil health and the revolutionary potential of conservation agriculture. In: Rethinking food and agriculture. Woodhead Publishing, pp 219–229

Mupangwa W, Twomlow S, Walker S (2012) Reduced tillage, mulching and rotational effects on maize (Zea mays L.), cowpea (Vigna unguiculata (Walp) L.) and sorghum (Sorghum bicolor L. (Moench)) yields under semi-arid conditions. Field Crop Res 132:139–148

Naab JB, Mahama GY, Yahaya I, Prasad PVV (2017) Conservation agriculture improves soil quality, crop yield, and incomes of smallholder farmers in North Western Ghana. Front Plant Sci 8:996

Nyasimi M, Kimeli P, Sayula G, Radeny M, Kinyangi J, Mungai C (2017) Adoption and dissemination pathways for climate-smart agriculture technologies and practices for climate-resilient livelihoods in Lushoto, Northeast Tanzania. Climate 5(3):63

Oppenheimer M, Campos M, Warren R, Birkmann J, Luber G, O'Neill B, Takahashi K et al (2015) Emergent risks and key vulnerabilities. In: Climate change 2014 impacts, adaptation and vulnerability: part A: global and sectoral aspects. Cambridge University Press, pp 1039–1100

O'Rourke SM, Angers DA, Holden NM, McBratney AB (2015) Soil organic carbon across scales. Glob Chang Biol 21(10):3561–3574

Page KL, Dang YP, Dalal RC, Reeves S, Thomas G, Wang W, Thompson JP (2019) Changes in soil water storage with no-tillage and crop residue retention on a vertisol: impact on productivity and profitability over a 50 year period. Soil Tillage Res 194:104319

Page KL, Dang YP, Dalal RC (2020) The ability of conservation agriculture to conserve soil organic carbon and the subsequent impact on soil physical, chemical, and biological properties and yield. Front Sustain Food Syst 4:31

Pheap S, Lefevre C, Thoumazeau A, Leng V, Boulakia S, Koy R, Hok L, Lienhard P, Brauman A, Tivet F (2019) Multi-functional assessment of soil health under conservation agriculture in Cambodia. Soil Tillage Res 194:104349

Powlson DS, Stirling CM, Jat ML, Gerard BG, Palm CA, Sanchez PA, Cassman KG (2014) Limited potential of no-till agriculture for climate change mitigation. Nat Clim Chang 4(8):678–683

Pramanick B, Choudhary S, Kumar M, Singh SK, Jha RK, Singh SK, Salmen SH, Ansari MJ, Hossain A (2024a) Can site-specific nutrient management improve the productivity and resource use efficiency of climate-resilient finger millet in calcareous soils in India? Heliyon 10(12):e32774. https://doi.org/10.1016/j.heliyon.2024.e32774

Pramanick B, Kumar M, Naik BM, Singh SK, Singh SV (2024b) Soil carbon-nutrient cycling, energetics, and carbon footprint in calcareous soils with adoption of long-term conservation tillage practices and cropping systems diversification. Sci Total Environ 912:169421. https://doi.org/10.1016/j.scitotenv.2023.169421

Preiti G, Romeo M, Bacchi M, Monti M (2017) Soil loss measure from Mediterranean arable cropping systems: effects of rotation and tillage system on C-factor. Soil Tillage Res 170:85–93

Sommer R, Bossio D (2014) Dynamics and climate change mitigation potential of soil organic carbon sequestration. J Environ Manag 144:83–87

Srinivasarao C, Venkateswarlu B, Lal R, Singh AK, Kundu S (2013) Sustainable management of soils of dryland ecosystems of India for enhancing agronomic productivity and sequestering carbon. Adv Agron 121:253–329

Stagnari F, Ramazzotti S, Pisante M (2010) Conservation agriculture: a different approach for crop production through sustainable soil and water management: a review. In: Organic farming, pest control and remediation of soil pollutants: organic farming, pest control and remediation of soil pollutants. Springer, pp 55–83

Steward PR, Dougill AJ, Thierfelder C, Pittelkow CM, Stringer LC, Kudzala M, Shackelford GE (2018) The adaptive capacity of maize-based conservation agriculture systems to climate stress in tropical and subtropical environments: a meta-regression of yields. Agric Ecosyst Environ 251:194–202

TerAvest D, Carpenter-Boggs L, Thierfelder C, Reganold JP (2015) Crop production and soil water management in conservation agriculture, no-till, and conventional tillage systems in Malawi. Agric Ecosyst Environ 212:285–296

Thierfelder C, Amézquita E, Stahr K (2005) Effects of intensifying organic manuring and tillage practices on penetration resistance and infiltration rate. Soil Tillage Res 82(2):211–226

Thierfelder C, Mwila M, Rusinamhodzi L (2013) Conservation agriculture in eastern and southern provinces of Zambia: long-term effects on soil quality and maize productivity. Soil Tillage Res 126:246–258

Turmel M-S, Speratti A, Baudron F, Verhulst N, Govaerts B (2015) Crop residue management and soil health: a systems analysis. Agric Syst 134:6–16

Wacker TS, Jensen LS, Thorup-Kristensen K (2022) Conservation agriculture affects soil organic matter distribution, microbial metabolic capacity and nitrogen turnover under Danish field conditions. Soil Tillage Res 224:105508

Watson R, Albon S, Aspinall R, Austen M, Bardgett B, Bateman I, Berry P et al (2011) UK National Ecosystem Assessment: understanding nature's value to society. Synthesis of key findings
Wheeler T, Von Braun J (2013) Climate change impacts on global food security. Science 341(6145):508–513
Wittwer RA, Franz Bender S, Hartman K, Hydbom S, Lima RAA, Loaiza V, Nemecek T et al (2021) Organic and conservation agriculture promote ecosystem multifunctionality. Sci Adv 7(34):eabg6995
WMO (2021) WMO Provisional Statement on the State of the Global Climate in 2020. Retrieved from: https://public.wmo.int/en/media/press-release/provisional-statement-on-state-of-global-climate-2020
Zani CF, Manning DAC, Abbott GD, Taylor JA, Cooper J, Lopez-Capel E (2023) Diversified crop rotations and organic amendments as strategies for increasing soil carbon storage and stabilisation in UK arable systems. Front Environ Sci 11:493
Zheng C, Jiang Y, Chen C, Sun Y, Feng J, Deng A, Song Z, Zhang W (2014) The impacts of conservation agriculture on crop yield in China depend on specific practices, crops and cropping regions. Crop J 2(5):289–296

Chapter 14
Organic Agriculture and Improved Ecosystem Services for Food and Nutritional Security

S. K. Sharma, B. S. S. Siddartha Naik, S. Sai Chandrika, M. Sunil Kumar, G. Krishna Reddy, K. Gurava Reddy, P. Gurumurthy, Dinesh Kumar, and Rupesh Tirunagiri

Abstract The ability for developing nations to secure food and nutrition often relies on the effective use of natural resources. Evaluations of food and nutrition preservation typically consider three elements: availability, access, and utilization. Ecosystems contribute to these elements both directly and indirectly by providing vital services that aid in sustainable agricultural production, profit generation, expansion, and the provision of energy for cooking. However, there can be instances where the domestic exploitation of these natural resources jeopardizes food security, hampers national initiatives to reduce poverty, and endangers the continuity of key ecosystem functions. Sometimes, the resources and services utilized to ensure access and utilization can adversely affect the ecosystem functions that guarantee food availability. These findings highlight the necessity of integrating ecosystem services into the approaches to food security and poverty alleviation in developing

---

S. K. Sharma (✉)
Indian Council of Agricultural Research, Pusa, New Delhi, India

B. S. S. Siddartha Naik · S. S. Chandrika · M. S. Kumar · G. K. Reddy
Department of Agronomy, Agricultural College, Acharya N.G. Ranga Agricultural University, Guntur, Andhra Pradesh, India

K. G. Reddy
Department of Agricultural Extension, Administrative Office, Acharya N.G. Ranga Agricultural University, Guntur, Andhra Pradesh, India

P. Gurumurthy
Department of Soil Science and Agricultural Chemistry, Agricultural College, Acharya N.G. Ranga Agricultural University, Naira, Andhra Pradesh, India

D. Kumar
Department of Agronomy, Indian Institute of Soil and Water Conservation Research Centre, Datia, Madhya Pradesh, India

R. Tirunagiri
Division of Science and Agricultural Chemistry, Indian Agricultural Research Institute, Pusa, New Delhi, India

B. Pramanick et al. (eds.), *Climate-Smart Agricultural Technologies*, Sustainability Sciences in Asia and Africa, https://doi.org/10.1007/978-981-96-7699-6_14

nations. Consumer preference for foods rich in beneficial compounds, in addition to essential nutrients, has seen a significant rise. The deficiency of vital minerals, vitamins, and essential proteins in the daily diets of approximately three billion people worldwide is the main reason for their nutritional inadequacy, as stated by Welch (Food Nutr Bull 26(4): 419–421, 2005). Overfertilization and intensive crop cultivation in India are causing soils to lose vital nutrients and organic carbon, leading to a decline in crop quality. This nutrient deficiency in major food crops, posing a malnutrition risk for infants and working women, can be counteracted by adopting agronomic practices such as organic farming.

Keywords Organic farming · Biocontrol · Food security · Ecosystem services · Food quality

14.1 Introduction

For nearly 79.8% of the global rural population, farming is the primary source of income, which until the first quarter of the previous century, leveraged biological diversity. Back then, the versatility of agriculture was a crucial factor in its sustainability. As demand for crops and other agricultural products grew over time, farming became more complex with the use of agrochemicals. Traditional farming depended solely on technologies driven by the supply of inputs and necessitated large quantities of chemical fertilizers, pesticides, irrigation, capital, and labor. Despite this, it expanded sufficiently to meet the needs of the global population a few decades ago (Smil 2000). The rest of the world, witnessing the success of chemical farming in developed countries, adopted this approach, leading to changes in land use (Goldewijk and Ramankutty 2004; Sandhu et al. 2010; Naik et al. 2022) and the onset of agroecosystem degradation (Tilman et al. 2002). Ensuring the sustainability of agricultural production and food and nutrition security involves providing everyone with sufficient nutritious food while also ensuring that the agricultural production system does not negatively impact the environment. In other words, sustainable agriculture is a production system that meets today's needs without exhausting natural resources, thereby securing the ability of future generations to meet their own needs. It is clear, however, that this expectation is not being met, and modern agriculture is falling short of these sustainability demands.

14.2 Organic Agriculture

Organic farming prioritizes establishing an ideal connection between humans and the planet. Over the last 30 years, there has been a global rise in consciousness and action toward environmental conservation, safeguarding of agroecosystems, and high-quality food production. Today's organic farming deviates from its roots in its aim to encourage healthy soil, food, and people. Various leading organizations have their own definitions of organic farming or organic agriculture, but core notions such as sustainability and ecosystem services remain key to these definitions (Fig. 14.1). Without sacrificing natural resources, it focuses on enabling future generations to fulfill their needs. This commitment is missing in modern agriculture, making it evident that it will not be able to satisfy the sustainability criteria.

The main pillars of organic farming are the following:

- Organic threshold specifications
- Reliable certification and regulatory affairs mechanisms
- Technological products
- An effective and viable market network.

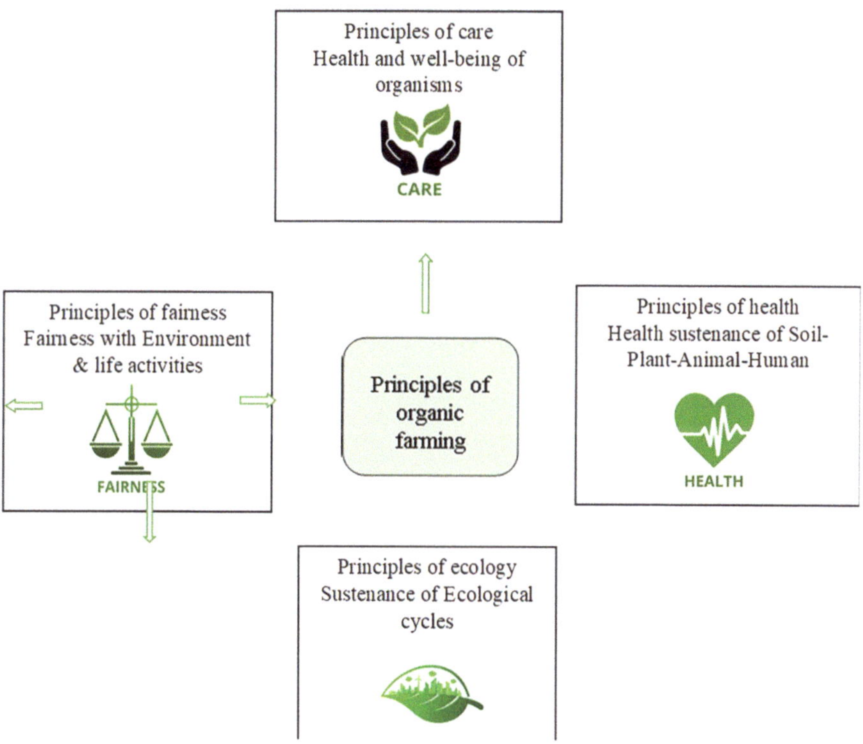

Fig. 14.1 Principles of organic farming and their importance

14.2.1 Principles of Organic Farming

Apart from these, the organic agricultural approach follows some basic principles of health, ecology, fairness, and care suggested by IFOAM. These principles are the core of organic agriculture that ensures sustainable development (Luttikholt 2007) (Fig. 14.2; Tables 14.1 and 14.2).

14.2.2 Reliable Services of the Ecosystem in Organic Farming

Healthy approaches to organic cultivation and other agrienvironmental services (AES) have been proposed as strategies to mitigate the negative effects of intensive agriculture. Organic farming practices can positively impact soil health, nutrient cycling, biodiversity enhancement, and the delivery of ecosystem services. Ecosystem services, deeply tied to land-based ecosystems, include biological pest control (targeting insects, plants, and diseases), the maintenance of crop and pollinator diversity, soil conservation, hydrogeochemical cycles, carbon sequestration, and cultural benefits (Rundlof and Smith 2006). As per Daily (1997), ecosystem services encompass both natural and modified ecosystems that uphold human life and well-being through diverse functions and processes. Presently, these services are recognized and incorporated in the Assessment of Millennium Ecosystem initiative backed by the United Nations. Agriculture-related ecosystem services are divided into four sections (Reid et al. 2005), and these are not only directly relevant to organic farming but also resonate strongly with the principles of organic cultivation (Fig. 14.3).

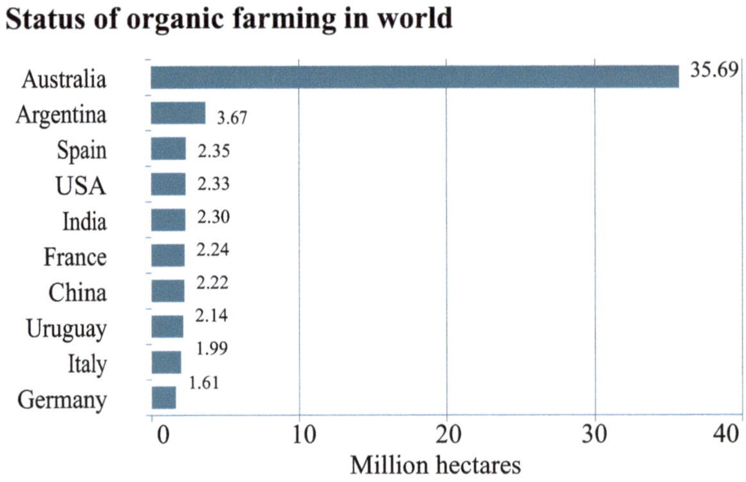

Fig. 14.2 World: The ten countries with the largest areas of organic agricultural land 2019. (Source: FiBL (2021))

Table 14.1 Organic agricultural land (including inconversion areas) and organic share of total agricultural land by region 2019

Region	Organic agricultural land [ha]	Share of total agricultural land
Africa	2′030′830	0.2%
Asia	5′911′622	0.4%
Europe	16′528′677	3.3%
Latin America	8′292′139	1.2%
North America	3′647′623	0.8%
Oceania	35′881′053	9.6%
World	**72′285′656**	**1.5%**

Source: FiBL survey (2021)

Table 14.2 Status of organic farming in India

The status of organic production in India (2020–2021)	
Overall acreage under certified organic cultivation	2657889.33 ha
Acreage of forest and wild for gathering of minor forest outgrowths	1681295.61 ha
The entire acreage under organic certification	4339184.94 ha
Total production	3468991.98 MT
Total quantity exported	888179.69 MT
Value of total export	7078.50 crores

Source: APEDA (2021)

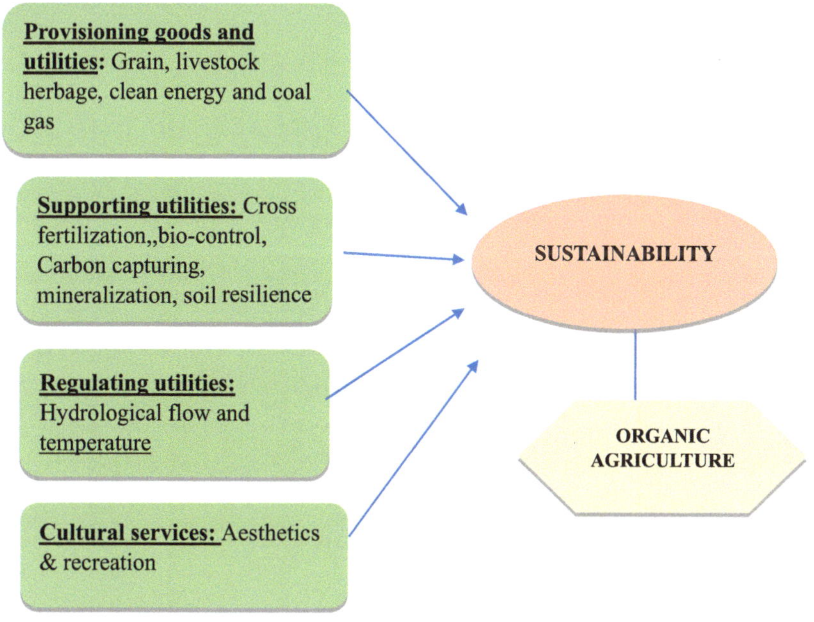

Fig. 14.3 Services of ecosystem in organic farming establishing sustainability

14.2.3 Extend of Services from Forest Ecosystem for Food and Nutritional Security

Based on the Non-legally Binding Instrument on All Types of Forests established by the UN in 2008, the seven criteria for Sustainable Forest Management (SFM) are as follows (Figs. 14.4, 14.5, and 14.6):

14.2.4 Types and Descriptions of Ecosystem Services (Fig. 14.7)

Conversely, two ecosystem services—the supply of food and fundamental materials—essentially function as stock-flow resources. These are a type of renewable resource with a finite supply, delivering a consistent stream of benefits through their use, consumption, or extraction. The unique attribute of stock-flow resources is the flexibility of their harvest rate, which can be adjusted based on available technology, and they can be either used immediately or preserved for future needs. Unlike fund-service resources, these services can be consumed (i.e., depleted), but they do not degrade or wear out. This distinction is significant as when a stock-flow resource like raw materials (such as fuelwood or timber from forests) is exploited, it reduces the resource stock and compromises the multiple fund services that forest ecosystems provide.

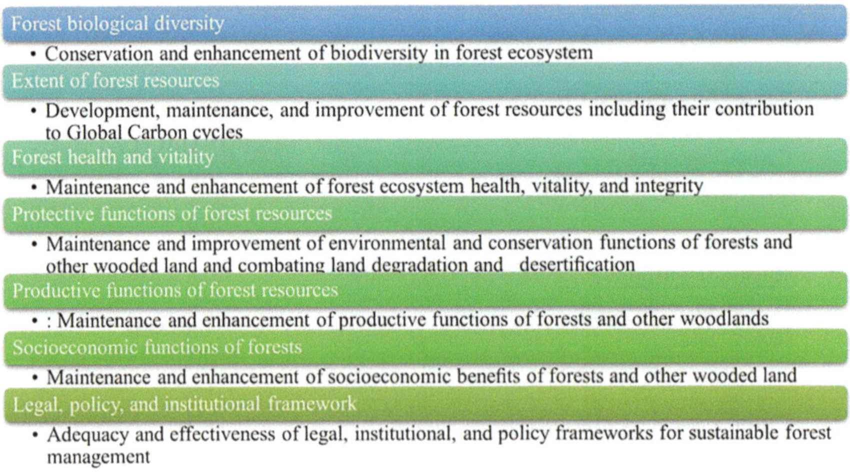

Fig. 14.4 The seven universally accepted and broadly implemented components of sustainable forest management (SFM), also known as criteria

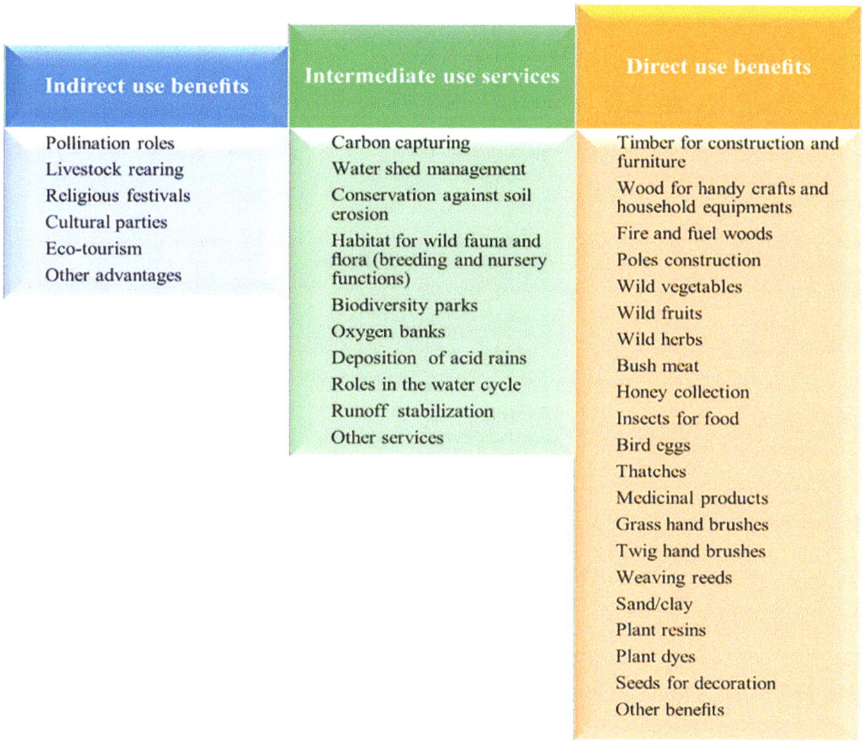

Fig. 14.5 Examples of direct use benefits, indirect use benefits, and intermediate use services from forest and allied ecosystems (Dlamini 2007; Dlamini and Samboko 2017)

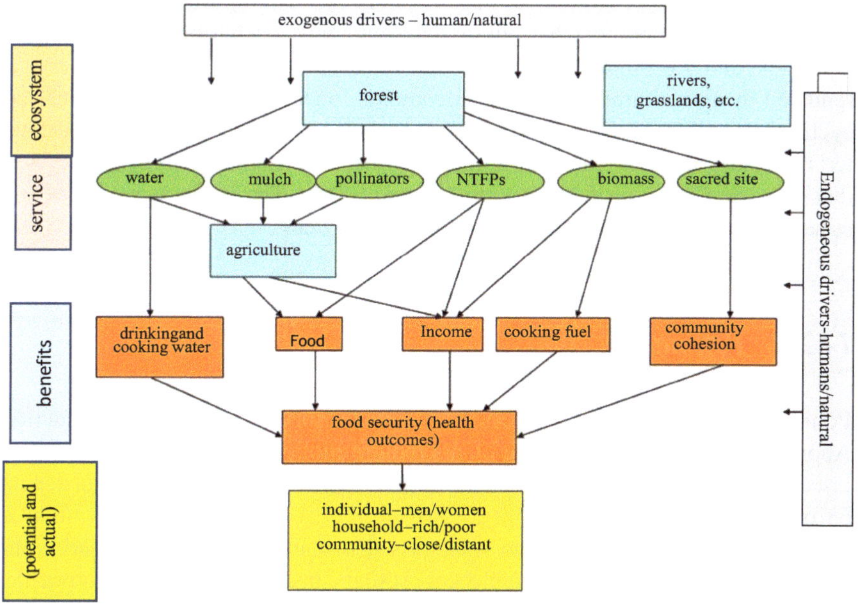

Fig. 14.6 Schematic of the direct and indirect routes by which ES and benefits contribute to food and nutritional outcomes (Poppy et al. 2014)

Fig. 14.7 Types and descriptions of services in the ecosystem (Costanza et al. 1997)

14.2.5 Services of Ecosystem from Sustainable Farming

Agricultural ecosystems play an indispensable role in human survival as they provide essential resources like food, fuel, and fiber. Accounting for 34% of the global land surface, they comprise 5.1 billion livestock units, 3.5 billion hectares of grazing lands, and 1.8 billion hectares of cultivated lands (WFP 2018). The services rendered by agricultural ecosystems encompass foundational services like nutrient cycling and soil formation; delivery services such as food, fuel, fiber, and freshwater; regulatory services that control flood, disease, and climate; and cultural services that offer aesthetic, spiritual, and recreational values. The implementation of sustainable agricultural methods can bolster these services.

14.2.6 Pollination

Pollination is an essential process for human survival and food production. Approximately 90% of flowering plants require biotic pollination, primarily from insects, for seed production. More than 200,000 species of flowering plants depend on over 100,000 different species to provide this ecosystem service. Without it, many interdependent species and processes that maintain ecological equilibrium would collapse. Honeybees, bumblebees, solitary bees, butterflies, and hoverflies are among the primary pollinators for most plants. Among the 128 most vital crops, 89 rely on pollinators to varying extents (Klein et al. 2007).

Insect pollination holds worldwide significance, delivering substantial economic, aesthetic, and cultural values to society, in addition to being crucial for ecological processes in land ecosystems by Rader et al. (2016). It aids in preserving genetic diversity within plant populations (Kearns et al. 1998) and enhances crop productivity, quality, and fecundity, resulting in strong seed production by Albrecht et al. (2012). Recent calculations by IPBES (2016) put the annual value of pollinators' contribution to ecosystem services between 243 and 588 billion USD.

Confined that numerous factors, including habitat fragmentation, land-use patterns, agrochemicals (pesticides), invasive species, predation and competition, and climate change events, can reduce insect pollinator populations (Kerr et al. 2015). Bee community alterations caused by management practices can in turn modify the pollination services offered to crops. Intensive agriculture, focusing on a few crops and their specific needs, simplifies the landscape, leading to high usage of chemical fertilizers and pesticides, monotonous crop rotations, and low crop diversity.

On the other hand, lower intensive methods of crop management like organic farming and crop diversification can enhance the diversity and abundance of insect pollinator species. Recently, Banaszak-Cibicka et al. (2019) showed that using organic composts and manures has increased the impact of increase in number of bumblebees and consequently the seed yield of many pulse and oil seed crops.

In summary, pollination serves as a vital ecosystem service for the sustainability of crop production. Intense agricultural activities, like repetitive crop rotations and excessive use of agrochemicals, are leading to a decrease in insect pollinators. However, adopting practices such as organic farming, diversification in agriculture, crop rotation, agroforestry, and reducing agrochemical application can augment the diversity and abundance of insect pollinators.

14.2.7 Pest Control Through Biological Approach

Major contributors to the stagnation or reduction in crop yields are pests and diseases (Kumar and Singh 2014). A lack of chemical weed control can result in up to 70% losses in field crop yields (Kumar and Singh 2014). However, the utilization of pesticides in farming is a contentious topic due to their detrimental impacts on the environment, human health, and soil microorganisms (Verger and Boobis 2013). Cycoń et al. (2017) reported that prolonged presence of agrochemicals in the soil obstructs the growth and metabolism of soil microbes. Campos et al. (2019) stated that overuse of inputs like pesticides and fertilizers leads to environmental pollution and develops resistance to pesticides.

Alam et al. (2017) reported that practices of sustainable farming methods boost the efficiency of pesticides through approaches like integrated pest management (IPM), integration of plant and animal systems, and material substitution. As per Jabran et al. (2015), the use of allelopathy for crop enhancement and pest control is gaining popularity. There is a growing trend toward employing botanical pesticides as biological control agents. Essential oils, which are secondary metabolites generated by plants, are the most frequently used botanical pesticides. These oils have a crucial role in the plant's life cycle by fascinating pollinators and seed dispersers.

14.3 The Interplay Between Organic and Sustainable Agriculture Through Ecosystem Services

For the benefit of future generations, sustainable agriculture must address environmental, economic, and social issues while preserving the efficiency of agricultural production systems. In many developing nations, poor farmers in marginal and arid regions often encounter barriers to accessing inputs and irrigation due to lack of capital, infrastructure, and a market. The focus on healthy biological cycles and reduced input costs in organic farming could provide significant opportunities to maximize agricultural productivity. Organic farming, which emphasizes the use of organic inputs, crop rotation, and biological production cycles, follows rigorous guidelines and standards throughout the supply chain, favoring practices that are nonpolluting and environmentally friendly. Hence, organic farming tends to contribute more notably to the preservation of crop environments and ecosystem services (Maitra et al. 2018; Pramanick et al. 2014). It plays a crucial role in aiding national efforts to tackle climate change. Typically, organic farmers command higher and more stable prices for their products (Valkila 2009), and it is often noted that organic inputs, which are generally cheaper, locally sourced, and derived from recycled waste materials, result in lower production costs (Eyhorn et al. 2007). If the production costs decrease and the product garners a premium price in any production system, then the profits will be higher, as is often seen in organic farming; thus, organic farming provides farmers with economic stability.

Organic agriculture demonstrates considerable stability in its social aspects, given its capacity to foster social capital and empower small-scale farmers via the establishment of cooperatives (Rice 2001). There is room within organic agriculture for utilizing local resources and integrating traditional and indigenous technical knowledge. If there is significant cluster activity, producers can opt for group certification under the participatory guarantee scheme (PGS) for organic production targeting local markets. As organic farming requires more labor than conventional farming, it can create employment opportunities, benefiting rural communities in developing nations (Bray 2002). Additionally, organic farming promotes a system-based approach, providing abundant opportunities for women workers and family laborers to engage in associated activities, thereby fostering social ties. This diversity can decrease vulnerability and promote sustainability within a system.

Lotter et al. (2003) stated that organic agriculture generally adheres to agroecological principles, undoubtedly yielding stable productivity and resilience to unpredictable climate conditions by eliminating agricultural workers' exposure to chemical inputs, organic farming offers considerable health benefits, safeguarding community health. A variety of ecosystem services are employed for food production, supported by soil fertility preservation, plant protection, and water regulation (Takatsuka et al. 2009; Pretty et al. 2003; Pretty 2005). Sustainable agriculture involves using nature's goods and services while conserving them for future generations. According to Lampkin and Measures (2001), organic agriculture is among the production systems that strive to achieve sustainability by using and preserving ecosystem services. It is widely acknowledged that organic farming contributes more environmental benefits compared to conventional intensive agriculture.

14.4 Significance of Ecosystem Services in Food and Nutritional Security

Research probing into the significance of ecosystems in augmenting the nutritional and food security highlights the intricate bond between sustainability of food security and environment. The function, structure, and services of ecosystems hold a level of complexity that remains to be fully understood by ecologists, economists, and other scientists, particularly in relation to the intricate exchanges of advantages and pressures between ecosystems and the dimensions of human food security. Nevertheless, various studies and evaluations have documented the advantages of certain systems and their contribution to socioeconomic well-being and food security. Pimentel et al. scrutinized studies regarding the contributions and values of forest resource utilization, concluding that preserving the integrity of forests is essential to global food security, primarily since the poor significantly depend on forest resources. To evaluate the role of forests and nontimber forest products (NTFPs) in the agricultural systems of developing countries, ten unique categories of forest uses were outlined (Fig. 14.8).

14.5 Food Security and Agriculture

The World Health Organisation (WHO) defines food and nutritional security as the condition "when all individuals consistently have access to enough safe and nutritious food to sustain a healthy and active life." The United Nations Committee on World Food Security adds that food security is realized when all people, at all times, have physical, social, and economic access to adequate, safe, and nutritious food that caters to their dietary preferences and requirements for leading an active, healthy life (Grebmer et al. 2020).

By the year 2050, the world population is projected to reach approximately 10 billion, creating a demand for 4.5 billion tons of cereals annually. Consequently, achieving food security has emerged as a key goal (Why Food Storage 2020).

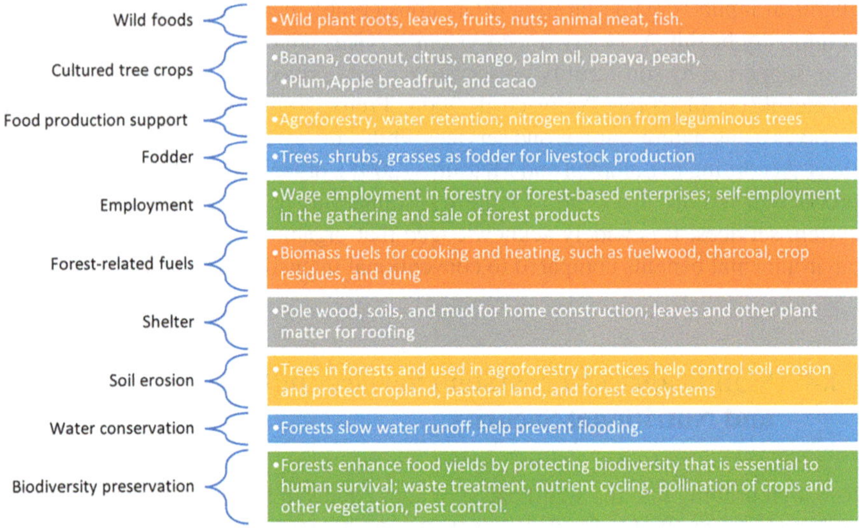

Fig. 14.8 Forest services and products that support global nutritional and food security

According to the FAO, over 60% of the global populace relies on agriculture for their livelihood. Around the world, 2.3 billion people are employed in agriculture, with 96% of them residing in developing countries. The 2008 World Development Report indicates that between 65% and 98% of rural households globally derive their income from agriculture. Paradoxically, countries where agriculture forms the main economic sector are often more exposed to food insecurity, hunger, and poverty. Despite the food surplus in the two Asian giants, China and India, as shown by growing buffer stocks, millions are still facing food insecurity and malnutrition. Yet due to the growth in the industrial and service sectors, the agricultural contribution to the gross domestic product has been decreasing. From 2009 to 2011, agriculture only contributed 2.8% to the rise in global GDP (Zavatta 2014), while the industrial sector and the services sector accounted for 24.6% and 73.2%, respectively.

The state of food security in each nation changes annually, largely due to variations in food prices and individual income levels. The Global Food Security Index (2019) evaluates the affordability, availability, and quality of food across 118 developing and developed countries. This index shows that while 91% of the countries have sufficient food supplies, roughly one-fourth of the population is malnourished.

Despite farming being the only means of feeding billions, the gravity of the threats it faces keeps intensifying. Challenges such as climate change, fluctuating market prices, lack of adequate infrastructure, population growth, and decreasing cultivable land urgently call for solutions to ensure the provision of affordable, hygiene food for individual. The recent COVID-19 pandemic has further complicated the agricultural landscape. It is still too early to predict future consequences in these unpredictable times.

14.6 Crucial Factors Judging Security of Food on Globe

The crucial issue relating to security of food revolves around securing access to affordable and healthy food for at-risk populations around the world. As some researchers suggest, the paramount challenge and focal point should be to enhance agricultural yield. Regrettably, any attempts to fortify the food system can only bear fruit if multiple key determinants are adequately tackled. In the following discussion, we will delve into the four principal factors that significantly influence global food security.

14.6.1 Food Costs

Over the previous 5 years, 26 nations have reported an increase in food prices by 5%, while 7 countries have experienced an inflation rate in food prices more than twice this level. The escalating cost of food is a distressing reality for those living in poverty. Conversely, farmers are faced with the inability to sell their crops at a fair rate, leading them into financial debt. Tragically, since 1998, more than 5000 cotton farmers in Andhra Pradesh, burdened by debt, have taken their own lives.

14.6.2 Food Nutrition

Although food security is usually associated with total caloric consumption, it currently involves the threefold challenge of malnutrition, hunger (inadequate energy intake from food), and severe illnesses resulting from under nutrition and pesticide pollution. At the same time, close to half of the world food production is dumped into vain (Anonymous 2018). One in five countries worldwide do not have nutritional policies in place for adults or children. It is of immediate necessity to implement nutritional monitoring, a crucial step for devising personalized, precise responses to under nutrition and obesity (Pramanick et al. 2023).

14.6.3 Food Storage Potential

Each year, a staggering quantity of food is wasted globally. Roughly 1.8 billion hectares of farmland, which is 34% of the world's total agricultural acreage, is utilized to cultivate food that is eventually discarded or lost. This translates to an area equivalent to the United States, India, and Egypt being dedicated to producing food that is never consumed by Corrado et al. (2019). Grebmer et al. (2020) reported that world discards about 1.3 billion tons of food suitable for mankind consumption, valued near 3.2 trillion pounds, with the lion's share originating from the United States, Europe, Australia, and Southeast Asia. However, a quarter of the globe's freshwater resources are exploited to cultivate crops that end up being wasted. This wasted food persists in the environment and poses adverse environmental impacts without contributing to nutrition. Considering global warming and the urgency to cut carbon emissions, the problem of inefficient food storage will increasingly worsen. From 2010 to 2016, worldwide food loss and waste contributed 8–10% of all greenhouse gas emissions induced by humans. It is projected that by 2035, production of edible grain will reach 358 million tons, while static grain storage will only be 261 million tons, indicating a shortfall of 98 million tons. This highlights the immense need for expanding grain storage capacity. In 2019, the total volume of refrigerated warehouses hit 640 million cubic meters, up by 2.9% from 2016. India leads the market, with states like Gujarat, Himachal Pradesh, and Pondicherry seeing substantial growth in refrigerated warehouse capacities. The United States and China are the second and third largest global markets, respectively (Why Food Storage 2020).

14.6.4 Food and Nutritional Security Can Be Structured Around Four Key Dimensions

Food security can be conceptualized as encompassing four aspects:

- Availability refers to the provision of adequate amounts of food of suitable quality, derived from both natural and farmed systems.
- Access pertains to the ability of individuals to acquire food consistently, whether through their own production, markets, or other sources.
- Utilization entails the ways in which individuals extract energy and nutrients from food.
- Stability implies the consistent provision of sufficient, accessible, and utilizable food on a sustainable basis (Fig. 14.9).

14.6.5 Contributions of Ecosystem to Food Security

Ecosystems, such as forests, play a role in every aspect of food security.

- Forests serve as direct food sources, through edible wild flora and fauna, as well as offering genetic materials for domestication and indirectly by generating income from forest products which can be used to buy food.
- Medicinal plants obtained from forests contribute to human health, thereby amplifying the positive effects of food intake.

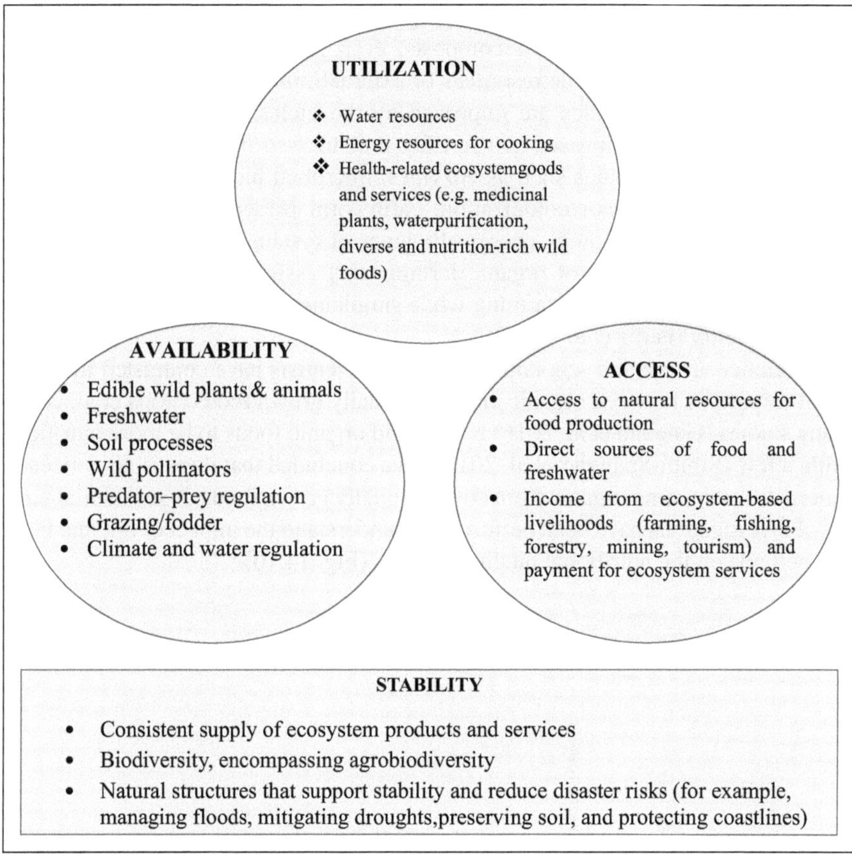

Fig. 14.9 Dimensions of food security

- Coastal forests, like mangroves, provide protection against floods for coastal regions, thereby ensuring consistent food production in adjacent fields and fish ponds.

14.7 Impact of Organic Farming on Nutritional Security

Numerous elements, like soil attributes (consistency, acidity, organic carbon, accessible nutrients, water retention capacity, etc.), weather conditions, nutrient stewardship, other farming methods, and harvest timing, affect the quality characteristics of plant-based products (Rembiałkowska 2007). Organic farming can help stop the ongoing degradation of soil health by rejuvenating, preserving, and enhancing the ecosystem's natural cycles (Kesavan and Swaminathan 2008). As indicated earlier, organic farming operations focus more on the buildup of soil organic matter using organic fertilizers like rotational cropping, green manuring, cover crops, FYM, digesters, special bioenhancers, composts, crop residues, etc. and lessen their dependence on nonrenewable resources or external inputs. Soil's physical, chemical, and biological qualities are improved by the inclusion of organic fertilizers (Dubey et al. 2020). Improved soil fertility is witnessed in organic systems, with improved quality attributes such as enzymes, microbial biomass carbon, development of mycorrhizal root colonization, earthworm presence, etc. (Mäder et al. 2002), and heightened biomes, energy efficiency of systems, and nutrient utilization efficiency are other traits of organic farming that assist in addressing the issues linked to high external input farming while simultaneously bettering parameters of our food quality (Pérez et al. 2019).

A number of research scholars and organic scientists have contrasted the nutritional disparities between organic and traditionally grown foods (Popa et al. 2019). Many studies (Galgano et al. 2016) have found organic foods to be more nutritious, while a few (Smith-Spangler et al. 2012) have concluded that there is no noticeable pattern differentiating organic from conventionally grown foods. A substantial number of investigations have been performed to understand the impact of organic farming methods on the nutritional quality of crops (Fig. 14.10).

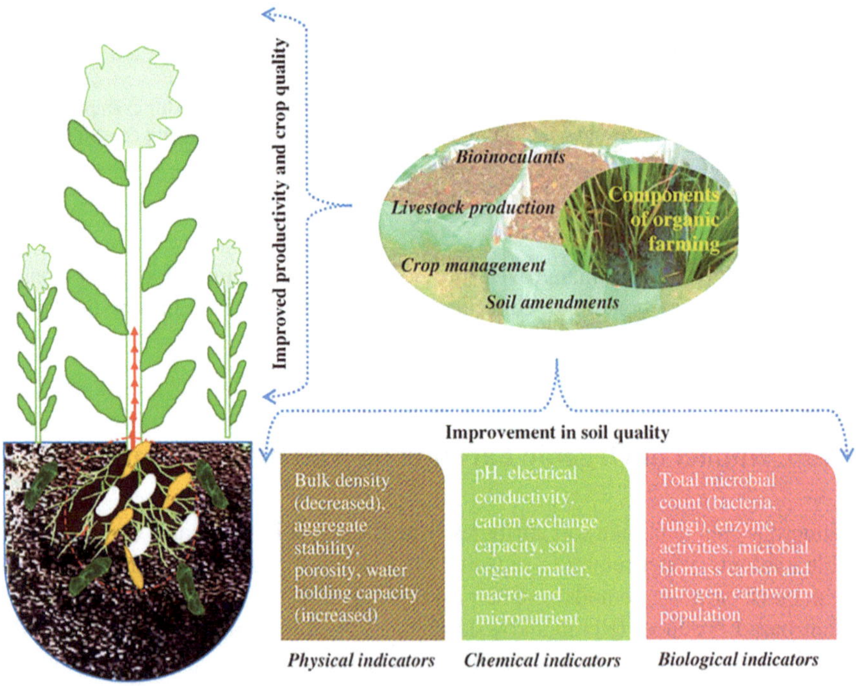

Fig. 14.10 The components of organic agriculture intervene on the physical, chemical, and biological levels of the soil and revitalize its health properties. The soil's vitality is a potential indicator of both increased crop yield and crop quality

14.8 Conclusion

The importance of ecosystem services and organic farming in supporting sustainable survival, notably in preserving the environment and ensuring worldwide food availability, is vital. Nonetheless, environmental sustainability and food security are not properly consolidated within the ecosystem service model. Typically, policies centered on food security overlook or fail to enhance the wellness and vigor of ecosystems, focusing mainly on agricultural productivity, trade, and microeconomic policies. Economic durability, driven by market forces, is often given precedence over ecological endurance, which hinges on the health of ecosystems. To successfully balance food availability and environmental conservation, both economic and ecological durability are key. If ecological sustainability is overlooked, the potential repercussions include severe ecosystem harm and ongoing food shortages, both of which can be significantly lessened through the application of organic farming techniques.

References

Alam MK, Salahin N, Islam S, Begum RA, Hasanuzzaman M, Islam MS, Rahman MM (2017) Patterns of change in soil organic matter, physical properties and crop productivity under tillage practices and cropping systems in Bangladesh. J Agric Sci 155:216–238

Albrecht M, Schmid B, Hautier Y, Müller CB (2012) Diverse pollinator communities enhance plant reproductive success. Proc R Soc B Biol Sci 279:4845–4852

Anonymous (2018) How organic agriculture can help achieve food security and SDG 2 organic without boundaries

APEDA (2021) Organic farming in India. Available at: http://apeda.gov.in/apedawebsite/organic/Organic_Products

Banaszak-Cibicka J, Weronika B, Twerd L (2019) Possible expansion of the range of Xylocopa violacea L. (Hymenoptera, Apiformes, Apidae) in Europe. Turk J Zool 43(6):14. https://doi.org/10.3906/zoo-1812-6

Bray DB (2002) Social dimensions of organic coffee production in Mexico: lessons for eco-labeling initiatives. Soc Nat Resour 15:429–446

Campos EV, Proença PL, Oliveira JL, Bakshi M, Abhilash PC, Fraceto LF (2019) Use of botanical insecticides for sustainable agriculture: future perspectives. Ecol Indic 105:483–495

Corrado S, Caldeira C, Eriksson M, Hanssen OJ, Hauser HE, Holsteijn F, Liu G, Östergren K, Parry A, Secondi L, Stenmarck Å, Sala S (2019) Food waste accounting methodologies: Challenges, opportunities, and further advancements. Glob Food Sec 20:93–100

Costanza R, d'Arge R, de Groot R, Farber S, Grasso M, Hannon B, Limburg K, Naeem S, O'Neill RV, Paruelo J, Raskin RG, Sutton P, van den Belt M (1997) The value of the world's ecosystem services and natural capital. Nature 387:253–260

Cycoń M, Mrozik A, Piotrowska-Seget Z (2017) Bioaugmentation as a strategy for the remediation of pesticide-polluted soil: a review. Chemosphere 172:52–71

Daily GC (1997) Introduction: What are ecosystem services? In: Daily GC, Ed, Nature's Services: Societal Dependence on Natural Ecosystems, Island Press, Washington DC, 1-10

Dlamini CS (2007) Towards the improvement of policy and strategy development for the sustainable management of non-timber Forest products: Swaziland-a case study. Ph.D. dissertation. University of Stellenbosch, South Africa

Dlamini CS, Samboko PC (2017) The value of non-timber forest products in Zambia: direct use benefits and their contribution to rural livelihoods. IAPRI Working Paper No. 118
Dubey PK, Singh A, Chaurasia R, Abhilash PC (2020) Climate adaptive agricultural interventions for food, nutritional, health and livelihood security. In: Dhyani S, Gupta A, Karki M (eds) Nature-based solutions for resilient ecosystems and societies, disaster resilience and green growth. Springer, Singapore. https://doi.org/10.1007/978-981-15-4712-6_16
Eyhorn F, Ramakrishnan M, Mader P (2007) The viability of cotton-based organic farming systems in India. Int J Agric Sustain 5(1):25–38
FiBL (2021) The World of Organic Agriculture 2021. Frick, Switzerland
Galgano F, Tolve R, Colangelo MA, Scarpa T, Caruso MC (2016) Conventional and organic foods: a comparison focused on animal products. Cogent Food Agric 2(1):1142818. https://doi.org/10.1080/23311932.2016.1142818
Global Food Security Index (2019) The percentage of cultivated land equipped for irrigation is inadequate to meet global need
Goldewijk KK, Ramankutty N (2004) Land cover change over the last three centuries due to human activities: the availability of new global data sets. GeoJournal 61:335–344
Grebmer K, Bernstein J, Alders R, Dar O, Kock R, Rampa F, Wiemers M, Acheampong K, Hanano A, Higgins B, NíChéilleachair R, Foley C, Gitter S, Ekstrom K, Fritschel H (2020) Global Hunger Index: one decade to zero hunger: linking health and sustainable food systems
IPBES (2016) The assessment report of the Intergovernmental Science-Policy Platform on Biodiversity and Ecosystem Services on pollinators, pollination and food production. Potts SG, Imperatriz-Fonseca VL, Ngo HT (eds). Secretariat of the Intergovernmental Science-Policy Platform on Biodiversity and Ecosystem Services, Bonn, Germany, pp. 552
Jabran K, Mahajan G, Sardana V, Chauhan BS (2015) Allelopathy for weed control in agricultural systems. Crop Prot 72:57–65
Kearns CA, Inouye DW, Waser NM (1998) Endangered mutualisms: the conservation of plant-pollinator interactions. Annu Rev Ecol Syst 29:83–112
Kerr JT, Pindar A, Galpern P, Packer L, Potts SG, Roberts SM, Rasmont P, Schweiger O, Colla SR, Richardson LL, Wagner DL (2015) Climate change impacts on bumblebees converge across continents. Science 349:177–180
Kesavan PC, Swaminathan MS (2008) Strategies and models for agricultural sustainability in developing Asian countries. Philos Trans R Soc B: Biol Sci 363(1492):877–891
Klein AM, Vaissiere BE, Cane JH, Steffan-Dewenter I, Cunningham SA, Kremen C, Tscharntke T (2007) Importance of pollinators in changing landscapes for world crops. Proc R Soc B Biol Sci 274:303–313
Kumar S, Singh A (2014) Biopesticides for integrated crop management: environmental and Regulatory aspects. J Biofertil Biopestic:5–e121
Lampkin N, Measures M (2001) Organic farm management handbook: organic farming research unit. Institute of Rural Studies, University of Wales, Aberystwyth
Lotter DW, Seidel R, Liebhardt W (2003) The performance of organic and conventional cropping systems in an extreme climate year. Am J Altern Agric 18(3):146–154
Luttikholt LWM (2007) Principles of organic agriculture as formulated by the International Federation of Organic Agriculture Movements. NJAS-Wageningen J Life Sci 54(4):347–360
Mäder P, Fließbach A, Dubois D, Gunst L, Fried P, Niggli U (2002) Soil fertility and biodiversity in organic farming. Science 296(5576):1694–1697
Maitra S, Mondal TK, Shankar T, Palai JB (2018) Combatting climate change by organic agriculture
Naik BSSS, Sharma SK, Pramanick B, Chaudhary R, Yadav SK, Tirunagari R, Gaber A, Hossain A (2022) Silicon in combination with farmyard manure improves the productivity, quality and nitrogen use efficiency of sweet corn in an organic farming system. Silicon 14:5733–5743. https://doi.org/10.1007/s12633-022-01818-0
Pérez IAV, Toral JN, Vázquez ATP, Hernández FG, Ferrer GJ, Cano DG (2019) Potential for organic conversion and energy efficiency of conventional livestock production in a humid tropical region of Mexico. J Clean Prod 241:118354

Popa ME, Mitelut AC, Popa EE, Stan A, Popa VI (2019) Organic foods contribution to nutritional quality and value. Trends Food Sci Technol 84:15–18. https://doi.org/10.1016/j.tifs.2018.01.003
Poppy GM, Chiotha S, Eigenbrod F, Harvey CA, Honzák M, Hudson MD, Jarvis A, Madise NJ, Schreckenberg K, Shackleton CM, Villa F, Dawson TP (2014) Source: Philosophical Transactions: Biological Sciences. Achieving food and environmental security: new approaches to close the gap. Philos Trans R Soc B 369(1639):1–13
Pramanick B, Brahmachari K, Ghosh A (2014) Efficacy of *Kappaphycus* and *Gracilaria* sap on growth and yield improvement of sesame in new alluvial soil. J Crop Weed 10(1):77–81
Pramanick B, Mahapatra BS, Datta D, Dey P, Singh SP, Kumar A, Paramanik B, Awasthi N (2023) An innovative approach to improve oil production and quality of mustard (*Brassica juncea* L.) with multi-nutrient-rich polyhalite. Heliyon 9(3):e13997. https://doi.org/10.1016/j.heliyon.2023.e13997
Pretty JN (2005) The pesticide detox: towards a more sustainable agriculture. Earthscan, London
Pretty JN, Morison JIL, Hine RE (2003) Reducing food poverty by increasing agricultural sustainability in developing countries. Agric Ecosyst Environ 95:217–234
Rader R, Bartomeus, I, Garibaldi LA, et al (2016) Non-bee insects are important contributors to global crop pollination. PNAS 113(1):146–151
Reid WV, Mooney HA, Cropper A, Capistrano D, Carpenter SR (2005) Millennium ecosystem assessment synthesis report. Island Press, Washington, DC
Rembiałkowska E (2007) Quality of plant products from organic agriculture. J Sci Food Agric 87(15):2757–2762
Rice RA (2001) Noble goals and challenging terrain: organic and fair-trade coffee movements in the global marketplace. J Agric Environ Ethics 14(1):39–66
Rundlof M, Smith H (2006) The effect of organic farming on butterfly diversity depends on landscape context. J Appl Ecol 43:1121–1127
Sandhu HS, Wratten SD, Cullen R (2010) Organic agriculture and ecosystem services. Environ Sci Policy 13:1–7
Smil V (2000) Feeding the World. MIT Press, Cambridge, MA
Smith-Spangler C, Brandeau ML, Hunter GE, Bavinger JC, Pearson M, Eschbach PJ et al (2012) Are organic foods safer or healthier than conventional alternatives? A systematic review. Ann Intern Med 157(5):348–366. https://doi.org/10.7326/0003-4819-157-5-201209040-00007
Takatsuka Y, Cullen R, Wilson M, Wratten S (2009) Using stated preference techniques to value four key ecosystem services on New Zealand arable land. Int J Agric Sustain 7:1–13
Tilman D, Cassman G, Matson PA, Naylor R, Polasky S (2002) Agricultural sustainability and intensive production practices. Nature 418:671–677
UN (2008) Resolution 62/98 establishing the non-legally binding instrument on all types of forests UN population prospects. 2017. Report by the UN Department of Economic and Social Affairs Population Division
Valkila J (2009) Fair trade organic coffee production in Nicaragua-Sustainable development or a poverty trap? Ecol Econ 68(12):3018–3025
Verger PJ, Boobis AR (2013) Reevaluate pesticides for food security and safety. Science 341:717–718
WFP (2018) World Food Programme- year in Review 2018. Rome RM, Italy
Why food storage lies at the core of future food security (2020) Israelagri. http://www.israelagri.com/?CategoryID=521andArticleID=1858
Zavatta G (2014) Agriculture remains central to the world economy. 60% of the population depends on agriculture for survival. http://www.expo2015.org/magazine/en/economy/agriculture-remains-central-to-the-world-economy.html

Chapter 15
Cropping System Approach in Climate-Resilient Food Crop Production

Sagar Maitra, Sumit Ray, Lalichetti Sagar, Masina Sairam, Biswajit Pramanick, Harun Gitari, D. T. Santosh, Gaikwad Dinkar Jagannath, Anjana Atapattu, and Luciana Raphael Chappa

Abstract The agriculture sector is facing many problems as a result of direct and indirect consequences of climate change, eventually disturbing the agricultural productivity and raising concerns on global food security. Therefore, there is an urgent need to cope up agriculture with the changing climatic conditions and sustain crop yields. In this juncture, the crop diversification and inclusion of stress-resistant varieties in the cropping system were gaining momentum. Crop diversification lowers the chance of total crop failure because a single crop under these unpredictable weather conditions might result in significant losses. On the other hand, climate-resilient varieties were developed either through breeding or through genetic modification to withstand biotic and abiotic stresses and produce stable yields. Further, when the strategies were administered agronomically using climate-adaptive techniques and technologies, in particular through sustainable agricultural practices like

S. Maitra · S. Ray · L. Sagar · M. Sairam
Department of Agronomy and Agroforestry, M. S. Swaminathan School of Agriculture, Centurion University of Technology and Management, Paralakhemundi, Odisha, India

B. Pramanick
Department of Agronomy, Dr. Rajendra Prasad Central Agricultural University, Pusa, Bihar, India

H. Gitari
Department of Agricultural Science and Technology, School of Agriculture and Environmental Sciences, Kenyatta University, Nairobi, Kenya

D. T. Santosh · G. D. Jagannath
Research Centre for Smart Agriculture, Centurion University of Technology and Management, Paralakhemundi, Odisha, India

A. Atapattu
Agronomy Division, Coconut Research Institute of Sri Lanka, Lunuwila, Sri Lanka

L. R. Chappa (✉)
Faculty of Natural Resources and Agricultural Sciences, Swedish University of Agricultural Science, Uppsala, Sweden

B. Pramanick et al. (eds.), *Climate-Smart Agricultural Technologies*, Sustainability Sciences in Asia and Africa, https://doi.org/10.1007/978-981-96-7699-6_15

conservation agriculture, agroforestry, etc., with the goal of improving water use efficiency, carbon restoration, and management of pests and diseases further reduced environmental impacts and induced system sustainability.

Keywords Climate change · Crop diversification · Climate resilience · Cropping system · Recarbonization

15.1 Introduction

Climate change is one of the most important challenges to modern agriculture. The agricultural sector is highly vulnerable to both direct and indirect impacts of climatic change. The direct impacts include biotic pressures and climatic anomalies such as temperature changes, irregular rainfall, frequency of extreme weather events, etc. (Nastis et al. 2012; Sagar et al. 2022, 2023). Indirectly, soil degradation, soil salinization, and introduction of exotic diseases, plants, insects, etc. have an adverse impact on global agricultural productivity (Bhadra et al. 2022; Alam 2014). There is an urgent need to mitigate these negative impacts of climate change to ensure global food security. Addressing, the aforementioned issues in the light of shrinking arable land owing to rapid population growth and industrial development is further challenging (Maitra et al. 2024).

Improving system productivity by changing the cropping systems associated with adoption of modern technology is the most promising option available to withstand the negative consequences of climate change in a sustainable manner (Rehman et al. 2016; Maitra et al. 2022). In this scenario, diversification of cropping system by replacing the traditional crops with those crops that are best suited to the agroclimatic setup of the region could be one of the key strategies (Santosh et al. 2024; De and Bodosa 2014) (Fig. 15.1). Since, with the increase in biodiversity of a region the pest and diseases could be efficiently managed (Ratnadass et al. 2012). This was mainly due to the reason that the occurrence of pests and diseases were mostly crop specific in nature. The impact of weed infestation was minimized by the choice of best suited crops inclusion in the existing cropping system (Shahzad et al. 2021). Moreover, introduction of climate-resilient varieties into the existing cropping systems could lead to withstand and perform better under extreme climatic conditions, namely, drought, heat, salinity, water logging, heavy metal stress, etc. (Wang et al. 2003, 2010).

In addition, intensification of cropping system with the inclusion of legumes is also gaining worldwide attention. The ability of legumes to fix atmospheric nitrogen in the soil attributed for its role in reducing the reliance on synthetic nitrogenous fertilizers which are the key contributors of greenhouse gas emission in the world (Abd-Alla et al. 2023). According to an estimate, around 0.7% of the global greenhouse gas emission was attributed by the production of nitrogenous fertilizers (FAO

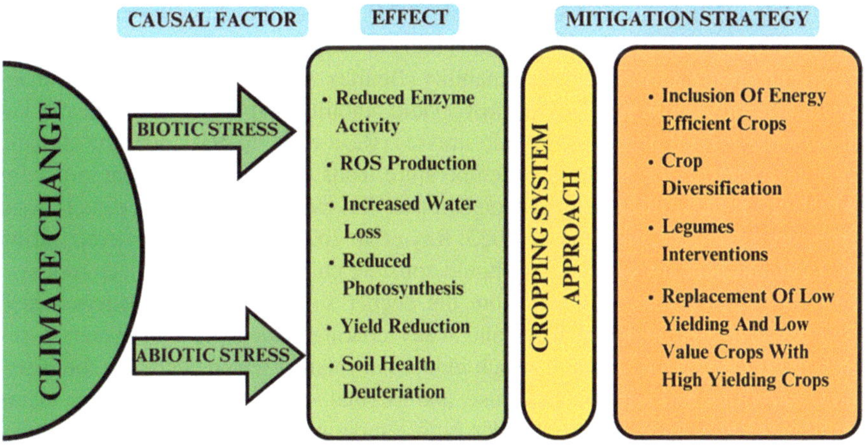

Fig. 15.1 Response of cropping systems approach to climate change

2018). Therefore, inclusion of legumes in the cropping system not only enhances the soil fertility but also reduces the greenhouse gas emission from agricultural sector and mitigates climate change (Yang et al. 2024).

In the light of above facts, it is essential to comprehend the ideas behind creating a climate-resilient cropping system for the production of food crops in order to guarantee both ecological health and global food security in a sustainable manner. All the necessary information required for investigating the topic was included in this chapter.

15.2 Crop Diversification: Choice of Climate-Resilient Crops

Green revolution during the 1960s aided in achieving food security but intensive agriculture disoriented the agricultural systems sustainability in a long run. The use of fossil fuel-based agricultural inputs contributed a major part to global warming and climate change affecting cropping system. A reorientation of cropping systems through crop diversification under the changing climatic scenario for achieving food, nutritional, and environmental security is the need of the hour. Crop diversification is a dynamic aspect of cropping systems which involves various crops and cropping patterns suitable to various agroecosystems for enhancing system productivity and sustainability (Yadav and Kumar 2023; Pramanick et al. 2024). Crop diversification helps in achieving the sustainable development goals while taking into account different aspects of gender equality, poverty, food, and nutritional security and climate change (Feliciano 2019). It provides an option to farmers for growing of suitable crops while reducing the risk of crop failure due to various abiotic reasons in a climate-resilient manner (Acevedo et al. 2020). Climate resilience can be explained as the ability or efficiency of a system to recover, mitigate, and adapt to climate related shocks (Ahmed et al. 2024). Under the changing climatic scenario, adopting crop diversification strategies while including climate-resilient crops can ensure agricultural sustainability (Benitez-Alfonso et al. 2023). A climate-resilient crop should not only improve yield stability and improve soil health but also tackle several biotic and abiotic stresses (Raza et al. 2019). Under the climatic change scenario, the increase in temperature leads to drought and heat stress for which crops like millets, cowpea, pigeon pea, etc. can be adopted in a cereal-based cropping system (Kaushik et al. 2023; Ray et al. 2021). The choice of crop during drought and heat stress tolerance should consider characteristics like deep root system for better access of water from the deeper soil profile, small leaves which reduces leaf area and leaf hairs and waxy coating which prevents water losses through transpiration, drought which induces dormancy which decreases the metabolic activities and reduces water use, and thermal tolerance which helps in maintaining cellular level function under high temperature, production of heat shock proteins, early maturity, small fruits and seeds, high water content for maintaining turgor pressure and better cooling through evapotranspiration, multilayer canopies for maintaining soil temperature and moisture retention, drought-induced flowering which helps in short vegetative and reproductive period, hormonal and antioxidant production, better microbial association, etc. (Fahad et al. 2017; Kim et al. 2020; Bodner et al. 2015; Li et al. 2022; Gobu et al. 2017).

The climate change not only increases temperature but also leads to erratic rainfall and which sometime leads to flooding. Flood-resilient crop should have characteristics like submergence tolerance where plant can withstand water stagnation for certain period (Singh et al. 2017). Furthermore, a heavy rainfall can cause waterlogging, and for such situation, waterlogging tolerance is important which means it should be able to withstand a situation where a partial flood submerges the roots but not the leaves; it should also be able to float due to the formation of aerenchyma tissue, should have adventitious roots to enhance the uptake of nutrients and oxygen, produce antioxidants, and faster recovery, and it should be able to withstand salinity (Fukao et al. 2019; Kim et al. 2024; Glenz et al. 2006; Kaur et al. 2020; Oladosu et al. 2020). Some examples include flood and excess water stress-tolerant crops such as rice, taro, lotus, water spinach, etc. Change in climate also leads to advancement of pest and diseases and makes plants more susceptible toward biotic stresses (Singh et al. 2023a, b). To tackle these pest and diseases, a crop with enhanced immune response, systemic-acquired resistance where defense mechanism activates through initial infection or pest attack, production of antimicrobial and insecticidal compounds, hormonal regulations, physical barriers like waxy coating, thick cuticle, epigenetic regulation, etc. should be taken into account (Al-Khayri et al. 2023; Fürstenberg-Hägg et al. 2013; Xiao et al. 2023). Crops like deep-rooted wheat, legumes perennial crops, cover crops, etc. increase yield stability and soil health (Shang et al. 2024). Crops with characteristics like better carbon sequestration, extensive root systems for better access to nutrient and water, soil microbes' association observed in between legumes and rhizobium, less soil compaction, better soil aeration, increase in biomass production, etc. lead to better soil management and enhanced productivity whereafter in long run it promotes sustainable food system (Koudahe et al. 2022; Wei et al. 2024; Gudla et al. 2023). In changing climatic scenario, the increase in sea level, changing precipitation pattern, warmer temperature, increased use of groundwater, etc. have led to salinity problem which threatens ecosystem health, freshwater resources, and agricultural productivity (Khondoker et al. 2023). While selection of crops under cereal-based cropping system, a crop with better salt tolerance ability, deep root, small leaves, succulent stem, wary or hairy coating, antioxidant production, hormonal regulation, primed stress response, etc. should be considered (Singh and Devi 2023). Along with these characters, crops with salt accumulation and salt-excluding trait are also considered. Heavy metal accumulation has also been observed with changing climatic condition. Increased weathering process, soil erosion, change in soil pH, altered soil microbial activity, atmospheric deposition, urbanization, wildfires due to climate change, improper nutrient cycling, etc. act as catalysts for increasing heavy metal stress to the crop (Mok et al. 2023). While choosing a crop in a cropping system under heavy metal stress characters like heavy metal tolerance, metal exclusion and accumulation, antioxidant and hormone production, deep and extensive root system, production of metal binding compounds like phytochilatins, etc. should be taken under consideration (Singh et al. 2016). Heavy metal stress can also be managed through including trees with the existing cropping system (Emamverdian et al. 2015; Islam and Sandhi 2022). Plants like eucalyptus helps in heavy metal bioremediation through absorbing and transporting heavy metals from the soil through the roots and shoots to the leaves (Madejón et al. 2017). The absorbed heavy metals are stored in vacuoles known as vacuolar compartmentalization, cell walls, and other organelles (Patil et al. 2023). Eucalyptus also produces phenolics, flavonoids, catalase, peroxidase, auxin, ethylenes, etc. to detoxify heavy metals and repair damages (Jańczak-Pieniążek et al. 2022). Eucalyptus has advantages when agroforestry is adopted, but it also has allelopathic effect as it releases volatile compounds like benzoic, cinnamic, and phenolic acids which hampers growth and development of other crops (Zhiqun et al. 2017; Zhang and Fu 2009). So while choosing climate resilient crops for a cropping system, several aspects like allelopathic effect also need to be taken under consideration. Hence, a proper choice of climate-resilient crop supports sustainable agriculture, farm resilience, food security, and mitigation of climate change and ensures productive and resilient system.

15.3 Choice of Stress-Tolerant Crop Varieties and Hybrids

The changing climatic conditions will force the future crops to grow in an unprecedented environment while facing new challenges. On the other hand, the demand for nutritious and quantity of food is increasing with increasing global population. To fulfil this demand along with climatic stress like drought, floods, heat and cold waves, etc. along with increasing incidence of pest and disease epidemic. A key element of contemporary agriculture is the development of crop types and hybrids that can withstand stress, as this allows farmers to adapt for worse environmental circumstances (Atlin et al. 2017) (Table 15.1). In the recent scenario of soil degradation, water shortages, and climate change, it is more important to adopt cropping systems with proper choice of crop varieties and hybrids lead to increased resilience within the system.

Moreover, while choosing a crop for drought-prone areas under a cropping system, special attention should be given to the varietal characters like deep root system, drought-induced gene expression, water efficiency, photosynthetic activity, yield stability, drought recovery, adaption to various soil type, early maturity, etc. (Panda et al. 2021). Triplett, droughtgard, pannaer of maize, kavir, chamran and ciano of wheat, sahbhagi dhan, IR64, Jyothi of rice, IT97K-499-35, IT93K-503-1 of

Table 15.1 Stress-resistant varieties/hybrids of different food crops

Crops	Varieties/hybrids	Stress resistance	References
Rice	Sahbhagi Dhan	Drought	Anantha et al. (2016)
	Swarnaprabha and Kattamodan	Drought	Swapna and Shylaraj (2017)
	Dasht	Salinity	Kamyab-Talesh et al. (2024)
	B55, Banjiemang, and Lijianghegu	Low temperature	Ye et al. (2009)
Wheat	JM418, HM19, and SM22	Drought	Bao et al. (2023)
	Pasban-90	Salinity	Irshad et al. (2022)
	HD 2967, WH 1249, and HI 1617	High temperature	Lamba et al. (2023)
	XC-45	Salinity	Zhang et al. (2020)
Sorghum	KWS Kallisto	Drought	Frantová et al. (2024)
	M-81E	Salinity	Zheng et al. (2021)
	Hongkezi	Low-temperature	Chopra et al. (2015)
	PYPS 8 and PYPS 2	Pest (shoot fly)	Nagesh Kumar et al. (2022)
Chick pea	D0091-10, K010-10, D0085-10, and K005-10	Drought	Shah et al. (2020)
	GNG1969, GNG1488, and PantG186	High temperature	Devi et al. (2022)
	H 04-75, H 08-75, H 12-26, H 09-96	High temperature	Sunil et al. (2023)
	IC 116783 and ICC 15200	Cold temperature	Mir et al. (2021)
Red gram	PRG58	Drought	Basha et al. (2014)
	ICC 4958	Drought	Shende et al. (2020)
	BRG-5	Drought	Mallikarjuna and Devaraja (2023)
	PUSA 991, MA6, and CORG 9701	Pest (pod borrer)	Das et al. (2024)
Green gram	LGG 583	Drought	Kanavi et al. (2023)

cowpea, etc. are some developed varieties for drought resistance (Bhattacharjee et al. 2023). Due to climatic aberrations, excess rainfall leads to flood, and in such situation, selection of a variety for flood tolerance is preferred. Swarna-Sub1, IR64-Sub1, Jyothi of rice, Flood-Tolerant Wheat (FTW), WL711 of wheat, Pusa 1121 for chickpea, ICRISAT's SPH 5044 for sorghum, etc. are some developed varieties for flood tolerance (Dar et al. 2013).

The changing climate also leads to heat and cold wave which decreases the crop productivity drastically. Under such stress condition, while selecting a variety, characteristics like high and low temperature tolerance, winter hardiness, short maturity, drought tolerance, etc. should be considered (Raza et al. 2019). Some examples for cold stress-tolerant varieties are Frostmaster and Norstar of wheat, DKC 9088 of maize, CTR, and Jyothi of rice (Hassan et al. 2021). Along with other stress, the problem of salinity has also increased due to increase in sea level, excessive use of fertilizer and groundwater, soil degradation, change in precipitation patterns, etc. So selection of varieties for saline stress characters like ion exclusion and compartmentalization, osmotic adjustment, antioxidant production, tolerance to high EC, SAR, ESP, deep root system, small leaf area, short maturity time, etc. should be taken into consideration (Gupta and Huang 2014). Some of the examples of salinity stress-tolerant varieties are Pokkali, CSR10, IR64-Sub1 for rice, Kharchia 65, KRL 1-4, WL 711 for wheat, Salt-Tolerant Maize (STM), DKC 9088 for maize, Salt-Tolerant Sorghum (STS), ICRISAT's SPH 5044 for sorghum, Salt-Tolerant Chickpea (STC), Pusa 1121 for chickpea, etc. (ICAR-CSSRI).

While selection of a variety for a cropping system, special attention should also be given to local adaptation and climatic condition, on farm testing and screening, crop management knowledge and awareness of farmers, varietal or breeding history, soil type, market demand, yield potential, etc. (Burgess et al. 2022). The cropping method in which a crop is cultivated has a direct impact on the selection of stress-tolerant crop types and hybrids. Farmers and breeders may choose cultivars and hybrids that not only withstand difficult environments but also support resilient and sustainable agricultural systems by taking into account variables including climate, soil type, moisture availability, and management techniques (Benitez-Alfonso et al. 2023). It is impossible to overestimate the significance of cropping systems in the selection of stress-tolerant crops, particularly in light of the global struggles with climate change and food security (Datta et al. 2022). Farmers may maximize crop production, lower losses, and advance sustainable agricultural practices by choosing hybrids and varieties that are suited to certain cropping systems. Stress-tolerant crops will become more and more important in cropping systems as the world's agricultural environment changes (Godoy et al. 2021). They are essential for maintaining food security, reducing the effects of climate change, and safeguarding natural resources for future generations.

15.4 Contingency Crop Planning Under Natural Calamities

Food security needs to be ensured for the growing global population. To ensure the food security, agriculture plays a vital role but the impact of climate change on agriculture in observed all over the world (Pawlak and Kołodziejczak 2020). Agriculture also acts as the fountain head of livelihood for major part of the population, which is also affected by weather aberrations. The natural calamities occurring in the current situation are no more guided by the nature where as it is the result of man-made disruptions and nonsustainable development (Chaudhary and Piracha 2021). Weather aberrations like drought, flood, erratic rainfall, frost, hailstorm, heat waves, cold waves, etc. affect crop production and productivity, increase pest and disease incidence, and destabilize the economic stability of small and marginal farmers while jeopardizing food security and sustainability (Skendžić et al. 2021). Regardless of prediction of extreme events by the use of technologies, the impact of climate change is hard to manage through a single management system (Bolan et al. 2024). In this context, contingency crop planning acts as an imperative strategy for mitigating the adverse effect of climate change. Strategic selection and management of crops for ensuring agricultural productivity and sustainability while mitigating the changing climatic action (Malhi et al. 2021). Contingency crop planning includes technology related to soil, land, water, crop, etc., institutional and policy support, previous and future meteorological data, time, etc. (Srinivasarao et al. 2020). Climatic conditions like erratic rainfall, long gap in rainfall, early and late onset of monsoon, early cessation of rain, and prolonged dry spells during crop season affect crop productivity (Baffour-Ata et al. 2021). Under such circumstances, management measures like cultivation of crops with low water requirement, short duration crops, supplemental irrigation especially during critical crop growth, higher seed rate for maintaining plant population, weeding, presowing irrigation, dry sowing, alternate crop selection, seed treatment, cultivation of pulses and millets, mulching, gap filling, thinning, ratooning of crops like pearl millet and sorghum, adoption of wider spacing, in situ rain water harvesting, stripping of leaves, intercropping, spraying of nutrients like 2% urea, etc. (Joshi and Kar 2001; Raza et al. 2019; Srinivasarao et al. 2020; Yadav et al. 2021; Gupta et al. 2021).

Drought is one of the most severe natural calamities which leads to significant reduction of crop yield. The recurring events of drought have increased due to climate change, while making it imperative for farmers to adopt effective strategy for its mitigation. Some of the mitigation strategies are selection of drought-resistant varieties, optimizing water use, use of cover crops, mulching, intercropping, short duration crops, contour ploughing, agroforestry, etc. (Fan et al. 2023; Shaffril et al. 2024). Calamities like hail storms, heat waves, cold waves, cyclones, intrusion of sea water, etc. add up to the losses of crop productivity. For mitigation of such unpredictable calamities, management strategies like historical metrological data analysis, mapping, wind breaks, drainage systems, efficient irrigation systems, netting, sea walls, leaching, farmers training, community involvement, etc. need to be adopted (Khan et al. 2023a, b; Grigorieva et al. 2023).

Cropping system plays a vital role in contingency crop planning especially under unpredictable extreme weather events. The selection and management strategies significantly enhances the resilience and sustainability of agricultural practices (Baraj et al. 2024). One of the dry components of cropping system is diversification and risk management. Adoption of variety of crops in a farming system substantially reduces the risk of crop failure (Wang et al. 2021). Various types of crops have different levels of tolerance which ensures survival of some crops without hampering the yield even under adverse climatic condition. Intercropping, mixed cropping, and crop rotation are some of such examples for better risk management. Planting cover crops or cover cropping not only reduces soil erosion but also improves soil fertility and enhances organic matter and water retention in soil. In addition to it, selection of climate-resistant crops with a shorter growing period helps in decreasing the adverse effect of climate change (Koudahe et al. 2022; Scavo et al. 2022). Cropping system also provides economic stability through market flexibility, multiple revenue stream, and value addition.

Research on the influence of rapid climate change and sensitivity on agriculture is gradually acknowledging the need of adaptation approaches in overcoming weather anomalies across the world. For such situation, there is a need to place a greater emphasis on adaptation, and mitigation techniques to extreme weather dangers should be accurately predicted (Tripathi and Mishra 2017). Short-term solutions to protect agricultural output and guarantee food security for households and the country should be implemented in tandem with long-term plans to lessen the effects of climate change that will have an impact on rural residents' livelihoods on a sectoral, temporal, and spatial level (Viana et al. 2022). Furthermore, in order to lessen the negative effects of unpredictable weather events, the implementation of real-time contingency plans and programs necessitates some sort of preparation for specific anticipated weather abnormalities based on long-term experiences or trends along with an actual response to the prevailing condition in an indispensable approach.

15.5 Agronomic Manipulation or Adaption to Avoid Adverse Impact of Stresses

Due to climate change, promoting global food production and climate-resilient crops to cope with abiotic stress is an important task in agriculture. Various abiotic stress like salinity, drought, high and low temperature, etc. has negative impact on crop production. So adaption of improved agricultural practices, genetic engineering, breeding technology, etc. can guarantee the food security under the changing climatic scenario, but most of the technologies are time taking process and needs proper skill for small-scale farmers to adapt (Nischalke 2015; Grigorieva et al. 2023). Under such scenario, agronomical practices or agronomic adaptation can help farmers as it is practical, adaptive, and sustainable. Agronomic manipulation refers to the use of various agricultural practices and techniques to minimize the adverse effect of climate change (O'Brien et al. 2021). Agronomic adaption is a nonresource intensive as it does not require significant investment and accessible to small-scale farmers, adaptive according to various climatic conditions, increases soil health, has low environmental impact, improves crop diversity, and integrates various factors like soil, water, crops, pest, etc. which in long run may mitigate climate change more efficiently without affecting world food demands (Tadesse and Ahmed 2023; Agrawal et al. 2021). Agronomic adaptation practices and techniques involve crop selection that involves selection of crops which can tolerate abiotic stress while maintaining the yield and reducing the chances of crop failure, soil management practices such as conservation tillage, mulching, and cover cropping which reduces soil disturbance, preserves soil and soil moisture, regulates temperature and weed growth, and provides conducive ecosystem for beneficial microorganisms (Mariani and Ferrante 2017; Manik et al. 2019; Ogunsola et al. 2021). It also involves irrigation and fertilizer management such as irrigation scheduling according to crop water requirement, weather forecasting, moisture availability, etc., use of precision irrigation techniques, use of balanced, slow-releasing and efficient fertilizer, and precision fertilizer application (Liang et al. 2020; Abioye et al. 2022). Biotic stress management like pest and disease management through crop rotation, biological control, and targeted pesticide application is also taken under consideration during agronomic modification. Planting dates adjustment such as early or late sowing according to the climatic condition, proper plant population for better air circulation and decreasing the risk of pest and disease attack, and optimizing orientation help minimize soil moisture loss and regulate soil temperature regulation that are easy agronomic adaptation which can be easily accepted by small-scale farmers (Hossain et al. 2021). Cover cropping and mulching techniques help in keeping weed population in check, maintaining soil moisture and regulating soil temperature which can be easily integrated in a cereal-based cropping system (Iqbal et al. 2020; El-Beltagi et al. 2022; Silva and Bagavathiannan 2023). Crop diversification and agroforestry practices are agronomic modifications which help in enhancing biodiversity and ecosystem services, reduce wind and water erosion, improve soil health, maintain temperature, and provide extra income to farmers (Fahad et al. 2022; Kay et al. 2019; Udawatta et al. 2019). In order to alleviate the negative effects of environmental stress on crop productivity and assure food security in the face of a rapidly changing climate, a synergistic combination of several agronomic approaches can offer a comprehensive answer.

15.6 Enhancement of Water Use Efficiency

The ever-growing global population needs to be fed while keeping sustainability as one of the focuses. Along with the increasing demand for food, the need for water also needs to be addressed, and agriculture is one of the biggest industries for food production as well as water consumption (Ingrao et al. 2023). In agriculture, crop production is mostly dependent upon rainfall as main source of water, but with the changing climatic scenario, the erratic rainfall patterns along with intensive agriculture have led to overexploitation of groundwater which in return depletes the water table while threatening long-run crop production and rural sustainability (Datta et al. 2022). Additionally, urbanization and industrialization also act as a catalyst for such situation (Mallareddy et al. 2023). To prevent the problem of water scarcity in crop production, enhancing the efficiency of available water can be a game changer. The concept of WUE was introduced by Briggs and Shantz in 1913 and can be defined as grain yield per unit water supplied or the relationship between productivity and water uses. The WUE can be affected by various reasons such as abiotic stresses, leaf level process, CO_2 concentration, genetic makeup, plant and canopy level responses, carbon and water dynamics, canopy level interaction, air temperature, solar radiation, choice of crops, cultural activities, etc. (Hatfield and Dold 2019). The changing climatic scenario also leads to more incidence of abiotic stress like heat and drought stresses. This stress, especially during pollination and grain filling, leads to decrease in productivity as water requirement for the crop is more during these critical periods. Moreover, the leaf level WUE are a complex activity as it is guided by various enzymatic activities which is drastically affected by even by small change in the environment which has been well recorded (Slattery et al. 2018; Baslam et al. 2020). Under such changing climatic scenario, cropping system can act as a mitigating strategy.

Cropping systems not only have a positive interaction with microclimate but also increase the efficiency of resource utilization. In a intercropping system, CO_2 concentration over the canopy decreases which leads to better carbon fixation and photosynthesis and hence increases productivity (Maitra et al. 2023; Burgess et al. 2022). A intercropping system not only increases WUE but also radiation use efficiency (RUE) and land use efficiency (LUE) (Zhou et al. 2021). The different height of crops in a cropping system provides shade to others which decreases the water losses through evapotranspiration and hence increasing WUE while providing a soothing microclimatic condition (Maitra et al. 2021). Adoption of conservation tillage with cropping system adds up to better WUE by permanent soil cover, minimum soil disturbance, and crop rotation strategies (Pramanick et al. 2022).

In a sustainable cropping system, crop selection is one of the important factors which determines the system efficiency. Selection of crops with less water requirement or drought-tolerant varieties can increase WUE (Seleiman et al. 2021). Crops like millets and legumes are more drought tolerant as compared to maize, wheat, etc. Along with that, reducing soil moisture depletion and optimizing water consumption can be achieved by rotating crops with varying water requirements and growing seasons (Das et al. 2021; Wang et al. 2024a, b). Water consumption can be balanced, for instance, by switching out high-water-demanding crops like maize with low-water-demanding crops like beans. Furthermore, in a cropping system, crop rotation with varying growth seasons can guarantee a consistent layer of soil cover, minimize soil erosion, and preserve soil moisture (Yadav et al. 2023). Reducing evaporation, holding onto soil moisture, and controlling soil temperature are all achieved by maintaining soil cover through mulching, cover crops, or residues. By acting as a barrier, soil cover lowers water loss through evaporation and stops direct sunlight from heating the soil. Additionally, by preserving soil organic matter, this technique enhances soil structure and water-holding ability (Demo and Bogale 2024).

Maintaining a proper plant density in a cropping system may be matched to water availability to minimize water waste and guarantee effective usage. While undercrowding might result in unused water resources, overcrowding can lead to greater competition among plants for water (Zhang et al. 2021). Farmers can guarantee that every plant gets enough water without wasting resources by maximizing plant density. By using drip irrigation or other precision irrigation techniques, water loss may be minimized and water can be applied directly to the roots and lowering evaporation, runoff, and soil compaction; drip irrigation makes sure that water gets to the plant roots, where it is needed. When compared to conventional flood irrigation techniques, this strategy can save up to 50% of the water needed (Levidow et al. 2014). Cropping system improves the organic matter content and soil structure which can help in increasing crop water availability, infiltration, and retention of water (Yang et al. 2020). Techniques like adding compost, manure, or cover crops can help achieve these goals. As a result, there is less need for regular watering since the soil can hold onto more water. By adding organic mulches, soil temperature, conserve moisture, and soil evaporation can be maintained (Iqbal et al. 2020). Mulch retains soil moisture and functions as an insulating layer to lessen temperature variations in the soil. Organic mulches like wood chips, grass, or straw contribute organic matter to the soil, enhancing soil health (Demo and Bogale 2024). Water consumption may be optimized and losses can be minimized by scheduling crop management and irrigation based on meteorological data. Farmers may better prepare for and anticipate weather extremes by using weather monitoring to modify irrigation plans. By ensuring that water is administered when needed, this strategy lowers waste and runoff (Kelly et al. 2021). Water waste may be avoided and effective utilization can be ensured by routinely evaluating crop water stress and modifying management techniques. In order to detect water stress in crops, crop monitoring entails analyzing crop growth, soil moisture, and weather patterns (Ahmad et al. 2021). By responding promptly to water stress, farmers can adjust irrigation schedules (Sumit et al. 2024; Sinha et al. 2024), apply targeted irrigation, or implement conservation measures to optimize water use.

15.7 Integrated Pest, Disease, and Weed Management

Along with changing climatic conditions, farmers have to face the threat of pest, disease, and weed which leads to significant yield loss. Chemical management approaches have negative effect on environment (Tudi et al. 2021). So integrated pest, diseases and weed management (IPDM) is the solution for management of pest, disease, and weed while keeping sustainability in focus. IPDM is a holistic approach which uses multiple techniques and strategies like physical, cultural, chemical, and biological approaches in a coordinated and synergistic way in an agricultural ecosystem (Deguine et al. 2021). IPDM not only focuses on environmental protection but also on ecosystem services, crop diversity, resistance management, cost-effectiveness, improved crop quality, human health, mitigation of climate change, etc. (Rao et al. 2020; Angon et al. 2023). One of the critical components of IPDM is a well-planned sustainable cropping system which directly influences the efficiency of the management practices while maintaining the population dynamics of pest, disease, and weeds (Maclaren et al. 2020; Angon et al. 2023). A well-planned or designed cropping system helps in breaking the life cycle of pest, disease, and weed. A sustainable cropping system for IPDM includes crop rotation, intercropping, crop spacing and density, crop selection, etc. Adoption of crop rotation breaks the life cycle of pest, diseases, and weed by removing the host crops and hence reducing the risk of carryover (Jalli et al. 2021). One such example is rotation of maize with legumes which decreases the risk of maize pest, disease, and weed population. Intercropping in a cropping system helps in growing of multiple crops together and hence increases ecological interaction. It also adds different crops from different family while keeping the population of pest, disease, and weed in check (Mupangwa et al. 2021). Some of the examples are as follows: planting of marigold along with tomatoes decreases the incidence of nematodes (Huang et al. 2022) and intercropping of maize and beans decreases the population of weeds like *Amaranthus blitoides*, *A. retroflexus*, *Setaria* sp., *Tribulus terrestris*, *A. blitoides*, *A. retroflexus*, *Echinochloa crus-galli*, *Solanum nigrum*, and *Convolvulus arvensis* (Bilalis et al. 2010). Selection of a crop with natural resistance or tolerance to pest and diseases and having allelopathic effect on weeds helps in increasing resilience to pest, diseases, and weed in a cropping system (Huss et al. 2022). Crop spacing and density also plays a vital role in IPDM as closer spacing increases easy spreading of pest and diseases whereas wider spacing decreases the spreading as well as increases the efficiency of different resources (Tudi et al. 2021). A cropping system also increases the soil organic matter, soil fertility, soil cover, water-holding capacity, ecological interaction, soil microbes' population, etc. (Yang et al. 2020). Increase in beneficial microbes also decreases the soil-borne pathogens, supports plant growth, and decomposes organic matter while indirectly increases population of beneficial insects like earthworms (Wang et al. 2024a, b). A sustainable cropping system for IPDM also requires farmers' knowledge and participation to ensure that a cropping system is planned in accordance with local demand, supply, and agroclimatic conditions. Regular monitoring can help identify thresholds of management practices. Adaptive management involves adjusting the cropping system in response to pest, disease, and weed dynamics.

The basis of IPDM is a well-planned cropping system, which creates the conditions for managing weeds, diseases, and pests. Farmers may break the cycles of disease and pests, stop the life cycles of weeds, boost the populations of beneficial organisms, improve the fertility and health of the soil, raise crop yields and quality, lessen their need on chemical inputs, and implement diversified and resilient cropping systems (Shah et al. 2021). Cropping systems are important in IPDM because they may act as a physical barrier against pest, disease, and weed, alter the microclimate to inhibit the growth of pests and diseases, encourage biodiversity and ecological balance, encourage the application of conservation agriculture techniques, and increase the overall resilience of agricultural ecosystems (Grasswitz 2019). Therefore, farmers, researchers, and policymakers may collaborate to create and promote sustainable agriculture practices that prioritize social responsibility, environmental stewardship, and economic viability by acknowledging the significance of cropping systems in IPDM. Establishing resilient and varied agricultural systems is essential to minimizing the effects of climate change and attaining sustainable food security.

15.8 Reduced Tillage, Conservation Agriculture, and Recarbonization of Soil

Increase in carbon dioxide (CO_2) emission is one of the serious issues in the world as it is the most important factor for increasing global temperature, that is, global warming. Around 10–12% of world's anthropogenic CO_2 emission is contributed by agriculture (Lynch et al. 2021). Intensification of agriculture especially fossil fuel-based intense mechanization and monoculture which is driven by worlds food demand leads to the increase in CO_2 efflux (Khan et al. 2023a, b). Under current situation especially after green revolution, maintaining crop yield while decreasing CO_2 efflux runs unparallelly with each other. The CO_2 emission from agriculture in return hampers the crop productivity due to unpredictable and sudden changes in climatic conditions. The amount of CO_2 emission is influenced by various factors, out of which soil, location, climate, characteristics, and management techniques play a significant role (Raza et al. 2019). The development of effective agricultural techniques that preserve or even boost crop yields while reducing CO_2 outflow is essential for sustainable agriculture. It is essential to implement land management techniques that minimize CO_2 losses from soils by conserving carbon deposits over the long term (Singh et al. 2024). This will maximize the global carbon cycle. One important component affecting the soil's net CO_2 exchange rate (NCER) is soil tillage. The physical characteristics of the soil, including temperature, moisture, oxygen concentration, microbial interaction with the substrate, and substrate dispersion, may all be changed by tillage (Gougoulias et al. 2014). A higher O_2 level in the soil due to intense plowing creates a more conducive environment for microbial activity and, as a result, raises NCER (Gougoulias et al. 2014). Land degradation and climatic change are the major challenges for agricultural sustainability and food security. In this context, a perfect cropping system while adapting conservation agriculture (CA) can act as a perfect strategy as it helps in improving soil health, enhancing crop productivity, and promoting sustainable agriculture while decreasing the CO_2 emission.

The goal of CA is to manage agroecosystems sustainably by emphasizing on soil conservation, efficient use of water, and maximum crop yields. The cropping system is the foundation of CA, and its design and management have a significant influence on the overall success of CA practices (Rodríguez et al. 2022). Cropping systems are essential to CA because they support biodiversity, preserve soil health, and reduce climate change. CA includes practices like crop rotation or crop diversification, permanent soil cover, and minimum disturbances to the soil. These practices improve soil fertility, enhance biodiversity of soil microbes, break disease and pest cycle, reduce soil erosion, retain moisture, improve crop resilience, improve soil health, enhance ecosystem services, etc. (Bhan and Behera 2014; Wolf et al. 2023). Recarbonization of soil is a critical aspect of CA as it helps in mitigation of climate change due to sequestration of carbon in soil (Rodrigues et al. 2023; Singh et al. 2024). Recarbonization of soil is done through plant residue and organic amendments, soil biota, and microbial activities and minimal disturbances of soil. Recarbonization of soil builds soil organic carbon, improves soil fertility, and increases water-holding capacity while reversing soil degradation and restoring soil health and productivity. It also decreases the CO_2 emission to the atmosphere which is one of the important greenhouse gasses contributing to the rapid climate change (Hannam 2024).

15.9 Adoption of Mixed Stand, Intercropping, and Agroforestry

Under the changing climatic scenario, it is challenging to ensure food security mitigating the extreme weather event. In this context, traditional farming system are no more efficient, sustainable, and climate resilient. Innovative approaches which include diversification and synergistic and ecosystem services need to be adopted for gaining prominence under such condition. Mixed stand, intercropping and agroforestry in cropping system, can be a promising solution in these innovative approaches (Pretty and Bharucha 2014). Integration of structures and multiple species helps in enhancement of soil health, ecosystem resilience, and biodiversity while improving crop yield, decreasing pest and disease attack and carbon sequestration (Huss et al. 2022).

A mixed stand in a cropping system means growing of multiple crop species in a same field simultaneously or in sequence to enhance synergistic effects and enhance ecological interaction (Curtright and Tiemann 2021). Mixed stand approach mimics the natural ecosystem diversity where coexistence and interaction of different species creates a productive and resilient environment. Mixed stand includes polyculture, intercropping, relay cropping, companion planting, etc. (Chamkhi et al. 2022). Growing multiple crops together in same field at the same time is known as polyculture whereas growing two or more crops together in the same field where one is main and other being the secondary crops is known as intercropping (Maitra et al. 2021). Relay cropping is planting of a second crop in the same field while one crop is in its harvesting stage. In this type of cropping, one crop growing stage overlaps growing stage of another (Lamichhane et al. 2023). Companion planting is planting of different crops together to enhance growth, health, and pest disease resistance through their synergistic effect such as maize, beans, and squash which is known as "Three sisters," marigold with tomatoes, etc. (Finch and Collier 2011). Companion planting helps each other species by providing nutrient, shade, weed control, pest and disease control, moisture retention, etc. A mixed stand helps in improved soil health through fulfilling nutrient requirement, reducing soil depletion and improved fertility, better ecosystem service by enhancing ecosystem services, climate regulation, pest control, etc., reduced pest and diseases attack, better crop yield, less use of chemical, increased farm resilience, etc. (Derpsch et al. 2024). While designing a mixed stand, certain characteristics like compatible crops, spatial arrangement, crop diversification and rotation, and optimization of inputs like water and nutrients need to be kept under consideration.

Agroforestry designs and manages cropping system that integrates trees and crops and sometime livestocks too for diversification, resilient farm, synergies, and ecological interaction (Low et al. 2023). Agroforestry helps in improving soil health and fertility, biodiversity, ecosystem services, crop yields, and quality, reduces soil erosion and water pollution, and provides additional income to farmers, carbon sequestration, sustainable agriculture, and rural development (Fahad et al. 2022). Some of the common agroforestry practices are silviculture (tress with crops), agrosilvipasture (trees with crop and livestock), forest farming (crops in forest ecosystem), permaculture (regenerative cropping system), etc. (Elevitch et al. 2018). Tree roots firmly hold soil, decreases erosion, adds organic matter, and improves soil structure and fertility. Agroforestry helps in diversification which encourages growth and development of beneficial microorganism and insects (Fahad et al. 2022). Shading effect of trees in a agroforestry system reduces evapotranspiration losses, decreasing soil temperature while promoting water cycle. It also decreases intensive tillage and monoculture practices. Trees absorb CO_2 from the atmosphere and store it in the soil; hence, it helps in carbon sequestration (Markwitz et al. 2020). Cropping systems that incorporate agroforestry, mixed stands, and intercropping enhance productivity, assist farmers create climate-resilient farms, and support international efforts to address climate change, biodiversity loss, and sustainable development. By implementing these, rural populations may be supported and the environment can be preserved, all while ensuring food security (Low et al. 2023).

15.10 Soil Health Improvement

One of the most essential and strategic resource for crop production is soil. From an ecocentric and anthropocentric standpoint, soil is a very precious resource and a very complicated ecosystem (Yang et al. 2020). In addition to providing food, fuel, and fiber, soil also aids in the decomposition of organic matter (such as dead plant and animal matter), recycling of necessary nutrients, detoxification of organic contaminants, sequestration of carbon, regulation of water quality and supply, provision of habitat for a wide variety of animals and microorganisms (soil is an important biodiversity reservoir), source of raw materials (clay, sand, gravel), etc. (Zhang et al. 2024).

Unfortunately, a variety of invasive anthropogenic activities in intensive agriculture have caused soil to deteriorate fast both globally and locally, with detrimental consequences for ecosystem and human health (Das et al. 2022). This is troubling because, at the human temporal scale, soil is a nonrenewable resource (i.e., soil loss and deterioration are not reversible during a human lifespan) (Di Stefano et al. 2023). In order to promote long-term sustainable cropping systems, the notion of soil health assessment has been centered on certain soil qualities and the soil's capacity to sustain a variety of biological activities in its suitable ecosystem since the 1990s (Lehmann et al. 2020). Soil health is therefore described as the soil's fitness to sustain crop development without degrading soil or otherwise damaging the environment (Acton and Gregorich 1995) or the soil's capacity to operate and deliver ecosystem services (Van Es and Karlen 2019). There has been interchangeability between the words "soil health" and "soil quality," with a focus mostly on agricultural productivity and a modicum of environmental sustainability (Doran et al. 1996). The current intensive agricultural situation has created a worrying situation for soil health and sustainability of production since it has led to nutrient imbalances and a decrease in soil fertility due to an extensive and poorly managed inorganic feeding system (Gamage et al. 2023). As a result, it is now essential to maintain soil fertility using a few straightforward yet unconventional methods, such as employing a few promising crops that are believed to boost soil health and restoring crop residue to the soil through proper crop diversification (Mosier et al. 2021). By implementing the best crop management techniques and creating an efficient cropping system, it is possible to ensure both soil fertility and profitability (Shah and Wu 2018).

Cropping system includes various spatial and temporal aspects which impacts soil health and quality along with crop diversification, crop rotation, intercropping, and other agronomical practices. The longstanding cropping systems were focused on maximization of yield, but the modern agriculture focuses on environmental sustainability without hampering the productivity. One of the key aspects of a sustainable cropping system is crop diversification (Rodriguez et al. 2021). Plant diversity in a cropping system is also known as crop diversification plays a vital role in increasing soil biodiversity, nutrient use efficiency, and reducing soil-borne pathogens. Crop diversity increases heterogeneity of soil chemical nutrients, population of beneficial microbes, and soil physical structure which leads to better soil health and hence increases productivity (Williams et al. 2023). Bainard et al. (2017) reported that crop diversity decreased the soil-borne diseases especially by introducing pulse crops in the rotation. Crop diversification along with new breeding techniques can increase the functionalities in an agroecosystem in multifold. Producing new cultivars with better nutrient use efficiency and better symbiotic relationship with beneficial microorganism by the use of modern breeding techniques can be helpful in crop diversification (Yang et al. 2020). Crop diversification for a sustainable cropping system also depends on adoption of different types of crops like C3, C4, legume, nonlegumes, etc. Better crop productivity, increased soil nutrient use efficiency, lower farm input costs, and less environmental impact from artificial chemical applications could all be achieved through properly planned crop diversification with enhanced capacity to interact with beneficial soil biota, resulting in improved soil health and sustainable agroecosystems (Waha et al. 2020). Crop rotation along with crop diversification can also act wonder for a sustainable cropping system. Crop rotation is a practical way for managing agroecosystem biodiversity by managing disease and pest outbreaks and hence increasing the productivity of the system. The efficiency of the rotation depends on rotation length, series, frequency, soil characteristic, agronomical history, etc. (Mortensen and Smith 2020). Soil health can be influenced by these factors in many ways. One such example is crop rotation might break the life cycle of soil-borne diseases linked with a particular crop or crop genotype, hence improving possibilities for some soil beneficial microorganisms to develop and reducing disease load (Panth et al. 2020). Selecting the ideal rotation sequence to optimize soil benefits can be challenging since certain crops perform better in rotation than others (Koyama et al. 2022). For instance, crop rotation with grain legumes can boost wheat productivity and protein content as the subsequent crop because the soil will have more accessible nitrogen from the legumes' biological fixation (Gan et al. 2003). According to Yang et al. (2012), varying genotypes of chickpeas or legume crops like pea and chickpea grown in rotation might alter the beneficial microbial communities in the soil and affect the productivity of pulse crops and the subsequent wheat crop. Intercropping is an additional strategy for enhancing soil health in a cropping system. By decreasing artificial chemical pollution (Lemaire et al. 2014), preventing soil disease (Vukicevich et al. 2016), improving soil nutrient and spatial use efficiency (Hinsinger et al. 2011), and fostering the bio-functionalities of soil microorganisms, intercropping practices can improve soil health.

Global soil degradation in agriculture is a result of numerous anthropogenic practices used in different cropping systems, including intensive tillage, the use of fossil fuels, the draining of wetlands, the adoption of heavy machinery in farming operations, fertilization, and pesticide management. Soil health is also threatened by other factors such as wind and water erosion, organic matter loss in peat and mineral soils, compaction, sealing, pollution, salinization, desertification, floods, landslides, and a decrease in biodiversity (Stolte et al. 2016). One of the most serious socioeconomic and environmental issues endangering human existence and well-being is soil degradation, particularly when it comes to food security and safety. In this regard, it is undeniable that one of the most pressing and unsettling issues facing our society in the twenty-first century will be feeding the world's rapidly expanding population, especially in light of the current state of climate change and its anticipated severe negative effects on food production (Gomiero 2016). In this context, maintaining the soil health while adopting a sustainable cropping system can show a profound impact on soil's chemical, biological, and physical properties.

15.11 Conclusion

It is essential to have a planned and comprehensive approach to handle the adverse impacts of climate change effectively. This chapter clearly highlighted the importance of crop diversification and adoption of stress-tolerant varieties to stabilize the yields and minimize crop failures under extreme climatic conditions. The adoption of efficient agronomic management practices provided the competitive advantage to the crop by enhancing water use efficiency and pest and disease resistance in the cropping system. Further, through adoption of sustainable agricultural practices such as conservation agriculture, recarbonization of soil, and agroforestry systems significantly contributed for the enhancement of soil health. Consequently, the strategies covered in this chapter form the foundation for creating a resilient cropping system in the face of climate change.

References

Abd-Alla MH, Al-Amri SM, El-Enany AWE (2023) Enhancing rhizobium-legume symbiosis and reducing nitrogen fertilizer use are potential options for mitigating climate change. Agriculture 13(11):2092

Abioye EA, Hensel O, Esau TJ, Elijah O, Abidin MSZ, Ayobami AS, Yerima O, Nasirahmadi A (2022) Precision irrigation management using machine learning and digital farming solutions. Agric Eng 4(1):70–103. https://doi.org/10.3390/agriengineering4010006

Acevedo M, Pixley K, Zinyengere N, Meng S, Tufan H, Cichy K, Bizikova L, Isaacs K, Ghezzi-Kopel K, Porciello J (2020) A scoping review of adoption of climate-resilient crops by small-scale producers in low- and middle-income countries. Nat Plants 6(10):1231–1241. https://doi.org/10.1038/s41477-020-00783-z

Acton DF, Gregorich LJ (1995) The health of our soils—towards sustainable agriculture in Canada. Centre for Land and Biological Resources Research, Research Branch, Agriculture and Agri-Food Canada, Ottawa, p 138. https://doi.org/10.5962/bhl.title.58906

Agrawal T, Hirons M, Gathorne-Hardy A (2021) Understanding farmers' cropping decisions and implications for crop diversity conservation: insights from Central India. Sustain Environ Res 3:100068. https://doi.org/10.1016/j.crsust.2021.100068

Ahmad U, Alvino A, Marino S (2021) A review of crop water stress assessment using remote sensing. Remote Sens 13(20):4155. https://doi.org/10.3390/rs13204155

Ahmed MM, Sharif U, Raza A, Safdar M, Ali W, Asim M, Muzammal H, Sattar J, Maqbool S, Zaheer M (2024) Climate-smart agriculture for resilience and profitability. In: Garwi J, Masengu R, Chiwaridzo O (eds) Practice, progress, and proficiency in sustainability, pp 27–50. https://doi.org/10.4018/979-8-3693-2011-2.ch002

Alam A (2014) Soil degradation: a challenge to sustainable agriculture. Int J Sci Res Agric Sci 1(4):50–55

Al-Khayri JM, Rashmi R, Toppo V, Chole PB, Banadka A, Sudheer WN, Nagella P, Shehata FW, Al-Mssallem QM, Alessa MF, Almaghasla IM, Rezk AA (2023) Plant secondary metabolites: the weapons for biotic stress management. Metabolites 13(6):716. https://doi.org/10.3390/metabo13060716

Anantha MS, Patel D, Quintana M, Swain P, Dwivedi JL, Torres RO, Henry A (2016) Trait combinations that improve rice yield under drought: Sahbhagi Dhan and new drought-tolerant varieties in South Asia. Crop Sci 56(1):408–421

Angon PB, Mondal S, Jahan I, Datto M, Antu UB, Ayshi FJ, Islam MS (2023) Integrated Pest Management (IPM) in agriculture and its role in maintaining ecological balance and biodiversity. Adv Agric:1–19. https://doi.org/10.1155/2023/5546373

Atlin GN, Cairns JE, Das B (2017) Rapid breeding and varietal replacement are critical to adaptation of cropping systems in the developing world to climate change. Glob Food Sec 12:31–37. https://doi.org/10.1016/j.gfs.2017.01.008

Baffour-Ata F, Antwi-Agyei P, Nkiaka E, Dougill AJ, Anning AK, Kwakye SO (2021) Effect of climate variability on yields of selected staple food crops in northern Ghana. J Agric Food Res 6:100205. https://doi.org/10.1016/j.jafr.2021.100205

Bainard LD, Navarro-Borrell A, Hamel C, Braun K, Hanson K, Gan Y (2017) Increasing the frequency of pulses in crop rotations reduces soil fungal diversity and increases the proportion of fungal pathotrophs in a semiarid agroecosystem. Agric Ecosyst Environ 240:206–214. https://doi.org/10.1016/j.agee.2017.02.020

Bao X, Hou X, Duan W, Yin B, Ren J, Wang Y, Liu X, Gu L, Zhen W (2023) Screening and evaluation of drought resistance traits of winter wheat in the North China Plain. Front Plant Sci 14:1194759. https://doi.org/10.3389/fpls.2023.1194759

Baraj B, Mishra M, Sudarsan D, Da Silva RM, Santos CG (2024) Climate change and resilience, adaptation, and sustainability of agriculture in India: a bibliometric review. Heliyon 10(8):e29586. https://doi.org/10.1016/j.heliyon.2024.e29586

Basha PA, Seenaiah R, Sreenath AB, Babu TM, Naik ST (2014) Photosynthetic studies in four red gram (*Cajanus cajan*) cultivars under drought conditions. Int J Plant Anim Environ Sci 4:1–8

Baslam M, Mitsui T, Hodges M, Priesack E, Herritt MT, Aranjuelo I, Sanz-Sáez Á (2020) Photosynthesis in a changing global climate: scaling up and scaling down in crops. Front Plant Sci 11. https://doi.org/10.3389/fpls.2020.00882

Benitez-Alfonso Y, Soanes BK, Zimba S, Sinanaj B, German L, Sharma V, Kolesnikova A, Dunn JA, Martin CA, Rahman MK, Saati-Santamaría Z, García-Fraile P, Ferreira AE, Frazão LA, Cowling WA, Siddique HMK, Pandey KM, Farooq M, Varshney RK, Foyer CH (2023) Enhancing climate change resilience in agricultural crops. Curr Biol 33(23):R1246–R1261. https://doi.org/10.1016/j.cub.2023.10.028

Bhadra P, Maitra S, Shankar T, Hossain A, Praharaj S, Aftab T (2022) Climate change impact on plants: plant responses and adaptations. In: Plant perspectives to global climate changes. Academic Press, pp 1–24

Bhan S, Behera U (2014) Conservation agriculture in India – problems, prospects and policy issues. Int Soil Water Conserv Res 2:1–12. https://doi.org/10.1016/s2095-6339(15)30053-8

Bhattacharjee B, Ali A, Rangappa K, Choudhury BU, Mishra VK (2023) A detailed study on genetic diversity, antioxidant machinery, and expression profile of drought-responsive genes in rice genotypes exposed to artificial osmotic stress. Sci Rep 13:45661. https://doi.org/10.1038/s41598-023-45661-8

Bilalis D, Papastylianou P, Konstantas A, Patsiali S, Karkanis A, Efthimiadou A (2010) Weed-suppressive effects of maize–legume intercropping in organic farming. Int J Pest Manag 56:173–181. https://doi.org/10.1080/09670870903304471

Bodner G, Nakhforoosh A, Kaul H (2015) Management of crop water under drought: a review. Agron Sustain Dev 35:401–442. https://doi.org/10.1007/s13593-015-0283-4

Bolan S, Padhye LP, Jasemizad T, Govarthanan M, Karmegam N, Wijesekara H, Amarasiri D, Hou D, Zhou P, Biswal BK, Balasubramanian R, Wang H, Siddique KHM, Rinklebe J, Kirkham MB, Bolan NS (2024) Impacts of climate change on the fate of contaminants through extreme weather events. Sci Total Environ 909:168388. https://doi.org/10.1016/j.scitotenv.2023.168388

Briggs LJ, Shantz HL (1913) The water requirement of plants. In: Bureau of Plant Industry Bulletin. US Department of Agriculture, Washington, DC, pp 282–285

Burgess AJ, Cano MEC, Parkes B (2022) The deployment of intercropping and agroforestry as adaptation to climate change. Crop Environ 1:145–160. https://doi.org/10.1016/j.crope.2022.05.001

Chamkhi I, Cheto S, Geistlinger J, Zeroual Y, Kouisni L, Bargaz A, Ghoulam C (2022) Legume-based intercropping systems promote beneficial rhizobacterial community and crop yield under stressing conditions. Ind Crop Prod 183:114958. https://doi.org/10.1016/j.indcrop.2022.114958
Chaudhary MT, Piracha A (2021) Natural disasters—origins, impacts, management. Encyclopedia 1:1101–1131. https://doi.org/10.3390/encyclopedia1040084
Chopra R, Burow G, Hayes C, Emendack Y, Xin Z, Burke J (2015) Transcriptome profiling and validation of gene based single nucleotide polymorphisms (SNPs) in sorghum genotypes with contrasting responses to cold stress. BMC Genomics 16:1040
Crop Improvement – ICAR-CSSRI:: Central Soil Salinity Research Institute. Available at https://cssri.res.in/crop-improvement/
Curtright AJ, Tiemann LK (2021) Intercropping increases soil extracellular enzyme activity: a meta-analysis. Agric Ecosyst Environ 319:107489. https://doi.org/10.1016/j.agee.2021.107489
Dar MH, De Janvry A, Emerick K, Raitzer D, Sadoulet E (2013) Flood-tolerant rice reduces yield variability and raises expected yield, differentially benefiting socially disadvantaged groups. Sci Rep 3:3315. https://doi.org/10.1038/srep03315
Das P, Pramanick B, Goswami SB, Maitra S, Ibrahim SM, Laing AM, Hossain A (2021) Innovative land arrangement in combination with irrigation methods improves the crop and water productivity of rice (*Oryza sativa* L.) grown with okra (Abelmoschus esculentus L.) under raised and sunken bed systems. Agronomy 11(10):2087
Das BS, Wani SP, Benbi DK, Muddu S, Bhattacharyya T, Mandal B, Santra P, Chakraborty D, Bhattacharyya R, Basak N, Reddy NN (2022) Soil health and its relationship with food security and human health to meet the sustainable development goals in India. Soil Secur 8:100071. https://doi.org/10.1016/j.soisec.2022.100071
Das BC, Patra S, Sarkar S, Pramanik S, Dhar PP, Samanta A (2024) Impacts of varieties and their biochemical parameters on pod damage by borer complex in pigeon pea in eastern India. AMA Agric Mech Asia Afr Lat Am 55(1):256–268
Datta P, Behera B, Rahut DB (2022) Climate change and Indian agriculture: a systematic review of farmers' perception, adaptation, and transformation. Environ Challenges 8:100543. https://doi.org/10.1016/j.envc.2022.100543
De UK, Bodosa K (2014) Crop diversification in Assam and use of essential modern inputs under changing climatic condition: indication of a retrograded option. J Climatol Weather Forecasting 2(2):1–14
Deguine JP, Aubertot JN, Flor RJ, Lescourret F, Wyckhuys KA, Ratnadass A (2021) Integrated pest management: good intentions, hard realities. A review. Agron Sustain Dev 41:38. https://doi.org/10.1007/s13593-021-00689-w
Demo AH, Bogale GA (2024) Enhancing crop yield and conserving soil moisture through mulching practices in dryland agriculture. Front Agron 6:1361697. https://doi.org/10.3389/fagro.2024.1361697
Derpsch R, Kassam A, Reicosky D, Friedrich T, Calegari A, Basch G, Gonzalez-Sanchez E, Santos DRD (2024) Nature's laws of declining soil productivity and conservation agriculture. Soil Secur 14:100127. https://doi.org/10.1016/j.soisec.2024.100127
Devi P, Jha UC, Prakash V, Kumar S, Parida SK, Paul PJ, Prasad PV, Sharma KD, Siddique KH, Nayyar H (2022) Response of physiological, reproductive function and yield traits in cultivated chickpea (Cicer arietinum L.) under heat stress. Front. Plant Sci 13:880519. https://doi.org/10.3389/fpls.2022.880519
Di Stefano C, Nicosia A, Pampalone V, Ferro V (2023) Soil loss tolerance in the context of the European Green Deal. Heliyon 9:e12869. https://doi.org/10.1016/j.heliyon.2023.e12869
Doran JW, Sarrantonio M, Liebig MA (1996) Soil health and sustainability. Adv Agron 56:1–54. https://doi.org/10.1016/S0065-2113(08)60178-9
El-Beltagi HS, Basit A, Mohamed HI, Ali I, Ullah S, Kamel EAR, Shalaby TA, Ramadan KMA, Alkhateeb AA, Ghazzawy HS (2022) Mulching as a sustainable water and soil saving practice in agriculture: a review. Agronomy 12:1881. https://doi.org/10.3390/agronomy12081881

Elevitch CR, Mazaroli DN, Ragone D (2018) Agroforestry standards for regenerative agriculture. Sustainability 10:3337. https://doi.org/10.3390/su10093337

Emamverdian A, Ding Y, Mokhberdoran F, Xie Y (2015) Heavy metal stress and some mechanisms of plant defense response. Sci World J 2015:756120. https://doi.org/10.1155/2015/756120

Fahad S, Bajwa AA, Nazir U, Anjum SA, Farooq A, Zohaib A, Sadia S, Nasim W, Adkins S, Saud S, Ihsan MZ, Alharby H, Wu C, Wang D, Huang J (2017) Crop production under drought and heat stress: plant responses and management options. Front Plant Sci 8:1147. https://doi.org/10.3389/fpls.2017.01147

FAO (2018) The future of food and agriculture – Alternative pathways to 2050. Summary version. Rome. 60 pp. Licence: CC BY-NC-SA 3.0 IGO.https://openknowledge.fao.org/server/api/core/bitstreams/2c6bd7b4-181e-4117-a90d-32a1bda8b27c/content

Fahad S, Chavan SB, Chichaghare AR, Uthappa AR, Kumar M, Kakade V, Pradhan A, Jinger D, Rawale G, Yadav DK, Kumar V, Farooq HT, Ali B, Sawant VA, Saud S, Chen S, Poczai P (2022) Agroforestry systems for soil health improvement and maintenance. Sustainability 14(22):14877. https://doi.org/10.3390/su142214877

Fan T, Li S, Zhao G, Wang S, Zhang J, Wang L, Dang Y, Cheng W (2023) Response of dryland crops to climate change and drought-resistant and water-suitable planting technology: a case of spring maize. J Integr Agric 22:2067–2079. https://doi.org/10.1016/j.jia.2022.08.044

Feliciano D (2019) A review on the contribution of crop diversification to Sustainable Development Goal 1 "No poverty" in different world regions. Sustain Dev 27:795–808. https://doi.org/10.1002/sd.1923

Finch S, Collier R (2011) The influence of host and non-host companion plants on the behaviour of pest insects in field crops. Entomol Exp Appl 142:87–96. https://doi.org/10.1111/j.1570-7458.2011.01191.x

Frantová N, Rábek M, Porčová L, Jovanović I, Širůček P, Lukas V, Hájek J, Elzner P, Holková L, Smutná P, Smutný V (2024) Monitoring drought tolerance mechanisms of sorghum and maize under unevenly distributed precipitation. Int J Plant Prod 18:49–67. https://doi.org/10.1007/s42106-023-00278-9

Fukao T, Barrera-Figueroa BE, Juntawong P, Peña-Castro JM (2019) Submergence and waterlogging stress in plants: a review highlighting research opportunities and understudied aspects. Front Plant Sci 10:340. https://doi.org/10.3389/fpls.2019.00340

Fürstenberg-Hägg J, Zagrobelny M, Bak S (2013) Plant defense against insect herbivores. Int J Mol Sci 14:10242–10297. https://doi.org/10.3390/ijms140510242

Gamage A, Gangahagedara R, Gamage J, Jayasinghe N, Kodikara N, Suraweera P, Merah O (2023) Role of organic farming for achieving sustainability in agriculture. Farm Syst 1:100005. https://doi.org/10.1016/j.farsys.2023.100005

Gan YT, Miller PR, McConkey BG, Zentner RP, Stevenson FC, McDonald CL (2003) Influence of diverse cropping sequences on durum wheat yield and protein in the semiarid Northern Great Plains. Agron J 95:245–252. https://doi.org/10.2134/agronj2003.2450

Glenz C, Schlaepfer R, Iorgulescu I, Kienast F (2006) Flooding tolerance of Central European tree and shrub species. For Ecol Manag 235:1–13. https://doi.org/10.1016/j.foreco.2006.05.065

Gobu OR, Baghel PBM, Chourasia KN (2017) Resistance/tolerance mechanism under water deficit (drought) condition in plants. Int J Curr Microbiol Appl Sci 6:66–78. https://doi.org/10.20546/ijcmas.2017.604.009

Godoy F, Olivos-Hernández K, Stange C, Handford M (2021) Abiotic stress in crop species: improving tolerance by applying plant metabolites. Plants 10:186. https://doi.org/10.3390/plants10020186

Gomiero T (2016) Soil degradation, land scarcity and food security: reviewing a complex challenge. Sustainability 8:281. https://doi.org/10.3390/su8030281

Gougoulias C, Clark JM, Shaw LJ (2014) The role of soil microbes in the global carbon cycle: tracking the below-ground microbial processing of plant-derived carbon for manipulating carbon dynamics in agricultural systems. J Sci Food Agric 94:2362–2371. https://doi.org/10.1002/jsfa.6577

Grasswitz TR (2019) Integrated pest management (IPM) for small-scale farms in developed economies: challenges and opportunities. Insects 10:179. https://doi.org/10.3390/insects10060179
Grigorieva E, Livenets A, Stelmakh E (2023) Adaptation of agriculture to climate change: a scoping review. Climate 11:202. https://doi.org/10.3390/cli11100202
Gudla SL, Puligadda LS, Devarakonda N, Ray S (2023) Assessing soil fertility dynamics and carbon sequestration potential in groundnut (Arachis hypogaea L.) cultivation areas of Sri Satya Sai District, Andhra Pradesh, India. Int J Environ Clim Change 13:4384–4393. https://doi.org/10.9734/ijecc/2023/v13i103115
Gupta B, Huang B (2014) Mechanism of salinity tolerance in plants: physiological, biochemical, and molecular characterization. Int J Genomics 2014:1–18. https://doi.org/10.1155/2014/701596
Gupta AK, Srivastava R, Hodam S, Singh S (2021) Handbook for district administration on disaster risk management in agriculture sector. National Institute of Disaster Management, New Delhi, pp 4–32
Hannam I (2024) Sustainable soil management and soil carbon sequestration. In: Ginzky H, Corrêa FDA, Dooley E, Heuser IL, Kameri-Mbote P, Kibugi R, Ruppel OC (eds) International yearbook of soil law and policy 2022. Springer, Cham. https://doi.org/10.1007/978-3-031-40609-6_1
Hassan MA, Xiang C, Farooq M, Muhammad N, Yan Z, Hui X, Yuanyuan K, Bruno AK, Lele Z, Jincai L (2021) Cold stress in wheat: plant acclimation responses and management strategies. Front Plant Sci 12:676884. https://doi.org/10.3389/fpls.2021.676884
Hatfield JL, Dold C (2019) Water-use efficiency: advances and challenges in a changing climate. Front Plant Sci 10:103. https://doi.org/10.3389/fpls.2019.00103
Hinsinger P, Betencourt E, Bernard L, Brauman A, Plassard C, Shen J, Tang X, Zhang F (2011) P for two, sharing a scarce resource: soil phosphorus acquisition in the rhizosphere of intercropped species. Plant Physiol 156:1078–1086. https://doi.org/10.1104/pp.111.175331
Hossain A, Skalicky M, Brestic M, Maitra S, Ashraful Alam M, Syed MA, Hossain J, Sarkar S, Saha S, Bhadra P, Shankar T (2021) Consequences and mitigation strategies of abiotic stresses in wheat (Triticum aestivum L.) under the changing climate. Agronomy 11(2):241
Huang F, Deng X, Gao L, Cai X, Yan D, Cai Y, Chen X, Yang M, Tong W, Yu L (2022) Effect of marigold (Tagetes erecta L.) on soil microbial communities in continuously cropped tobacco fields. Sci Rep 12:23517. https://doi.org/10.1038/s41598-022-23517-x
Huss CP, Holmes KD, Blubaugh CK (2022) Benefits and risks of intercropping for crop resilience and pest management. J Econ Entomol 115:1350–1362. https://doi.org/10.1093/jee/toac045
Ingrao C, Strippoli R, Lagioia G, Huisingh D (2023) Water scarcity in agriculture: an overview of causes, impacts and approaches for reducing the risks. Heliyon 9:e18507. https://doi.org/10.1016/j.heliyon.2023.e18507
Iqbal R, Raza MAS, Valipour M, Saleem MF, Zaheer MS, Ahmad S, Toleikiene M, Haider I, Aslam MU, Nazar MA (2020) Potential agricultural and environmental benefits of mulches—a review. Bull Natl Res Cent 44:75. https://doi.org/10.1186/s42269-020-00290-3
Irshad A, Ahmed RI, Rehman SU, Sun G, Ahmad F, Sher MA, Aslam MZ, Hassan MM, Qari SH, Aziz MK, Khan Z (2022) Characterization of salt tolerant wheat genotypes by using morpho-physiological, biochemical, and molecular analysis. Front Plant Sci 13:956298. https://doi.org/10.3389/fpls.2022.956298
Islam M, Sandhi A (2022) Heavy metal and drought stress in plants: the role of microbes—a review. Gesunde Pflanzen 75:695–708. https://doi.org/10.1007/s10343-022-00762-8
Jalli M, Huusela E, Jalli H, Kauppi K, Niemi M, Himanen S, Jauhiainen L (2021) Effects of crop rotation on spring wheat yield and pest occurrence in different tillage systems: a multi-year experiment in Finnish growing conditions. Front Sustain Food Syst 5:647335. https://doi.org/10.3389/fsufs.2021.647335
Jańczak-Pieniążek M, Cichoński J, Michalik P, Chrzanowski G (2022) Effect of heavy metal stress on phenolic compounds accumulation in winter wheat plants. Molecules 28:241. https://doi.org/10.3390/molecules28010241
Joshi NNL, Kar NA (2001) Contingency crop planning for dryland areas in relation to climate change. Indian J Agron 54:237–243. https://doi.org/10.59797/ija.v54i2.4787

Kamyab-Talesh F, Mousavi SF, Asadi R, Rezaei M, Khaledian MR (2024) Evaluation of some rice cultivars' response to salinity stress using resistance indices. Arch Agron Soil Sci 60(9):1303–1314.https://doi.org/10.1080/03650340.2014.891730

Kanavi MSP, Govind G, Somu G, Krishna PB (2023) Identification of promising drought tolerant genotypes in green gram [*Vigna radiata* (L.)]. Int J Trop Agric 41:139–144

Kaur G, Singh G, Motavalli PP, Nelson KA, Orlowski JM, Golden BR (2020) Impacts and management strategies for crop production in waterlogged or flooded soils: a review. Agron J 112:1475–1501. https://doi.org/10.1002/agj2.20093

Kaushik N, Siddique KH, Ahmed MS, Hakimi SS (2023) Climate-smart agriculture strategies for South Asia to address the challenges of climate change: identification of climate-resilient agriculture practices for India, Bangladesh, and Afghanistan. APN Sci Bull 13:241–262. https://doi.org/10.30852/sb.2023.2349

Kay S, Rega C, Moreno G, Herder MD, Palma JH, Borek R, Crous-Duran J, Freese D, Giannitsopoulos M, Graves A, Jager M, Lamersdorf N, Memedemin D, Mosquera-Losada R, Pantera A, Paracchini ML, Paris P, Roces-Diaz JV, Rolo V, Rosati A, Herzog F (2019) Agroforestry creates carbon sinks whilst enhancing the environment in agricultural landscapes in Europe. Land Use Policy 83:581–593. https://doi.org/10.1016/j.landusepol.2019.02.025

Kelly T, Foster T, Schultz DM, Mieno T (2021) The effect of soil-moisture uncertainty on irrigation water use and farm profits. Adv Water Resour 154:103982. https://doi.org/10.1016/j.advwatres.2021.103982

Khan R, Alabsi AAN, Muda I (2023a) Comparing the effects of agricultural intensification on CO2 emissions and energy consumption in developing and developed countries. Front Environ Sci 10:1065634. https://doi.org/10.3389/fenvs.2022.1065634

Khan SM, Shafi I, Butt WH, De La Torre Diez I, Flores MAL, Galán JC, Ashraf I (2023b) A systematic review of disaster management systems: approaches, challenges, and future directions. Land 12:1514. https://doi.org/10.3390/land12081514

Khondoker M, Mandal S, Gurav R, Hwang S (2023) Freshwater shortage, salinity increase, and global food production: a need for sustainable irrigation water desalination—a scoping review. Earth 4:223–240. https://doi.org/10.3390/earth4020012

Kim Y, Chung YS, Lee E, Tripathi P, Heo S, Kim K (2020) Root response to drought stress in rice (Oryza sativa L.). Int J Mol Sci 21:1513. https://doi.org/10.3390/ijms21041513

Kim Y, Webber H, Adiku SG, Nóia DS, Júnior R, Deswarte J, Asseng S, Ewert F (2024) Mechanisms and modelling approaches for excessive rainfall stress on cereals: waterlogging, submergence, lodging, pests and diseases. Agric For Meteorol 344:109819. https://doi.org/10.1016/j.agrformet.2023.109819

Koudahe K, Allen SC, Djaman K (2022) Critical review of the impact of cover crops on soil properties. Int Soil Water Conserv Res 10:343–354. https://doi.org/10.1016/j.iswcr.2022.03.003

Koyama A, Dias T, Antunes PM (2022) Application of plant–soil feedbacks in the selection of crop rotation sequences. Ecol Appl 32:e2501. https://doi.org/10.1002/eap.2501

Lamba K, Kumar M, Singh V, Chaudhary L, Sharma R, Yashveer S, Dalal MS (2023) Heat stress tolerance indices for identification of the heat tolerant wheat genotypes. Sci Rep 13(1):1–13. https://doi.org/10.1038/s41598-023-37634-8

Lamichhane JR, Alletto L, Cong W, Dayoub E, Maury P, Plaza-Bonilla D, Reckling M, Saia S, Soltani E, Tison G, Debaeke P (2023) Relay cropping for sustainable intensification of agriculture across temperate regions: crop management challenges and future research priorities. Field Crop Res 291:108795. https://doi.org/10.1016/j.fcr.2022.108795

Lehmann J, Bossio DA, Rillig MC (2020) The concept and future prospects of soil health. Nat Rev Earth Environ 1:544–553. https://doi.org/10.1038/s43017-020-0080-8

Lemaire G, Franzluebbers A, De Faccio Carvalho PC, Dedieu B (2014) Integrated crop–livestock systems: strategies to achieve synergy between agricultural production and environmental quality. Agric Ecosyst Environ 190:4–8. https://doi.org/10.1016/j.agee.2013.08.009

Levidow L, Zaccaria D, Maia R, Vivas E, Todorovic M, Scardigno A (2014) Improving water-efficient irrigation: prospects and difficulties of innovative practices. Agric Water Manag 146:84–94

Low G, Dalhaus T, Miranda PM, Meuwissen MPM (2023) Mixed farming and agroforestry systems: a systematic review on value chain implications, Agric Syst 206:103606. https://doi.org/10.1016/j.agsy.2023.103606

Li B, Zhang X, Morita S, Sekiya N, Araki H, Gu H, Han J, Yang L, Liu X (2022) Are crop deep roots always beneficial for combating drought: a review of root structure and function, regulation and phenotyping. Agric Water Manag 271:107781. https://doi.org/10.1016/j.agwat.2022.107781

Liang Z, Liu X, Xiong J, Xiao J (2020) Water allocation and integrative management of precision irrigation: a systematic review. Water 12:3135. https://doi.org/10.3390/w12113135

Lynch J, Cain M, Frame D, Pierrehumbert R (2021) Agriculture's contribution to climate change and role in mitigation is distinct from predominantly fossil CO_2-emitting sectors. Front Sustain Food Syst 4. https://doi.org/10.3389/fsufs.2020.518039

MacLaren C, Storkey J, Menegat A, Metcalfe H, Dehnen-Schmutz K (2020) An ecological future for weed science to sustain crop production and the environment: a review. Agron Sustain Dev 40:24. https://doi.org/10.1007/s13593-020-00631-6

Madejón P, Marañón T, Navarro-Fernández CM, Domínguez MT, Alegre JM, Robinson B, Murillo JM (2017) Potential of Eucalyptus camaldulensis for phytostabilization and biomonitoring of trace-element contaminated soils. PLoS One 12:e0180240. https://doi.org/10.1371/journal.pone.0180240

Maitra S, Hossain A, Brestic M, Skalicky M, Ondrisik P, Gitari H, Brahmachari K, Sankar T, Bhadra P, Palai JB, Jena J, Bhattacharya U, Duvvada SK, Lalichetti S, Sairam M (2021) Intercropping—a low input agricultural strategy for food and environmental security. Agronomy 11:343. https://doi.org/10.3390/agronomy11020343

Maitra S, Mondal T, Hossain A, Kalasare RS, Praharaj S, Joy JMM, Shankar T, Pramanick B, Battaglia ML, Bhattacharya U, Zupanič M (2022) Impact of climate change on growth and productivity of major field crops. In: Response of Field Crops to Abiotic Stress Choudhury S, Moulick D, CRC Press, pp 217–226. https://doi.org/10.1201/9781003258063-18

Maitra S, Sahoo U, Sairam M, Gitari HI, Rezaei-Chiyaneh E, Battaglia ML, Hossain A (2023) Cultivating sustainability: a comprehensive review on intercropping in a changing climate. Res Crops 24(4):702–715

Maitra S, Shankar T, Hossain A, Sairam M, Sagar L, Sahoo U, Gaikwad DJ, Pramanick B, Mandal TK, Sarkar S, Gitari HI (2024) Climate-smart millets production in future for food and nutritional security. In: Adapting to climate change in agriculture-theories and practices: Approaches for adapting to climate change in agriculture in India, Springer Nature Switzerland, pp. 11-41 https://doi.org/10.1007/978-3-031-28142-6_2

Malhi GS, Kaur M, Kaushik P (2021) Impact of climate change on agriculture and its mitigation strategies: a review. Sustainability 13:1318. https://doi.org/10.3390/su13031318

Mallareddy M, Thirumalaikumar R, Balasubramanian P, Naseeruddin R, Nithya N, Mariadoss A, Eazhilkrishna N, Choudhary KA, Deiveegan M, Subramanian E, Padmaja B, Vijayakumar S (2023) Maximizing water use efficiency in rice farming: a comprehensive review of innovative irrigation management technologies. Water 15:1802. https://doi.org/10.3390/w15101802

Mallikarjuna BO, Devaraja TN (2023) Frontline demonstration a tool to study drought tolerant and high yielding red gram variety for Davanagere District. Int J Curr Microbiol Appl Sci 12(8):100–105. https://doi.org/10.20546/ijcmas.2023.1208.012

Manik SMN, Pengilley G, Dean G, Field B, Shabala S, Zhou M (2019) Soil and crop management practices to minimize the impact of waterlogging on crop productivity. Front Plant Sci 10. https://doi.org/10.3389/fpls.2019.00140

Mariani L, Ferrante A (2017) Agronomic management for enhancing plant tolerance to abiotic stresses—drought, salinity, hypoxia, and lodging. Horticulturae 3:52. https://doi.org/10.3390/horticulturae3040052

Markwitz C, Knohl A, Siebicke L (2020) Evapotranspiration over agroforestry sites in Germany. Biogeosciences 17:5183–5208. https://doi.org/10.5194/bg-17-5183-2020

Mir AH, Bhat MA, Dar SA, Sofi PA, Bhat NA, Mir RR (2021) Assessment of cold tolerance in chickpea (Cicer spp.) grown under cold/freezing weather conditions of North-Western Himalayas of Jammu and Kashmir, India. Physiol Mol Biol Plants 27(5):1105–1118. https://doi.org/10.1007/s12298-021-00997-1

Mok WJ, Ghaffar MA, Noor MIM, Lananan F, Azra MN (2023) Understanding climate change and heavy metals in coastal areas: a macroanalysis assessment. Water 15:891. https://doi.org/10.3390/w15050891

Mortensen DA, Smith RG (2020) Confronting barriers to cropping system diversification. Front Sustain Food Syst 4:564197. https://doi.org/10.3389/fsufs.2020.564197

Mosier S, Córdova SC, Robertson GP (2021) Restoring soil fertility on degraded lands to meet food, fuel, and climate security needs via perennialization. Front Sustain Food Syst 5:706142. https://doi.org/10.3389/fsufs.2021.706142

Mupangwa W, Nyagumbo I, Liben F, Chipindu L, Craufurd P, Mkuhlani S (2021) Maize yields from rotation and intercropping systems with different legumes under conservation agriculture in contrasting agro-ecologies. Agric Ecosyst Environ 306:107170. https://doi.org/10.1016/j.agee.2020.107170

Nagesh Kumar MV, Ramya V, Govindaraj M, Dandapani A, Maheshwaramma S, Ganapathy KN, Kavitha K, Goverdhan M, Jagadeeshwar R (2022) India's rainfed sorghum improvement: three decades of genetic gain assessment for yield, grain quality, grain mold and shoot fly resistance. Front Plant Sci 13:1056040. https://doi.org/10.3389/fpls.2022.1056040

Nastis SA, Michailidis A, Chatzitheodoridis F (2012) Climate change and agricultural productivity. Afr J Agric Res 7(35):4885–4893

Nischalke SM (2015) Adaptation options to improve food security in a changing climate in the Hindu Kush-Himalayan region. In: Leal Filho W (ed) Handbook of climate change adaptation. Springer, Berlin, Heidelberg, pp 1423–1442. https://doi.org/10.1007/978-3-642-38670-1_103

O'Brien P, Kral-O'Brien K, Hatfield JL (2021) Agronomic approach to understanding climate change and food security. Agron J 113:4616–4626. https://doi.org/10.1002/agj2.20693

Ogunsola OA, Adeniyi OD, Adedokun VA (2021) Soil management and conservation: an approach to mitigate and ameliorate soil contamination. In: IntechOpen eBooks. https://doi.org/10.5772/intechopen.94526

Oladosu Y, Rafii MY, Arolu FC, Chukwu SC, Muhammad I, Kareem I, Arolu IW (2020) Submergence tolerance in rice: review of mechanism, breeding and future prospects. Sustainability 12:1632. https://doi.org/10.3390/su12041632

Panda D, Mishra SS, Behera PK (2021) Drought tolerance in rice: focus on recent mechanisms and approaches. Rice Sci 28:119–132. https://doi.org/10.1016/j.rsci.2021.01.002

Panth M, Hassler SC, Baysal-Gurel F (2020) Methods for management of soilborne diseases in crop production. Agriculture 10:16. https://doi.org/10.3390/agriculture10010016

Patil G, Divya MP, Shah P, Maitry A (2023) Metal accumulation ability of different eucalyptus species at the early stage. Forestist 74:26–34. https://doi.org/10.5152/forestist.2023.22078

Pawlak K, Kołodziejczak M (2020) The role of agriculture in ensuring food security in developing countries: considerations in the context of the problem of sustainable food production. Sustainability 12:5488. https://doi.org/10.3390/su12135488

Pramanick B, Kumar M, Naik BM, Kumar M, Singh SK, Maitra S, Naik BS, Rajput VD, Minkina T (2022) Long-term conservation tillage and precision nutrient management in maize-wheat cropping system: effect on soil properties, crop production, and economics. Agronomy 12(11):2766

Pramanick B, Kumar M, Naik BM, Singh SK, Singh SV (2024) Soil carbon-nutrient cycling, energetics, and carbon footprint in calcareous soils with adoption of long-term conservation tillage practices and cropping systems diversification. Sci Total Environ 912:169421. https://doi.org/10.1016/j.scitotenv.2023.169421

Pretty J, Bharucha ZP (2014) Sustainable intensification in agricultural systems. Ann Bot 114:1571–1596. https://doi.org/10.1093/aob/mcu205

Rao A, Singh RG, Mahajan G, Wani S (2020) Weed research issues, challenges, and opportunities in India. Crop Prot 134:104451. https://doi.org/10.1016/j.cropro.2018.02.003

Ratnadass A, Fernandes P, Avelino J, Habib R (2012) Plant species diversity for sustainable management of crop pests and diseases in agroecosystems: a review. Agron Sustain Dev 32:273–303

Ray S, Umesha C, Meshram MR (2021) Response of irrigation and hydrogel on growth and yield of pearl millet (*Pennisetum glaucum* L.) under eastern UP condition. Biol Forum 13:404–407

Raza A, Razzaq A, Mehmood S, Zou X, Zhang X, Lv Y, Xu J (2019) Impact of climate change on crops adaptation and strategies to tackle its outcome: a review. Plants 8:34. https://doi.org/10.3390/plants8020034

Rehman A, Jingdong L, Khatoon R, Hussain I, Iqbal MS (2016) Modern agricultural technology adoption: its importance, role, and usage for the improvement of agriculture. Life Sci J 14(2):70–74

Rodrigues CID, Brito LM, Nunes LJR (2023) Soil carbon sequestration in the context of climate change mitigation: a review. Soil Syst 7:64. https://doi.org/10.3390/soilsystems7030064

Rodriguez C, Dimitrova Mårtensson L, Zachrison M, Carlsson G (2021) Sustainability of diversified organic cropping systems—challenges identified by farmer interviews and multi-criteria assessments. Front Agron 3:698968. https://doi.org/10.3389/fagro.2021.698968

Rodríguez BC, Durán-Zuazo VH, Rodríguez MS, García-Tejero IF, Ruiz BG, Tavira SC (2022) Conservation agriculture as a sustainable system for soil health: a review. Soil Syst 6:87. https://doi.org/10.3390/soilsystems6040087

Sagar L, Praharaj S, Singh S, Attri M, Pramanick B, Maitra S, Hossain A, Shankar T, Palai JB, Sahoo U (2022) Drought and heat stress tolerance in field crops: consequences and adaptation strategies. In: Response of field crops to abiotic stress. CRC Press, pp 91–102

Sagar L, Singh S, Sharma A, Maitra S, Attri M, Sahoo RK, Ghasil BP, Shankar T, Gaikwad DJ, Sairam M, Sahoo U, Hossain A, Roy S (2023) Role of soil microbes against abiotic stresses induced oxidative stresses in plants. In: Microbial symbionts and plant health: trends and applications for changing climate. Springer

Santosh DT, Debnath S, Maitra S, Sairam M, Sagar L, Hossain A, Moulick D (2024) Alleviation of climate catastrophe in agriculture through adoption of climate-smart technologies. In: Brestic M, Skalicky M, Ondrisik P (eds) Climate crisis: adaptive approaches and sustainability. Springer Nature Switzerland, Cham, pp 307–332

Scavo A, Fontanazza S, Restuccia A, Pesce GR, Abbate C, Mauromicale G (2022) The role of cover crops in improving soil fertility and plant nutritional status in temperate climates: a review. Agron Sustain Dev 42:5. https://doi.org/10.1007/s13593-022-00825-0

Seleiman MF, Al-Suhaibani N, Ali N, Akmal M, Alotaibi M, Refay Y, Dindaroglu T, Abdul-Wajid HH, Battaglia ML (2021) Drought stress impacts on plants and different approaches to alleviate its adverse effects. Plants 10:259. https://doi.org/10.3390/plants10020259

Shaffril HAM, Samah AA, Samsuddin SF, Ahmad N, Tangang F, Sidique SF, Rahman AH, Burhan SAN, Shah JA, Khalid NA (2024) Diversification of agriculture practices as a response to climate change impacts among farmers in low-income countries: a systematic literature review. Clim Serv 35:100508. https://doi.org/10.1016/j.cliser.2024.100508

Shah F, Wu W (2018) Soil and crop management strategies to ensure higher crop productivity within sustainable environments. Sustainability 11:1485. https://doi.org/10.3390/su11051485

Shah TM, Imran M, Atta BM et al (2020) Selection and screening of drought tolerant high yielding chickpea genotypes based on physio-biochemical indices and multi-environmental yield trials. BMC Plant Biol 20:171. https://doi.org/10.1186/s12870-020-02381-9

Shah KK, Modi B, Pandey HP, Subedi A, Aryal G, Pandey M, Shrestha J (2021) Diversified crop rotation: an approach for sustainable agriculture production. Adv Agric 2021:1–9. https://doi.org/10.1155/2021/8924087

Shahzad M, Jabran K, Hussain M, Raza MAS, Wijaya L, El-Sheikh MA, Alyemeni MN (2021) The impact of different weed management strategies on weed flora of wheat-based cropping systems. PLoS One 16(2):e0247137

Shang Y, Olesen JE, Lærke PE, Manevski K, Chen J (2024) Perennial cropping systems increased topsoil carbon and nitrogen stocks over annual systems—a nine-year field study. Agric Ecosyst Environ 365:108925. https://doi.org/10.1016/j.agee.2024.108925

Shende RT, Sengar RS, Kumar M, Gupta S, Singh G, Singh SK (2020) Evaluation of yield and yield related traits of chickpea (*Cicer arietinum* L.) genotypes under water stress condition. Int J Curr Microbiol Appl Sci 9(12):1960–1979

Silva GC, Bagavathiannan M (2023) Mechanisms of weed suppression by cereal rye cover crop: a review. Agron J 115:1571–1585. https://doi.org/10.1002/agj2.21347

Singh RB, Devi R (2023) Achieving salinity-tolerance in cereal crops: major insights into genomics-assisted breeding (GAB). In: IntechOpen eBooks. https://doi.org/10.5772/intechopen.112570

Singh S, Parihar P, Singh R, Singh VP, Prasad SM (2016) Heavy metal tolerance in plants: role of transcriptomics, proteomics, metabolomics, and bionomics. Front Plant Sci 6:1143. https://doi.org/10.3389/fpls.2015.01143

Singh A, Septiningsih EM, Balyan HS, Singh NK, Rai V (2017) Genetics, physiological mechanisms and breeding of flood-tolerant rice (Oryza sativa L.). Plant Cell Physiol 58:1951–1964. https://doi.org/10.1093/pcp/pcw206

Singh BK, Delgado-Baquerizo M, Egidi E, Guirado E, Leach JE, Liu H, Trivedi P (2023a) Climate change impacts on plant pathogens, food security and paths forward. Nat Rev Microbiol 21:640–656. https://doi.org/10.1038/s41579-023-00900-7

Singh SV, Chaturvedi S, Nayak P, Dhyani VC, Nadipamu TMK, Singh DK, Gudapaty P, Mathyam P, Srinivasarao K, Gavindaraju K (2023b) Carbon offset potential of biochar based straw management under rice – wheat system along Indo-Gangetic Plains of India. Sci Total Environ 897:165176. https://doi.org/10.1016/j.scitotenv.2023.165176

Singh S, Kiran BR, Mohan SV (2024) Carbon farming: a circular framework to augment CO2 sinks and to combat climate change. Environ Sci Adv 3:522. https://doi.org/10.1039/d3va00296a

Sinha A, Ray S, Majumder S, Dash B, Barik BR, Reddy MD (2024) Assessment of wheat crop water requirement for South Bihar by FAO CROPWAT 8.0. Agric Sci Dig Res J. https://doi.org/10.18805/ag.d-6020

Skendžić S, Zovko M, Živković IP, Lešić V, Lemić D (2021) The impact of climate change on agricultural insect pests. Insects 12:440. https://doi.org/10.3390/insects12050440

Slattery RA, Walker BJ, Weber AP, Ort DR (2018) The impacts of fluctuating light on crop performance. Plant Physiol 176:990–1003. https://doi.org/10.1104/pp.17.01234

Srinivasarao C, Rao K, Gopinath K, Prasad Y, Arunachalam A, Ramana DBV, Chary GR, Gangaiah B, Venkateswarlu B, Mohapatra T (2020) Agriculture contingency plans for managing weather aberrations and extreme climatic events: development, implementation and impacts in India. Adv Agron:35–91. https://doi.org/10.1016/bs.agron.2019.08.002

Stolte J, Tesfai M, Oygarden L, Kvaerno S, Keizer J, Verheijen F, Pano P, Ballabio C, Hessel R (2016) Soil threats in Europe: status, methods, drivers and effects on ecosystem services: deliverable 2.1 RECARE project. European Commission DG Joint Research Centre

Sumit R, Bandana RB, Sitabhra M, Anamika S, Devender MR, Lalichetti S, Supradip S (2024) Assessment of crop water requirement using FAO CROPWAT 8.0 for groundnut in South Odisha. Plant Sci Today. https://doi.org/10.14719/pst.3438

Sunil R, Chhabra AK, Yadav RK, Kumar S (2023) Assessment of chickpea (*Cicer arietinum* L.) genotypes under normal and late sown environments using stress indices. Agric Sci Dig 43(6):807–811

Swapna S, Shylaraj KS (2017) Screening for osmotic stress responses in rice varieties under drought condition. Rice Sci 24(5):253–263

Tadesse B, Ahmed M (2023) Impact of adoption of climate smart agricultural practices to minimize production risk in Ethiopia: a systematic review. J Agric Food Res 13:100655. https://doi.org/10.1016/j.jafr.2023.100655

Tripathi A, Mishra AK (2017) Knowledge and passive adaptation to climate change: an example from Indian farmers. Clim Risk Manag 16:195–207. https://doi.org/10.1016/j.crm.2016.11.002
Tudi M, Ruan HD, Wang L, Lyu J, Sadler R, Connell D, Cordia C, Phung DT (2021) Agriculture development, pesticide application and its impact on the environment. Int J Environ Res Public Health 18:1112. https://doi.org/10.3390/ijerph18031112
Udawatta RP, Rankoth L, Jose S (2019) Agroforestry and biodiversity. Sustainability 11:2879. https://doi.org/10.3390/su11102879
van Es HM, Karlen DL (2019) Reanalysis validates soil health indicator sensitivity and correlation with long-term crop yields. Soil Sci Soc Am J 83:721–732. https://doi.org/10.2136/sssaj2018.09.0338
Viana CM, Freire D, Abrantes P, Rocha J, Pereira P (2022) Agricultural land systems importance for supporting food security and sustainable development goals: a systematic review. Sci Total Environ 806:150718. https://doi.org/10.1016/j.scitotenv.2021.150718
Vukicevich E, Lowery T, Bowen P, Úrbez-Torres JR, Hart M (2016) Cover crops to increase soil microbial diversity and mitigate decline in perennial agriculture: a review. Agron Sustain Dev 36:38. https://doi.org/10.1007/s13593-016-0385-7
Waha K, Dietrich JP, Portmann FT, Siebert S, Thornton PK, Bondeau A, Herrero M (2020) Multiple cropping systems of the world and the potential for increasing cropping intensity. Glob Environ Chang 64:102131. https://doi.org/10.1016/j.gloenvcha.2020.102131
Wang W, Vinocur B, Altman A (2003) Plant responses to drought, salinity and extreme temperatures: towards genetic engineering for stress tolerance. Planta 218:1–14
Wang Q, Li Y, Alva A (2010) Cropping systems to improve carbon sequestration for mitigation of climate change. J Environ Prot 1(03):207
Wang T, Jin H, Fan Y, Obembe O, Li D (2021) Farmers' adoption and perceived benefits of diversified crop rotations in the margins of U.S. Corn Belt. J Environ Manag 293:112903. https://doi.org/10.1016/j.jenvman.2021.112903
Wang B, Wang G, Van Dam J, Yang X, Ritsema C, Siddique KH, Du T, Kang S (2024a) Diversified crop rotations improve crop water use and subsequent cereal crop yield through soil moisture compensation. Agric Water Manag 294:108721. https://doi.org/10.1016/j.agwat.2024.108721
Wang X, Chi Y, Song S (2024b) Important soil microbiota's effects on plants and soils: a comprehensive 30-year systematic literature review. Front Microbiol 15. https://doi.org/10.3389/fmicb.2024.1347745
Wei X, Xie B, Wan C, Song R, Zhong W, Xin S, Song K (2024) Enhancing soil health and plant growth through microbial fertilizers: mechanisms, benefits, and sustainable agricultural practices. Agronomy 14(3):609. https://doi.org/10.3390/agronomy14030609
Williams A, Birt HWG, Raghavendra A, Dennis GP (2023) Cropping system diversification influences soil microbial diversity in subtropical dryland farming systems. Microb Ecol 85:1473–1484. https://doi.org/10.1007/s00248-022-02074-w
Wolf MK, Wiesmeier M, Macholdt J (2023) Importance of soil fertility for climate-resilient cropping systems: the farmer's perspective. Soil Secur 13:100119. https://doi.org/10.1016/j.soisec.2023.100119
Xiao Z, Yao X, Bai S, Wei J, An S (2023) Involvement of an enhanced immunity mechanism in the resistance to Bacillus thuringiensis in Lepidopteran pests. Insects 14(2):151. https://doi.org/10.3390/insects14020151
Yadav KS, Kumar A (2023) Crop diversification for climate resilient agriculture: opportunity and challenges. Indian Farm 73(02):12–14.2
Yadav VL, Yadav VP, Yadav GN, Yadav SK (2021) Contingent crop planning for aberrant weather conditions in dryland agriculture. Pharm Innov J 10:1223–1226. https://www.thepharmajournal.com/archives/2021/vol10issue8s/PartQ/S-10-8-31-146.pdf
Yadav D, Singh D, Babu S, Madegowda M, Singh D, Mandal D, Rathore AC, Sharma VK, Singhal V, Kumawat A, Yadav DA, Yadav RK, Kumar S (2023) Intensified cropping reduces soil erosion and improves rainfall partitioning and soil properties in the marginal land of the Indian Himalayas. Int Soil Water Conserv Res 12:521. https://doi.org/10.1016/j.iswcr.2023.10.002

Yang C, Hamel C, Gan Y, Vujanovic V (2012) Pyrosequencing reveals how pulses influence rhizobacterial communities with feedback on wheat growth in the semiarid Prairie. Plant Soil 367:493–505. https://doi.org/10.1007/s11104-012-1485-z

Yang T, Siddique KH, Liu K (2020) Cropping systems in agriculture and their impact on soil health: a review. Global Ecol Conserv 23:e01118. https://doi.org/10.1016/j.gecco.2020.e01118

Yang X, Xiong J, Du T, Ju X, Gan Y, Li S, Butterbach-Bahl K (2024) Diversifying crop rotation increases food production, reduces net greenhouse gas emissions, and improves soil health. Nat Commun 15(1):198

Ye C, Fukai S, Godwin I, Reinke R, Snell P, Schiller J, Basnayake J (2009) Cold tolerance in rice varieties at different growth stages. Crop Pasture Sci 60(4):328–338

Zhang C, Fu S (2009) Allelopathic effects of eucalyptus and the establishment of mixed stands of eucalyptus and native species. Forest Ecol Manag 258:1391–1396. https://doi.org/10.1016/j.foreco.2009.06.045

Zhang K, Tang J, Wang Y, Kang H, Zeng J (2020) The tolerance to saline–alkaline stress was dependent on the roots in wheat. Physiol Mol Biol Plants 26(5):947–954. https://doi.org/10.1007/s12298-020-00799-x

Zhang Y, Xu Z, Li J, Wang R (2021) Optimum planting density improves resource use efficiency and yield stability of rainfed maize in semiarid climate. Front Plant Sci 12:752606. https://doi.org/10.3389/fpls.2021.752606

Zhang Y, Jiang S, Quan D, Fang K, Wang B, Ma Z. (2024) Properties of Sustainable Earth Construction Materials:a State-of-the-Art Review. Sustainability 16(2):670. https://doi.org/10.3390/su16020670

Zheng H, Sun X, Li J, Song Y, Song J, Wang F, Liu L, Zhang X, Sui N (2021) Analysis of N6-methyladenosine reveals a new important mechanism regulating the salt tolerance of sweet sorghum. Plant Sci 304:110801

Zhiqun T, Jian Z, Junli Y, Chunzi W, Danju Z (2017) Allelopathic effects of volatile organic compounds from Eucalyptus grandis rhizosphere soil on Eisenia fetida assessed using avoidance bioassays, enzyme activity, and comet assays. Chemosphere 173:307–317. https://doi.org/10.1016/j.chemosphere.2017.01.004

Zhou H, Zhou G, Zhou L, Lv X, Ji Y, Zhou M (2021) The interrelationship between water use efficiency and radiation use efficiency under progressive soil drying in maize. Front Plant Sci 12. https://doi.org/10.3389/fpls.2021.794409

Chapter 16
Holistic Approaches to Enhancing Climate Resilience and Biodiversity Conservation for Food Security

Sk Asraful Ali, Tony Manoj Kumar Nandipamu, Rashmi Jha, K. M. Krishna, Kadagonda Nithinkumar, Nilutpal Saikia, Unti Miiri Ezing, T. H. V. Meghana, Ajay Anto Soy, Dendukuri Udaya Lakshmi, Doppalapudi Vijaya Rani, Ateka Sultana, Shashank Shekher Singh, and Shiv Vendra Singh

Abstract Climate change creates problems in agriculture, disrupts crop yields, and threatens global food security. Rising temperatures, erratic rainfall, and extreme weather events induce losses in staple crops like maize and wheat. Mitigation strategies in agriculture, crucial for climate change combat, include reducing greenhouse gas emissions through practices like conservation agriculture and efficient livestock management. Sustainable land management and carbon sequestration methods,

---

S. A. Ali · K. M. Krishna · K. Nithinkumar · N. Saikia · U. M. Ezing · A. A. Soy
ICAR-Indian Agricultural Research Institute, Pusa Campus, New Delhi, India

T. M. K. Nandipamu
G.B. Pant University of Agriculture and Technology, Pantnagar, Uttarakhand, India

R. Jha
Tamil Nadu Agricultural University, Coimbatore, Tamil Nadu, India

T. H. V. Meghana
University of Agricultural Sciences, GKVK, Bangalore, Karnataka, India

D. U. Lakshmi
Banaras Hindu University, Varanasi, Uttar Pradesh, India

D. V. Rani
Sardar Vallabhbhai Patel University of Agriculture and Technology, Meerut, Uttar Pradesh, India

A. Sultana
Midnapore City College, Paschim Medinipur, West Bengal, India

S. S. Singh
Department of Agronomy NIAS, Gopal Narayan Singh University, Sasaram, Rohtas, Bihar, India

S. V. Singh (✉)
Rani Lakshmi Bai Central Agricultural University, Jhansi, Uttar Pradesh, India

B. Pramanick et al. (eds.), *Climate-Smart Agricultural Technologies*, Sustainability Sciences in Asia and Africa, https://doi.org/10.1007/978-981-96-7699-6_16

such as agroforestry and afforestation, serve as carbon sinks, enhancing soil health and biodiversity while mitigating climate impacts. Adaptation strategies in agriculture include crop diversification, resilient varieties, improved water management, sustainable agroecological practices, and risk reduction through early warning systems. Enhancing biodiversity conservation in agriculture through agroecological approaches is crucial for climate resilience. Ecosystem-based adaptation strategies harness biodiversity to mitigate climate impacts. Integrating these strategies fosters climate-smart agriculture, balancing productivity with sustainability for future food security. Addressing socio-economic dimensions, advancing genomic technologies, and enhancing policy and governance are crucial for fostering climate-resilient agriculture. This present chapter focuses on the aspects of mitigation adoption strategies to impart climate resilience besides ensuring national food security and ecological biodiversity.

Keywords Climate change · Climate change mitigation and adaptation · Biodiversity conservation · Climate-smart agriculture · Food security

16.1 Introduction

Agriculture is the most comprehensive word used to denote the many ways in which crop plants and domestic animals sustain the global human population by providing food and other products (Harris and Fuller 2014). The agricultural sector, which plays a crucial role in ensuring global food security, is increasingly confronted with issues arising from climate change. Increasing temperatures, modified precipitation patterns, and severe weather events pose a hazard to the production of crops, the health of animals, and total agricultural output (Smith et al. 2008). Efforts are now being made to improve the ability of agriculture to withstand climate change by implementing a comprehensive strategy that includes reducing greenhouse gas emissions, adapting to changing conditions, and preserving biodiversity. Mitigation measures seek to diminish the role of agriculture in generating greenhouse gas emissions, which are responsible for driving climate change. These include improving livestock management practices and promoting sustainable land use (Smith et al. 2007). Adopting practices like precision agriculture, agroforestry, and enhanced livestock management may effectively capture carbon, decrease emissions, and improve soil health (Balafoutis et al. 2017; Nikzaad and Nusrathali 2023). On the other hand, adaptation strategies focus on adjusting agricultural practices to the changing climate. Examples include developing drought-resistant crop varieties and improving water management (Grigorieva et al. 2023; Akinnagbe and Irohibe 2014). In order to increase agricultural resilience to present and future climate challenges, adaptation strategies are crucial. Crop diversification, water conservation, and the utilization of resilient crop varieties are climate-smart practices that farmers can implement (Okoronkwo et al. 2024). In addition, enhanced meteorological predictions and advanced alert systems may assist farmers in anticipating and addressing climate-related hazards, therefore minimizing crop damage and protecting their means of living.

Biodiversity protection is essential for improving the ability of agriculture to withstand and recover from challenges. Diverse ecosystems provide essential ecological services, including pollination, pest control, and soil fertility, that are crucial for agricultural output (Power 2010). Biodiversity in agriculture, or agrobiodiversity, refers to all crops and animal breeds, their wild relatives, and other species (e.g., pollinators, symbionts, pests, parasites, predators, decomposers, and competitors) that coexist and interact within croplands and/or their surrounding environments. It includes populations of variable and adaptable landraces, as well as wild and weedy relatives, from which the entire range of domestic crops is derived components of agrobiodiversity, including genes, populations, species, communities, and ecosystems, as well as the landscapes in which agroecosystems are embedded (Altieri and Rogé 2009). Maintaining diverse ecosystems can improve soil health, pest control, and crop resilience to climate change (Montoya et al. 2020). Biodiversity is fundamental to agricultural productivity and resilience to climate change. Therefore, to safeguard food production, agriculture sustainability, and food security, there is a need to support the advancement of biodiversity conservation in agricultural landscapes to improve functionality and productivity through ecosystem services the biodiversity provides (Mamabolo et al. 2020). Conservation initiatives, such as the preservation of indigenous habitats and the promotion of agricultural techniques that are conducive to biodiversity, may bolster the ability of ecosystems to withstand and recover from disturbances while also facilitating the production of food in a sustainable manner. Climate-smart agriculture aims to increase productivity, enhance resilience, and reduce greenhouse gas emissions in agricultural systems (Lipper et al. 2014). There is a need for sustainable intensification of agriculture, which involves maximizing yields while minimizing negative environmental impacts (Tilman et al. 2011). It is important to balance biodiversity conservation with food production to ensure long-term sustainability (Brussaard et al. 2010). Therefore, there is a need to integrate mitigation and adaptation strategies to address the dual challenges of climate change and food security (Wilbanks and Sathaye 2007). This book chapter explores the interconnected nature of mitigation, adaptation, and biodiversity conservation in enhancing climate resilience in agriculture. It highlights innovative strategies and best practices for sustainable agricultural development in a changing climate. By integrating these approaches, we can build a more resilient agricultural sector capable of ensuring food security for future generations.

16.2 Climate Change Impacts on Agriculture

The fields, which were once a kaleidoscope of colorful crops, are now haunted by the quiet danger of climate change. The basis of our farming systems quakes under the intense heat and unpredictable showers. Changes in temperature, rainfall patterns, and extreme weather events are having a disastrous effect on crop output, upsetting entire agricultural systems and casting a long shadow on global food security (Gitarskiy 2019). Climate change's uneven distribution across different regions has both positive and negative effects on crop yields. For example, losses in maize crops have been observed in North America, Mexico, Africa, and China; gains in wheat crops have been observed in South America, Mexico, and South Asia; and losses in soybean and rice crops have been observed in the USA, Brazil, and Southeast Asia (Falconnier et al. 2020). While there may be possible improvements in higher latitudes, there are also major risks of losses in low-latitude areas, which might have an impact on many smallholder farmers' livelihoods and food security (Field and Barros 2014).

Global food security is seriously threatened by rising temperatures. According to studies, important crops including wheat, rice, and soybeans might see output decreases of 6–12% with just a 1 °C increase. Li et al. (2022) found a 10% reduction in maize yields with a 2 °C temperature increase. The fact that the warmest years on record have been after 2000 and that future estimates indicate significantly higher temperatures possibly reaching 4.8 °C by 2100 exacerbates this worry. This emphasizes how urgently research and adaptation plans are needed to lessen the effects of climate change on agricultural productivity and guarantee the security of the world's food supply. The rain's once-reliable cadence has become dissonant. Unpredictable weather patterns, such as floods and droughts, cause havoc in agricultural systems. Lesk et al. (2016) pointed out that droughts parch the land, leaving crops parched and thirsty. Conversely, floods submerge fields, killing crops and interfering with the cycles of planting and harvesting. According to a study by Döll et al. (2009), changes in precipitation brought on by climate change could result in a 20% decrease in world food output. Another layer of disturbance is brought about by extreme weather events, which are the pounding drum of this climate crisis. Heat waves, such as the one that ravaged Europe in 2018, have the power to wilt and hinder the growth of crops. The World Meteorological Organization (WMO) (2018) predicted that by the end of the century, the likelihood of these heatwaves could increase by four to five times. Furthermore, severe floods and storms can cause severe damage to infrastructure, making it difficult to store and transport agricultural goods, thereby endangering food security.

Climate change emerges as a major biophysical driver, with droughts and other shifts increasingly recognized as key forces shaping land use patterns (Paudel et al. 2020). Water scarcity further compels farmers to adapt their land use, often toward drought-resistant crops (Shiferaw et al. 2014). Additionally, infrastructure and market access play a role in influencing crop choices and land management practices (Sunderland and Rowland 2019; WoldeYohannes et al. 2018). Land use changes hold significant power to shape our food security and environmental well-being. Understanding how agricultural practices adapt to challenges like droughts, particularly in diverse geographical contexts, is crucial for building resilient rural communities (Paudel et al. 2019). Studies reveal a complex interplay of drivers influencing these changes, often categorized as socio-economic and biophysical (Shiferaw et al. 2019). On the one hand, agricultural expansion and intensification can boost food production, but at the potential cost of straining resources and ecosystems (Garg et al. 2020). Similarly, population growth drives the demand for more food and land, sometimes leading to unsustainable practices like deforestation (Zhou et al. 2020). Climate change, particularly droughts, stands out as a major concern due to its significant impact on rural livelihoods, especially in developing countries (Azadi et al. 2021). The navigation of complex dance between land use, food security, and environmental health helps to understand these diverse drivers and is essential for designing effective adaptation strategies and building resilient communities in the face of climate challenges (Tahmasebi et al. 2020). In the 1990s, water scarcity led to the abandoning of water-intensive crops. From 2005 to 2012, drought-tolerant pomegranate trees replaced 60% of farmland. By 2018, government initiatives and market forces diversified orchards to 90% of the land, with pomegranate, olive, and fig replacing traditional crops (Tahmasebi et al. 2020) (Fig. 16.1).

An important prerequisite for a healthy food production system is soil fertility, which is seriously threatened by rising temperatures. Lal (2009) issued a warning about possible deterioration and a drop in organic matter content, which could impair the structure of the soil and limit its ability to absorb nutrients. This presents a dire picture for crop production, as does the increased frequency and severity of extreme weather events like floods and droughts (Li et al. 2022). There is a siege against the delicate balance of life on our Earth. A key threat to soil health and fertility is the complicated dance of soil processes being disrupted by rising atmospheric carbon dioxide (CO_2), rising temperatures, and creeping salinity—all characteristics of climate change. Lal (2009) and Bardgett et al. (2021) have both drawn attention to the cascade impact and its potentially significant effects on agriculture and food security. Elevated CO_2 can directly promote plant development, but it also modifies the dynamics of soil moisture and might cause increased evapotranspiration, which could dry out soils in some areas, especially those that are already vulnerable to drought (Xu et al. 2019). Plants may become stressed as a result and find it more difficult to use the extra CO_2 for growth. Besides, raising the temperature can hasten the organic matter's breakdown in the soil, releasing nutrients more quickly (Suseela et al. 2012). Although this may appear advantageous at first, over time, it may result in the loss of vital nutrients, decreasing soil fertility. Furthermore, the variety and activity of soil microbes—the microscopic workhorses in charge of nutrient cycling and decomposition—can be adversely affected by rising temperatures (Liu et al. 2021). The availability of nutrients for plants may be further hampered by this disturbance in the microbial population. Moreover, salt might also damage soil microorganisms, which would further jeopardize the cycling of nutrients and the general health of the soil (Wang et al. 2021).

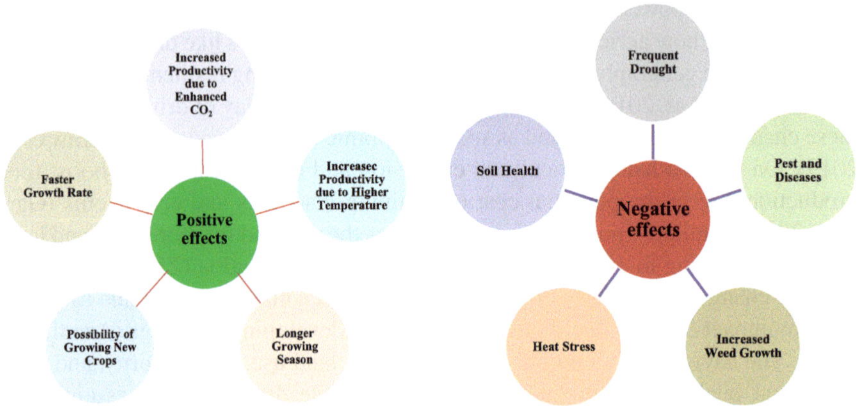

Fig. 16.1 Impacts of climate change on agriculture

Global climate change is expected to have a major impact on agricultural productivity, food security, and biodiversity (Morton 2007). The shift of plants latitudinally and altitudinally in search of favorable conditions is one predicted effect (Davis and Shaw 2001). The distribution of plants is already changing, and this tendency has already been noted in a number of locations, including Alaska, the Mediterranean, Scandinavia, and the Alps (Vermeulen et al. 2012). While species at the contracting edge of their range fear population decline and even extinction, those expanding at the leading edge of their range gain access to fresh area (Walther 2003). Due to competition and rising temperatures, species that are restricted by geography may have an improvement in reproduction and survival at their poleward and upper altitudinal limits as a result of rising temperatures, but they may face threats at their equatorial and lower altitudinal limits (Muluneh 2021). Oceanic and continental habitats may react differently, and certain species may become extinct as a result of their inability to adapt. On the other hand, different migration rates are causing new groups to emerge (Walther 2003). Even while historical distributions indicate that adaptation has not had much of an impact, continuous population differentiation emphasizes how strongly climatic change is exerting selective pressure (Izah et al. 2023). Temperature fluctuations altering precipitation patterns, variations in ocean temperature, and sea level rise are all direct effects of climate change on species and ecosystems. Changes in wildfire frequency and severity are examples of indirect effects. Due to this fragility, species become extinct in their existing habitats and colonize new areas, which may upend entire communities (Muluneh 2021). Due to their rich biodiversity and varied terrestrial ecosystems, the Melanesian islands, for instance, have recently experienced increased vulnerability to habitat degradation, fragmentation, and climate change. Physiology, behavior, and appearance are among the many changes that climate change is bringing about in species (Lavergne et al. 2010). Examples include the quicker growth and maturity of painted turtles in warmer years and the bodyweight loss of North American wood mice as a result of rising temperatures (Sheridan and Bickford 2011). Productivity and species composition have also been found to be impacted by variations in disturbance kinds, intensities, and frequencies (Kulakowski et al. 2011). Ecosystems are likewise impacted by extreme weather events and their aftereffects, such as landslides and wildfires. These effects are further intensified by human activity (Muluneh 2021).

The threat of starvation is quite real due to the relentless progress of climate change. Godfray et al. (2010) soberingly pointed out that the negative impacts of the global crisis on agricultural productivity pose a severe impediment to feeding the world's rising population and jeopardize the food security of millions of people globally. Extreme weather, broken agricultural systems, and diminishing crop yields are the threads that make up this menace. Our capacity to produce food is being severely hampered by rising temperatures, unpredictable rainfall patterns, and an increase in the frequency and severity of heatwaves, floods, and droughts. Crop yields can be considerably lowered by even a small increase in temperature, with staple crops like wheat and maize being especially vulnerable (Lobell et al. 2011). Floods submerge fields, killing crops and upsetting planting and harvesting cycles, whereas droughts parch the ground, leaving crops parched and withered (Lesk et al. 2016). When these factors come together, there may be severe food shortages, especially in areas where food insecurity is already a problem. When we take into account the expanding global population, the issue gets much more complicated. According to UN estimates, there will be 9.7 billion people on the planet by 2050, meaning that the food supply will need to rise significantly to keep up with demand. But these estimates are thrown off by climate change, which makes it harder and harder to close the gap between food production and food consumption. Serious repercussions from this mismatch could include widespread starvation, malnutrition, and societal upheaval (Arias et al. 2021). This impending food crisis is not affecting everyone evenly. The most vulnerable groups are probably going to be affected the most, especially those in emerging nations with little resources and a reliance on rain-fed agriculture. According to the Food and Agriculture Organization (FAO) (2019), millions of people are currently experiencing food insecurity as a result of climate change, and this figure is predicted to increase in the years to come.

16.3 Mitigation Strategies in Agriculture

Addressing the impact of greenhouse gas emissions in agriculture is essential to combat climate change and ensure the sustainability of our food systems. We can work toward a more resilient and environmentally friendly agricultural sector by implementing strategies to reduce emissions and promote sustainable practices. The significance of minimizing greenhouse gas emissions in agriculture cannot be overstated. Mitigating greenhouse gas emissions can result in a more resilient and environmentally friendly agricultural sector, making it essential for us to implement strategies aimed at reduction while also promoting sustainable practices. Given that agriculture is a significant contributor to greenhouse gas emissions, accounting for a substantial portion of the overall emissions globally, it is imperative to implement comprehensive and effective measures to reduce its environmental impact. For instance, the use of synthetic fertilizers in agriculture releases nitrous oxide, a potent greenhouse gas, into the atmosphere, contributing to the exacerbation of climate change. Therefore, adopting sustainable agricultural practices that focus on slow-release/controlled-release fertilizers coupled with precision farming techniques can significantly reduce emissions and promote a more sustainable food production system for future generations. In addition, reducing greenhouse gas emissions in agriculture can also have positive impacts on soil health, water quality, and overall ecosystem resilience.

16.3.1 Reducing Greenhouse Gas Emissions in Agriculture

Lowering greenhouse gas emissions in agriculture is vital for reducing climate change and protecting biodiversity (Pasricha and Ghosh 2023). Agriculture is a substantial contributor to global greenhouse gas emissions, with agriculture-related emissions accounting for around 18% of total world emissions (Ritchie et al. 2020). A staggering 21% increase in GHG emissions from agriculture was reported over the last three decades (International Energy Agency (IEA) 2021). Climate-smart agriculture (CSA) is a method that tries to boost production, strengthen resilience, and minimize greenhouse gas emissions. CSA encompasses a range of methods and technology customized to specific agroecological circumstances and socio-economic situations, such as conservation agricultural techniques, agroforestry, precision farming, water management measures, and enhanced livestock management. Tillage techniques have a major influence on greenhouse gas (GHG) emissions in cropping systems, with varied percentage changes seen in different studies (Zaman et al. 2021). No-tillage strategies have been associated with an 8.64% decrease in GHG emissions compared to conventional tillage methods in a rice-wheat cropping system (Choudhary et al. 2020) (Fig. 16.2).

Conversely, extensive tillage practices can lead to a considerable rise in GHG emissions owing to soil disturbance and organic matter decomposition under conventional practices (Behnke et al. 2018). Studies have found a 20% drop in GHG emissions in maize-producing systems using decreased tillage methods. Additionally, decreased tillage has been related to a 15% reduction in nitrous oxide (N_2O) emissions compared to traditional tillage practices (Valujeva et al. 2022). These findings underline the necessity of adopting appropriate tillage strategies to decrease GHG emissions and promote sustainable agriculture systems. Zero tillage helps to maintain a protective layer of crop residue on fields, reducing evaporation and runoff, which helps reduce soil erosion and conserve soil moisture. The lack of frequent soil interruption under the no-till system prevents the oxidation of soil organic carbon over traditional tillage systems. Li et al. (2019) documented the reduction in methane emissions in direct-seeded rice under a no-till system upon converting from puddling in China. A significant reduction in CH_4 emissions by 10–14% was reported when the alternate wet and dry method was adopted over conventional

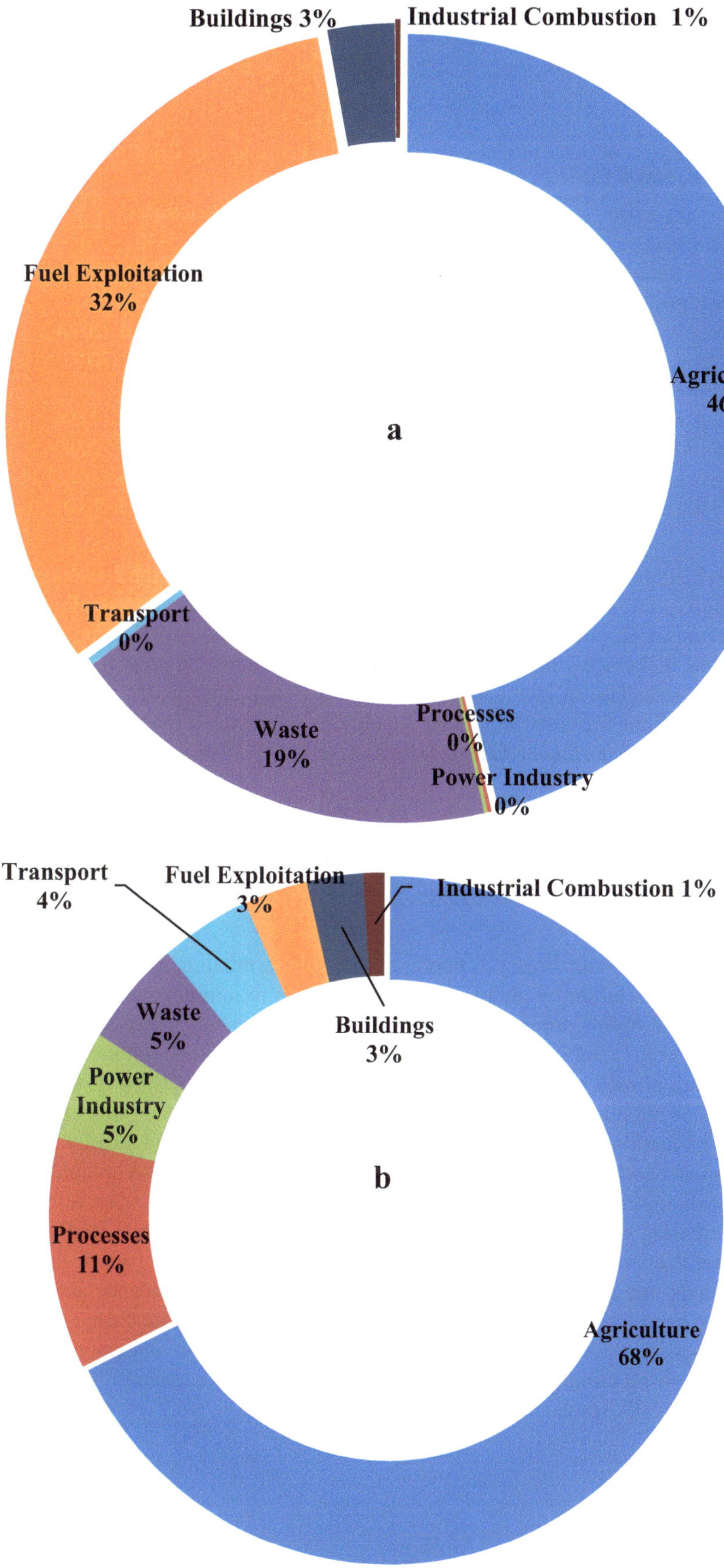

Fig. 16.2 Global methane (**a**) and nitrous oxide (**b**) emissions by sector

Table 16.1 Effects of different tillage systems on CH_4 emissions (kg CH_4-C ha^{-1}) and N_2O emissions (kg N_2O-N ha^{-1})

Crop/cropping system	Conventional tillage	No tillage	References
CH_4 emissions (kg CH_4-C/ha)			
Rice (South Korea)	0.038	0.028	Ali et al. (2009)
Late rice (China)	0.010	0.003	Li et al. (2011)
Rice (India)	0.66	0.52	Ahmad et al. (2009)
Rice-soybean (Brazil)	0.25	0.40	Metay et al. (2007)
N_2O emissions (kg N_2O-N/ha)			
Corn-soybean-wheat (USA)	1.20	1.32	Grandy et al. (2006)
Soybean-wheat (Brazil)	1.16	1.19	Jantalia et al. (2008)
Corn-soybean-alfalfa (USA)	7.58	7.74	Parkin and Kaspar (2006)
Forage grass-rice (Brazil)	8.69	6.12	Passianoto et al. (2003)

flooding in Tamil Nadu of South India (Oo et al. 2018). Similarly, the conversion of crop residues into biochar through pyrolysis and its soil addition successfully offset the N_2O and CH_4 emissions by 10.9% and 28.4%, respectively, over no residue application in the rice-wheat system of Indo-Gangetic plains (Singh et al. 2023b). A significant reduction in methanotrophic microbial abundance under alternate wet and dry conditions resulted in aerobic conditions unfavorable for the microbiome and reduced methane fluxes (Gangopadhyay et al. 2023) (Table 16.1).

16.3.2 Sustainable Land Management and Soil Carbon Sequestration

Land use management strategies play a crucial role in minimizing climate change impacts on agriculture by giving sustainable options to minimize greenhouse gas emissions and boost carbon sequestration. Practices such as conservation agriculture, agroforestry, and organic farming are helpful in limiting emissions of carbon dioxide and nitrous oxide through strategies including minimal soil disturbance, crop residue retention, and diverse cropping systems. These methods not only aid in lowering emissions but also contribute to enhancing soil health and fertility, resulting in greater carbon storage in soils. Efficient water management techniques, such as drip irrigation and rainwater collection, contribute to maximizing water resources and decreasing energy use for irrigation, hence reducing the carbon footprint of agricultural operations. Additionally, boosting biodiversity conservation through agroecology and integrated pest control increases ecosystem resilience and minimizes the need for chemical inputs, further contributing to climate change mitigation efforts. Sustainable land management methods, present a comprehensive approach toward climate-friendly agriculture that is both ecologically sustainable and economically viable.

The land use management strategies including conservation agriculture, organic farming, and agroforestry by adopting crop diversification into the systems would be a stepping stone in the process of climate change mitigation. A long-term study conducted by Alletto et al. (2022) demonstrated that crop diversification within a field can increase soil organic carbon stocks by 8% and enhance soil health by 45%, a high spot to amplify the claims of crop diversification. Additionally, diversified crop rotations can reduce net CO_2-eq emissions by 106.8 ± 31.7 million tonnes annually, offsetting 5.6% of the annual GHG emissions associated with China's food system (Yang et al. 2024). Diversified cropping systems can enhance soil health, increase resilience to climate variability, and reduce greenhouse gas emissions. Additionally, sustainable land use systems can improve food security by ensuring reliable water supplies and soil integrity, especially in the face of extreme climate events like droughts (Reay 2020). These practices are essential for meeting climate targets and reducing the conversion of natural ecosystems, particularly tropical deforestation and degradation.

Carbon sequestration in agriculture has a dual function in reducing climate change and boosting food security. By absorbing and storing carbon dioxide from the atmosphere in soil and plants, carbon sequestration helps to decrease greenhouse gas emissions, therefore helping to mitigate climate change. This method not only serves to offset carbon emissions but also improves soil health, fertility, and water retention capacity, making agricultural systems more adaptable to climatic unpredictability. Enhanced soil fertility resulting from carbon sequestration measures leads to enhanced crop production and better agricultural outputs, eventually boosting food security. Studies have indicated that organic farming techniques may lead to a modest rate of carbon sequestration at 0.07 Mg C ha^{-1} yr^{-1}, while turning a section of grassland into arable land can result in a higher rate of 0.18 Mg C ha^{-1} yr^{-1}, both domestically and internationally. Over the years, agricultural greenhouse gas emissions have dropped by 108 Mt CO_2-eq, illustrating the potential influence of carbon sequestration strategies in agriculture. Sustainable agriculture methods including cover crops, decreased tillage, and plant variety are known to enhance carbon absorption in soil, leading to emission reductions. The US agriculture sector's emissions account for around 9% of all US GHG emissions, stressing the need to develop techniques that promote carbon sequestration to prevent climate change consequences. Globally, agriculture has the technological capacity to sequester up to around 5.3 Gt CO_2 per year by 2030, demonstrating the enormous role agriculture may play as a net-negative emission sector.

By sequestering carbon in the soil, biochar may help remove 300–660 million tonnes of carbon dioxide from the atmosphere worldwide by 2050, with studies indicating an increase of up to 20% in organic carbon in the soil when employing biochar. Additionally, biochar may lower nitrous oxide emissions from the soil by 12–50%, significantly improving the environment (Karan et al. 2023). The use of biochar not only assists in carbon sequestration but also promotes crop yields by up to 42%, improves soil fertility, raises nutrient levels, and increases water retention capacity (Gupta et al. 2020). Furthermore, biochar has the potential to trap one billion tonnes of carbon yearly if all agricultural leftovers worldwide were turned into

Table 16.2 Carbon sequestration rates in different production systems using biochar

Biochar feedstock	Pyrolysis temp. (°C)	System	CO_2 sequestration increased by	References
Wheat straw	350–550	Rice-based cropping system	14–57%	Zhang et al. (2010)
Bamboo leaf	500	Chinese chestnut forest	9.4% (1-year period)	Singh et al. (2020)
Invasive weed	450	Coal mine degraded land	13–91%	Ghosh and Maiti (2023)
Wood chips	500	Wheat-based cropping system	50% (10-year period)	Lehmann et al. (2021)
Sugarcane bagasse	400–800	Calcaric dark red soil	50–70%	Kameyama et al. (2012)

biochar, offsetting nearly 80% of greenhouse gas emissions from agriculture. These results underline the essential importance of biochar in sustainable agricultural techniques for climate change mitigation. Biochar's long-term stability in soil enables it to resist decomposition and persist in the soil for longer periods, functioning as a carbon sink by keeping CO_2 out of the atmosphere for over a thousand years (Glaser et al. 2009). Additionally, biochar promotes soil fertility, improves water retention capacity, raises nutrient levels, and increases crop yields, making it a great tool for sustainable agricultural techniques (Singh et al. 2018, 2023a, b) (Table 16.2).

16.3.3 Efficient Livestock and Manure Management

Efficient livestock and manure management play a significant role in boosting food security and reducing climate change. Livestock supply chains account for 7.1 Gt CO_2-eq, equal to 14.5% of worldwide anthropogenic greenhouse gas emissions, with cattle (beef, milk) accounting for nearly two-thirds of that amount, mostly owing to methane emissions originating from rumen fermentation (Shields and Orme-Evans 2015). Enteric methane emissions contribute 30% of worldwide methane emissions. To address climate change caused by animal agriculture, a variety of mitigation techniques have been suggested, including enhancing the productivity and efficiency of livestock production via improved nutrition and genetics. Improved animal health and lifespan, as well as improved feed management, may contribute to higher output and fewer emissions. Manure management strategies that guarantee the collection and recycling of nutrients might further contribute to mitigation. Sourcing low-emission intensity inputs (feed and energy in particular) is another approach for minimizing emissions. Livestock supply chains contribute to 14.5% of worldwide anthropogenic greenhouse gas emissions, with cattle being a significant contributor owing to methane emissions from rumen fermentation (Grossi et al. 2019). Manure storage accounts for around 10% of total emissions, and improved manure management may lead to large reductions in methane and nitrous oxide emissions by up to 55% and 41%, respectively (EPA 2022). Strategies such as minimizing manure storage time, composting, daily spreading, and pasture-based management may considerably decrease methane emissions from livestock operations (Frimawaty et al. 2023). Livestock production systems create methane (CH_4) and nitrous oxide (N_2O) emissions from manure, with the relative significance of these gases dependent on manure composition, management strategies, and climatic conditions (Petersen et al. 2013). Strategies such as composting, manure drying, semipermeable coverings, and shorter manure storage periods may considerably reduce methane emissions from animal operations (Rivera and Chará 2021). These measures may lead to reductions of up to 50% in CH_4 or N_2O emissions from manure, contributing to climate change mitigation efforts. Additionally, using animal manure correctly may lead to decreased natural gas use and imports, helping climate change mitigation efforts. Implementing good manure management procedures is vital for lowering greenhouse gas emissions from animal operations and developing sustainable agriculture practices.

16.3.4 Agroforestry and Afforestation as Carbon Sinks

Agroforestry and afforestation serve as essential techniques in mitigating climate change by serving as carbon sinks and lowering greenhouse gas emissions. Agroforestry systems, which integrate trees with crops and cattle, have been demonstrated to retain more atmospheric carbon in plant parts and soil than traditional farming. Studies suggest that agroforestry may considerably enhance soil organic carbon, with transitions from agriculture to agroforestry resulting in an average 34% increase in soil organic carbon. Similarly, the conversion from pasture/grassland to agroforestry leads to around a 10% increase in soil organic carbon (Kay et al. 2019). Agroforestry methods not only play a key part in global carbon sequestration but also help in the long-term storage of atmospheric carbon dioxide, thereby contributing to limiting global warming. Correspondingly, afforestation, the purposeful introduction of trees into agricultural landscapes, has been recognized as a significant natural climate solution. Afforestation may successfully sequester carbon in both aboveground biomass and soil, absorbing carbon dioxide from the atmosphere and storing it for the long term (Tefera et al. 2019). By increasing carbon sequestration, afforestation systems provide various advantages such as increased soil health, biodiversity protection, better ecosystem services, and sustainable agricultural output. These activities emphasize the tremendous potential of agroforestry and afforestation in addressing climate change and promoting sustainable land use practices.

16.4 Adaptation Strategies in Agriculture

16.4.1 Crop Diversification and Resilient Varieties

Crop diversity is a significant method to prevent climate change impacts and promote food security, as proven by many study results. Douyon et al. (2022) noted that crop diversity may increase household food security by supplying various crops that farmers may not access owing to cost or other restrictions, hence boosting resilience in rural communities depending on local production. Additionally, research by Ijaz et al. (2019) and Adjimoti and Kwadzo (2018) underlined the favorable influence of crop diversity on household food security status and its dynamic role in providing sustained food security. Furthermore, research by Mapemba et al. (2018) revealed that crop variety increases agricultural productivity, food consumption, and nutritional outcomes, particularly in smallholder agriculture. Moreover, crop diversity has been demonstrated to promote technical efficiency, lessen climate change implications for farmers, and lead to better health and nutrition outcomes (Omotoso and Omotayo 2024). While the economic costs of diversifying agricultural systems might occasionally exceed ecological advantages, policies encouraging diversified and sustainable production systems are vital for guaranteeing ecological, economic, and social sustainability (Mzyece and Ng'ombe 2021; Vernooy 2022). The impact of crop diversity in enhancing livelihoods, health, nutrition, household food security, climate resilience, and ecological sustainability is substantial and highlights the necessity of supporting varied agricultural systems at many levels. The outcome underlines numerous advantages of crop variety in tackling climate change issues and maintaining sustainable food production systems. Resilient crop types are vital in minimizing the consequences of climate change and boosting food security, as stated in different study papers. The creation of climate-resilient crop varieties is vital to meet the difficulties presented by climate change on global food security and agricultural productivity (Dhankher and Foyer 2018). These robust cultivars are distinguished by features like high-temperature tolerance, decreased water needs, resistance to pests and diseases, and higher nutritional content by biofortification. Studies underline that heat stress and lower water availability due to climate change may lead to decreased agricultural yields and quality, making the development of robust crops a priority for sustainable agriculture. Notable examples are the Tutti apple variety produced for hot-temperature locations and Green Super Rice (GSR) intended to manage climate concerns while assuring high production potential and resilience (Benitez-Alfonso et al. 2023). The redesigning of crop varieties to resist climate-related pressures is vital in guaranteeing agricultural sustainability and food security in the face of changing climatic conditions (Pixley et al. 2023; Benitez-Alfonso et al. 2023). These creative breeding initiatives attempt to generate crops that can tolerate climate-related challenges and retain production under changing environmental circumstances. Redesigning crop varieties to respond to climate challenges is vital for guaranteeing agricultural sustainability and food security in the face of developing climatic circumstances (Kopeć 2024). By integrating

Table 16.3 Crop varietal traits developed to mitigate climate change and enhance food security

S. no.	Crop	Description
1	Tutti apple	A hot climate apple variety developed for areas with high temperatures, known for its flavor and sustainability
2	Green super Rice (GSR)	High-yielding rice varieties designed to address climate challenges, combining drought tolerance and disease resistance
3	Climate-resilient maize	Maize varieties with enhanced drought tolerance to mitigate yield losses in water-scarce regions
4	Heat-tolerant wheat	Wheat varieties developed for heat resistance to maintain productivity under rising temperature scenarios
5	Zinc-enriched rice	Rice varieties biofortified with zinc to combat nutrient deficiencies and improve human health outcomes
6	Iron-fortified beans	Beans enriched with iron to address malnutrition and enhance public health outcomes
7	Drought-tolerant sorghum	Sorghum varieties bred for drought tolerance to ensure stable yields in water-stressed conditions
8	Pest-resistant soybeans	Soybean varieties resistant to pests and diseases, ensuring sustainable production under changing pest dynamics
9	Climate-resilient barley	Barley varieties with improved heat tolerance and disease resistance for sustainable cultivation
10	Nutrient-enhanced maize	Maize varieties biofortified with essential nutrients to improve nutritional content and combat malnutrition
11	Disease-resistant tomatoes	Tomato varieties bred for disease resistance, ensuring stable production despite changing disease pressures
12	Climate-adaptive potatoes	Potatoes adapted to changing climates, maintaining productivity under varying temperature and water conditions
13	Resilient quinoa	Quinoa varieties resilient to heat stress and water scarcity, ensuring stable yields in challenging environments
14	Disease-tolerant groundnut	Groundnut varieties resistant to diseases, ensuring sustainable production despite changing disease dynamics
15	Climate-adapted cabbage	Cabbage varieties adapted to changing climates, maintaining quality and yield under varying environmental conditions
16	Resilient lentils	Lentil varieties resilient to heat stress and pests, ensuring stable production in challenging climates
17	Pest-resistant maize	Maize varieties resistant to pests, ensuring sustainable yields even under changing pest pressures
18	Climate-adaptive carrots	Carrot varieties adapted to climate changes, maintaining quality and yield under varying environmental conditions

characteristics that boost nutritional content and tolerance to environmental pressures, biofortified crop varieties provide a complete strategy for tackling food security concerns in places sensitive to climate change impacts (Table 16.3). The coupling of nutritional improvement with resilience characteristics underlines the promise of contemporary crop breeding approaches in promoting sustainable agriculture, boosting food security, and enhancing human health outcomes within a changing climatic scenario.

16.4.2 Improved Water Management and Irrigation Techniques

Improved water management strategies play a crucial role in minimizing the consequences of climate change on agriculture. At the agricultural level, improved land and water management practices may boost carbon storage in soils, adding considerably to climate change mitigation efforts. Water availability is a vital aspect of agricultural output, and variations in water access considerably affect agriculture. Globally, irrigated agriculture accounts for the bulk of freshwater withdrawals, underscoring the significance of effective water management methods to meet water shortages aggravated by climate change (Frimpong et al. 2023). Efficient soil conservation measures, rainwater gathering, and enabling farmer groups to maintain irrigation systems are critical parts of sustainable water management in agriculture. Water-smart crop farming is a critical method that focuses on improving water resource usage while boosting agricultural output. This method comprises choosing drought-resistant agricultural types, adopting modern irrigation techniques like drip systems, and encouraging nature-based solutions to promote climate resilience. Smallholders are urged to employ modern irrigation methods like drip or sprinkler systems to avoid water loss and enhance water efficiency (He and Rosa 2023). Moreover, sustainable water management strategies are vital for adjusting to climate change concerns in agriculture. Implementing techniques such as additional irrigation during important cropping times, adopting drought-resistant crop types, and managing sustainable water sources like rainwater collecting are essential steps toward guaranteeing agricultural sustainability against changing climatic circumstances (Chartzoulakis and Bertaki 2015). Collaborative efforts among stakeholders and the implementation of climate-smart farming methods are crucial for long-term success in minimizing the consequences of climate change on agriculture.

Pressurized irrigation systems play a crucial role in minimizing climate change effects on agriculture by boosting water efficiency and sustainability. Studies suggest that moving to current pressured systems like drip and spray irrigation may lead to better water productivity, lower energy consumption, and improved agricultural yields per unit of water used (Rosa 2022). Sustainable irrigation methods not only lower water consumption but also help facilitate agricultural intensification and global food security (Nikolaou et al. 2020). By enhancing irrigation efficiency using pressured systems, agriculture can adjust to changing climatic conditions and boost resistance to water shortage issues. Drip irrigation systems are helpful in climate change mitigation in agriculture by boosting water efficiency and crop output. Research reveals that drip irrigation may greatly boost agricultural yields compared to conventional systems, with some instances exhibiting twice or triple the output. Additionally, drip irrigation saves water use by 20–60% while delivering accurate water distribution to plants, reducing resource utilization (El-Fakharany and Salem 2021). This sustainable technique not only enhances water productivity but also adds to environmental sustainability by avoiding water loss and runoff. By utilizing drip irrigation, farmers boost crop yields, minimize water use, and adjust to changing climatic circumstances efficiently. Similarly, sprinkler irrigation systems serve a significant role in minimizing climate change effects on agriculture by providing regulated water application to crops, comparable to rainfall. Compared to gravity irrigation, sprinkler systems give greater control over water scheduling and levels, leading to increased water efficiency and sustainability (Frisvold and Bai 2016). Research indicates that sustainable irrigation approaches, including sprinkler systems, may help reduce heat stress, enhance water production, and boost soil moisture, all crucial climate adaptation strategies in a water-scarce world (Rosa 2022). By adopting sprinkler irrigation in the thrust areas, farmers can optimize resource utilization, boost crop yields, and react to changing climatic circumstances effectively.

Advanced irrigation techniques involving artificial intelligence (AI) through real-time soil matric potential monitoring are helpful tools in reducing climate change consequences on agriculture. By continually tracking fluctuations in soil parameters, such as moisture content, and different contaminants, real-time monitoring allows rapid decision-making for system control and management. This technique revolutionizes field measurements by giving critical information for improving agricultural operations and boosting soil health. Studies have demonstrated that real-time continuous soil monitoring (RTCSM) may update strategies for soil process models, control techniques, digital soil mapping, and climate change mitigation, adding considerably to sustainable agricultural practices (Maia et al. 2022). Machine learning methods have been applied to assess soil matric potential in the root zone based on remotely sensed evapotranspiration data (Fan et al. 2022). These models can learn the connections between soil, crop, and weather-related elements to forecast soil moisture levels correctly. By integrating information and communication technology (ICT) in agriculture and installing IoT soil moisture sensors for irrigation scheduling, farmers may maximize water usage efficiency and boost crop output. Remote sensing methods paired with machine learning algorithms provide a potential strategy to monitor soil moisture levels in large-scale agricultural operations efficiently.

16.4.3 Sustainable Agroecological Practices

Implementing sustainable agroecological techniques is vital for limiting climate change implications in agriculture. Organic agriculture, defined by the elimination of synthetic fertilizers and pesticides, has been found to dramatically lower greenhouse gas (GHG) emissions compared to traditional agricultural practices. The latest research by Franzluebbers and Hendrickson (2024) indicated that organic farming methods may cut carbon dioxide (CO_2) emissions by up to 40% and nitrous oxide (N_2O) emissions by 70%. Additionally, biochar incorporation in agriculture has attracted attention as a possible approach to trap carbon in soil and prevent climate change (Singh et al. 2018). Studies by Lehmann and Joseph (2015) and Goswami et al. (2021) indicated that biochar application may improve soil carbon sequestration by up to 200% relative to nonamended soils. Furthermore, integrating agricultural leftovers as mulch may boost soil organic carbon content and minimize soil erosion, helping with climate change mitigation. Research done by Lal (2004) found that applying agricultural leftovers as mulch may enhance soil organic carbon by 15–40%, resulting in better soil fertility and lower GHG emissions. These agro-ecological approaches have various advantages beyond climate change mitigation. Organic agriculture boosts biodiversity, soil health, and water conservation (Mäder et al. 2002; Sapinas and Abbott 2020), whereas biochar application enhances soil fertility, nitrogen retention, and crop yield (Spokas et al. 2012; Roy et al. 2022). Similarly, applying agricultural leftovers as mulch increases soil structure, moisture retention, and weed control (Giller et al. 2009; Corbeels et al. 2020). By incorporating these techniques into agricultural systems, farmers may contribute to climate change mitigation while boosting the resilience and sustainability of their enterprises. However, widespread adoption of these approaches needs supporting policies, extension services, and financial incentives to overcome constraints such as restricted availability of organic inputs, technical expertise, and upfront investment expenditures (Pretty et al. 2012; Muhie 2022). Additionally, other studies are required to maximize the adoption of these practices across varied agroecological settings and quantify their long-term implications on GHG emissions, soil health, and agricultural output.

16.4.4 Risk Reduction and Early Warning Systems

Risk reduction and early warning systems play a significant role in minimizing the implications of climate change in agriculture. Disaster risk reduction entails utilizing risk analysis, weather warnings, and good agricultural practices to guide farm management, assist response and recovery initiatives, and build resilience in farming operations. Integrating disaster risk reduction into agricultural development ensures that measures, responsible agencies, and budget allocations are linked to connect humanitarian relief with long-term development goals (FAO 2020). Climate change worsens climate-related dangers such as severe heat, floods, droughts, and storms, hurting food production significantly, particularly in underdeveloped countries. Early warning systems are vital for limiting the detrimental consequences of severe climatic events by providing timely information for decision-making processes. However, there is an agreement that these systems are not completely reaching their promise owing to problems in decision-making processes rather than early warnings themselves (Smith 2015).

Diversifying food sources, developing early warning systems for extreme weather events, increasing organic carbon in the soil for better water retention, promoting education on food preservation techniques, investing in climate-resilient food crops, and providing smallholder farmers with access to credit and services are all critical strategies for enhancing food security in the face of climate change. These activities not only decrease risks connected with climate-related catastrophes but also contribute to strengthening resilience in agricultural systems (Kirubanandan

2023). Investing in disaster risk reduction not only saves lives but also aids sustainable agricultural growth by minimizing rehabilitation expenses post-disasters. Disaster risk reduction links with climate change adaptation strategies to minimize exposure and susceptibility and build resistance to severe climatic impacts. Strengthening early warning systems involves efficient communication, customized information for diverse stakeholders, investment in strengthening decision-making processes, and local capacity development to guarantee prompt action in response to climate extremes (UNEP 2023). By applying these measures and investing in early warning systems with efficient decision-making processes, the agricultural industry can greatly strengthen its resistance to climate change effects and help to guarantee food security despite changing climatic circumstances.

16.5 Integrating Biodiversity Conservation

16.5.1 The Role of Biodiversity in Enhancing Resilience

Biodiversity plays a crucial role in enhancing climate resilience in agriculture as it promotes various ecosystem services such as pollination of crops, biological pest control, maintenance of soil structure and fertility, and nutrient cycling, all of which are essential for agricultural productivity and resilience to climate change. Benefits of biodiversity in agriculture include pest and disease control where a biodiverse system with diverse fauna, fungi, and microorganisms helps regulate pest populations naturally, reducing the need for synthetic pesticides; soil health, where biodiversity in soil microbiomes enhances nutrient cycling, soil structure, and resilience, leading to healthier and more productive soil, higher yields, and better crop quality; and genetic material repository, where biodiversity fosters genetic diversity in crops, enabling adaptation to environmental challenges and facilitating the development of new crop varieties suited to local conditions. Climate change affects agriculture in a number of ways, with farmers facing increasingly erratic rainfall, altered seasonal weather patterns, and more frequent adverse weather events such as floods and droughts, all of which have a direct impact on land productivity by affecting agricultural crops and an indirect impact via water and soil resources, exacerbating soil erosion. Practices that strengthen resilience to these stresses show the uses of biodiversity at three different scales: the landscape, the farming system, and the species or variety (Mijatović et al. 2013). At the landscape scale, agricultural activities form diverse land patterns, including home gardens, fields, orchards, forests, fishing spots, and pastures; ecosystem degradation undermines climate resilience due to population pressure, overuse, and expanding agricultural areas. Thus, protecting and restoring ecosystems enhance resilience against droughts, floods, and extreme weather, with coastal ecosystem restoration like mangroves helping counteract extreme weather risks and reduce saltwater intrusion and rehabilitating water and soil resources mitigating drought impacts, as seen in traditional practices combined with modern methods that proved successful in landscape restoration in Ethiopia's Tigray region (Edwards et al. 2010). At the farming system level, resilience-building practices aim to minimize yield losses by diversifying crop species, integrating crops and livestock, and implementing adjustments such as crop rotation, irrigation, and sustainable soil management; diversified farming often includes low-input methods for soil fertility and water conservation, with adaptation strategies like adjusting cropping calendars (Wajih 2008) and utilizing crops like tamarind leaves to combat soil salinity in regions like Jaffna, Sri Lanka (Vakeesan et al. 2008). At the level of species/variety, resilience is enhanced by cultivating fast-growing and stress-tolerant crop varieties while safeguarding diversity in agricultural systems, such as in Niger and Mali where farmers increasingly opt for fast-growing varieties like pearl millet and sorghum to cope with erratic rainfall (Kouressy et al. 2003; Bezançon et al. 2009); in India, community seed banks are established to access flood- and salt-tolerant rice varieties in flood-prone areas, with communities engaging in selection and breeding projects to improve crop stress tolerance; in Nepal, traditional rice landraces, including drought-resistant varieties, are reintroduced through a participatory plant breeding program conserving 69 rice varieties (Ulsrud et al. 2008). Therefore, preserving and promoting agricultural biodiversity are fundamental for building climate resilience in the agricultural sector.

16.5.2 Agroecological Approaches to Conserve Biodiversity

Agroecological approaches play a significant role in conserving biodiversity in agriculture, emphasizing the interrelatedness of all agroecosystem components and the complex dynamics of ecological processes. Agroecology builds biodiversity into agriculture and food systems, making it a key strategy for global biodiversity conservation. By incorporating practices such as agroforestry, intercropping, and diversified cropping systems, agroecological approaches enhance agricultural biodiversity, provide habitat for beneficial species, and contribute to the sustainable use of biodiversity while supporting productivity and resilience. Some of the agroecological approaches that conserve biodiversity include crop rotations, conservation agriculture with minimum tillage and crop residue mulch, cover crops grown instead of maintaining bare fallow, organic crop production prohibiting the use of pesticides and synthetic fertilizers, integrated pest management, integration of livestock and crops in cropping systems, and field margin practices like conservation headlands, hedgerows, and wildflower strips. Pasture-fed livestock systems, multi-paddock grazing, organic grazing, tree intercropping, multistrata agroforestry, permaculture, silvopasture, rewilding, and agricultural land abandonment also contribute to biodiversity conservation by minimizing soil tillage, increasing plant diversity, and promoting ecological regeneration, thus highlighting the importance of agroecology in sustainable agriculture and biodiversity conservation efforts (Burgess et al. 2024).

16.5.3 Ecosystem-Based Adaptation Strategies

Ecosystem-based adaptation (EbA) entails utilizing biodiversity and ecosystem services within adaptation strategies to counter the adverse impacts of climate change. Healthy ecosystems contribute significantly to reducing vulnerability to climate change and disaster risks by fulfilling two key functions: Physical protection: Ecosystems, such as wetlands, forests, and coastal areas, act as natural barriers or buffers, reducing exposure to hazards and mitigating their impacts. Socio-economic support: Beyond their protective functions, ecosystems sustain human livelihoods by providing essential resources like food, medicine, and construction materials, thereby enhancing resilience to disasters. Examples of ecosystem-based disaster risk reduction measures include the restoration of coastal vegetation like mangroves to shield shorelines from storm surges, management of invasive species to prevent land degradation and protect food security, and strategic ecosystem management to complement and extend the lifespan of built infrastructure investments. Many ecosystem-based disaster risk reduction initiatives align with EbA efforts. For instance, enhancing the functionality of protective forests is a key component of climate protection programs in various countries. Given forests' crucial role in mitigating natural hazards, these programs aim to enhance forest stability, promote species diversity, encourage natural regeneration, prevent wildfires, and control pests and diseases to reduce risks effectively (Lo 2016).

16.5.4 Synergies Between Agriculture, Biodiversity, and Food Security

Outlined are various recent strategies to enhance food security and preserve biodiversity, along with an examination of science's potential role in bridging the gap between them (Brussaard et al. 2010). Scientific efforts to bolster food production primarily reside in production ecology, where disciplines like crop ecophysiology, plant breeding, nutrition, and crop protection, buoyed by technological advancements like mechanization, have substantially augmented light, water, and nutrient-use efficiencies, expanding cropping intensities for major crops globally, epitomized by the green revolution of the mid-twentieth century (Tilman et al. 2002). Concurrently, research on livestock physiology, breeding, nutrition, and disease management, as well as on fish and aquaculture, has yielded comparable advancements. Despite the availability of scientific knowledge and technologies, realizing potential increases in food production remains elusive under field conditions in most countries, hindered by factors like deficient education and health systems, dysfunctional markets, political instability, and the absence of alternative livelihood options for transitioning from subsistence to market-oriented agriculture. Conversely, in biodiversity conservation, significant scientific strides have emerged from conservation biology, focusing on habitat requirements, emblematic species, and Red List species management, alongside prioritizing conservation actions, especially in biodiversity hotspots (Sodhi et al. 2010). However, despite the existence of over 130,000 protected areas covering nearly 14% of the Earth's surface, biodiversity decline persists due to habitat fragmentation, resource overexploitation, pollution, climate change, and invasive species proliferation, partly stemming from poor management and commercial expansion. To mitigate persistent hunger and biodiversity loss, two contrasting approaches emerge: an intensive agriculture model, emphasizing enhanced resource-use efficiency and mechanization to expand food production while conserving biodiverse areas, and an eco-agriculture approach, emphasizing landscape-level biodiversity's centrality in sustaining agricultural production and ecosystem services, advocating for diversified, extensive, and multifunctional agriculture with regionalization over globalization (Scherr and McNeely 2008).

16.5.5 Breeding for Climate-Resilient Crops

Breeding resilient crops to combat climate change is becoming increasingly important. Traditional breeding methods are slow, but advancements in genome sequencing now enable swift manipulation of crops. CRISPR/Cas, particularly, holds promise for revolutionizing crop development by targeting specific genome sites. Other approaches like mutation breeding, marker-assisted selection (MAS), genetic engineering, and RNA interference (RNAi) are also employed (Gaba et al. 2021). Genome editing, including techniques like zinc finger nucleases (ZFNs) and CRISPR/Cas, presents significant opportunities for improving crop tolerance to environmental stresses. Advanced phenotyping methods coupled with genome-wide analysis allow for better adjustment of physiological traits to unstable environments. CRISPR/Cas9 presents opportunities for enhancing tolerance to water deficit and heat stress in cereal crops, as highlighted by Massel et al. (2021). While gene editing holds promise, it is still in its early stages for crop improvement. Crop wild relatives, with natural tolerance to various stresses, offer genetic variability crucial for resilience in cultivated crops. Root architecture, critical for water and nutrient absorption, is gaining importance in breeding for stress tolerance. However, measuring root traits remains challenging. Despite this, advances in phenotyping offer avenues for holistic selection strategies, especially in enhancing water-use efficiency (WUE) (Ober et al. 2021). As climate change intensifies, breeders must adapt to increased environmental variability, including extreme weather events and shifts in pest and pathogen distributions. Utilizing genomic resources and bioinformatics platforms aids in identifying genes underlying stress responses, opening new avenues for crop improvement through genome editing. Despite challenges, there are ample opportunities to accelerate genetic gains in crop improvement programs. The adoption of impactful technologies by breeders and farmers, along with regulatory and financial support from governments, will be crucial in meeting the increasing demands for food production in the coming years (Langridge et al. 2022).

16.5.6 Genetic Diversity and Climate Adaptation

Exploring gene banks (Blair et al. 2012) and uncovering biodiversity in nature (Boyle et al. 2017) are crucial for addressing food shortages caused by rising temperatures and drought. Integrating climate data and genomics can accelerate the identification of new sources of tolerance. By incorporating advanced analytical frameworks like GP and ML, we can unlock these variants and transfer them into cultivated gene pools for farmers' use. Gene bank research should focus on studying adaptation and diversification mechanisms in crops and exploring the genetic potential of wild relatives to withstand abiotic stresses (Miska and Ferguson-Smith 2016). Genomic signatures of adaptation in dry habitats aid in germplasm characterization and identification of variation sources, facilitating future genome-assisted selection and crop improvement. The integration of quantitative genetics with big data analytics, particularly ML, promises to enhance parent selection accuracy and reinforce the relevance of crops and their wild relatives in addressing evolutionary biology and ecological genetics questions. Looking ahead, understanding the molecular footprint of selection and adaptation in wild crop relatives and early landraces can benefit plant breeding and climate adaptation by bridging the gap between phenotyping and genotyping in gene banks (Watson et al. 2018).

16.5.7 Disease-Resistant and Stress-Tolerant Crop Varieties

16.5.7.1 Plant Biotic Stress Resistance

Traditional methods of breeding for plant resistance to pathogens have involved extracting "R" (resistance) genes from related plant hosts and integrating them into high-yielding cultivars. However, this process is time-consuming, pathogen-specific, and short-lived due to the continuous evolution of new pathogen races that eventually overcome plant resistance conferred by R genes (Roberts and Mattoo 2018). Antimicrobial peptides (AMPs) offer a promising alternative to R genes for plant resistance to pathogens (Epand 2016). AMPs, small proteins with diverse structures, can provide durable immunity to plants against pathogens. Their ability to interact with pathogen membranes through electrostatic interactions and disrupt cellular functions makes them effective against a broad spectrum of pathogens. AMPs can be modified to enhance potency and produced in higher amounts using genetic biotechnology. They are not harmful to host cells, making them suitable for disease management in plants. AMPs are also utilized by microorganisms for defense purposes, highlighting their versatility in plant defense strategies (Paiva and Breukink 2012).

16.5.7.2 Plant Abiotic Stress Resistance

To create stress-tolerant crops, novel strategies are required, as traditional breeding methods have struggled, likely due to complex multigenic traits. Genetic transformation involving genes conferring tolerance to abiotic stresses is essential. Techniques like comparative transcriptomics have identified transcription factors and other molecules crucial for stress tolerance, including proteins like heat shock proteins (HSPs), osmolytes like proline, and hormones such as abscisic acid (Kaur et al. 2017). Genetic engineering has successfully enhanced stress tolerance in plants, for example, by overexpressing transcription factors like TaHsfA2d from wheat and improving proteins like rubisco activase (Chauhan et al. 2013). Small molecules like glycine betaine and polyamines also play significant roles in stress responses (Kusano et al. 2008). Transgenic approaches have yielded promising results, demonstrating enhanced tolerance to various abiotic stresses (Kasukabe et al. 2004). These advancements provide optimism for field trials of engineered crop plants, aiming to develop high-yielding crops resilient to harsh environmental conditions and meet global food demands.

16.6 Policy and Governance for Resilient Agriculture

Agriculture has the potential to both aggravate and alleviate climate change, depending on the management practices adopted. In spite of its significant (17%) contribution to global greenhouse gas emissions (FAO 2021), agriculture is still considered a potential mitigator of climate change, due to its carbon capture and storage capacity. Hence, climate-smart agriculture is crucial for curbing this cataclysm. Therefore, many international and national policies are being formulated and implemented to promote the adoption of the same. The first global treaty to explicitly address climate change was the United Nations Framework Convention on Climate Change (UNFCCC), signed in 1992 by 154 states at the Earth Summit. UNFCCC established an annual forum, namely, the Conference of the Parties (COP), for international negotiations on mitigating and adapting to climate change. In the 23rd session of the COP (COP-23), an agreement, namely, Koronivia Joint Work on Agriculture (KJWA), was made among the countries to work together for the development of the agriculture sector and to ensure both food security and climate change mitigation, simultaneously (Ramifehiarivo et al. 2022). In COP 27, held in the year 2022, another agreement, namely, "Sharm el-Sheikh joint work on implementation of climate action on agriculture and food security," was adopted based on the outcomes of KJWA, to make climate-smart agriculture practices a reality. Its progress and outcomes will be reported by the subsidiary bodies (SBs) of UNFCCC at COP 31. At COP 28, held in UAE, 159 countries signed "the declaration on sustainable agriculture, resilient food systems and climate action," to work together to transform the agriculture and food systems urgently, to adapt to climate change. Nations committed to integrating agriculture and food systems into their Nationally Determined

Contributions, National Adaptation Plans, National Biodiversity Strategies and Action Plans, long-term strategies, and other related strategies. The nations Brazil, Cambodia, Norway, Rwanda, and Sierra Leone unitedly launched the "Alliance of Champions for Food System Transformation," through a "whole of government" approach, to transfer their national food systems to become more efficient and make them models for implementation across the world. "Action Agenda on Regenerative Landscapes" was also launched at COP 28 in cooperation with the World Business Council for Sustainable Development (WBCSD), the Boston Consulting Group (BCG), and the UN High-Level Climate Champions (HLCC), with an aim to transfer more agricultural lands to regenerative landscapes, by the year 2030. Huge financial investments are also being made to make agriculture and food systems more climate-resilient. Consultative Group for International Agricultural Research (CGIAR) received an 890 million USD investment to augment agricultural research and extension, as food security and climate change concerns increase. The Agriculture Innovation Mission for Climate (AIM for Climate) doubled (from USD 8 billion at COP27 to USD 17 billion at COP 28) its investment in climate-smart agriculture. The Food and Agriculture Organization (FAO) finances climate actions through funding channels such as the Green Climate Fund (GCF), Global Environmental Facility (GEF), and the Adaptation Fund, through bilateral funding and supporting the operationalization of the FAST (Food and Agriculture for Sustainable Transformation) Initiative. FAO Strategic Framework 2022–2031, FAO Strategy on Climate Change 2022–2031, and FAO Action Plan 2022–2025 aim to achieve better production, nutrition, environment, and life, through interventions at global and regional, country, and local levels. World Bank finances almost USD 3 billion annually for climate-smart agriculture. All these international policies help the beneficiary nations to achieve sustainable development and reduced emissions, thus promoting global resilience to climate change while assuring food and nutritional security.

India, being an agrarian country, is highly impacted by climate change. An assessment made by Rama Rao et al. (2019) proved that 193 districts of the country are "highly" or "very highly" vulnerable to the adverse effects of climate change. Considering the tremendous contribution that India makes to global food security, various international and national organizations are implementing different policies to promote climate-smart agriculture in the country (Aryal et al. 2020). Indian Council of Agricultural Research (ICAR) launched a network project, namely, National Innovations on Climate Resilient Agriculture (NICRA) in 2011, with the objective of increasing the flexibility of Indian agriculture so as to withstand climatic unpredictability, through strategic research and technology demonstrations (Suresh and Viswanathan 2022). NICRA promotes the idea of climate-resilient villages (CRVs), through developing and demonstrating location-specific climate-resilient technologies (CRTs) in the farmers' fields, with the help of Krishi Vigyan Kendras. CRTs demonstrated include weather-smart, water-smart, seed/breed-smart, carbon-smart, nutrient-smart, energy-smart, institution-smart, and knowledge-smart agricultural practices (Aryal et al. 2020; Dinesh et al. 2017). Climate-smart villages are being established in the country in collaboration with the

CGIAR Research Program on Climate Change, Agriculture and Food Security (CCAFS), International Maize and Wheat Improvement Center (CIMMYT), International Food Policy Research Institute (IFPRI), and various ICAR institutes. The Indian state of Haryana has highly benefited from the CSV project. Under the All India Coordinated Project (AICRP) on Integrated Farming Systems, it is also aimed to develop climate-smart integrated farming system (CS-IFS) villages that cover arid and coastal ecosystems, those that are most vulnerable to climate vagaries (Babu et al. 2023). ICAR-Central Research Institute for Dryland Agriculture (CRIDA) has developed District Agricultural Contingency Plans (DACPs) for 650 districts in collaboration with natural resource management divisions of other ICAR institutes, AICRP on Dryland Agriculture, AICRP on Agrometeorology, and SAUs. These DACPs help the farmers to increase their preparedness, real-time response, relief, and rehabilitation under climatic vagaries. With the aim of making Indian agriculture more productive, sustainable, remunerative, and climate-resilient, the government introduced the National Mission for Sustainable Agriculture (NMSA) in 2014–2015 under the National Action Plan on Climate Change (NAPCC).

Farm Water Management (FWM), Soil Health Management (SHM), and Soil Health Card Scheme are the components of NMSA that help to improve the input-use efficiency and soil fertility, paving the way to climate resilience in agriculture. The national government also promotes crop diversification in the country, which helps in stabilizing production under adverse climatic conditions and also reduces GHG emissions. The Crop Diversification Programme (CDP), a sub-scheme of Rashtriya Krishi Vikas Yojana (RKVY), is being implemented in the states of Punjab, Haryana, and Western Uttar Pradesh to promote the adoption of pulses, oilseeds, coarse cereals, nutri-cereals, cotton, etc. instead of water-demanding rice. National Food Security Mission (NFSM), Mission for Integrated Development of Horticulture (MIDH), National Mission on Sustainable Agriculture (NMSA), National Agroforestry Policy (NAP), Sub-Mission on Agroforestry (SMAF), Watershed Development Fund, etc. are few other schemes that promote crop diversification. As soil health, especially soil organic carbon, is the key to climate change mitigation and adaptation, many policies are being formulated to ensure the same in the country. Balanced plant nutrition, through integrated use of organic and inorganic fertilizers, is being promoted by different agencies. National Biogas and Manure Management Programme; National Project on Organic Farming; National Policy on Crop Residue Management; National Soil Health Mission; the introduction of customized fertilizers, micronutrient sources, nanofertilizers, and neem-coated urea; and the promotion of conservation agriculture, agroforestry, precision agriculture, etc. encourage the farmers to adopt soil-friendly agricultural practices. To increase preparedness against climate vagaries, the government provides agrometeorological services to the farmers through Gramin Krishi Mausam Seva (GKMS), a scheme by the Indian Meteorological Department (IMD), in collaboration with the State Agricultural University (SAU) and ICAR. Crop and region-specific advisory services are made based on the climatic conditions, twice a week. DAAS (District Agromet Advisory Services), a project that involves SAUs, ICAR, KVKs, Department of Agriculture and Cooperation, state departments of agriculture/horticulture/animal husbandry, nongovernmental organizations (NGOs), media agencies, etc., provides a spectrum of services to the farmers including meteorological, agricultural, extension, and information dissemination (Mohapatra and Sharma 2020). Meghdoot is a mobile application which is initiated by IMD, Indian Institute of Tropical Meteorology (IITM), and ICAR that gives district-wise weather-based advisory in crop and livestock management, to help farmers make weather-sensitive decisions such as sowing, agrochemical application, scheduling of irrigation, and vaccination of livestock. To increase economic resilience of the farmers to climate change, the government of India has introduced various national missions and insurance schemes such as National Agricultural Insurance Scheme (NAIS), Weather Based Crop Insurance Scheme (WBCIS), Modified National Agricultural Insurance Scheme (MNAIS), and Pradhan Mantri Fasal Bima Yojana (PMFBY), to provide micro-insurance to farmers to cover the loss by natural calamities (Swain 2023). The government also provides various incentives to promote climate-smart practices. It provides subsidies on drip irrigation (45–55% of the indicative unit cost) and laser land levelers to increase water-use efficiency and thus increase drought resilience. In western India, diversification of transplanted rice-based systems is being promoted as rice is known to deplete the water table and also contribute considerably to GHG emissions. The Punjab government is incentivizing direct-seeded rice (DSR), at the rate of Rs. 1500/- per hectare, while the Haryana government declared Rs. 7000/- per acre for crops other than rice. Government subsidy of 75–80% is declared on types of machinery such as happy seeder, super seeder, zero tillage/seed cum fertilizer drill, rotary mulcher, and straw baler that shall promote straw incorporation and mulching rather than burning, which ensures not only a cleaner atmosphere, but also improved soil health and climate resilience. Recently, the Indian government launched an initiative named the Green Credit Program (GCP) which incentivizes plantation on degraded wasteland through the formation of green credits, which can be marketed by the planter to generate income.

The private sector also makes a significant contribution to making Indian agriculture climate smart. Recently, many private companies such as Grow Indigo and Varaha have been helping Indian farmers in accumulating carbon credits and trading them in the international market. The Indian government partners with IBM and Microsoft to develop AI-based crop yield prediction models and sensors, respectively, to increase the resilience and productivity of Indian agriculture. Public-private partnership is also contributing a lot to the application of drones in agriculture. Bayer, a German multinational company, as a part of its "DirectAcres" program, is planning to bring one million hectares under direct-seeded rice in India by 2030. This is expected to decrease water use (40%), GHG emission (45%), and labor requirement (50%), compared to transplanted rice. Several rules enforced by the government on agricultural practices indirectly help in climate change mitigation. The promotion of neem coating of urea, which increases its efficiency, also decreases N_2O emissions from the field (Bordoloi et al. 2020). Prohibition of rice residue burning helps in preserving soil organic carbon (Dutta et al. 2022). Prevention of rice transplanting before June 15 in the Punjab area helps preserve groundwater and thus increases climate resilience. The National Institution for Transforming India

(NITI Aayog), in collaboration with the Ministry of Agriculture and Farmers' Welfare (MoA & FW), Government of India, and FAO, launched an investment forum, namely, "Investment Forum for Advancing Climate Resilient Agrifood Systems in India" on January 18, 2024, at New Delhi. The forum aims to develop strategies for investments and partnerships among government agencies, private sectors, farmers' organizations, and financial institutions in India, to enhance climate resilience in the agrifood system. In its Nationally Determined Contributions (INDCs), India has committed to UNFCCC, to bring about a 33–35% reduction in the intensity of its GHG emissions per unit of GDP by 2050, compared to the emission levels in 2005 (Babu et al. 2023). Most of the approaches followed by the government to achieve this goal are of a top-down nature, which fails to get implemented by the farmers, who are the real stakeholders of CRA (Goswami et al. 2023). Therefore, more improvements and additions in the policy frameworks are required, to achieve the sustainable development goals and to meet the country's commitments to UNFCCC.

16.7 Economics of Climate-Resilient Agriculture

Various economic approaches, such as cost-effectiveness analysis, multi-criteria analysis, social return on investment, and willingness to pay, can be utilized to evaluate the costs and benefits of CSA technologies (Mutenje et al. 2019; Chaudhury et al. 2016). According to a study conducted by Ng'ang'a et al. (2021), climate-smart agricultural (CSA) practices, such as improved livestock housing, supplementary feeding, improved nutrient management, crop rotation, mixed cropping, minimum tillage, and improved varieties, have positive net present values (NPVs) and higher internal rate of returns (IRRs). Among these practices, improved livestock housing showed the highest NPV (USD 6042 per hectare), while supplementary feeding and improved nutrient management had the highest IRR values (272% and 324%, respectively). The study also found that the packages containing three CSA strategies of improved maize varieties, soil and water conservation, and cereal-legume diversification generate positive economic benefits in terms of NPV, IRR, and payback period for smallholder farming systems in South Africa (Mutenje et al. 2019). Mogaka et al. (2022) conducted a study to evaluate the comparative profitability and relative risk of implementing climate-smart soil (CSS) practices among farmers in Kakamega, Siaya, and Bungoma counties in Western Kenya. The CSS practices assessed included agroforestry, intercropping, liming, organic manure use, inorganic fertilizer, and improved hybrid seeds, based on the climate-smart agriculture (CSA) pillars, which are production, adaptation, and mitigation. The study found that agroforestry was the most profitable, with a net present value of USD 16,071 per hectare. Intercropping was the second most profitable, with USD 10,487 per hectare. Improved hybrid seeds were the least profitable, with a net present value of USD 881 per hectare.

16.8 Future Directions and Research Needs

Addressing the socio-economic dimensions of climate resilience is imperative, focusing on gender, youth, and marginalized groups' roles in building community-level resilience. Fostering knowledge sharing and capacity building among stakeholders can accelerate the adoption of climate-smart agricultural practices. Integrating climate resilience into national agricultural policies and strategies, alongside robust monitoring and evaluation frameworks, will be essential. Developing climate risk management strategies, including early warning systems and weather-indexed insurance, can help farmers mitigate climate-related shocks. Multi-stakeholder partnerships between governments, civil society, academia, and the private sector can drive the adoption of climate-smart practices and enhance resilience at scale. Advancements in genomic technologies will enable a more precise and efficient selection of desirable traits, focusing on developing crops that can thrive in diverse environmental conditions, including drought, heat, and salinity. There will also be a growing emphasis on improving the nutritional quality of crops, increasing micronutrient content, and reducing anti-nutritional factors. Pan-genomics and population genomics will play a crucial role in utilizing genetic diversity to develop more resilient and productive varieties. Integrating various omics technologies will provide a comprehensive understanding of crop traits, facilitating more targeted breeding efforts. Gene editing technologies, such as CRISPR-Cas9, will continue to revolutionize crop breeding by enabling precise modifications of target genes to introduce desirable traits. Improving policy and governance will be necessary in the future to enhance climate-resilient agriculture. This involves providing incentives for climate-smart activities and promoting collaborations between government agencies, commercial sectors, and farmers' organizations. Investment and partnerships will be essential in promoting the development of climate-resilient agrifood systems, with a focus on collaboration and the formulation of strategic plans. The adoption of climate-smart agriculture technologies and the exploration of innovative solutions for soil health and water management will play a crucial role in addressing the challenges ahead. Capacity building and knowledge sharing are crucial for fostering the implementation of climate-resilient practices, necessitating training and the spread of information. Preserving habitats and implementing biodiversity-friendly agricultural methods will be essential for the conservation of biodiversity. By combining climate mitigation and adaptation techniques, as well as research and development endeavors, we can strengthen agricultural resistance to climate change and ensure food security for future generations. Further research and development of digital agriculture technologies are needed to enhance their effectiveness in mitigating climate change and ensuring food security. This could include the integration of advanced data analytics, AI, and automation technologies to optimize resource use and improve crop yields (Balasundram et al. 2023).

16.9 Conclusion

Climate change threatens global food security through disrupted agricultural systems, reduced crop yields, and extreme weather events. Rising temperatures, unpredictable rainfall, and population growth exacerbate the challenge. Vulnerable communities, reliant on rainfed agriculture, face keen risks of starvation and malnutrition, demanding urgent action for adaptation and mitigation strategies. Efficient mitigation strategies in agriculture, such as reduced tillage, sustainable land management, and efficient livestock practices, offer pathways to combat climate change. However, other techniques like agroforestry and afforestation serve as crucial carbon sinks. Adaptation strategies in agriculture, like crop diversification, resilient varieties, improved water management, sustainable agroecological practices, and risk reduction measures, are crucial for mitigating climate change impacts. Effective implementation and investment in these strategies are essential for building resilience in agricultural systems. Integrating biodiversity conservation practices, agroecological approaches, ecosystem-based adaptation strategies, and breeding for climate-resilient crops is essential for enhancing agricultural resilience to climate change. Harnessing scientific advancements and traditional knowledge is critical for sustainable agricultural development in a changing climate. Addressing the socioeconomic dimensions of climate resilience, fostering knowledge sharing, and integrating climate-smart practices into national policies are imperative. Advancements in genomic technologies, gene editing, and omics will drive the development of resilient crop varieties. Improved policy and governance, alongside investment and partnerships, are crucial for promoting climate-resilient agrifood systems. Capacity building, biodiversity conservation, and digital agriculture advancements will further enhance agricultural resilience to climate change, ensuring food security for future generations.

References

Adjimoti GO, Kwadzo GTM (2018) Crop diversification and household food security status: evidence from rural Benin. Agric Food Secur 7(1):1–12

Ahmad S, Li C, Dai G, Zhan M, Wang J, Pan S, Cao C (2009) Greenhouse gas emission from direct seeding paddy field under different rice tillage systems in central China. Soil Tillage Res 106(1):54–61

Akinnagbe OM, Irohibe IJ (2014) Agricultural adaptation strategies to climate change impacts in Africa: a review. Bangladesh J Agric Res 39:407–418

Ali MA, Lee CH, Lee YB, Kim PJ (2009) Silicate fertilization in no-tillage rice farming for mitigation of methane emission and increasing rice productivity. Agric Ecosyst Environ 132(1–2):16–22

Alletto L, Vandewalle A, Debaeke P (2022) Crop diversification improves cropping system sustainability: an 8-year on-farm experiment in South-Western France. Agric Syst 200:103433

Altieri MA, Rogé P (2009) The ecological role and enhancement of biodiversity in agriculture. In: Agriculture, biodiversity and markets. Routledge, pp 15–32

Arias P, Bellouin N, Coppola E, Jones R, Krinner G, Marotzke J et al (2021) Climate change 2021: the physical science basis. Contribution of Working Group I to the Sixth Assessment Report of the Intergovernmental Panel on Climate Change; technical summary. IPCC

Aryal JP, Sapkota TB, Rahut DB, Jat ML (2020) Agricultural sustainability under emerging climatic variability: the role of climate-smart agriculture and relevant policies in India. Int J Innov Sustain Dev 14(2):219–245

Azadi H, Moghaddam SM, Burkart S, Mahmoudi H, Van Passel S, Kurban A, Lopez-Carr D (2021) Rethinking resilient agriculture: from climate-smart agriculture to vulnerable-smart agriculture. J Clean Prod 319:128602

Babu S, Das A, Singh R, Mohapatra KP, Kumar S, Rathore SS et al (2023) Designing an energy efficient, economically feasible, and environmentally robust integrated farming system model for sustainable food production in the Indian Himalayas. Sustain Food Technol 1(1):126–142

Balafoutis A, Beck B, Fountas S, Vangeyte J, Van der Wal T, Soto I et al (2017) Precision agriculture technologies positively contributing to GHG emissions mitigation, farm productivity and economics. Sustainability 9(8):1339

Balasundram SK, Shamshiri RR, Sridhara S, Rizan N (2023) The role of digital agriculture in mitigating climate change and ensuring food security: an overview. Sustainability 15(6):5325

Bardgett RD, Bullock JM, Lavorel S, Manning P, Schaffner U, Ostle N et al (2021) Combatting global grassland degradation. Nat Rev Earth Environ 2(10):720–735

Behnke GD, Zuber SM, Pittelkow CM, Nafziger ED, Villamil MB (2018) Long-term crop rotation and tillage effects on soil greenhouse gas emissions and crop production in Illinois, USA. Agric Ecosyst Environ 261:62–70

Benitez-Alfonso Y, Soanes BK, Zimba S, Sinanaj B, German L, Sharma V et al (2023) Enhancing climate change resilience in agricultural crops. Curr Biol 33(23):R1246–R1261

Bezançon G, Pham JL, Deu M, Vigouroux Y, Sagnard F, Mariac C et al (2009) Changes in the diversity and geographic distribution of cultivated millet (Pennisetum glaucum (L.) R. Br.) and sorghum (Sorghum bicolor (L.) Moench) varieties in Niger between 1976 and 2003. Genet Resour Crop Evol 56:223–236

Blair MW, Pantoja W, Carmenza Muñoz L (2012) First use of microsatellite markers in a large collection of cultivated and wild accessions of tepary bean (Phaseolus acutifolius A. Gray). Theor Appl Genet 125:1137–1147

Bordoloi N, Baruah KK, Hazarika B (2020) Fertilizer management through coated urea to mitigate greenhouse gas (N2O) emission and improve soil quality in agroclimatic zone of Northeast India. Environ Sci Pollut Res 27(11):11919–11931

Boyle EA, Li YI, Pritchard JK (2017) An expanded view of complex traits: from polygenic to omnigenic. Cell 169(7):1177–1186

Brussaard L, Caron P, Campbell B, Lipper L, Mainka S, Rabbinge R et al (2010) Reconciling biodiversity conservation and food security: scientific challenges for a new agriculture. Curr Opin Environ Sustain 2(1–2):34–42

Burgess PJ, Redhead J, Girkin N, Deeks L, Harris JA, Staley J (2024) Evaluating agroecological farming practices. Department for Environment, Food & Rural Affairs, London

Chartzoulakis K, Bertaki M (2015) Sustainable water management in agriculture under climate change. Agric Agric Sci Procedia 4:88–98

Chaudhury AS, Helfgott A, Thornton TF, Sova C (2016) Participatory adaptation planning and costing. Applications in agricultural adaptation in western Kenya. Mitig Adapt Strateg Glob Chang 21:301–322

Chauhan H, Khurana N, Agarwal P, Khurana JP, Khurana P (2013) A seed preferential heat shock transcription factor from wheat provides abiotic stress tolerance and yield enhancement in transgenic Arabidopsis under heat stress environment. PLoS One 8(11):e79577

Choudhary M, Panday SC, Meena VS, Singh S, Yadav RP, Pattanayak A et al (2020) Long-term tillage and irrigation management practices: strategies to enhance crop and water productivity under rice-wheat rotation of Indian mid-Himalayan Region. Agric Water Manag 232:106067

Corbeels M, Naudin K, Whitbread AM, Kühne R, Letourmy P (2020) Limits of conservation agriculture to overcome low crop yields in sub-Saharan Africa. Nat Food 1(7):447–454

Davis MB, Shaw RG (2001) Range shifts and adaptive responses to Quaternary climate change. Science 292(5517):673–679

Dhankher OP, Foyer CH (2018) Climate resilient crops for improving global food security and safety. Plant Cell Environ 41(5):877–884

Dinesh D, Aggarwal PK, Khatri-Chhetri A, Loboguerrero Rodriguez AM, Mungai C, Radeny MA et al (2017) The rise in Climate-Smart Agriculture strategies, policies, partnerships and investments across the globe. Agric Dev 30:4–9

Döll P, Fiedler K, Zhang J (2009) Global-scale analysis of river flow alterations due to water withdrawals and reservoirs. Hydrol Earth Syst Sci 13(12):2413–2432

Douyon A, Worou ON, Diama A, Badolo F, Denou RK, Touré S et al (2022) Impact of crop diversification on household food and nutrition security in southern and central Mali. Front Sustain Food Syst 5:751349

Dutta A, Bhattacharyya R, Chaudhary VP, Sharma C, Nath CP, Kumar SN, Parmar B (2022) Impact of long-term residue burning versus retention on soil organic carbon sequestration under a rice-wheat cropping system. Soil Tillage Res 221:105421

Edwards S, Egziabher TBG, Araya H (2010) Successes and challenges in ecological agriculture experiences from Tigray, Ethiopia. Tigray Project

El-Fakharany ZM, Salem MG (2021) Mitigating climate change impacts on irrigation water shortage using brackish groundwater and solar energy. Energy Rep 7:608–621

EPA (2022) Practices to reduce methane emissions from livestock manure management. United States Environmental Protection Agency. Retrieved from https://www.epa.gov/agstar/practices-reduce-methane-emissions-livestock-manure-management

Epand RM (ed) (2016) Host defense peptides and their potential as therapeutic agents. Springer, Cham

Falconnier GN, Corbeels M, Boote KJ, Affholder F, Adam M, MacCarthy DS et al (2020) Modelling climate change impacts on maize yields under low nitrogen input conditions in sub-Saharan Africa. Glob Chang Biol 26(10):5942–5964

Fan Y, Wang X, Funk T, Rashid I, Herman B, Bompoti N et al (2022) A critical review for real-time continuous soil monitoring: advantages, challenges, and perspectives. Environ Sci Technol 56(19):13546–13564

FAO (2019) The state of food and agriculture 2019: moving forward on food loss and waste reduction. UN

FAO (2020) Disaster risk reduction and the impact on agriculture and food security. Food and Agriculture Organization. Retrieved on 1 Mar 2024, from https://www.fao.org/climate--smart-agriculture-sourcebook/enabling-frameworks/module-c5-climate-resilience/chapter-c5-1/en/

FAO (2021) FAOSTAT: emissions shares. In: FAO.org [online]. https://www.fao.org/faostat/en/#data/EM

Field CB, Barros VR (eds) (2014) Climate change 2014–impacts, adaptation and vulnerability: regional aspects. Cambridge University Press

Franzluebbers AJ, Hendrickson JR (2024) Should we consider integrated crop–livestock systems for ecosystem services, carbon sequestration, and agricultural resilience to climate change? Agron J 116:415

Frimawaty E, Ilmika A, Sakina NA, Mustabi J (2023) Climate change mitigation and adaptation through livestock waste management. Glob J Environ Sci Manag 9(4):691–706

Frimpong F, Asante MD, Peprah CO, Amankwaa-Yeboah P, Danquah EO, Ribeiro PF et al (2023) Water-smart farming: review of strategies, technologies, and practices for sustainable agricultural water management in a changing climate in West Africa. Front Sustain Food Syst 7:1110179

Frisvold G, Bai T (2016) Irrigation technology choice as adaptation to climate change in the Western United States. J Contemp Water Res Educ 158(1):62–77

Gaba Y, Pareek A, Singla-Pareek SL (2021) Raising climate-resilient crops: journey from the conventional breeding to new breeding approaches. Curr Genomics 22(6):450
Gangopadhyay S, Chowdhuri I, Das N, Pal SC, Mandal S (2023) The effects of no-tillage and conventional tillage on greenhouse gas emissions from paddy fields with various rice varieties. Soil Tillage Res 232:105772
Garg KK, Anantha KH, Nune R, Akuraju VR, Singh P, Gumma MK et al (2020) Impact of land use changes and management practices on groundwater resources in Kolar district, Southern India. J Hydrol Reg Stud 31:100732
Ghosh D, Maiti SK (2023) Invasive weed-based biochar facilitated the restoration of coal mine degraded land by modulating the enzyme activity and carbon sequestration. Restor Ecol 31(3):e13744
Giller KE, Witter E, Corbeels M, Tittonell P (2009) Conservation agriculture and smallholder farming in Africa: the heretics' view. Field Crop Res 114(1):23–34
Gitarskiy ML (2019) The refinement to the 2006 IPCC guidelines for national greenhouse gas inventories. Fundam Appl Climatol 2:5–13
Glaser B, Parr M, Braun C, Kopolo G (2009) Biochar is carbon negative. Nat Geosci 2(1):2–2
Godfray HCJ, Beddington JR, Crute IR, Haddad L, Lawrence D, Muir JF et al (2010) Food security: the challenge of feeding 9 billion people. Science 327(5967):812–818
Goswami M, Pant G, Mansotra DK, Sharma S, Joshi PC (2021) Biochar: a carbon negative technology for combating climate change. In: Advances in carbon capture and utilization. Springer, Singapore, pp 251–272
Goswami M, Gupta AK, Kishan R, Baidya S, Khan YI, Prakash S et al (2023) An evaluation of climate resilient agricultural practices in India: a narrative synthesis of literature. Environ Sustain 6(1):7–23
Grandy AS, Loecke TD, Parr S, Robertson GP (2006) Long-term trends in nitrous oxide emissions, soil nitrogen, and crop yields of till and no-till cropping systems. J Environ Qual 35(4):1487–1495
Grigorieva E, Livenets A, Stelmakh E (2023) Adaptation of agriculture to climate change: a scoping review. Climate 11(10):202
Grossi G, Goglio P, Vitali A, Williams AG (2019) Livestock and climate change: impact of livestock on climate and mitigation strategies. Anim Front 9(1):69–76
Gupta DK, Gupta CK, Dubey R, Fagodiya RK, Sharma G, Noor Mohamed MB et al (2020) Role of biochar in carbon sequestration and greenhouse gas mitigation. In: Biochar applications in agriculture and environment management. Springer, Cham, pp 141–165
Harris DR, Fuller DQ (2014) Agriculture: definition and overview. In: Encyclopedia of global archaeology. Springer, Cham, pp 104–113
He L, Rosa L (2023) Solutions to agricultural green water scarcity under climate change. PNAS Nexus 2(4):pgad117
Ijaz M, Nawaz A, Ul-Allah S, Rizwan MS, Ullah A, Hussain M et al (2019) Crop diversification and food security. In: Hasanuzzaman M (ed) Agronomic crops: Volume 1: Production technologies. Springer, Singapore
International Energy Agency (IEA) (2021) GHG emissions from energy. Retrieved on 22 Dec 2021, from IEA Energy Data Centre
Izah SC, Iyiola AO, Yarkwan B, Richard G (2023) Impact of air quality as a component of climate change on biodiversity-based ecosystem services. In: Visualization techniques for climate change with machine learning and artificial intelligence. Elsevier, pp 123–148
Jantalia CP, dos Santos HP, Urquiaga S, Boddey RM, Alves BJ (2008) Fluxes of nitrous oxide from soil under different crop rotations and tillage systems in the south of Brazil. Nutr Cycl Agroecosyst 82:161–173
Kameyama K, Miyamoto T, Shiono T, Shinogi Y (2012) Influence of sugarcane bagasse-derived biochar application on nitrate leaching in calcaric dark red soil. J Environ Qual 41(4):1131–1137

Karan SK, Woolf D, Azzi ES, Sundberg C, Wood SA (2023) Potential for biochar carbon sequestration from crop residues: a global spatially explicit assessment. GCB Bioenergy 15(12):1424–1436

Kasukabe Y, He L, Nada K, Misawa S, Ihara I, Tachibana S (2004) Overexpression of spermidine synthase enhances tolerance to multiple environmental stresses and up-regulates the expression of various stress-regulated genes in transgenic Arabidopsis thaliana. Plant Cell Physiol 45(6):712–722

Kaur B, Handa AK, Mattoo AK (2017) Genetic engineering of tomato to improve nutritional quality, resistance to abiotic and biotic stresses, and for non-food applications. In: Achieving sustainable cultivation of tomatoes. Burleigh Dodds Science Publishing, pp 261–304

Kay S, Rega C, Moreno G, Den Herder M, Palma JH, Borek R, Crous-Duran J, Freese D, Giannitsopoulos M, Graves A, Jäger M, Lamersdorf N, Memedemin D, Mosquera-Losada R, Pantera A, Paracchini ML, Paris P, Roces-Díaz JV, Rolo V et al (2019) Agroforestry creates carbon sinks whilst enhancing the environment in agricultural landscapes in Europe. Land Use Policy 83:581–593. https://doi.org/10.1016/j.landusepol.2019.02.025

Kirubanandan S (2023) How to mitigate the effects of climate change on global food security. World Economic Forum. Retrieved on 2 Mar 2024, from https://www.weforum.org/agenda/2023/04/mitigate-climate-change-food-security/

Kopeć P (2024) Climate change—the rise of climate-resilient crops. Plan Theory 13(4):490

Kouressy M, Bazile D, Vaksmann M, Soumaré M, Doucouré COT, Sidibé A (2003) La dynamique des agroécosystèmes: un facteur explicatif de l'érosion variétale du sorgho: le cas de la zone Mali-sud. CNEARC-SAGERT

Kulakowski D, Bebi P, Rixen C (2011) The interacting effects of land use change, climate change and suppression of natural disturbances on landscape forest structure in the Swiss Alps. Oikos 120(2):216–225

Kusano T, Berberich T, Tateda C, Takahashi Y (2008) Polyamines: essential factors for growth and survival. Planta 228:367–381

Lal R (2004) Soil carbon sequestration impacts on global climate change and food security. Science 304(5677):1623–1627

Lal R (2009) Soils and food sufficiency: a review. In: Sustainable agriculture. Springer, pp 25–49

Langridge P, Alaux M, Almeida NF, Ammar K, Baum M, Bekkaoui F et al (2022) Meeting the challenges facing wheat production: the strategic research agenda of the global wheat initiative. Agronomy 12(11):2767

Lavergne S, Mouquet N, Thuiller W, Ronce O (2010) Biodiversity and climate change: integrating evolutionary and ecological responses of species and communities. Annu Rev Ecol Evol Syst 41:321–350

Lehmann J, Joseph S (2015) Biochar for environmental management: an introduction. In: Biochar for environmental management. Routledge, pp 1–13

Lehmann J, Cowie A, Masiello CA, Kammann C, Woolf D, Amonette JE et al (2021) Biochar in climate change mitigation. Nat Geosci 14(12):883–892

Lesk C, Rowhani P, Ramankutty N (2016) Influence of extreme weather disasters on global crop production. Nature 529(7584):84–87

Li D, Liu M, Cheng Y, Wang D, Qin J, Jiao J et al (2011) Methane emissions from double-rice cropping system under conventional and no tillage in southeast China. Soil Tillage Res 113(2):77–81

Li H, Guo HQ, Helbig M, Dai SQ, Zhang MS, Zhao M et al (2019) Does direct-seeded rice decrease ecosystem-scale methane emissions?—a case study from a rice paddy in southeast China. Agric For Meteorol 272:118–127

Li K, Pan J, Xiong W, Xie W, Ali T (2022) The impact of 1.5 C and 2.0 C global warming on global maize production and trade. Sci Rep 12(1):17268

Lipper L, Thornton P, Campbell BM, Baedeker T, Braimoh A, Bwalya M et al (2014) Climate-smart agriculture for food security. Nat Clim Chang 4(12):1068–1072

Liu W, Liu L, Yang X, Deng M, Wang Z, Wang P et al (2021) Long-term nitrogen input alters plant and soil bacterial, but not fungal beta diversity in a semiarid grassland. Glob Chang Biol 27(16):3939–3950

Lo V (2016) Synthesis report on experiences with ecosystem-based approaches to climate change adaptation and disaster risk reduction. Secretariat of the Convention on Biological Diversity, Montreal

Lobell DB, Bänziger M, Magorokosho C, Vivek B (2011) Nonlinear heat effects on African maize as evidenced by historical yield trials. Nat Clim Chang 1(1):42–45

Mäder P, Fliessbach A, Dubois D, Gunst L, Fried P, Niggli U (2002) Soil fertility and biodiversity in organic farming. Science 296(5573):1694–1697

Maia RF, Lurbe CB, Hornbuckle J (2022) Machine learning approach to estimate soil matric potential in the plant root zone based on remote sensing data. Front Plant Sci 13:931491

Mamabolo E, Makwela MM, Tsilo TJ (2020) Achieving sustainability and biodiversity conservation in agriculture: importance, challenges and prospects. Eur J Sustain Dev 9(3):616–616

Mapemba L, Makate C, Mango N, Sopo M (2018) The role of crop diversification in improving household food security in central Malawi. Agric Food Secur 7:7

Massel K, Lam Y, Wong AC, Hickey LT, Borrell AK, Godwin ID (2021) Hotter, drier, CRISPR: the latest edit on climate change. Theor Appl Genet 134(6):1691–1709

Metay A, Oliver R, Scopel E, Douzet JM, Moreira JAA, Maraux F et al (2007) N2O and CH4 emissions from soils under conventional and no-till management practices in Goiânia (Cerrados, Brazil). Geoderma 141(1–2):78–88

Mijatović D, Van Oudenhoven F, Eyzaguirre P, Hodgkin T (2013) The role of agricultural biodiversity in strengthening resilience to climate change: towards an analytical framework. Int J Agric Sustain 11(2):95–107

Miska EA, Ferguson-Smith AC (2016) Transgenerational inheritance: models and mechanisms of non–DNA sequence–based inheritance. Science 354:59–63

Mogaka BO, Karanja Ng'ang'a S, Bett HK (2022) Comparative profitability and relative risk of adopting climate-smart soil practices among farmers. A cost-benefit analysis of six agricultural practices. Clim Serv 26:100287

Mohapatra L, Sharma P (2020) Effectiveness of the SMS advisory service under Gramin Krishi Mausam Sewa (GKMS). Int J Farm Sci 10(3&4):12–17

Montoya D, Gaba S, de Mazancourt C, Bretagnolle V, Loreau M (2020) Reconciling biodiversity conservation, food production and farmers' demand in agricultural landscapes. Ecol Model 416:108889

Morton JF (2007) The impact of climate change on smallholder and subsistence agriculture. Proc Natl Acad Sci 104(50):19680–19685

Muhie SH (2022) Novel approaches and practices to sustainable agriculture. J Agric Food Res 10:100446

Muluneh MG (2021) Impact of climate change on biodiversity and food security: a global perspective—a review article. Agric Food Secur 10(1):1–25

Mutenje MJ, Farnworth CR, Stirling C, Thierfelder C, Mupangwa W, Nyagumbo I (2019) A cost-benefit analysis of climate-smart agriculture options in southern Africa: balancing gender and technology. Ecol Econ 163:126–137

Mzyece A, Ng'ombe JN (2021) Crop diversification improves technical efficiency and reduces income variability in Northern Ghana. J Agric Food Res 5:100162

Ng'ang'a SK, Miller V, Girvetz EH (2021) Is investment in Climate-Smart-agricultural practices the option for the future? Cost and benefit analysis evidence from Ghana. Heliyon 7:e06653

Nikolaou G, Neocleous D, Christou A, Kitta E, Katsoulas N (2020) Implementing sustainable irrigation in water-scarce regions under the impact of climate change. Agronomy 10(8):1120

Nikzaad RM, Nusrathali N (2023) Integrating livestock and crop systems for enhanced productivity and grassland conservation in developing countries. In: Grasslands – conservation and development. IntechOpen

Ober ES, Alahmad S, Cockram J, Forestan C, Hickey LT, Kant J et al (2021) Wheat root systems as a breeding target for climate resilience. Theor Appl Genet 134(6):1645–1662

Okoronkwo D, Ozioko R, Ugwoke RU, Nwagbo UV, Nwobodo C, Ugwu CH et al (2024) Climate smart agriculture? Adaptation strategies of traditional agriculture to climate change in sub-Saharan Africa. Front Clim 6:1272320

Omotoso AB, Omotayo AO (2024) Enhancing dietary diversity and food security through the adoption of climate-smart agricultural practices in Nigeria: a micro level evidence. Environ Dev Sustain:1–18. https://doi.org/10.1007/s10668-024-04681-8

Oo AZ, Sudo S, Inubushi K, Mano M, Yamamoto A, Ono K et al (2018) Methane and nitrous oxide emissions from conventional and modified rice cultivation systems in South India. Agric Ecosyst Environ 252:148–158

Paiva AD, Breukink E (2012) Antimicrobial peptides produced by microorganisms. In: Antimicrobial peptides and innate immunity. Springer, Basel, pp 53–95

Parkin TB, Kaspar TC (2006) Nitrous oxide emissions from corn–soybean systems in the Midwest. J Environ Qual 35(4):1496–1506

Pasricha NS, Ghosh PK (2023) Agriculture-related green house gas emissions and mitigation measures. Adv Agron 179:257–376

Passianoto CC, Ahrens T, Feigle BJ, Steudler PA, do Carmo JB, Melillo JM (2003) Emissions of CO2, N2O and NO in conventional and no-till management practices in Rondonia, Brazil. Biol Fertil Soils 38:200–208

Paudel B, Zhang Y, Yan J, Rai R, Li L (2019) Farmers' perceptions of agricultural land use changes in Nepal and their major drivers. J Environ Manag 235:432–441

Paudel B, Fall M, Daneshfar B (2020) GIS-based assessment of debris flow hazards in Kulekhani Watershed, Nepal. Nat Hazards 101:143–172

Petersen SO, Blanchard M, Chadwick D, Del Prado A, Edouard N, Mosquera J, Sommer SG (2013) Manure management for greenhouse gas mitigation. Animal 7(s2):266–282

Pixley KV, Cairns JE, Lopez-Ridaura S, Ojiewo CO, Dawud MA, Drabo I et al (2023) Redesigning crop varieties to win the race between climate change and food security. Mol Plant 16:1590

Power AG (2010) Ecosystem services and agriculture: tradeoffs and synergies. Philos Trans R Soc B 365(1554):2959–2971

Pretty JN, Williams S, Toulmin C (2012) Sustainable intensification: increasing productivity in African food and agricultural systems. Routledge

Rama Rao CA, Raju BMK, Islam A, Subbarao AVM, Rao KV et al (2019) Risk and vulnerability assessment of Indian agriculture to climate change. ICAR-Central Research Institute for Dryland Agriculture, Hyderabad

Ramifehiarivo N, Chevallier T, Defrance D, Brossard M, Chotte JL (2022) Framing the future of the Koronivia Joint Work on Agriculture from science-based evidence. A review. Agron Sustain Dev 42(5):102

Reay DS (2020) Land use and agriculture: pitfalls and precautions on the road to net zero. Front Clim 2:4

Ritchie H, Roser M, Rosado P (2020) CO_2 and greenhouse gas emissions. Our World in Data

Rivera JE, Chará J (2021) CH4 and N2O emissions from cattle excreta: a review of main drivers and mitigation strategies in grazing systems. Front Sustain Food Syst 5:657936

Roberts DP, Mattoo AK (2018) Sustainable agriculture—enhancing environmental benefits, food nutritional quality and building crop resilience to abiotic and biotic stresses. Agriculture 8(1):8

Rosa L (2022) Adapting agriculture to climate change via sustainable irrigation: biophysical potentials and feedbacks. Environ Res Lett 17(6):063008

Roy A, Chaturvedi S, Singh SV, Govindaraju K, Dhyani VC, Pyne S (2022) Preparation and evaluation of two enriched biochar-based fertilizers for nutrient release kinetics and agronomic effectiveness in direct-seeded rice. Biomass Convers Biorefinery 14:1–12. https://doi.org/10.1007/s13399-022-02488-z

Sapinas IP, Abbott LK (2020) Soil fertility management based on certified organic agriculture standards–a review. Sustain Agric Res 9(2):1–16

Scherr SJ, McNeely JA (2008) Biodiversity conservation and agricultural sustainability: towards a new paradigm of "ecoagriculture" landscapes. Philos Trans R Soc B 363(1491):477–494
Sheridan JA, Bickford D (2011) Shrinking body size as an ecological response to climate change. Nat Clim Chang 1(8):401–406
Shields S, Orme-Evans G (2015) The impacts of climate change mitigation strategies on animal welfare. Animals 5(2):361–394
Shiferaw B, Tesfaye K, Kassie M, Abate T, Prasanna BM, Menkir A (2014) Managing vulnerability to drought and enhancing livelihood resilience in sub-Saharan Africa: technological, institutional and policy options. Weather Clim Extrem 3:67–79
Shiferaw H, Bewket W, Alamirew T, Zeleke G, Teketay D, Bekele K et al (2019) Implications of land use/land cover dynamics and Prosopis invasion on ecosystem service values in Afar Region, Ethiopia. Sci Total Environ 675:354–366
Singh SV, Chaturvedi S, Dhyani VC, Datta D (2018) Biochar: an ecofriendly residue management approach. Indian Farm 69(8):27–29. https://epubs.icar.org.in/index.php/IndFarm/article/view/94471
Singh SV, Chaturvedi S, Dhyani VC, Kasivelu G (2020) Pyrolysis temperature influences the characteristics of rice straw and husk biochar and sorption/desorption behaviour of their biourea composite. Bioresour Technol 314:123674
Singh S, Luthra N, Mandal S, Kushwaha DK, Pathak SO, Datta D, Sharma R, Pramanick B (2023a) Distinct behavior of biochar modulating biogeochemistry of salt-affected and acidic soil: a review. J Soil Sci Plant Nutr 23(3):2981–2997. https://doi.org/10.1007/s42729-023-01370-9
Singh SV, Chaturvedi S, Nayak P, Dhyani VC, Nadipamu TMK, Singh DK, Gudapaty P, Mathyam P, Srinivasarao K, Gavindaraju K (2023b) Carbon offset potential of biochar based straw management under rice – wheat system along Indo-Gangetic Plains of India. Sci Total Environ 897:165176. https://doi.org/10.1016/j.scitotenv.2023.165176
Smith J (2015) Crops, crop pests and climate change–why Africa needs to be better prepared. CCAFS Working Paper
Smith P, Martino Z, Cai D (2007) Agriculture. In: Climate change 2007: mitigation. Cambridge University Press, Cambridge
Smith P, Martino D, Cai Z, Gwary D, Janzen H, Kumar P et al (2008) Greenhouse gas mitigation in agriculture. Philos Trans R Soc B 363(1492):789–813
Sodhi NS, Posa MRC, Lee TM, Bickford D, Koh LP, Brook BW (2010) The state and conservation of Southeast Asian biodiversity. Biodivers Conserv 19:317–328
Spokas KA, Cantrell KB, Novak JM, Archer DW, Ippolito JA, Collins HP et al (2012) Biochar: a synthesis of its agronomic impact beyond carbon sequestration. J Environ Qual 41(4):973–989
Sunderland TC, Rowland D (2019) Forests, land use, and challenges to climate stability and food security. In: Sustainable food and agriculture. Academic Press, pp 95–116
Suresh A, Viswanathan PK (2022) Building climate resilience in Indian farm households: an analysis of national and state policies and initiatives. Arab Econ Bus J 14(1):62–69
Suseela V, Conant RT, Wallenstein MD, Dukes JS (2012) Effects of soil moisture on the temperature sensitivity of heterotrophic respiration vary seasonally in an old-field climate change experiment. Glob Chang Biol 18(1):336–348
Swain M (2023) Redesigning crop insurance for coping with climate change in India. Indian J Appl Econ Bus 5(1):107–128
Tahmasebi T, Karami E, Keshavarz M (2020) Agricultural land use change under climate variability and change: drivers and impacts. J Arid Environ 180:104202
Tefera Y, Hailu Y, Siraj Z (2019) Potential of agroforestry for climate change mitigation through carbon sequestration. Agric Res Technol 22:556196
Tilman D, Cassman KG, Matson PA, Naylor R, Polasky S (2002) Agricultural sustainability and intensive production practices. Nature 418(6898):671–677
Tilman D, Balzer C, Hill J, Befort BL (2011) Global food demand and the sustainable intensification of agriculture. Proc Natl Acad Sci 108(50):20260–20264

Ulsrud K, Sygna L, O'Brien K (2008) More than rain: identifying sustainable pathways for climate adaptation and poverty reduction. Report prepared for the Development Fund, Norway

UNEP (2023) Climate information and early warning systems. United Nations Environment Programme. Retrieved on 2 Mar 2024, from http://www.unep.org/topics/climate-action/climate-transparency/climate-information-and-early-warning-systems

Vakeesan A, Nishanthan T, Mikunthan G (2008) Green manures: nature's gift to improve soil fertility. LEISA-LEUSDEN 24(2):16

Valujeva K, Pilecka-Ulcugaceva J, Skiste O, Liepa S, Lagzdins A, Grinfelde I (2022) Soil tillage and agricultural crops affect greenhouse gas emissions from Cambic Calcisol in a temperate climate. Acta Agric Scand Sect B Soil Plant Sci 72(1):835–846

Vermeulen SJ, Campbell BM, Ingram JS (2012) Climate change and food systems. Annu Rev Environ Resour 37:195–222

Vernooy R (2022) Does crop diversification lead to climate-related resilience? Improving the theory through insights on practice. Agroecol Sustain Food Syst 46(6):877–901

Wajih SA (2008) Adaptive agriculture in flood affected areas. LEISA Mag 24(4):24–25

Walther GR (2003) Plants in a warmer world. Perspect Plant Ecol Evol Syst 6(3):169–185

Wang M, Zhao S, Wang L, Chen S, Li S, Lei X et al (2021) Salt stress-induced changes in microbial community structures and metabolic processes result in increased soil cadmium availability. Sci Total Environ 782:147125

Watson A, Ghosh S, Williams MJ, Cuddy WS, Simmonds J, Rey MD et al (2018) Speed breeding is a powerful tool to accelerate crop research and breeding. Nat Plants 4(1):23–29

Wilbanks TJ, Sathaye J (2007) Integrating mitigation and adaptation as responses to climate change: a synthesis. Mitig Adapt Strateg Glob Chang 12:957–962

WMO (2018) WMO statement on the state of the global climate in 2017. World Meteorological Organisation (WMO)

WoldeYohannes A, Cotter M, Kelboro G, Dessalegn W (2018) Land use and land cover changes and their effects on the landscape of Abaya-Chamo Basin, Southern Ethiopia. Land 7(1):2

Xu HJ, Wang XP, Zhao CY, Zhang XX (2019) Responses of ecosystem water use efficiency to meteorological drought under different biomes and drought magnitudes in northern China. Agric For Meteorol 278:107660

Yang X, Xiong J, Du T, Ju X, Gan Y, Li S et al (2024) Diversifying crop rotation increases food production, reduces net greenhouse gas emissions and improves soil health. Nat Commun 15(1):198

Zaman M, Kleineidam K, Bakken L, Berendt J, Bracken C, Butterbach-Bahl K et al (2021) Greenhouse gases from agriculture. In: Measuring emission of agricultural greenhouse gases and developing mitigation options using nuclear and related techniques: applications of nuclear techniques for GHGs. Springer, pp 1–10

Zhang A, Cui L, Pan G, Li L, Hussain Q, Zhang X et al (2010) Effect of biochar amendment on yield and methane and nitrous oxide emissions from a rice paddy from Tai Lake plain, China. Agric Ecosyst Environ 139(4):469–475

Zhou Y, Li X, Liu Y (2020) Land use change and driving factors in rural China during the period 1995-2015. Land Use Policy 99:105048

Chapter 17
Harnessing Nanotechnology for Sustainable Plant Protection Approach

Shivali Sharma, Dhanapati Keerthana, Jonnada Likhita, Gaurav Sharma, Shubham A. Durgude, Susheel Kumar Singh, Ruchi Rajpoot, and Shiv Vendra Singh

Abstract Nanotechnology leads agricultural innovation, offering solutions for plant protection and nutrition amid climate change. This chapter examines how nanotechnology provides climate-smart strategies to enhance agriculture. Nano-fertilizers revolutionize nutrient management by delivering precise nutrients to crops, boosting yields, and reducing fertilizer runoff, which contributes to water pollution and greenhouse gas emissions. Similarly, nano-pesticides offer effective pest control with less chemical use, aligning with climate-smart practices by lowering the carbon footprint of crop protection. Nanosensors are vital for adaptive agriculture, enabling real-time monitoring of plant health, soil, and water needs, allowing farmers to make informed decisions, optimize resources, and increase resilience to climate variability. The chapter also highlights nanotechnology's role in water conservation through advanced irrigation and purification methods, essential in areas facing climate-induced water scarcity. Despite these benefits, integrating nanotechnology into agriculture poses challenges, including regulatory and safety concerns. This chapter addresses these issues while showcasing nanotechnology's potential to transform agricultural practices. By focusing on plant protection and nutrition, it emphasizes nanotechnology's promise to foster a sustainable and efficient future for food production.

Keywords Nanomaterials · Nano-pesticides · Nano-fertilizers · Sustainable production · Food security

---

S. Sharma · G. Sharma
College of Horticulture and Agroforestry, Rani Lakshmi Bai Central Agricultural University, Jhansi, India

D. Keerthana · J. Likhita · S. K. Singh · R. Rajpoot · S. V. Singh
College of Agriculture, Rani Lakshmi Bai Central Agricultural University, Jhansi, India

S. A. Durgude (✉)
Borlaug Institute of South Asia-CIMMY, Bihar, India
e-mail: s.durgude@cgiar.org

B. Pramanick et al. (eds.), *Climate-Smart Agricultural Technologies*, Sustainability Sciences in Asia and Africa, https://doi.org/10.1007/978-981-96-7699-6_17

17.1 Introduction

In the face of unprecedented global challenges, agriculture stands at a critical juncture. As the world's population is projected to reach 9.7 billion by 2050, a 34% increase from current levels, the imperative to increase food production by a staggering 60% looms large. The twenty-first century has ushered in unprecedented environmental challenges, with climate change emerging as a formidable threat to agricultural productivity. Recent studies indicate that global crop yields could decline by 3–7% for every degree Celsius increase in average temperature (Zhao et al. 2018). Simultaneously, demographic projections suggest a surge in global food demand, necessitating a 60% increase in agricultural output by mid-century. This climatic upheaval exacerbates existing challenges in crop protection. Pests, a perennial nemesis of farmers worldwide, continue to wreak havoc on crop yields. The statistics are sobering: Rice production suffers a 25% loss, wheat 5–10%, pulses 30%, oilseeds 35%, sugarcane 20%, and cotton a staggering 50% due to pest infestations (Dhaliwal et al. 2010; Joseph and Morrison 2006). These figures underscore the urgent need for innovative crop protection strategies that can withstand the dual pressures of climate change and pest proliferation.

Traditionally, agriculture has leaned heavily on chemical interventions to combat these challenges. For over five decades, farmers have relied on a triumvirate of solutions: synthetic fertilizers, pesticides, and genetically modified crops (Yadav et al. 2013). While these approaches have undeniably boosted productivity, they have also left in their wake a host of environmental concerns. Soil degradation, water pollution, and the emergence of pesticide-resistant pests stand as testament to the unsustainability of these conventional methods. These converging pressures underscore the urgency of innovative solutions in farming practices and crop management. Unfortunately, less consumption of nutritious food, especially in the developing world, is one of the ten risk factors for mortality. Nanotechnology, an interdisciplinary field that merges the principles of nanotechnology with biological systems, offers innovative solutions to these complex issues. Nano-formulations provide selective, targeted, and controlled release of pesticides, hence ecologically viable (Rajna et al. 2019) with added advantages of efficient management over conventional agrochemicals as well as environment sustainability (Kah et al. 2018). This cutting-edge field, operating at the nanoscale (1–100 nanometers), presents a unique opportunity to address agricultural challenges with precision and efficiency previously thought impossible along with immobilizing pollutants in the ecosystem (Singh et al. 2023). Nanotechnology's potential in agriculture spans a wide spectrum, from enhancing nutrient delivery systems to revolutionizing pest management strategies. One of the most promising applications of nanotechnology in crop protection lies in the development of nano-formulations for pesticide delivery. These advanced systems offer selective, targeted, and controlled release of active ingredients, marking a significant departure from conventional broad-spectrum applications. The result is a more ecologically viable approach to pest management, one that promises efficient control while minimizing environmental impact (Kah et al.

2018; Rajna et al. 2019). Beyond pest management, nanotechnology holds the key to optimizing nutrient delivery in crops. With current estimates suggesting that only 30% of applied fertilizers are utilized by plants (Durgude et al., 2022), there is vast room for improvement. Nano-enabled nutrient delivery systems have the potential to dramatically increase nutrient use efficiency, addressing the twin challenges of crop nutrition and environmental sustainability. Perhaps most excitingly, nanotechnology offers tools to develop crops with enhanced resilience to both biotic and abiotic stresses, a critical asset in the face of climate change.

Plant protection is an application of nanoscience with its scopes to enhance the efficiency of agriculture practices and contribute toward sustainable farming. Figure 17.1 demonstrates the targeted application of synthesized nanomaterials in plant protection. Nano-pesticides have proven to reduce the use of pesticides by 50% while providing better efficacy against pests. Pesticides are assumed to be more targeted that entails both cost-effectiveness and impact on the environment (Giraldo et al. 2014). Nano-fertilizers also improve nutrient uptake by 30–50%, which helps reduce waste nutrient runoff and optimum utilization of applied fertilizers (Cai et al. 2014). Other biological delivery systems that use nanotechnology improve biocontrol efficacy by 70%, which enhances accurate pest targeting for crop protection without causing undesirable effects on other beneficial organisms (Pradhan et al. 2020). Soil remediation is another area where nanoscience has

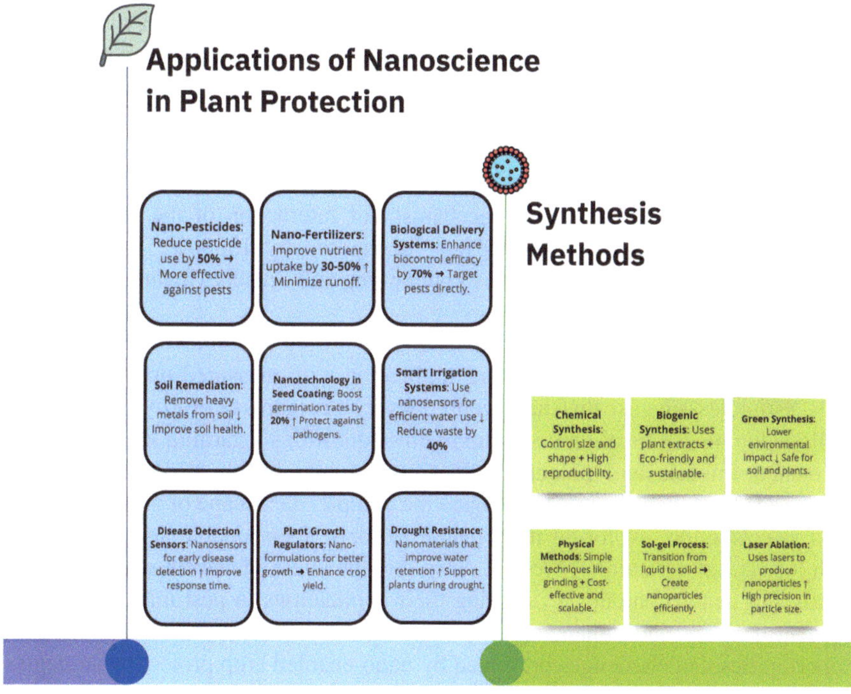

Fig 17.1 Applications of nanoscience as targeted solutions for sustainable farming

played a very crucial role, especially in the removal of heavy metals from the soil, hence enhancing soil health and reducing pollution. Nanotechnology for seed coating contributes positively to germination by 20%, with additional protection from pathogens, thereby aiding crop establishment (Cheloni et al. 2016). These smart irrigation systems, equipped with nanosensors to monitor the use of water, save around 40% of water otherwise wasted and also make irrigation more efficient (Davarpanah et al. 2016). Several mechanisms are involved in nanocomposites, which help deliver efficiency and resilience in agriculture. An essential mechanism is that of increased bioavailability of nutrients, meaning its effectiveness on the plant can be increased by as much as 90% (Giraldo et al. 2014). This facilitates greater availability from nutrient uptake by the plant, hence making the plants grow healthy.

Another benefit is enhanced uptake of nutrients through improved absorption in the nut, elevating the nutrient uptake capacity by 50%, thereby ensuring well-nourished crops (Cai et al. 2014). The controlled release of nutrients is an integral feature of nanocomposites and therefore extends the nutrient availability time in the soil up to several weeks. This resultant mechanism of nutrient sustaining depletes fewer nutrients and ensures their availability with time (Hossain et al. 2011). Furthermore, stress mitigation provided by nanocomposites enhances survival percentages by about 20–30% under conditions of drought stress, thus providing an essential benefit in regions experiencing water scarcity (Davarpanah et al. 2016). Another major development is the targeted delivery of pesticides through nanocomposites, which reduces the application of pesticides by 30–70% thereby resulting in less environmental contamination and enhanced effective pest management (Cheloni et al. 2016). Nanocomposites also activate plant defense mechanisms, thereby increasing the resistance level by 40%, which thus boosts the innate ability of crops to fight diseases (Pradhan et al. 2020). Nanocomposites also exhibit antimicrobial properties which lower fungal infections by 60%, hence keeping the crops healthy (Bortolin et al. 2013). The final mechanism of interaction of the soil is that it improves the nutrient retention capacity in the soil by 15–20%. Therefore, the growth of plants is enhanced besides maintaining soil fertility (Inyang et al. 2016) (Fig. 17.2).

However, as we stand on the brink of this nano-agricultural revolution, it is crucial to approach its development and application with measured optimism. While early estimates suggest that nanotechnology could increase crop yields by 30% while reducing chemical inputs by up to 50%, questions regarding long-term environmental impacts and potential effects on human health must be rigorously addressed. This chapter aims to navigate the complex landscape of nanotechnology in crop protection, exploring its potential to forge a sustainable and eco-friendly path forward in agriculture. We will delve into the latest advancements in nanomaterials and nanotechniques, examining their applications in pest management and nutrient delivery and enhancing crop resilience. By critically assessing both the opportunities and challenges presented by nano-enabled crop protection strategies, we seek to provide a comprehensive overview of how this cutting-edge field can contribute to more sustainable and resilient agricultural practices.

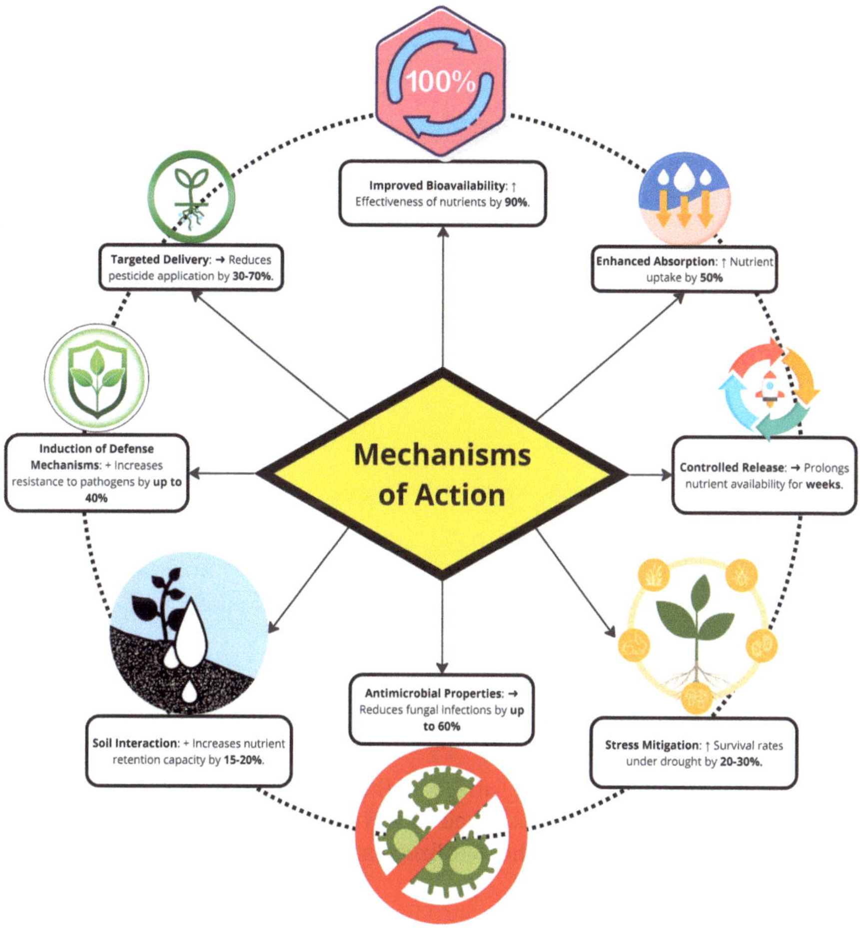

Fig. 17.2 Nanocomposites enhancing nutrient efficiency, plant protection, and soil and ecosystem resilience

17.2 Nanomaterial Synthesis Pathway

The synthesis process for the production of nanocomposites intended for use in agriculture starts with the chemical impregnation process (Fig. 17.3). In this process, urea undergoes adsorption in a porous structure of nanomaterials such as mesoporous silica or graphene oxide. This ensures the impregnation of nutrients within the nanomaterial in a stable and controlled manner to form a nanocomposite necessary for controlled nutrient release (Cai et al. 2014). The solution is then thermally treated under adequate conditions for the uniform deposition of nutrients inside the nanomaterial. This is done to minimize aggregation and ensure uniform quality (Giraldo et al. 2014). For enhanced nutrient retention, another technique that is used

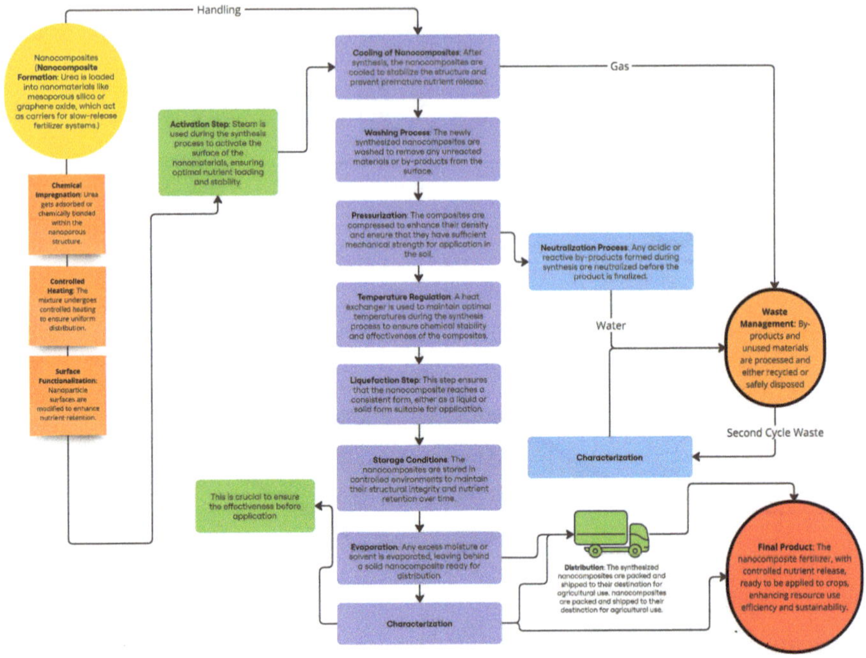

Fig. 17.3 Synthesis process of nanocomposites: from raw materials to sustainable agricultural solutions

is surface functionalization. The surface of the nanomaterials is thus changed to enhance its compatibility in the soil and root system of a plant. Therefore, there is proper nutrient uptake when the fertilizers are applied in the fields (Hossain et al. 2011).

17.2.1 Control, Activation, and Treatment

After the first synthesis, steam activation is already performed on the nanocomposites. This ensures the activation of the surface of the nanomaterial so that nutrient loading stability will be maximized whenever nutrients are captured. The nanocomposites are then cooled to ensure that the material will not freely release the encapsulated nutrient before it undergoes any testing or application process. Second, washing procedures are conducted to eliminate unreacted materials and impurities within. It removes impurities and leaves behind only a clean and efficient product (Bortolin et al. 2013).

17.2.2 Pressurization, Temperature Regulation, and Liquefaction

In the coming steps, compression of nanocomposites will be subjected to pressurization in order to achieve a sufficient density and mechanical strength for proper application of the material in soils. Through temperature regulation, a heat exchanger is attached to regulate steady conditions which ensure chemical stability during synthesis. In liquefaction, further consistency in formulation will be achieved whether in liquid or solid state that will be acceptable for various application requirements. However, proper storage conditions are crucial to ensure the integrity of the nanocomposites, thereby guaranteeing effectiveness until application in the field (Inyang et al. 2016).

17.2.3 Final Steps: Evaporation, Waste Treatment, and Product Distribution

Evaporation is the final step used to remove excess solvents or moisture to obtain a ready-to-distribute solid nanocomposite material as a by-product of completing the synthesis process. Characterization tests characterize the final product, checking whether it complies with requirements for quality and efficiency in agriculture (Cheloni et al. 2016). Reactive byproducts are neutralized safely by using appropriate neutralization in order to achieve as small an environmental effect as possible. Proper disposal of residual materials involves recycling and safe disposal of residual materials (Davarpanah et al. 2016). Finally, the product is a fertilizer nanocomposite which is packed and distributed to provide an answer that promotes resource use efficiency by controlling nutrient release and assuring sustainable farming methods (Pradhan et al. 2020).

17.3 Nano-formulations for Targeted Delivery of Pesticides

17.3.1 Nanoemulsions

Nanoemulsions are also the most important formulations to enhance the solubility and dissolution properties of poorly water-soluble substances (Wang et al. 2009). They have a small droplet size (20–200 nm) and are also referred to as mini-emulsions or ultrafine emulsions (Fernandes et al. 2014). These formulations enhance the solubility and bioavailability of poorly water-soluble pesticides, hence allowing enhanced penetration and distribution in plant tissues (Adak et al. 2012).

17.3.2 Nanocapsules

Nanocapsules are submicroscopic colloidal drug carriers with sizes ranging from 10 to 1000 nm, consisting of a liquid or solid core surrounded by a thin polymer membrane. Nanocapsules have diverse applications, including drug delivery, cosmetics, and agriculture. Nanocapsules can be engineered so that their contents are released upon specific triggers, such as pH changes or enzymatic activity (Pérez-de-Luque 2017).

17.3.3 Nanosuspension

Nanosuspensions are formulations of colloidal dispersion that contain active ingredient nanoparticles. Nanosuspension consists of the pure poorly water-soluble drug without any matrix material suspended in dispersion. For effective uptakes and systemic distribution of pesticides within plants in the control of pests and diseases, the use of nanosuspensions is quite important (Kah et al. 2018).

17.3.4 Nanoencapsulated Biopesticides

The coating of various substances within another material at various sizes in the nano-range and through which a chemical is slowly but efficiently released is called nanoencapsulation (Ragaei and Sabry 2014). Li et al. (2016) stated that nanocapsules exhibited a higher release rate than microcapsules because materials with a smaller size have larger surface areas being exposed to the surroundings. In other words, nanoencapsulated biopesticides are an evolutionarily safe and sustainable alternative to synthetic chemicals (Ghormade et al. 2011).

17.3.5 Lipid-Based Nanoparticles

Lipid-based nanoparticles (LBNPs) have emerged as a promising technology in agriculture and medicine. In agriculture, LBNPs can penetrate plant leaves, translocate bidirectionally, and deliver nutrients more effectively than conventional methods, potentially addressing nutrient deficiencies and increasing crop yields. These nanoparticles can be designed to carry various agrochemicals, including pesticides, herbicides, and fertilizers, offering improved crop protection and soil enhancement. LBNPs have advantages such as high stability, loading capacity, and cost-effectiveness.

17.3.6 Nanoencapsulation

This implies coating active substances with a nanoscale material, which enables the slow and efficient release of chemicals. According to Ragaei and Sabry (2014), with nanoencapsulation, the pesticide is only released in certain specified environments such as specific pH, temperature, moisture, and ultrasound frequency or in the presence of specific compounds (Urkude 2019). Since nanocapsule surface area relative to their size is large compared to microcapsules, it results in a higher rate of release. This makes them a very critical means of technique in protecting plants from pests (Li et al. 2016). This method has phenomenal applicability in the management of insect pests.

17.3.7 Nanogels

Nanogels can be defined as the aqueous dispersion of hydrogel particles that are, in turn, formed by cross-linked polymer networks on a nanoscale. Since they are more chemically, thermally, and mechanically stable than their natural pheromone counterparts, nanogel formulations can be highly useful in protecting the active ingredient from evaporation and environmental degradation (Raju 2019). Bhagat et al. (2013) demonstrated that nano-gelled pheromones successfully managed *Bactrocera dorsalis*, a key pest of fruit like guava—by providing long-lasting residual activity and high efficacy even in rainy seasons.

17.3.8 Metallic Nanoparticles

Only a few metallic and metal oxide nanoparticles are known to possess the potential for insecticidal activity. Nanostructured alumina, gold, silver, and cadmium sulfide have been reported to show toxicity against the pests (Ziaee and Hamzavi 2013). For example, Debnath et al. (2012) showed the amorphous lipophilic silica nanoparticles to be toxic to a grain pest called *Tribolium castaneum*. Moreover, nano-silica has been found to show insecticidal toxicity against the mosquito species which are vectors for various diseases (Goswami et al. 2010). DNA-tagged gold nanoparticles exhibited strong activity against *Spodoptera litura* (Chakravarthy et al. 2012), and the silver nanoparticle was reported to be more toxic to *Sitophilus oryzae* than silver nitrate plus *Euphorbia prostrata* extract, since less of the former was required to kill the adult insect (Zahir et al. 2012).

17.3.9 Nano-botanical Pesticides

Consumers are increasingly insisting on safe products; hence, there come uses for botanical-based insecticides. However, its use is limited owing to environmental instability (Forim et al. 2013). Nanotechnology enhances the stability and effectiveness of those botanicals (Ghormade et al. 2011). Nanostructured botanicals, like azadirachtin, rotenone, carvacrol, thymol, and curcumin, are very effective in controlling pests (Shah et al. 2016). Forim et al. (2013) observed that nanoparticles loaded with neem powder showed 100% larval mortality of diamondback moths and offered greater UV stability.

17.4 Mechanism of Nanoparticle Interaction with Plant Pathogens

17.4.1 Uptake and Translocation Mechanism

The uptake and translocation of nanoparticles in plants is a multistep mechanism which is pretty complex. Nanoparticles enter the cell membrane and cell wall of the epidermis of the root, and after passing through the stele and xylem, they reach the leaves (Tripathi et al. 2017a, b). Nanoparticles cross the intact cell membrane through size-specific pores for crossing the intact cell membrane (Rico et al. 2011). They move passively through the apoplast of the endodermis before reaching the stele. Nanoparticle uptake often employs active transport, with the use of cellular events like signaling, recycling, and plasma membrane regulation (Etxeberria et al. 2009; Tripathi et al. 2017a, b). Nanoparticles enter plant roots according to the forces of osmotic pressure, capillary forces, and cell wall pores, moving through plasmodesmatal connections or symplastic pathways or binding on carrier proteins, crossing through ion channels or aquaporins or entering by new pore formation via endocytosis (Liu et al. 2010; Gao et al. 2011). Once inside the plant cells, they move through the apoplast or symplast between the cells via plasmodesmata (Hauck et al. 2008). Crossing of the nanoparticles over the cell wall depends on the size of the particle. The smaller particles easily pass over the cell wall, while the larger particles enter via stomata, hydathodes, or flower stigmas (Fleischer et al. 1999; Hossain et al. 2016). The particles greater than 40 nm are transferred through stomatal pores and tend to accumulate there; they are then conducted throughout the other parts of the plant via the phloem (Tripathi et al. 2017a, b). Nanoparticles also penetrate seeds through parenchymatous spaces of the seed coat. Aquaporins regulate this entry into seeds (Lee et al. 2010; Abu-Hamdah et al. 2004).

17.4.2 Defence Mechanism

When plants come in contact with metallic nanoparticles, they may undergo oxidative stress, which brings about the onset of ROS and at the same time the activation of the antioxidant defense system (Rico et al. 2011). The antioxidant defense system is constituted by enzymatic elements, such as ascorbate peroxidase, catalase, superoxide dismutase, guaiacol peroxidase, and glutathione reductase, and also nonenzymatic antioxidants like glutathione, ascorbate, thiols, and phenolics (Rico et al. 2011; Singh and Lee 2016). Catalase and guaiacol peroxidase together with superoxide dismutase are involved in obviating the effects of the ROS and peroxy radicals, while Rico et al. (2011) reported about the conversion of superoxide to hydrogen peroxide via superoxide dismutase. Toxicity of titanium dioxide (TiO2) nanoparticles can be a result of the generation of ROS in the process of photocytotoxicity, but they are also signaling molecules that may activate the plants' antioxidant defense system, which detoxifies the radicals (Yin et al. 2012). In the case of ROS, ascorbate peroxidase takes water from hydrogen peroxide (H_2O_2) (Rico et al. 2011). The plant species have developed several biochemical mechanisms for the detoxification of ROS generated by nanoparticles as a result of oxidative stress (Wei and Wang 2013). Different types of nanoparticles (NMs) induce the activation of antioxidant enzymes in the following ways: nFe2O4, nCeO2, and nCo3O4 cause catalase to be triggered; nFe_3O_4, nCeO_2, nMnO_2, nCuO, and nAu cause guaiacol peroxidase to be increased; and nCeO_2, nPt, and fullerene cause superoxide dismutase activity to be increased (Tripathi et al. 2017a, b). Supplementation with nTiO2 in spinach intensified the activity of superoxide dismutase, catalase, ascorbate peroxidase, and guaiacol peroxidase proteins (Lei et al. 2008). Furthermore, the study of Song et al. (2012) showed that the activity of superoxide dismutase and catalase was increased with guaiacol peroxidase. The use of TiO_2 nanoparticles at lower concentrations led to increased levels of chlorophyll, peroxidase, catalase, malondialdehyde, and superoxide dismutase through the removal of ROS.

17.5 Influence of Nanomaterials in Plants

On the other hand, it has been noted that plants might experience various types of morphological changes according to the NMs used and their concentrations (Siddiqi et al. 2018). The influence of NMs on plants is either positive or adverse. Nano-growth stimulants have been proven to enhance seed germination and plant growth at later developmental stages (Nadiminti et al. 2013; Aslani et al. 2014). For instance, nano-TiO_2 and nano-SiO_2 seed treatment in soybean raised nitrate reductase activity that promoted seed germination with a more direct positive effect with the combined application (Lu et al. 2002). In spinach (*Spinacia oleracea* L.), the assimilation of nitrogen and photosynthesis was increased by TiO_2 nanoparticles at 0.25% concentration, and it increased dry weight by about 44% as compared to the control

(Zheng et al. 2005; Yang et al. 2006). Presoaking wheat seeds in MWCNT solution at a concentration of 40–160 μg L^{-1} increased root elongation and biomass production, although some experiments revealed no effects (Wang et al. 2012). Seeds of Indian mustard pre-treated with the oxidized MWCNT solution at the range of 2.3–46.0 μg L^{-1} increased seed germination percentage and uniformity, as well as root and shoot development (Mondal et al. 2011). Even similar results were obtained for the soaking of watermelon (*Citrullus lanatus*) seeds in Fe_2O_3 nanoparticles which improved germination, growth, and fruiting (Li et al. 2013). Similarly, low concentrations of SiO_2 nanoparticles improved tomato seed germination (Siddiqui and Al-Whaibi 2014).

NMs can be used in plant cultures in vitro. Indeed, ZnO nanomaterials had more pronounced effects on the growth and physiological properties of tobacco calli compared to other forms of zinc because the protein content was higher and the growing ability was stronger (Mazaheri-Tirani and Dayani 2020). Foliar treatment of ZnO NPs at a dose of 10 mg L^{-1} enhanced the amount of chlorophyll, soluble protein, and phosphorus in cluster beans (*Cyamopsis tetragonoloba*) (Raliya and Tarafdar 2013). NMs are reported to enhance the seed germination percentage and nitrogen-fixing ability in plants (Yang et al. 2006). Foliar application of SiO2 nanoparticles on sugarcane (*Saccharum officinarum* L.) promoted growth under cold stress by maintaining effective quantum yield in photosynthesis and supplying photoprotection (Elsheery et al. 2020).

Though many authors have written about the positive impacts of NMs (Mondal et al. 2011; Raliya and Tarafdar 2013; Aslani et al. 2014), they also affect plants in a phytotoxic manner (Mazaheri-Tirani and Dayani 2020). The degree of these impacts depends upon the factors used for exposure such as NM dosage, size, nature, exposure time, and prevailing environmental conditions (Mazaheri-Tirani and Dayani 2020; Noori et al. 2020). For instance, while Yusefi-Tanha et al. (2020) showed that smaller-sized CuO nanoparticles (25 nm) were more lethal to soybeans than larger-sized particles (50 nm and 250 nm), the concentration response for smaller-sized particles was linear, while it was nonlinear for larger-sized particles. The toxicity of the nanoparticle is predominantly described by its germination impact, biomass, leaf number, and root elongation (Lin and Xing 2007; Doshi et al. 2008). Some other cytotoxic effects of OS include reduced seed germination, growth inhibition, or even death of plants (Yang et al. 2017), slow growth (Colman et al. 2013), altered subcellular metabolism (Zheng et al. 2005), oxidation damage to cell membranes (Noori et al. 2020), reduced photosynthesis (Barhoumi et al. 2015), chromosomal abnormalities (Raskar and Laware 2014), impaired transport of water (Martínez-Fernández et al. 2016), and disrupted gene expression (García-Sánchez et al. 2015).

Moreover, nanoparticles may hinder energy and electron transport pathways due to the suppression of genes encoding for transcription (Hossain et al. 2016). On the other hand, some nanoparticles may trigger stress response–related genes together with genes that code for water channels (Tripathi et al. 2017a, b). For instance, MWCNTs induce genes related to water transport, cell wall formation, and cell division (Khodakovskaya et al. 2012). Toxicity may be reported at high NM concentrations. For example, CeO2 nanoparticles (2000 and 4000 mg L^{-1}) have been found to cause DNA damage in soybean (López-Moreno et al. 2010), while TiO2 nanoparticles caused DNA damage in tobacco (Ghosh et al. 2019) and maize (Ruffini Castiglione et al. 2011). Such results make it necessary to test NMs on some plant species before applying them in the fields or selling them. Various nanoparticles and their active ingredient against different pests and pathogens are listed in Table 17.1.

Table 17.1 Nanoparticles and their active ingredients used against pests

Sl. no.	Nanomaterial	Crop species	Response	References
1	AgNPs	Tomato	Increased growth and resistance to pathogens	Khan et al. (2019)
2	Chitosan NPs	Rice	Increased germination and disease resistance	Divya et al. (2018)
3	TiO_2 NPs	Wheat	Improved photosynthesis and growth	Shah et al. (2016)
4	ZnO NPs	Maize	Enhanced growth and yield	Rameshraddy et al. (2017)
5	CuNPs	Potato	Reduced bacterial and fungal infections	Ismail (2021)
6	AuNPs	*Arabidopsis thaliana*	Increased root growth and nutrient uptake	Geisler-Lee et al. (2013)
7	rGO NPs	Lettuce	Improved growth and reduced pathogen infection	Begum, et al. (2011)
8	CeO_2 NPs	Corn (*Zea mays*)	Increased resistance to oxidative stress	Rico et al. (2011)
9	Mo NPs	Pea	Enhanced nitrogen fixation and growth	Sharma et al. (2016)
13	TiO_2 and SiO_2	*Oryza sativa*	Mitigated Cd toxicity and improved growth by stimulating antioxidant potential and inhibiting Cd translocation	Rizwan et al. (2019)
14	CuO	*Spinacia oleracea*	Improved photosynthesis and biomass production	Wang et al. (2019)
15	Ag NPs	*Vigna unguiculata*	Showed no phytotoxicity, but could inhibit the growth of *Xanthomonas axonopodis* pv. *malvacearum* and *Xanthomonas campestris* pv. *campestris* in vitro	Vanti et al. (2019)
16	Fe/SiO_2	*Arachis hypogaea, Zea mays*	Enhanced plant growth and biomass accumulation	Disfani et al. (2017)
17	MWCNTs	*Wheat, Maize, Garlic*	Improved and rapid germination, increased biomass accumulation, and water absorption potential of seeds	Srivastava and Rao (2014)

17.6 Effectivity of Nanocomposites in Plant Protection

Advanced solutions provided by nanotechnology in enhancing the detection, monitoring, and control of pathogens make plant disease management possible. Some of these include nanosensors and nano-biosensors, which allow for early detection through low levels of pathogens, such as gold nanoparticles and quantum dots that can bind specifically to a specific number of pathogens, offering high sensitivity (Rai et al. 2012; Zhang et al. 2018). Imaging is done through magnetic nanoparticles (MNPs) that visualize infected plant tissues (Huang et al. 2015). Nano-pesticides make pesticides more effective, reduce the adverse impact on the environment, and provide antibacterial activity as seen in silver nanoparticles (AgNPs) (Gogos et al. 2012). Nanocarriers in controlled-release formulations reduce cycles of pesticide application and limit contamination (Khot et al. 2012). Nano-fertilizers based on ZnO and TiO_2 improve plant growth and resistance to pathogens (Servin et al. 2015). These advances bring effective products to plant disease management for sustainable outputs (Table 17.2). Nano-pesticides offer numerous benefits over conventional pesticides, particularly in terms of reducing environmental impact and improving efficacy. These advantages stem from their unique properties, including enhanced delivery mechanisms, targeted action, and reduced usage amounts. The significant findings on the impact of nanocomposites that led to efficient plant protection are highlighted in Table 17.2.

Nano-pesticides have a high advantage over traditional pesticides because they reduce or show less environmental impact and much higher efficiency. Due to the small size of nanoparticles, they increase the surface area to volume ratio; thus, the active ingredients enhance their solubility and bioavailability. This way, uptake by pests or plants is raised at lower doses, further enhancing effectiveness (Li et al. 2016). Secondly, among various nano-based delivery systems, nanoencapsulation is a process that allows the slow release of pesticides, so the active ingredients will remain effective for longer periods with the occasion of less frequent application (Ragaei and Sabry 2014). Nanoparticles also reach better parts of the plant tissues compared to conventional pesticides and are faster in targeting pests (Chhipa 2019). They assist in keeping the active ingredients at the appropriate concentration to ensure reduced possibilities of the development of resistance to pests and prolonging the pesticide's effectiveness (Kah et al. 2018). Moreover, some of the nano-pesticide formulation degrades faster upon delivering their payload, thus leaving behind less residue that lessens the long-term ecological impacts often associated with conventional pesticide residues (Kookana et al. 2014).

Table 17.2 Effect of nanomaterials in plant protection: insect and pest management

Sl. no.	Nanostructure	Active ingredient	Biopolymer	Effect/target phytopathogen	References
1	Silver nanoparticles	Silver (Ag)	Chitosan	Antifungal and antibacterial against *Alternaria solani* and *Xanthomonas campestris*	Siddiqi and Husen (2022)
2	Magnesium oxide nanoparticles	Magnesium oxide (MgO)	Chitosan	Antibacterial against *Agrobacterium tumefaciens*	Choudhary and Kumar (2015)
3	Gold nanoparticles	Gold (Au)	Gelatin	Antifungal and antibacterial against *Rhizoctonia solani* and *Erwinia carotovora*	Shabana et al. (2020)
4	Silica nanoparticles	Silica (SiO2)	Chitosan	Antifungal against *Botrytis cinerea*	Qi et al. (2017)
5	Copper nanoparticles	Copper (Cu)	Alginate	Antibacterial against *Pseudomonas syringae*	Azam et al. (2017)
6	Zinc oxide nanoparticles	Zinc oxide (ZnO)	Gelatin	Antifungal against *Fusarium oxysporum*	Tarafdar et al. (2020)
7	Gold nanoparticles	Gold (Au)	Polyvinyl alcohol	Antiviral effect on Tobacco mosaic virus	Vargas-Hernandez et al. (2020)
8	Titanium dioxide (TiO2)	Titanium (Ti)	Pectin	Antifungal effect on *Alternaria dauci*	Ahamad and Siddiqui (2021)
9	Magnesium oxide (MgO)	Magnesium (Mg)	Carrageenan	Antibacterial effect on *Erwinia amylovora*	Kralova and Jampilek (2023)
10	Iron oxide (Fe2O3)	Iron (Fe)	Guar gum	Antifungal effect on *Botrytis cinerea*	Tiwari et al. (2023)
11	Nickel oxide (NiO)	Nickel (Ni)	Xanthan gum	Antibacterial effect on *Agrobacterium tumefaciens*	Sudheer et al. (2022)
12	Graphene oxide (GO)	Carbon (C)	Chitosan	Broad-spectrum antimicrobial activity	Gu et al. (2021)
13	Selenium nanoparticles	Selenium (Se)	Cellulose	Antifungal effect on *Aspergillus niger*	Islam et al. (2022)
14	Copper oxide (CuO)	Copper (Cu)	Gum arabic	Antibacterial effect on *Ralstonia solanacearum*	El-Batal et al. (2020)
15	Calcium carbonate (CaCO3)	Calcium (Ca)	Sodium alginate	Enhanced resistance against downy mildew	Tayeh et al. (2014)
16	Zinc sulfide (ZnS)	Zinc (Zn)	Chitosan	Antibacterial effect on *Xanthomonas axonopodis*	Choudhary et al. (2023)

17.7 Limitations and Future Prospects

The limitations identified in the application of nanotechnology for plant protection underscore the complexity of integrating these advanced materials into agricultural systems. However, these challenges also serve as catalysts for innovation and future research directions.

- Difficulty in ensuring uniform quality and stability of nanocomposites during production, including issues with aggregation and controlled nutrient release.
- Achieving the appropriate density and mechanical strength of nanocomposites for effective soil application remains challenging.
- Maintaining the integrity and effectiveness of nano-formulations during storage and transportation poses significant hurdles.
- Concerns about the long-term fate and persistence of nanomaterials in ecosystems, with potential unforeseen environmental impacts.
- Ensuring safe disposal and recycling of reactive byproducts and residual materials from nanomaterial synthesis processes.
- Lack of comprehensive regulatory frameworks and guidelines for the registration, approval, and use of nano-enabled agricultural products.
- Incomplete understanding of potential effects on human health and the environment, necessitating extensive safety assessments.
- Higher production costs compared to conventional agricultural inputs, potentially limiting widespread adoption, especially in developing countries.
- Difficulties in scaling up production from laboratory to commercial levels while maintaining efficacy and cost-effectiveness.
- Limited long-term studies on the interactions between nanomaterials and complex agricultural ecosystems, hindering risk assessment and management strategies.

Limitations in applying nanotechnology to plant protection are significant, and they are not insurmountable. By acknowledging these limitations and actively working to overcome them, we can harness the full potential of nanotechnology to transform agriculture, enhance food security, and adapt to the challenges posed by climate change. Nanotechnology has already demonstrated significant potential in transforming agriculture, and the future prospects outlined provide a roadmap for further advancements. By focusing on enhanced nano-formulations, precision agriculture, and biodegradable materials, we can address current challenges and unlock new opportunities for sustainable and efficient agricultural practices. These future directions underscore the importance of continued research and collaboration to fully realize the transformative impact of nanotechnology in agriculture.

- Future research could focus on developing nano-formulations with improved encapsulation efficiency and controlled release properties. This could involve exploring novel materials such as dendrimers and liposomes, which offer precise delivery mechanisms and enhanced stability of active ingredients.

- The advancement of nanosensors with higher sensitivity and specificity could revolutionize precision agriculture. These sensors could be integrated with IoT (Internet of things) platforms to provide real-time data on soil moisture, nutrient levels, and pest presence, enabling farmers to make informed decisions and optimize resource use.
- Developing biodegradable nanomaterials, such as those based on natural polymers like chitosan and cellulose, could address environmental concerns. These materials would degrade into non-toxic byproducts, reducing the ecological footprint of nanotechnology applications in agriculture.
- Nanotechnology could be integrated with biological control methods to enhance pest management strategies. For example, nanoencapsulated pheromones could be used to disrupt pest mating patterns, while nanocarriers could deliver biopesticides more effectively to target pests.
- Research into nanomaterials that enhance plant resilience to abiotic stresses, such as drought and salinity, could support climate adaptation strategies. This could involve the development of nanocomposites that improve water retention in soil or enhance the expression of stress-responsive genes in plants.
- Establishing comprehensive regulatory frameworks will be crucial to ensure the safe deployment of nanotechnologies in agriculture. This includes developing standardized testing protocols for assessing the safety and efficacy of nanomaterials, as well as guidelines for their use and disposal.
- Increasing public awareness and education about the benefits and risks of nanotechnology could facilitate its acceptance and adoption. This could involve outreach programs, workshops, and collaborations with educational institutions to disseminate knowledge and address public concerns.
- Encouraging collaboration between scientists, policymakers, and industry stakeholders could accelerate the innovation and implementation of nanotechnologies in agriculture. This could involve forming interdisciplinary research consortia and public-private partnerships to address complex challenges and drive technological advancements.

17.8 Conclusion

The chapter provides a comprehensive overview of the transformative role of nanotechnology in agriculture, emphasizing its potential to address critical challenges such as climate change, food security, and sustainable crop protection. Quantitatively, the use of nano-enabled crop protection technologies can reduce pesticide usage by up to 50%, significantly lowering environmental impact while maintaining effective pest control. Nano-formulations, such as nanoemulsions and nanocapsules, enhance the solubility and bioavailability of active ingredients, leading to a 30% increase in efficacy compared to traditional methods. In nutrient management, nano-fertilizers have demonstrated the ability to improve nutrient use efficiency by 40%, reducing fertilizer runoff and optimizing crop nutrition. This is achieved through mechanisms like increased bioavailability and controlled release, which ensure precise nutrient delivery to crops. Nanosensors play a crucial role in adaptive agriculture, providing real-time monitoring of plant health, soil, and water conditions. This technology enables farmers to make data-driven decisions, enhancing crop resilience to biotic and abiotic stresses and increasing adaptability to climate variability. Nanotechnology also offers advanced solutions for sustainable water management, with innovations in irrigation and water purification methods addressing climate-induced water scarcity. However, the chapter highlights the need to address regulatory and safety concerns, as well as potential long-term environmental impacts, to ensure the responsible integration of nanotechnology in agriculture. In conclusion, the chapter illustrates that by harnessing the power of nanotechnology, we can achieve a more sustainable and efficient future for agriculture, effectively tackling the dual challenges of climate change and food security. The quantitative benefits underscore the potential of these technologies, but a measured and responsible approach is essential to ensure their safe and effective integration into agricultural systems.

References

Abu-Hamdah R, Cho W-J, Cho S-J, Jeremic A, Kelly M, Ilie AE, Jena BP (2004) Regulation of the water channel aquaporin-1: isolation and reconstitution of the regulatory complex. Cell Biol Int 28:7–17

Adak T, Kumar J, Shakil NA, Walia S, Singh AK (2012) Development of controlled release formulations of imidacloprid employing novel nano-ranged carriers. J Environ Sci Health B 47(3):217–225

Ahamad L, Siddiqui ZA (2021) Effects of silicon dioxide, zinc oxide and titanium dioxide nanoparticles on *Meloidogyne incognita*, *Alternaria dauci* and *Rhizoctonia solani* disease complex of carrot. Exp Parasitol 230:108176

Aslani F, Bagheri S, Muhd Julkapli N, Juraimi AS, Hashemi FSG, Baghdadi A (2014) Effects of engineered nanomaterials on plants growth: an overview. Sci World J 2014:641759

Azam A, Ahmed AS, Oves M, Khan MS, Habib SS, Memic A (2017) Antimicrobial activity of metal oxide nanoparticles against Gram-positive and Gram-negative bacteria: a comparative study. Int J Nanomedicine 7:6003–6009. https://doi.org/10.2147/IJN.S35347

Barhoumi L, Oukarroum A, Taher LB, Smiri LS, Abdelmelek H, Dewez D (2015) Effects of superparamagnetic iron oxide nanoparticles on photosynthesis and growth of the aquatic plant *Lemna gibba*. Arch Environ Contam Toxicol 68:510–520

Bhagat D, Samanta SK, Bhattacharya S (2013) Efficient management of fruit pests by pheromone nanogels. Sci Rep 3(1):1–8

Bortolin A, Aouada F, Mattoso LHC, Ribeiro C (2013) Nanocomposite PAAm/methyl cellulose/montmorillonite hydrogel: slow release of fertilizers. J Agric Food Chem 61(31):7431–7439

Cai D, Wu Z, Jiang J, Wu Y, Feng H, Brown IG, Chu PK (2014) Controlling nitrogen migration through micro-nano networks. Sci Rep 4:3665

Chakravarthy AK, Bhattacharyya A, Shashank PR, Epidi TT, Doddabasappa B (2012) DNA-tagged nano gold: a new tool for the control of the armyworm, *Spodoptera litura* Fab. (Lepidoptera: Noctuidae). Afr J Biotechnol 11(38):9295–9301

Cheloni G, Marti E, Slaveykova VI (2016) Effects of copper oxide nanoparticles on green alga *Chlamydomonas reinhardtii*. Aquat Toxicol 170:120–128

Chhipa H (2019) Nanofertilizers and nanopesticides for agriculture. Environ Chem Lett 17:1141–1166

Choudhary P, Kumar R (2015) Synthesis and antibacterial activity of magnesium oxide nanoparticles. ScienceDirect, pp 26–38
Choudhary M, Pereira J, Davidson EB, Colee J, Santra S, Jones JB, Paret ML (2023) Improved persistence of bacteriophage formulation with Nano N-acetylcysteine–zinc sulfide and tomato bacterial spot disease control. Plant Dis 107(12):3933–3942
Colman BP, Arnaout CL, Anciaux S, Gunsch CK, Hochella MF Jr, Kim B, Lowry GV, McGill BM, Reinsch BC, Richardson CJ (2013) Low concentrations of silver nanoparticles in biosolids cause adverse ecosystem responses under realistic field scenario. PLoS One 8:e57189
Davarpanah S, Tehranifar A, Davarynejad G, Abadía J, Khorasani R (2016) Effects of foliar applications of zinc and boron nano-fertilizers on pomegranate yield and quality. Sci Hortic 210:57–64
Debnath N, Das S, Patra P, Mitra S, Goswami A (2012) Toxicological evaluation of entomotoxic silica nanoparticle. Toxicol Environ Chem 94(5):944–951
Dhaliwal GS, Jindal V, Dhawan AK (2010) Insect pest problems and crop losses: changing trends. Indian J Ecol 37(1):1–7
Disfani M, Mikhak A, Kassaee MZ, Maghari A (2017) Effects of nano Fe/SiO2 fertilizers on germination and growth of barley and maize. Archives of Agronomy and Soil Science 63(6):817–826
Divya K, Smitha V, Jisha MS (2018) Antifungal, antioxidant and cytotoxic activities of chitosan nanoparticles and its use as an edible coating on vegetables. International journal of biological macromolecules 114:572–577
Doshi R, Braida W, Christodoulatos C, Wazne M, O'Connor G (2008) Nano-aluminum: transport through sand columns and environmental effects on plants and soil communities. Environ Res 106:296–303
Durgude SA, Ram S, Kumar R, Singh SV, Singh V, Durgude AG, Hossain A (2022) Synthesis of Mesoporous Silica and Graphene-Based FeO and ZnO Nanocomposites for Nutritional Biofortification and Sustained the Productivity of Rice (*Oryza sativa L.*). J Nanobiotechnol 1:5120307
El-Batal AI, Balabel NM, Attia MS, El-Sayyad GS (2020) Antibacterial and antibiofilm potential of mono-dispersed stable copper oxide nanoparticles-streptomycin nano-drug: implications for some potato plant bacterial pathogen treatment. J Clust Sci 31:1021–1040
Elsheery NI, Sunoj VSJ, Wen Y, Zhu JJ, Muralidharan G, Cao KF (2020) Foliar application of nanoparticles mitigates the chilling effect on photosynthesis and photoprotection in sugarcane. Plant Physiol Biochem 149:50–60
Etxeberria E, Gonzalez P, Pozueta J (2009) Evidence for two endocytic transport pathways in plant cells. Plant Sci 177:341–348
Fernandes CP, de Almeida FB, Silveira AN, Gonzalez MS, Mello CB, Feder D, Falcão DQ (2014) Development of an insecticidal nanoemulsion with Manilkara subsericea (Sapotaceae) extract. J Nanobiotechnol 12(1):1–9
Fleischer A, O'Neill MA, Ehwald R (1999) The pore size of non-graminaceous plant cell walls is rapidly decreased by borate ester cross-linking of the pectic polysaccharide rhamnogalacturonan II. Plant Physiol 121:829–838
Forim MR, Costa ES, Da Silva MFDGF, Fernandes JB, Mondego JM, Boica Junior AL (2013) Development of a new method to prepare nano-/microparticles loaded with extracts of Azadirachta indica, their characterization and use in controlling Plutella xylostella. J Agric Food Chem 61(38):9131–9139
Gao J, Wang Y, Folta KM, Krishna V, Bai W, Indeglia P, Georgieva A, Nakamura H, Koopman B, Moudgil B (2011) *Polyhydroxy fullerenes* (fullerols or fullerenols): beneficial effects on growth and lifespan in diverse biological models. PLoS One 6:e19976
García-Sánchez S, Bernales I, Cristobal S (2015) Early response to nanoparticles in the Arabidopsis transcriptome compromises plant defence and root-hair development through salicylic acid signalling. BMC Genomics 16:341

García-Pinel B, Porras-Alcalá C, Ortega-Rodríguez A, Sarabia F, Prados J, Melguizo C, López-Romero JM (2019) Lipid-based nanoparticles: application and recent advances in cancer treatment. Nanomaterials 9(4):638
Geisler-Lee J, Wang Q, Yao Y, Zhang W, Geisler M, Li K, Ma X (2013) Phytotoxicity, accumulation and transport of silver nanoparticles by Arabidopsis thaliana. Nanotoxicology 7(3):323–337
Ghormade V, Deshpande MV, Paknikar KM (2011) Perspectives for nanobiotechnology enabled protection and nutrition of plants. Biotechnol Adv 29(6):792–803
Ghosh I, Sadhu A, Moriyasu Y, Bandyopadhyay M, Mukherjee A (2019) Manganese oxide nanoparticles induce genotoxicity and DNA hypomethylation in the moss Physcomitrella patens. Mutat Res Genet Toxicol Environ Mutagen 842:146–157
Giraldo JP, Landry MP, Faltermeier SM, McNicholas TP, Iverson NM, Boghossian AA, Reuel NF, Hilmer AJ, Sen F, Brew JA (2014) Plant nanobionics approach to augment photosynthesis and biochemical sensing. Nat Mater
Gogos A, Knauer K, Bucheli TD (2012) Nanomaterials in plant protection and fertilization: current state, foreseen applications, and research priorities. J Agric Food Chem 60(39):9781–9792
Goswami A, Roy I, Sengupta S, Debnath N (2010) Novel applications of solid and liquid formulations of nanoparticles against insect pests and pathogens. Thin Solid Films 519(3):1252–1257
Gu B, Jiang Q, Luo B, Liu C, Ren J, Wang X, Wang X (2021) A sandwich-like chitosan-based antibacterial nanocomposite film with reduced graphene oxide immobilized silver nanoparticles. Carbohydr Polym 260:117835
Hauck TS, Ghazani AA, Chan WC (2008) Assessing the effect of surface chemistry on gold nanorod uptake, toxicity, and gene expression in mammalian cells. Small 4:153–159
Hossain MK, Strezov V, Chan KY, Ziolkowski A, Nelson PF (2011) Influence of pyrolysis temperature on production and nutrient properties of wastewater sludge biochar. J Environ Manag
Hossain Z, Mustafa G, Sakata K, Komatsu S (2016) Insights into the proteomic response of soybean towards Al2O3, ZnO, and Ag nanoparticles stress. J Hazard Mater 304:291–305
Huang Y, Yu F, Park YS, Wang J, Shin Y (2015) Application of magnetic nanoparticles in plant biology and agriculture. Front Plant Sci 6:121
Inyang MI, Gao B, Yao Y, Xue Y, Zimmerman A, Mosa A, Pullammanappallil P, Ok YS, Cao X (2016) Biochar as a low-cost adsorbent for aqueous heavy metal removal. Crit Rev Environ Sci Technol
Islam SN, Naqvi SMA, Raza A, Jaiswal A, Singh AK, Dixit M, Ahmad A (2022) Mycosynthesis of highly fluorescent selenium nanoparticles from *Fusarium oxysporum*, their antifungal activity against black fungus *Aspergillus niger*, and in-vivo biodistribution studies. 3 Biotech 12(11):309
Ismail AM (2021) Efficacy of copper oxide and magnesium oxide nanoparticles on controlling black scurf disease on potato. Egypt J Phytopathol 49(2):116–130
Joseph T, Morrison M (2006). Nanotechnology in agriculture and food: a nanoforum report. Nanoforum.org
Kah M, Kookana RS, Gogos A, Bucheli TD (2018) A critical evaluation of nanopesticides and nanofertilizers against their conventional analogues. Nat Nanotechnol 13(8):677–684
Khan I, Saeed K, Khan I (2019) Nanoparticles: Properties, applications and toxicities. Arabian journal of chemistry 12(7):908–931
Khodakovskaya MV, De Silva K, Biris AS, Dervishi E, Villagarcia H (2012) Carbon nanotubes induce growth enhancement of tobacco cells. ACS Nano 6:2128–2135
Khot LR, Sankaran S, Maja JM, Ehsani R, Schuster EW (2012) Applications of nanomaterials in agricultural production and crop protection: a review. Crop Prot 35:64–70
Kookana RS, Boxall AB, Reeves PT, Ashauer R, Beulke S, Chaudhry Q, Clackett M (2014) Nanopesticides: guiding principles for regulatory evaluation of environmental risks. J Agric Food Chem 62(19):4227–4240
Kralova K, Jampilek J (2023) Applications of nanomaterials in plant disease management and protection. In: Nanotechnology in agriculture and agroecosystems. Elsevier, pp 239–296

Lee CW, Mahendra S, Zodrow K, Li D, Tsai YC, Braam J, Alvarez PJ (2010) Developmental phytotoxicity of metal oxide nanoparticles to *Arabidopsis thaliana*. Environ Toxicol Chem 29:669–675

Lei Z, Mingyu S, Xiao W, Chao L, Chunxiang Q, Liang C, Hao H, Xiaoqing L, Fashui H (2008) Antioxidant stress is promoted by nano-anatase in spinach chloroplasts under UV-B radiation. Biol Trace Elem Res 121:69–79

Li J, Chang PR, Huang J, Wang Y, Yuan H, Ren H (2013) Physiological effects of magnetic iron oxide nanoparticles towards watermelon. J Nanosci Nanotechnol 13:5561–5567

Li D, Liu B, Yang F, Wang X, Shen H, Wu D (2016) Preparation of uniform starch microcapsules by premix membrane emulsion for controlled release of avermectin. Carbohydr Polym 136:341–349

Lin D, Xing B (2007) Phytotoxicity of nanoparticles: inhibition of seed germination and root growth. Environ Pollut 150:243–250

Liu Q, Zhao Y, Wan Y, Zheng J, Zhang X, Wang C, Fang X, Lin J (2010) Study of the inhibitory effect of water-soluble fullerenes on plant growth at the cellular level. ACS Nano 4:5743–5748

López-Moreno ML, de la Rosa G, Hernández-Viezcas JÁ, Castillo-Michel H, Botez CE, Peralta-Videa JR, Gardea-Torresdey JL (2010) Evidence of the differential biotransformation and genotoxicity of ZnO and CeO2 nanoparticles on soybean (*Glycine max*) plants. Environ Sci Technol 44:7315–7320

Lu C, Zhang C, Wen J, Wu G, Tao M (2002) Research of the effect of nanometer materials on germination and growth enhancement of *Glycine max* and its mechanism. Soybean Sci 21:168–171

Martínez-Fernández D, Barroso D, Komárek M (2016) Root water transport of Helianthus annuus L. under iron oxide nanoparticle exposure. Environ Sci Pollut Res 23:1732–1741

Mazaheri-Tirani M, Dayani S (2020) In vitro effect of zinc oxide nanoparticles on *Nicotiana tabacum* callus compared to ZnO microparticles and zinc sulfate (ZnSO4). Plant Cell Tissue Organ Cult 140:279–289

Mondal A, Basu R, Das S, Nandy P (2011) Beneficial role of carbon nanotubes on mustard plant growth: an agricultural prospect. J Nanopart Res 13:4519

Nadiminti PP, Dong YD, Sayer C, Hay P, Rookes JE, Boyd BJ, Cahill DM (2013) Nanostructured liquid crystalline particles as an alternative delivery vehicle for plant agrochemicals. ACS Appl Mater Interfaces 5:1818–1826

Noori A, Donnelly T, Colbert J, Cai W, Newman LA, White JC (2020) Exposure of tomato (*Lycopersicon esculentum*) to silver nanoparticles and silver nitrate: physiological and molecular response. Int J Phytoremediation 22:40–45

Pérez-de-Luque A (2017) Interaction of nanomaterials with plants: what do we need for real applications in agriculture? Front Environ Sci 5:12

Pradhan S, Durgama M, Mailapalli DR (2020) Urea loaded hydroxyapatite nanocarrier for efficient delivery of plant nutrients in rice. Arch Agron Soil Sci

Qi K, Cheng B, Yu J, Ho W (2017) Review on the improvement of the photocatalytic and antibacterial activities of ZnO. ScienceDirect

Ragaei M, Sabry AKH (2014) Nanotechnology for insect pest control. Int J Sci Environ Technol 3(2):528–545

Rai M, Ingle AP, Pandit R, Paralikar P, Gupta I, Chaud M, Medici S (2012) Advances in nanotechnology for diagnosis and treatment of tuberculosis. Curr Nanosci 8(2):186–195

Rajna S, Paschapur AU, Raghavendra KV (2019) Nanopesticides: its scope and utility in pest management. Indian Farmer 6:17–21

Raju SVS (2019) Recent trends in insect pest management. AkiNik Publications

Raliya R, Tarafdar JC (2013) ZnO nanoparticle biosynthesis and its effect on phosphorous-mobilizing enzyme secretion and gum contents in Clusterbean (*Cyamopsis tetragonoloba* L.). Agric Res 2:48–57

Rameshraddy Pavithra GJ, Rajashekar Reddy BH, Salimath M, Geetha KN, Shankar AG (2017) Zinc oxide nano particles increases Zn uptake, translocation in rice with positive effect on growth, yield and moisture stress tolerance. Indian Journal of Plant Physiology 22:287–294

Raskar S, Laware S (2014) Effect of zinc oxide nanoparticles on cytology and seed germination in onion. Int J Curr Microbiol App Sci 3:467–473

Rico CM, Majumdar S, Duarte-Gardea M, Peralta-Videa JR, Gardea-Torresdey JL (2011) Interaction of nanoparticles with edible plants and their possible implications in the food chain. J Agric Food Chem 59:3485–3498

Rizwan M, Ali S, Zia-ur-Rehman M, Malik S, Qayyum M, Alamri S, Alyemeni M, Ahmad P (2019) Effect of foliar applications of silicon and titanium dioxide nanoparticles on growth, oxidative stress and cadmium accumulation by rice (Oryza sativa). Acta Physiologiae Plantarum 41. https://doi.org/10.1007/s11738-019-2828-7

Ruffini Castiglione M, Giorgetti L, Geri C, Cremonini R (2011) The effects of nano-TiO2 on seed germination, development and mitosis of root tip cells of *Vicia narbonensis* L. and *Zea mays* L. J Nanopart Res 13:2443–2449

Servin A, Elmer W, Mukherjee A, De La Torre-Roche R, Hamdi H, White JC, Bindraban P (2015) A review of the use of engineered nanomaterials to suppress plant disease and enhance crop yield. J Nanopart Res 17(2):92

Shabana T, Parveen S, Akhtar MS (2020) Biogenic synthesis of gold nanoparticles using plant extract and their antifungal activity against phytopathogens. ACS Nano. Retrieved from https://pubs.acs.org/doi/abs/10.1021/acsnano.0c03610

Shah MA, Wani SH, Khan AA (2016) Nanotechnology and insecticidal formulations. J Food Bioeng Nanoprocess 1:285–310

Sharma A, Mehta SK, Singh S, Gupta S (2016) Synthesized colloidal-supported Pt and bimetallic Pt–Mo nanoparticles as electrocatalyst in oxidation of methanol in alkaline solution. Journal of Applied Electrochemistry 46:27–38

Siddiqi KS, Husen A, Rao RAK (2018) A review on biosynthesis of silver nanoparticles and their biocidal properties. J Nanobiotechnol 16(1):1–28

Siddiqui MH, Al-Whaibi MH (2014) Role of nano-SiO2 in germination of tomato (*Lycopersicum esculentum* seeds Mill.). Saudi J Biol Sci 21:13–17

Siddiqi KS, Husen A (2022) Plant response to silver nanoparticles: a critical review. Critical Reviews in Biotechnology 42(7):973–990

Singh J, Lee B-K (2016) Influence of nano-TiO2 particles on the bioaccumulation of Cd in soybean plants (*Glycine max*): a possible mechanism for the removal of Cd from the contaminated soil. J Environ Manag 170:88–96

Singh SV et al (2023) Nanomaterials for inorganic pollutants removal from contaminated water. In: Nano-bioremediation for water and soil treatment: an eco-friendly approach. Apple Academic Press. https://doi.org/10.1201/9781003414964-6

Song G, Gao Y, Wu H, Hou W, Zhang C, Ma H (2012) Physiological effect of anatase TiO2 nanoparticles on Lemna minor. Environ Toxicol Chem 31:2147–2152

Srivastava A, Rao DP (2014) Enhancement of seed germination and plant growth of wheat, maize, peanut and garlic using multiwalled carbon nanotubes. Eur Chem Bull 3(5):502–504

Sudheer WN, Praveen N, Al-Khayri JM, Jain SM (2022) Role of plant tissue culture medium components. In: Advances in plant tissue culture. Academic Press, pp 51–83

Tarafdar JC, Sharma S, Raliya R (2020) Zinc oxide nanoparticles induce defense responses and toxicity in the *arbuscular mycorrhizal* fungus. NCBI

Tayeh C, Siah A, Randoux B, Halama P, Walters DR, Reignault P (2014) Topical application of inducers for disease control. In: Induced resistance for plant defense. Wiley, pp 193–231

Tiwari S, Sharma B, Singh H, Biswas P, Kumari A (2023) 11 Nanoparticles in pest management. In: Advances in nanotechnology for smart agriculture: techniques and applications. CRC Press, p 221

Tripathi DK, Shweta, Singh S, Singh S, Pandey R, Singh VP, Sharma NC, Prasad SM, Dubey NK, Chauhan DK (2017a) An overview on manufactured nanoparticles in plants: uptake, translocation, accumulation and phytotoxicity. Plant Physiol Biochem 110:2–12

Tripathi DK, Singh S, Singh S, Pandey R, Singh VP, Sharma NC, Prasad SM, Dubey NK, Chauhan DK (2017b) An overview on manufactured nanoparticles in plants: uptake, translocation, accumulation and phytotoxicity. Plant Physiol Biochem 110:2–12

Urkude R (2019) Application of nanotechnology in insect pest management. Int Res J Sci Eng 7(6):151–156

Vanti GL, Nargund VB, NBK, Vanarchi R, Kurjogi M, Mulla SI, Patil RR (2019) Synthesis of Gossypium hirsutum-derived silver nanoparticles and their antibacterial efficacy against plant pathogens. Applied Organometallic Chemistry 33(1):e4630

Vargas-Hernandez M, Macias-Bobadilla I, Guevara-Gonzalez RG, Rico-Garcia E, Ocampo-Velazquez RV, Avila-Juarez L, Torres-Pacheco I (2020) Nanoparticles as potential antivirals in agriculture. Agriculture 10(10):444

Wang L, Dong J, Chen J, Eastoe J, Li X (2009) Design and optimization of a new self- nanoemulsifying drug delivery system. J Colloid Interface Sci 330(2):443–448

Wang X, Han H, Liu X, Gu X, Chen K, Lu D (2012) Multi-walled carbon nanotubes can enhance root elongation of wheat (*Triticum aestivum*) plants. J Nanopart Res 14:841

Wang Y, Liu H, Yu J, Hu B, Zhao H, Tsiakaras P, Song S (2019) Copper oxide derived nanostructured self-supporting Cu electrodes for electrochemical reduction of carbon dioxide. Electrochimica Acta 328:135083

Wei H, Wang E (2013) Nanomaterials with enzyme-like characteristics (nanozymes): next-generation artificial enzymes. Chem Soc Rev 42:6060–6093

Yadav SK, Babu S, Yadav MK, Singh K, Yadav GS, Pal S (2013) A review of organic farming for sustainable agriculture in Northern India. Int J Agron 2013:1–8

Yang F, Hong F, You W, Liu C, Gao F, Wu C, Yang P (2006) Influence of nano-anatase TiO2 on the nitrogen metabolism of growing spinach. Biol Trace Elem Res 110:179–190

Yang X, Pan H, Wang P, Zhao F-J (2017) Particle-specific toxicity and bioavailability of cerium oxide (CeO2) nanoparticles to *Arabidopsis thaliana*. J Hazard Mater 322:292–300

Yin L, Colman BP, McGill BM, Wright JP, Bernhardt ES (2012) Effects of silver nanoparticle exposure on germination and early growth of eleven wetland plants. PLoS One 7:e47674

Yusefi-Tanha E, Fallah S, Rostamnejadi A, Pokhrel LR (2020) Particle size and concentration dependent toxicity of copper oxide nanoparticles (CuONPs) on seed yield and antioxidant defense system in soil grown soybean (*Glycine max* cv. Kowsar). Sci Total Environ 136994

Zahir AA, Bagavan A, Kamaraj C, Elango G, Rahuman AA (2012) Efficacy of plant-mediated synthesized silver nanoparticles against *Sitophilus oryzae*. J Biopest 5:95

Zhang H, Wang R, Chen H, Wang Y (2018) Quantum dots for bioimaging applications. J Mater Chem B 6(18):2855–2871

Zheng L, Hong F, Lu S, Liu C (2005) Effect of nano-TiO2 on strength of naturally aged seeds and growth of spinach. Biol Trace Elem Res 104:83–91

Zhao C, Liu B, Piao S, Wang X, Lobell DB, Huang Y, Asseng S (2018) Temperature increase reduces global yields of major crops in four independent estimates. Proceedings of the National Academy of sciences 114(35):9326–9331

Ziaee M, Hamzavi F (2013) Application of nanoparticles in pest management programs-a review. In: Proceedings of International Conference on Green AgroIndustry (ICGAI), pp 386–393

Chapter 18
Multifarious Approaches of Agri-Based Carbon Credit Generation for Green Economy

Rashida Hameed, Adeel Abbas, Rashmi Sharma, and Shiv Vendra Singh

Abstract This chapter explores the diverse strategies within agriculture that contribute to carbon credit generation, fostering a sustainable and eco-friendly economic framework. Beginning with an overview of carbon credits and their pivotal role in mitigating climate change, the discussion highlights how these credits incentivize reductions in greenhouse gas emissions. It then delves into soil carbon sequestration, emphasizing techniques such as no-till farming, cover cropping, and crop rotation, which enhance soil health and maximize carbon storage. Agroforestry systems are examined for their ability to integrate trees and shrubs into agricultural landscapes, sequestering carbon while simultaneously improving biodiversity, soil stability, and water management. The chapter addresses methane reduction in agriculture management, focusing on practices like improved livestock feeding strategies and manure management to lower methane emissions. It also discusses biochar utilization as a method to sequester carbon and improve soil fertility by converting organic waste into stable charcoal. Moreover, the integration of renewable energy sources in agricultural operations is considered, showcasing how solar, wind, and bioenergy can reduce carbon footprints and contribute to a greener economy. The chapter concludes by synthesizing these approaches, underscoring their collective impact on creating sustainable agricultural systems that not only mitigate climate change but also enhance economic resilience and environmental health. By understanding and implementing these multifaceted strategies, agriculture can play a pivotal role in the transition toward a green economy, generating carbon credits that support both ecological and economic goals.

R. Hameed · A. Abbas (✉)
School of Environment and Safety Engineering, Institute of Environment and Ecology, Jiangsu University, Zhenjiang, China

R. Sharma
Department of Agronomy, Govind Ballabh Pant University of Agriculture and Technology, Pantnagar, Uttarakhand, India

S. V. Singh
Department of Agronomy, Rani Lakshmi Bai Central Agricultural University, Jhansi, Uttar Pradesh, India

B. Pramanick et al. (eds.), *Climate-Smart Agricultural Technologies*, Sustainability Sciences in Asia and Africa, https://doi.org/10.1007/978-981-96-7699-6_18

Keywords Carbon credit · Agroforestry · Biochar · Green economy · Renewable energy · Carbon trading

18.1 Introduction

As the world faces the growing threat of climate change, there is an urgent need to explore and implement sustainable practices across all sectors, including agriculture (Adisa et al. 2024; Eruaga 2024). Agriculture, traditionally seen as a significant source of greenhouse gas emissions, holds immense potential to become a vital component of climate change mitigation through carbon sequestration and emission reduction strategies (Dahai et al. 2024; Kamyab et al. 2024). This chapter delves into the multifarious approaches to generating carbon credits through agricultural practices, highlighting their significance in fostering a green economy (Baral et al. 2024; Yang et al. 2024a). Agriculture is a double-edged sword in the context of climate change. On one hand, it contributes substantially to greenhouse gas emissions, primarily through methane from livestock, nitrous oxide from fertilizers, and carbon dioxide from soil management practices (Jordon et al. 2024; Kollar 2023; Thamarai et al. 2024). Methane, for instance, is produced during the digestion process of ruminant animals like cows, while nitrous oxide emissions are a by-product of synthetic fertilizers and manure (Giamouri et al. 2023; Króliczewska et al. 2023). On the other hand, agriculture offers unique opportunities for carbon sequestration and emission reductions that can transform it into a net positive for the environment (Feng et al. 2023; Kamyab et al. 2024). Practices such as no-till farming, cover cropping, and crop rotation enhance the soil's ability to capture and store carbon, while integrating trees into agricultural systems (agroforestry) can significantly increase carbon sequestration (Morya et al. 2023; Pandao et al. 2023).

Carbon credits are market-based instruments that provide economic incentives for reducing greenhouse gas emissions (Braccio 2024; Kumar 2024). Each carbon credit represents one ton of carbon dioxide (or an equivalent amount of other greenhouse gases) that has been reduced, avoided, or sequestered (Michaelowa et al. 2023; Singh et al. 2024b). These credits can be traded in carbon markets, where entities that need to offset their emissions purchase them from those who have implemented sustainable practices (Laine et al. 2023; Trouwloon et al. 2023). By participating in carbon markets, farmers and agricultural enterprises can generate additional income while contributing to climate change mitigation efforts (Barbato and Strong 2023; Tumushabe et al. 2023). This economic benefit can be particularly valuable for smallholder farmers in developing countries, providing them with new revenue streams and promoting sustainable agricultural practices (Adisa et al. 2024; Amede et al. 2023). Agricultural practices have the potential to generate carbon credits through a variety of methods (Barbato and Strong 2023; Paul et al. 2023). Enhancing soil carbon sequestration involves practices like reduced tillage, which minimizes soil disturbance, and the use of cover crops, which protect the soil and add organic matter (Francaviglia et al. 2023). Integrating agroforestry systems combines trees and shrubs with crops and livestock, creating a more diverse and resilient agricultural system that sequesters carbon above and below ground (Nwaogu et al. 2023; Vinodhini et al. 2023). Reducing methane emissions from livestock can be achieved through dietary modifications, improved manure management, and the use of anaerobic digesters, which capture methane for use as a renewable energy source (Aluwong et al. 2011; Sikiru et al. 2024). Utilizing biochar, a form of charcoal produced from plant matter, can improve soil health and sequester carbon for hundreds to thousands of years (Pandian et al. 2024). Incorporating renewable energy sources, such as solar and wind power, can reduce the carbon footprint of agricultural operations and generate additional carbon credits (Chen et al. 2024; Kantwa et al. 2024) (Fig. 18.1).

Transitioning to a green economy involves adopting sustainable practices that promote economic growth while reducing environmental impacts (Hutajulu et al. 2024; Xia et al. 2024). Agri-based carbon credit generation is a key pathway toward this goal. It encourages sustainable farming practices, provides financial incentives for farmers, and supports global efforts to combat climate change (Byrne et al. 2024; De Corato et al. 2024). Moreover, these practices contribute to broader environmental and social benefits, such as enhanced biodiversity, improved water quality, and increased resilience to climate extremes (Ghadiri et al. 2024). By understanding and implementing these multifaceted strategies, agriculture can play a pivotal role in fostering a green economy, ensuring food security, and safeguarding the planet for future generations (Adisa et al. 2024). The adoption of these practices not only mitigates climate change but also enhances the sustainability and productivity of agricultural systems, ultimately benefiting both the environment and society.

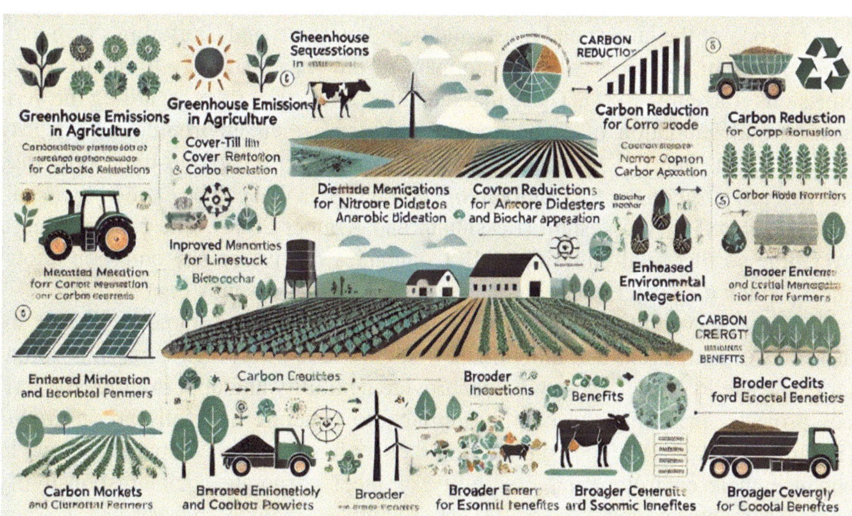

Fig. 18.1 Sustainable agricultural practices that mitigate climate change by reducing emissions and sequestering carbon. It showcases methods, carbon credit incentives, and environmental and societal benefits

18.2 Understanding Carbon Credits

Carbon credits are tradable certificates representing the right to emit a certain amount of carbon dioxide or an equivalent amount of other greenhouse gases (GHGs) (Braccio 2024). Each carbon credit typically equates to one ton of carbon dioxide that has been either reduced, avoided, or sequestered (Zakkour et al. 2024). The fundamental purpose of carbon credits is to provide a financial incentive for entities, such as companies and governments, to reduce their greenhouse gas emissions (Grau Ruiz 2024). By assigning a monetary value to carbon reductions, carbon credits create an economic motivation for adopting environmentally friendly practices and technologies (Gao et al. 2024). This mechanism is crucial for mitigating the adverse effects of climate change, as it encourages industries to innovate and implement strategies that lower their carbon footprint (Liu and Zhang 2024). Additionally, carbon credits help direct funding to projects that might otherwise lack the financial support necessary to reduce emissions, such as reforestation, renewable energy installations, and improved agricultural practices (Braccio 2024). The revenue generated from selling carbon credits can make a significant difference in financing sustainable development projects, particularly in developing countries where resources are often limited (Rana et al. 2024).

The market dynamics of carbon credits are characterized by two primary types: compliance markets and voluntary markets (Bleuel and Müller 2024). Compliance markets are regulated by mandatory national, regional, or international policies that set a cap on the total level of greenhouse gas emissions allowed (Friedman 2024). Entities that exceed their emission allowances must purchase carbon credits to offset their excess emissions, creating a demand-driven market. Examples of compliance markets include the European Union Emissions Trading System (EU ETS), which is one of the largest and most established markets, and California's Cap-and-Trade Program, which is a key component of the state's strategy to reduce GHG emissions (Lazenby 2024). These markets often have strict rules regarding the verification and validation of carbon credits to ensure their environmental integrity (Pande 2024). The EU ETS, for instance, operates on a cap-and-trade principle, where a cap is set on the total amount of certain GHGs that can be emitted by installations covered by the system (Lessmann and Kramer 2024). The cap is reduced over time so that total emissions fall. Companies receive or buy emission allowances, which they can trade with one another as needed (Ivanov et al. 2024).

In contrast, voluntary markets operate outside regulatory frameworks and are driven by organizations and individuals who choose to offset their emissions voluntarily (de Mello 2024). These markets provide flexibility and opportunities for various projects, including those in agriculture, to generate and sell carbon credits (Baumber et al. 2024). The voluntary market is often used by companies aiming to enhance their environmental credibility or meet corporate social responsibility goals (Almontaser and Gerged 2024). Examples include the Verified Carbon Standard (VCS) and the Gold Standard, which certify projects that demonstrate real and additional emission reductions (Salzman and Weisbach 2024). Projects in the voluntary market can range from forest conservation and reforestation to methane capture from landfills and livestock manure management (Malec et al. 2024). The voluntary carbon market has grown significantly in recent years, driven by increased corporate commitments to sustainability and net-zero targets (Van Tam et al. 2024). Companies are increasingly recognizing the importance of addressing their environmental impact and are using voluntary carbon credits as part of their broader sustainability strategies (Grau Ruiz 2024).

Both types of markets play a vital role in driving investments into sustainable practices and technologies, thereby contributing to the global effort to combat climate change (Kawamori 2024). Compliance markets ensure that major emitters are held accountable for their emissions and incentivize large-scale reductions, while voluntary markets enable a broader range of participants to contribute to emission reduction efforts (Verschuuren et al. 2024). The interaction between these markets also helps to stabilize carbon prices and ensure a steady demand for high-quality carbon credits (Zhang et al. 2024). Moreover, as more companies and countries commit to net-zero targets, the demand for carbon credits is expected to rise, further driving innovation and investment in low-carbon technologies and sustainable practices (Evro et al. 2024). The development of robust monitoring, reporting, and verification (MRV) systems is essential for maintaining the credibility of carbon credits and ensuring that emission reductions are real and permanent (Baiz 2024). These systems provide transparency and accountability, fostering trust in the carbon credit market and encouraging greater participation from a wide range of stakeholders (Suryanarayana and Rao 2024).

18.3 Soil Carbon Sequestration

Soil carbon sequestration is a critical component of climate change mitigation strategies, offering a way to capture and store atmospheric carbon dioxide in the soil (Singh et al. 2024b). This process not only helps to reduce greenhouse gas concentrations in the atmosphere but also improves soil health and agricultural productivity (Kamyab et al. 2024). By enhancing the soil's ability to store carbon, farmers can play a significant role in mitigating climate change while reaping multiple agricultural benefits (Naeem et al. 2023). One of the most effective techniques for soil carbon sequestration is no-till farming, which involves minimal soil disturbance (Mattila and Vihanto 2024). Traditional plowing or tilling breaks up soil structure and exposes organic matter to oxygen, accelerating its decomposition and releasing carbon dioxide (Filipović et al. 2024). No-till farming, on the other hand, allows crop residues to remain on the field, protecting the soil and promoting the accumulation of organic matter. This method not only increases carbon sequestration but also improves soil moisture retention and reduces erosion (Mattila and Vihanto 2024). By maintaining a protective layer of organic matter on the soil surface, no-till farming fosters a healthier soil ecosystem that supports microbial activity and nutrient cycling (Singh et al. 2024a).

Cover cropping is another essential technique in soil carbon sequestration. This practice involves planting certain crops, such as legumes, grasses, or clovers, during off-seasons when main crops are not grown (Grohs et al. 2024; Saniga et al. 2024). These cover crops provide ground cover, reducing soil erosion and enhancing soil structure. More importantly, they add organic matter to the soil when they decompose, increasing the soil's carbon content (Fontana et al. 2024). Cover crops also improve soil fertility by fixing nitrogen and can suppress weeds, reducing the need for chemical herbicides (Debaeke et al. 2024). The diverse root structures of cover crops enhance soil porosity and microbial diversity, contributing to a more robust and resilient agricultural system (Gentsch et al. 2024).

Crop rotation, the practice of growing different types of crops in the same area across a sequence of growing seasons, is another effective strategy (Yang et al. 2024b). This technique prevents soil depletion, as different crops have varying nutrient requirements and pest resistance (Mohanty et al. 2024). Crop rotation contributes to soil carbon sequestration by diversifying the organic residues returned to the soil, enhancing microbial activity, and improving soil structure (Liu et al. 2024). For example, rotating deep-rooted crops with shallow-rooted ones can help distribute organic matter more evenly throughout the soil profile. This practice not only boosts soil health but also reduces the risk of pest and disease cycles, thereby promoting sustainable farming (Vanlauwe et al. 2024). Agroforestry systems, which integrate trees and shrubs into agricultural landscapes, can significantly boost soil carbon sequestration (Bhardwaj et al. 2024). Trees and shrubs add organic matter through leaf litter and root biomass, and their deep root systems help stabilize soil and store carbon at greater depths (Witzgall et al. 2024). Agroforestry practices, such as alley cropping (planting rows of trees with crops in between) or silvopasture (integrating trees with livestock grazing), create a diverse ecosystem that enhances biodiversity, improves water retention, and increases resilience to climate extremes (Arshad et al. 2024). By combining agricultural and forestry practices, agroforestry systems optimize land use and provide multiple ecological and economic benefits (Veste et al. 2024) (Fig. 18.2).

Applying compost and other organic amendments to agricultural fields is another effective way to increase soil carbon content (Farooqi et al. 2024). These materials add organic matter and nutrients to the soil, enhancing microbial activity and soil structure (Yao et al. 2024). Composting agricultural residues and manure not only recycles nutrients but also reduces methane emissions from decomposition (Bremaghani 2024). Enhanced root systems can also contribute to soil carbon sequestration (Forfora et al. 2024). Deep-rooted plants, such as certain perennial grasses, transfer carbon deeper into the soil, where it is more likely to be stored long term (Patra et al. 2024). These plants also improve soil structure and increase the stability of soil organic matter. Soil carbon sequestration offers a multitude of benefits beyond carbon storage (Bera et al. 2024). Enhanced soil organic matter improves soil fertility, water retention, and crop yields. Healthy soils are more resilient to droughts and floods, reducing the impact of extreme weather events on agricultural productivity (Kundu and Kumar 2024). Additionally, practices that promote soil carbon sequestration often reduce the need for synthetic fertilizers and

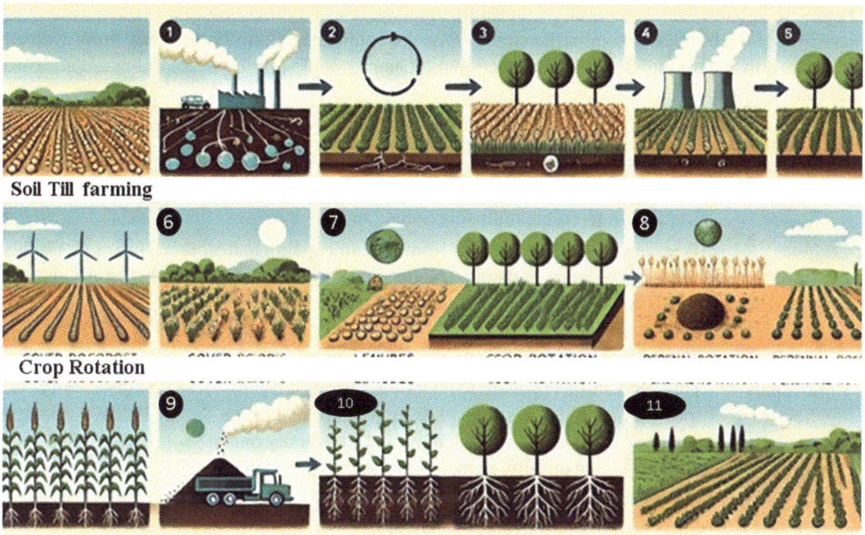

Fig. 18.2 Soil carbon sequestration techniques in agriculture, including no-till farming with crop residues, cover crops like legumes and grasses, crop rotation with different crops, agroforestry with trees and crops together, compost application, and deep-rooted plants like perennial grasses, all set against a clear sky and healthy landscape

pesticides, leading to lower input costs and less environmental pollution (Srivastav et al. 2024). Implementing soil carbon sequestration practices requires knowledge, commitment, and sometimes initial investment. However, the long-term benefits, including increased farm profitability, sustainability, and climate resilience, make these practices worthwhile (Gillespie 2024). As part of a broader strategy to combat climate change, soil carbon sequestration not only mitigates greenhouse gas emissions but also supports sustainable agricultural systems that can feed a growing global population (Jebari et al. 2024) (Table 18.1).

18.4 Agroforestry Systems

Agroforestry systems integrate trees and shrubs into agricultural landscapes, combining forestry and agricultural production to create multifunctional land use systems. This approach is particularly effective for sequestering carbon, as trees absorb carbon dioxide from the atmosphere and store it in their biomass and soil. The deep root systems of trees also help stabilize the soil, reduce erosion, and enhance water infiltration, further contributing to carbon storage and improving overall soil health. By incorporating trees into croplands and pastures, agroforestry systems transform conventional farming practices into more sustainable and resilient operations, providing a natural solution to climate change mitigation. The benefits of agroforestry

Table 18.1 Detailed overview of agroforestry systems, highlighting their integration into agricultural landscapes, carbon sequestration benefits, enhancement of biodiversity, improvement of soil health, microclimate regulation, economic benefits, and various successful examples from different regions

Aspect	Description	References
Integration	Agroforestry systems integrate trees and shrubs into agricultural landscapes, combining forestry and agricultural production to create multifunctional land use systems	Jose et al. (2021) and Veste et al. (2024)
Carbon sequestration	Trees absorb carbon dioxide from the atmosphere and store it in their biomass and soil. The deep root systems of trees stabilize the soil, reduce erosion, and enhance water infiltration. This process contributes to carbon storage and improves overall soil health	Diop et al. (2022) and Priori et al. (2020)
Biodiversity	Trees and shrubs provide habitats for various wildlife species, promoting ecological diversity and fostering a balanced ecosystem. This increase in biodiversity helps control pests naturally, reducing the need for chemical pesticides and contributing to a healthier environment	Akinsorotan et al. (2023)
Soil health	Trees improve soil health by adding organic matter through leaf litter and root biomass. This boosts soil fertility and water retention, leading to higher crop yields and more resilient farming systems. Trees also enhance soil structure and microbial activity	Nair et al. (2021)
Microclimate regulation	Trees regulate microclimates by offering shade and windbreaks, protecting crops and livestock from extreme weather conditions. This reduces heat stress on crops, improves water efficiency, and enhances animal welfare, leading to better weight gain and overall health of livestock	Saeed (2023)
Economic benefits	Agroforestry provides additional income streams through the production of timber, fruits, nuts, resins, and other tree-related products. This diversification of income sources improves the economic stability of farming communities and reduces their vulnerability to market fluctuations and climate variability	Gabay et al. (2020) and Muys (2021)
Examples	*Sahel, Africa*: FMNR technique revitalizes degraded lands *Central and South America*: Alley cropping with trees and crops *Temperate regions*: Silvopasture integrates trees with livestock grazing *India*: Combining fruit trees with traditional crops improves food security and diversifies incomes	Ahimbisibwe et al. (2024) and Hemida (2023)
Overall impact	Agroforestry enhances sustainability, mitigates climate change, improves biodiversity and soil health, and provides diversified and stable income sources	Khan et al. (2021)

extend far beyond carbon sequestration. One of the most significant advantages is the enhancement of biodiversity. Trees and shrubs provide habitats for various wildlife species, promoting ecological diversity and fostering a balanced ecosystem. This increase in biodiversity can help control pests naturally, reducing the need for chemical pesticides and contributing to a healthier environment. Additionally, the presence of trees improves soil health by adding organic matter through leaf litter and root biomass, which boosts soil fertility and water retention. This, in turn, can lead to higher crop yields and more resilient farming systems. Trees also play a crucial role in regulating microclimates, offering shade and windbreaks that protect crops and livestock from extreme weather conditions. Agroforestry also offers farmers additional income streams through the production of timber, fruits, nuts, resins, and other tree-related products. This diversification of income sources can improve the economic stability of farming communities and reduce their vulnerability to market fluctuations and climate variability.

There are numerous successful examples of agroforestry systems implemented in different regions and climates around the world. In the Sahel region of Africa, for instance, the Farmer Managed Natural Regeneration (FMNR) technique has revitalized degraded lands by encouraging the regrowth of native trees and shrubs. This practice has improved soil fertility, increased crop yields, and enhanced local biodiversity, all while providing much-needed resources such as firewood and fodder. In Central and South America, alley cropping, which involves planting rows of trees with crops in between, has been widely adopted. This method not only sequesters carbon but also provides shade for crops, reducing heat stress and improving water efficiency. Coffee and cocoa plantations often use shade trees to enhance production quality and biodiversity. In temperate regions, silvopasture, which integrates trees with livestock grazing, has gained popularity. This system enhances forage production, provides shelter for animals, and creates additional sources of income through timber and non-timber forest products. Silvopasture also improves animal welfare by providing shade and wind protection, leading to better weight gain and overall health of livestock. In India, the practice of combining fruit trees with traditional crops (such as mango or guava orchards intercropped with vegetables) has not only improved food security but also diversified farmers' incomes and resilience to climate shocks.

These examples demonstrate the versatility and effectiveness of agroforestry systems in addressing environmental and economic challenges across various landscapes. By adopting agroforestry practices, farmers can create more sustainable agricultural systems that contribute to climate change mitigation, enhance biodiversity, improve soil health, and provide diversified and stable income sources. The success of these systems across different regions and climates underscores their potential as a global strategy for achieving sustainable agricultural development and resilience in the face of changing climate conditions.

18.5 Methane Reduction in Agriculture Management

Methane (CH_4) is a potent greenhouse gas, significantly contributing to climate change, with agriculture being a primary source (Bray and Poston 2024). Methane emissions in agriculture mainly stem from enteric fermentation in ruminant animals like cows and sheep and from manure management (Lambo et al. 2024). Enteric fermentation produces methane as microorganisms in the stomachs of these animals break down food, while manure management generates methane during the anaerobic decomposition of organic matter in manure storage systems (Faizan et al. 2024). Methane's global warming potential is 25 times greater than carbon dioxide over a century, making its reduction critical for climate mitigation. To tackle methane emissions, several innovative practices have emerged (Kamyab et al. 2024). Dietary modifications, such as incorporating fats, oils, and feed additives like seaweed, can reduce methane production in livestock (Cheong et al. 2023). Improving forage quality also enhances digestion efficiency, lowering emissions (Santander et al. 2023). Effective manure management strategies, including anaerobic digesters, which convert methane into biogas, solid-liquid separation, and aerobic composting, further mitigate methane emissions (El Mashad et al. 2023). Anaerobic digesters not only manage manure but also process other organic wastes, producing renewable energy and nutrient-rich digestate (Chojnacka and Moustakas 2024).

The economic viability of these practices is essential for widespread adoption. Initial investments can be high, particularly for anaerobic digesters, along with ongoing operational costs for maintenance and feed additives (Nleya et al. 2023). However, these costs can be offset by revenues from biogas production, carbon credits, and improved livestock productivity (van Kooten and Zanello 2023). Long-term savings arise from reduced energy and fertilizer costs, lower veterinary expenses, and potential subsidies or incentives for sustainable practices (Neethirajan 2023). Additionally, the environmental and social benefits, including enhanced air quality, soil health, and climate change mitigation, add significant value (Agathokleous et al. 2023). While implementing methane reduction practices in agriculture requires financial and operational commitment, the long-term economic and environmental benefits make them a sustainable and profitable choice for farmers (Vogel et al. 2023). With appropriate support and incentives, these practices can lead to a more sustainable and resilient agricultural sector (Amede et al. 2023).

18.6 Biochar Utilization

Biochar is a form of charcoal produced from plant matter through a process called pyrolysis, where organic materials like wood, crop residues, or manure are heated in the absence of oxygen (Singh et al. 2024; Chaturvedi et al. 2021; Maniraj et al. 2023). This process stabilizes the carbon within the biochar, preventing it from being released into the atmosphere as carbon dioxide (Li and Tasnady 2023). When applied to soil, biochar can remain stable for hundreds to thousands of years, effectively sequestering carbon and reducing greenhouse gas emissions (Elkhlifi et al. 2023; Singh et al. 2023b). This long-term carbon storage makes biochar a potent tool in the fight against climate change, as it locks away carbon that would otherwise contribute to global warming (Werner 2024). Beyond its role in carbon sequestration, biochar offers numerous agricultural benefits. Its porous structure significantly enhances soil fertility by improving nutrient retention and availability

(Sarma et al. 2024). This structure increases the soil's cation exchange capacity, which allows it to hold onto essential nutrients like nitrogen, phosphorus, and potassium more effectively (Singh et al. 2023a, b; Antonangelo et al. 2024). As a result, plants have better access to these nutrients, leading to healthier growth and more productive crops (Chandrasekaran et al. 2024). Additionally, biochar aids in moisture retention by improving the soil's ability to retain water (Khan et al. 2024). This is particularly beneficial in arid and semiarid regions where water scarcity is a significant concern, as it helps plants withstand periods of drought and reduces the need for frequent irrigation (Alharbi et al. 2024).

The combined effects of enhanced nutrient availability and improved moisture retention lead to increased crop yields (Jiang et al. 2024). Studies have shown that biochar can significantly boost agricultural productivity, making it a valuable addition to farming practices (Khan et al. 2024). Furthermore, biochar improves soil structure by reducing compaction and increasing aeration, which promotes root growth and enhances plant resilience (Sultan et al. 2024). Biochar also boosts microbial activity in the soil, fostering a healthier soil ecosystem that supports plant growth (Pagliaccia et al. 2024). Overall, biochar utilization presents a sustainable and multifaceted approach to improving agricultural productivity and combating climate change (Upadhyay et al. 2024). By stabilizing carbon through pyrolysis and enhancing soil properties, biochar provides long-term benefits for both the environment and agriculture (Das and Ghosh 2024). Its ability to improve soil fertility, retain moisture, boost crop yields, and contribute to carbon sequestration makes it an essential tool for farmers looking to enhance productivity while reducing their carbon footprint (Han et al. 2024).

18.7 Renewable Energy Integration

Integrating renewable energy sources into agricultural practices offers a sustainable approach to meeting farms' energy needs while reducing environmental impact (Kumar et al. 2024). Solar panels, wind turbines, and biogas systems are prominent renewable energy options providing significant benefits for agriculture (Panda et al. 2024). Solar energy, harnessed through panels installed on farm buildings, greenhouses, or open fields, provides a reliable and cost-effective way to power irrigation systems, heating, cooling, and other operations, especially in sunny regions (Ogunboyo and Ogunlade 2024). Excess solar energy can be stored in batteries for use during cloudy days or nighttime, ensuring a consistent energy supply (Falope et al. 2024). Wind turbines, strategically placed on farms, capture wind energy to power activities such as water pumping, grain drying, and machinery operation, particularly in areas with consistent wind patterns (Manesh et al. 2024). Farms can even feed excess wind energy into the grid, creating an additional revenue stream. Biogas production, which involves the anaerobic digestion of organic waste like animal manure and crop residues, generates methane-rich biogas used for electricity, heat, or vehicle fuel (Vasan et al. 2024). Biogas systems help manage waste and reduce methane emissions from manure storage, providing a renewable energy source and contributing to greenhouse gas reduction (Kabeyi et al. 2024). Renewable energy projects on farms also generate carbon credits by reducing greenhouse gas emissions or capturing and storing carbon, providing an additional revenue stream (Pretty et al. 2002). For instance, a dairy farm in California installed solar panels on its rooftops and open land, meeting energy needs for milking machines, cooling systems, and lighting (Mohsenimanesh et al. 2021). By switching to solar energy, the farm significantly reduced its carbon emissions and earned carbon credits sold to local businesses seeking to offset their carbon footprint (Anukwonke and Abazu 2022). The additional revenue helped cover the initial installation costs and provided ongoing financial benefits (Lee and Shepley 2020). Similarly, a crop farm in Iowa integrated wind turbines to harness wind energy, which powered irrigation systems and grain dryers, reducing energy costs. The wind energy project generated carbon credits sold in the voluntary carbon market, enhancing the farm's sustainability profile and attracting environmentally conscious customers (Rosales et al. 2021).

18.8 Conclusion

The integration of renewable energy in agriculture, through solar panels, wind turbines, and biogas systems, presents significant economic and social benefits. Economically, farmers can achieve substantial financial gains from carbon credits and improved farm productivity. Renewable energy projects reduce energy costs, generate additional revenue from excess energy fed into the grid, and create valuable carbon credits that can be sold in carbon markets. Improved soil fertility, moisture retention, and enhanced crop yields further boost farm profitability. Socially, renewable energy integration enhances rural livelihoods by creating new jobs and promoting sustainable farming communities. It supports food security by ensuring reliable and efficient farming practices, contributing to the overall well-being of farming populations. However, several challenges must be addressed to maximize these benefits. Barriers to adoption include high initial investment costs, lack of technical knowledge, and limited access to financing. Solutions to these challenges involve providing financial incentives, such as grants and low-interest loans, to help farmers cover the initial costs of renewable energy systems. Offering training and technical support can empower farmers with the knowledge and skills needed to implement and maintain these technologies. Additionally, successful models of renewable energy integration should be scaled and replicated through government and private sector partnerships, fostering an environment conducive to widespread adoption. In conclusion, renewable energy integration in agriculture offers a multifaceted approach to economic and social development. By addressing the challenges and leveraging the economic and social benefits, renewable energy can play a crucial role in building sustainable and resilient farming communities, enhancing productivity, and contributing to global food security.

References

Adisa O, Ilugbusi BS, Adewunmi O, Franca O, Ndubuisi L (2024) A comprehensive review of redefining agricultural economics for sustainable development: overcoming challenges and seizing opportunities in a changing world. World J Adv Res Rev 21:2329–2341

Agathokleous E, Frei M, Knopf OM, Muller O, Xu Y, Nguyen TH, Gaiser T, Liu X, Liu B, Saitanis CJ (2023) Adapting crop production to climate change and air pollution at different scales. Nat Food 4:854–865

Ahimbisibwe V, Zhunusova E, Kassa H, Günter S (2024) Technical efficiency drivers of farmer-led restoration strategies, and how substantial is the unrealised potential for farm output? Agric Syst 213:103799

Akinsorotan OA, Akinsorotan AM, Adewale RO, Akande AB (2023) Detrimental effects of agrochemical-based agricultural intensification on biodiversity: evidence from some past studies. In: One Health implications of agrochemicals and their sustainable alternatives. Springer, pp 275–298

Alharbi SF, Aldakil M, Felemban A, Abdelrahim A (2024) Agricultural and technology-based strategies to improve water use efficiency in Arid and Semiarid areas. Water 16(13):1842

Almontaser TS, Gerged AM (2024) Insights into corporate social responsibility disclosure among multinational corporations during host-country political transformations: evidence from the Libyan oil industry. Corp Soc Responsib Environ Manag 31:1818–1837

Aluwong T, Wuyep P, Allam L (2011) Livestock-environment interactions: methane emissions from ruminants. Afr J Biotechnol 10:1265–1269

Amede T, Konde AA, Muhinda JJ, Bigirwa G (2023) Sustainable farming in practice: building resilient and profitable smallholder agricultural systems in sub-Saharan Africa. Sustainability 15:5731

Antonangelo JA, Culman S, Zhang H (2024) Comparative analysis and prediction of cation exchange capacity via summation: influence of biochar type and nutrient ratios. Front Soil Sci 4:1371777

Anukwonke CC, Abazu CI (2022) Green business through carbon credits. In: Climate change alleviation for sustainable progression. CRC Press, pp 315–332

Arshad MA, Rouf S, Abbas RN, Shahbaz Z, Aleem K, Shahbaz H, Pervaiz R, Sarwar A, ur Rehman H (2024) Navigating synergies: a comprehensive review of agroforestry system and agronomy crops. Haya Saudi J Life Sci 9:97–113

Baiz P (2024) Blockchain and carbon markets: standards overview. arXiv preprint arXiv:2403.03865

Baral SK, Goel R, Singh T, Kumar R (2024) Green metaverse for greener economies. CRC Press

Barbato CT, Strong AL (2023) Farmer perspectives on carbon markets incentivizing agricultural soil carbon sequestration. npj Clim Action 2:26

Baumber A, Cross R, Ampt P, Waters C, Ringbauer J, Bowdler I, Scott A, Gordon L, Sutton A, Metternicht G (2024) Soil-based carbon farming: opportunities for collaboration. J Rural Stud 108:103268

Bera T, Samui S, Dey A, Ankireddypalli J (2024) Soil carbon sequestration in the context of climate change. In: Climate change impacts on soil-plant-atmosphere continuum. Springer, pp 63–106

Bhardwaj KK, Yadav R, Goyal V, Sharma MK, Ahlawat KS (2024) Role of agroforestry systems in enrichment of soil organic carbon and nutrients: a review. Environ Conserv J 25:289–296

Bleuel S, Müller C (2024) Unlocking the potential: expert insights on the long-term compatibility of forest carbon credits with the EU ETS. Forest Policy Econ 162:103185

Braccio S (2024) The market of carbon credits and biofuels. Politecnico di Torino

Bray D, Poston TM (2024) The methane majors: climate change and animal agriculture in US courts. Bray, Daina, and Thomas M. Poston

Bremaghani A (2024) Utilization of organic waste in compost fertilizer production: implications for sustainable agriculture and nutrient management. Law Econ 18:86–98

Byrne N, McCarthy O, Ryan-Doyle M (2024) Leveraging the potential of cooperative agri-advisory services in the transition to sustainable and landscape-based agriculture. Int Food Agribus Manag Rev 1:1–29

Chandrasekaran P, Ashokkumar N, Ashok S (2024) Plant nutrient balance for robust vegetable cultivation and healthier human nutrition. Vigyan Varta 5(6):43–48

Chaturvedi S, Singh SV, Dhyani VC, Govindaraju K, Mandal S (2021) Characterization, bio-energy value and thermal stability of biochar derived from diverse agriculture and forestry lignocellulosic waste. Biomass Convers Biorefinery 13:1–14. https://doi.org/10.1007/s13399-020-01239-2

Chen L, Hu Y, Wang R, Li X, Chen Z, Hua J, Osman AI, Farghali M, Huang L, Li J (2024) Green building practices to integrate renewable energy in the construction sector: a review. Environ Chem Lett 22:751–784

Cheong K-L, Zhang Y, Li Z, Li T, Ou Y, Shen J, Zhong S, Tan K (2023) Role of polysaccharides from marine seaweed as feed additives for methane mitigation in ruminants: a critical review. Polymers 15:3153

Chojnacka K, Moustakas K (2024) Anaerobic digestate management for carbon neutrality and fertilizer use: a review of current practices and future opportunities. Biomass Bioenergy 180:106991

Dahai H, Zhihong Y, Lin Q, Yuhong L, Lei T, Jiang L, Liandong Z (2024) The application of magical microalgae in carbon sequestration and emission reduction: removal mechanisms and potential analysis. Renew Sustain Energ Rev 197:114417

Das SK, Ghosh GK (2024) Conversion of biomass into low-cost biochar along with organic manure improved soil hydro-physical environment through technological intervention for sandy soil restoration. Biomass Convers Biorefinery 14:5373–5385

De Corato U, Viola E, Keswani C, Minkina T (2024) Impact of the sustainable agricultural practices for governing soil health from the perspective of a rising agri-based circular bioeconomy. Appl Soil Ecol 194:105199

de Mello FP (2024) Voluntary carbon offset programs in aviation: a systematic literature review, vol 147. Transp Policy, p 158

Debaeke P, Perronne R, Colbach N, Moreau D, Barre P, Lecouviour F, Durand-Tardif M (2024) Non-chemical weed management: which crop functions and traits to improve through breeding? Crop Prot:106631

Diop M, Chirinda N, Beniaich A, El Gharous M, El Mejahed K (2022) Soil and water conservation in Africa: state of play and potential role in tackling soil degradation and building soil health in agricultural lands. Sustainability 14:13425

El Mashad HM, Barzee TJ, Franco RB, Zhang R, Kaffka S, Mitloehner F (2023) Anaerobic digestion and alternative manure management technologies for methane emissions mitigation on Californian dairies. Atmosphere 14:120

Elkhlifi Z, Iftikhar J, Sarraf M, Ali B, Saleem MH, Ibranshahib I, Bispo MD, Meili L, Ercisli S, Torun Kayabasi E (2023) Potential role of biochar on capturing soil nutrients, carbon sequestration and managing environmental challenges: a review. Sustainability 15:2527

Eruaga MA (2024) Policy strategies for managing food safety risks associated with climate change and agriculture. Int J Sch Res Rev 4(1):21–32

Evro S, Oni BA, Tomomewo OS (2024) Global strategies for a low-carbon future: lessons from the US, China, and EU's pursuit of carbon neutrality. J Clean Prod:142635

Faizan M, Rehman M, Fiaz M, Tauqir N, Aamir J (2024) Ruminants enteric methane production: its deleterious effect on environment and mitigation strategies-a review. Pak J Sci 76:16

Falope T, Lao L, Hanak D, Huo D (2024) Hybrid energy system integration and management for solar energy: a review. Energy Convers Manag X:100527

Farooqi ZUR, Qadir AA, Khalid S, Murtaza G, Ashraf MN, Javed W, Waqas MA, Xu M (2024) Greenhouse gas emissions, carbon stocks and wheat productivity following biochar, compost and vermicompost amendments: comparison of non-saline and salt-affected soils. Sci Rep 14:7752

Feng J-C, Sun L, Yan J (2023) Carbon sequestration via shellfish farming: a potential negative emissions technology. Renew Sust Energ Rev 171:113018
Filipović A, Perčin A, Hadžiabulić A, Mandić A (2024) Transformation of organic matter and impact on the ecosystem. In: Agroforestry for carbon and ecosystem management. Elsevier, pp 311–329
Fontana MB, Novelli LE, Sterren MA, Uhrich WG, Rondán GA, Barbagelata PA, Benintende SM (2024) Cover crop benefit bacteria and increase aggregate-associate soil C and N storage. Geoderma Reg 36:e00743
Forfora N, Azuaje I, Vivas KA, Vera RE, Brito A, Venditti R, Kelley S, Tu Q, Woodley A, Gonzalez R (2024) Evaluating biomass sustainability: why below-ground carbon sequestration matters. J Clean Prod 439:140677
Francaviglia R, Almagro M, Vicente-Vicente JL (2023) Conservation agriculture and soil organic carbon: principles, processes, practices and policy options. Soil Syst 7:17
Friedman A (2024) The international maritime organization's revised greenhouse gas strategy: a political signal of shipping's regulatory future. Int J Mar Coast Law 1:1–27
Gabay M, Oldekop JA, Humphreys D, Kamoto J, Mutta DN, Rai N, Song C, Timko J, Rasmussen LV, Cheek JZ (2020) Contextual factors shaping forest-poverty dynamics. International Union of Forest Research Organizations (IUFRO)
Gao Y, Liu G, Meng F, Hao Y, Chen C, Casazza M (2024) A carbon responsibility allocation approach with incentives mechanism based on carbon emissions and carbon offsets accounting. J Clean Prod 434:139814
Gentsch N, Riechers FL, Boy J, Schweneker D, Feuerstein U, Heuermann D, Guggenberger G (2024) Cover crops improve soil structure and change organic carbon distribution in macroaggregate fractions. Soil 10:139–150
Ghadiri M, Krawchenko T, Newell R (2024) Applying a climate-biodiversity-health framework to support integrated food systems planning and policy. J Environ Manag 358:120769
Giamouri E, Zisis F, Mitsiopoulou C, Christodoulou C, Pappas AC, Simitzis PE, Kamilaris C, Galliou F, Manios T, Mavrommatis A (2023) Sustainable strategies for greenhouse gas emission reduction in small ruminants farming. Sustainability 15:4118
Gillespie S (2024) Practical regeneration: realistic strategies for climate smart agriculture. FriesenPress
Grau Ruiz MA (2024) International Fiscal Cooperation to better integrate public and private efforts on sustainability: the case of carbon offset credits. In: Sustainable finances and the law: between public and private solutions. Springer, pp 281–294
Grohs M, Giacomini SJ, da Rosa CA, Neu GRF, Dorneles AB, Lüdtke FL, Mc Mannis GS (2024) Seasonal and annual methane and nitrous oxide emissions affected by tillage and cover crops in flood-irrigated rice. Agric Ecosyst Environ 359:108747
Han Y, Tan Q, Zhang T, Wang S, Zhang T, Zhang S (2024) Development of an assessment-based planting structure optimization model for mitigating agricultural greenhouse gas emissions. J Environ Manag 349:119322
Hemida MAA (2023) The potentiality of agroforestry practices as essential land use option for Forest rehabilitation and livelihood improvement, case study of Nabag Forest Reserve, Kordofan State, Sudan. soe
Hutajulu H, Tahir U, Utami EY, Harsono I, Yuniarti T (2024) Green growth strategy and environmental policy integration in sustainable economic development: a review of bibliometric analysis. West Sci Soc Humanit Stud 2:474–482
Ivanov IT, Kruttli MS, Watugala SW (2024) Banking on carbon: corporate lending and cap-and-trade policy. Rev Financ Stud 37:1640–1684
Jebari A, Pereyra-Goday F, Kumar A, Collins AL, Rivero MJ, McAuliffe GA (2024) Feasibility of mitigation measures for agricultural greenhouse gas emissions in the UK. A systematic review. Agron Sustain Dev 44:2

Jiang M, Dong C, Bian W, Zhang W, Wang Y (2024) Effects of different fertilization practices on maize yield, soil nutrients, soil moisture, and water use efficiency in northern China based on a meta-analysis. Sci Rep 14:6480

Jordon MW, Buffet J-C, Dungait JA, Galdos MV, Garnett T, Lee MR, Lynch J, Röös E, Searchinger TD, Smith P (2024) A restatement of the natural science evidence base concerning grassland management, grazing livestock and soil carbon storage. Proc R Soc B 291:20232669

Jose S, Garrett HEG, Gold MA, Lassoie JP, Buck LE, Current D (2021) Agroforestry as an integrated, multifunctional land use management strategy. North Am Agrofor:1–25

Kabeyi MJB, Olanrewaju OA, Akpan J (2024) Biogas production and process control improvements. In: From biomass to biobased products. IntechOpen

Kamyab H, SaberiKamarposhti M, Hashim H, Yusuf M (2024) Carbon dynamics in agricultural greenhouse gas emissions and removals: a comprehensive review. Carbon Lett 34:265–289

Kantwa SR, Choudhary M, Agrawal RK, Dixit AK, Kumar S, Chary GR (2024) Reducing energy and carbon footprint through diversified rainfed cropping systems. Energy Nexus 14:100306

Kawamori K (2024) Capitalism created the climate crisis and capitalism will solve it: the market forces catalyzing a climate technology renaissance. Wiley

Khan N, Jhariya MK, Raj A, Banerjee A, Meena RS, Bargali SS, Yadav SK, Kumawat A (2021) Agroforestry and its services for soil management and sustainability. In: Sustainable intensification for agroecosystem services and management. Springer Singapore, pp 353–377

Khan S, Irshad S, Mehmood K, Hasnain Z, Nawaz M, Rais A, Gul S, Wahid MA, Hashem A, Abd_Allah EF (2024) Biochar production and characteristics, its impacts on soil health, crop production, and yield enhancement: a review. Plants 13:166

Kollar AJ (2023) Bridging the gap between agriculture and climate: mitigation of nitrous oxide emissions from fertilizers. Environ Prog Sustain Energy 42:e14069

Króliczewska B, Pecka-Kiełb E, Bujok J (2023) Strategies used to reduce methane emissions from ruminants: controversies and issues. Agriculture 13:602

Kumar S (2024) Will the green credit programme incentivize positive environmental actions? Ecol Econ Soc INSEE J 7:3–11

Kumar P, Raj A, Kumar VA (2024) Approach to reduce agricultural waste via sustainable agricultural practices. In: Valorization of biomass wastes for environmental sustainability: green practices for the rural circular economy. Springer, pp 21–50

Kundu B, Kumar R (2024) Enhancing crop resilience to climate change through biochar: a review. Int J Environ Clim Change 14:170–184

Laine A, Ahonen H-M, Pakkala A, Laininen J, Kulovesi K, Mäntylä I (2023) Guide to good practices for voluntary carbon markets: supporting voluntary mitigation action with carbon credits. Finnish Government

Lambo M, Ma H, Liu R, Dai B, Zhang Y, Li Y (2024) Mechanism, effectiveness, and the prospects of medicinal plants and their bioactive compounds in lowering ruminants' enteric methane emission. Animal:101134

Lazenby R (2024) Evaluating additionality analyses: California cap and trade and the regional greenhouse gas initiative. In: Sustainable finances and the law: between public and private solutions. Springer, pp 91–102

Lee J, Shepley MM (2020) Benefits of solar photovoltaic systems for low-income families in social housing of Korea: renewable energy applications as solutions to energy poverty. J Build Eng 28:101016

Lessmann C, Kramer N (2024) The effect of cap-and-trade on sectoral emissions: evidence from California. Energy Policy 188:114066

Li S, Tasnady D (2023) Biochar for soil carbon sequestration: Current knowledge, mechanisms, and future perspectives. C 9:67

Liu X-M, Zhang Y-Q (2024) RETRACTED ARTICLE: Digital transformation, green innovation, and carbon emission reduction performance of energy-intensive enterprises. Sci Rep 14:3905

Liu D, Gong H, Li J, Liu Z, Wang L, Ouyang Z, Xu L, Wang T (2024) Continuous crop rotation increases soil organic carbon stocks in river deltas: a 40-year field evidence. Sci Total Environ 906:167749

Malec K, Babu SC, Tohirzoda S (2024) Global methane pledge in Tajikistan: assessment report. International Food Policy Research Institute

Manesh MHK, Davadgaran S, Rabeti SAM (2024) New procedure for designing an innovative biomass-solar-wind polygeneration system for sustainable production of power, freshwater, and ammonia using 6E analyses, carbon footprint, water footprint, and multi-objective optimization. Renew Energy 230:120802

Maniraj J, Ramesh M, Kumar SG, Sahayaraj AF (2023) Introduction of biochar: sources, composition, and recent updates. In: Biochar and its composites: fundamentals and applications. Springer, pp 1–17

Mattila TJ, Vihanto N (2024) Agricultural limitations to soil carbon sequestration: plant growth, microbial activity, and carbon stabilization. Agric Ecosyst Environ 367:108986

Michaelowa A, Honegger M, Poralla M, Winkler M, Dalfiume S, Nayak A (2023) International carbon markets for carbon dioxide removal. PLOS Clim 2:e0000118

Mohanty LK, Singh N, Raj P, Prakash A, Tiwari AK, Singh V, Sachan P (2024) Nurturing crops, enhancing soil health, and sustaining agricultural prosperity worldwide through agronomy. J Exp Agric Int 46:46–67

Mohsenimanesh A, LeRiche EL, Gordon R, Clarke S, MacDonald RD, MacKinnon I, VanderZaag AC (2021) Dairy farm electricity use, conservation, and renewable production—a global perspective. Appl Eng Agric 37:977–990

Morya R, Bhargava A, Lalitha GR, Vamshi M, Verma S, Rastogi M, Raj S (2023) Soil management practices to enhance carbon sequestration rates-a review. Int J Environ Clim Change 13:3762–3776

Muys B (2021) Forest ecosystem services. In: Life on land. Springer, pp 386–395

Naeem S, Jilani MI, Kazerooni EA, Zahoor I (2023) Carbon farming: a solution to climate change. Int J Chem Biochem Sci 24(12):294–314

Nair PR, Kumar BM, Nair VD, Nair PR, Kumar BM, Nair VD (2021) Soil organic matter (SOM) and nutrient cycling. In: An introduction to agroforestry: four decades of scientific developments. Springer, pp 383–411

Neethirajan S (2023) Innovative strategies for sustainable dairy farming in Canada amidst climate change. Sustainability 16:265

Nleya Y, Young B, Nooraee E, Baroutian S (2023) Opportunities and challenges for anaerobic digestion of farm dairy effluent. ChemBioEng Rev 10:924–940

Nwaogu C, Fagbami OA, Olushola B, Cherubin MR (2023) Integrated agricultural systems for enhancing carbon stocks and climate change mitigation: Nigeria and Brazil. In: Handbook of nature-based solutions to mitigation and adaptation to climate change. Springer, pp 1–29

Ogunboyo PT, Ogunlade O (2024) Review of photovoltaic energy access for sustainable growth in the agricultural sector: economic, market and employment opportunities for rural communities of sub-Saharan African. Appl Photovolt Technol 1

Pagliaccia D, Ortiz M, Rodriguez MV, Abbott S, De Francesco A, Amador M, Lavagi V, Maki B, Hopkins F, Kaplan J (2024) Enhancing soil health and nutrient availability for Carrizo citrange (X Citroncirus sp.) through bokashi and biochar amendments: an exploration into indoor sustainable soil ecosystem management. Sci Hortic 326:112661

Panda S, Kumar R, Panda B, Panda B, Raj A (2024) Combining solar panels with plants for sustainable energy and food production: state of the art. Int J Appl 13:434–441

Pandao M, Jejal AD, Shukla EP, Prabhumitrareddy S, Rout S, Baral K, Bhadani M (2023) Unlocking the benefits of carbon sequestration for enhancing soil health. Int J Environ Clim Change 13:1349–1359

Pande R (2024) Can the market in voluntary carbon credits help reduce global emissions in line with Paris Agreement targets? American Association for the Advancement of Science, p eadp5223

Pandian K, Vijayakumar S, Mustaffa MRAF, Subramanian P, Chitraputhirapillai S (2024) Biochar–a sustainable soil conditioner for improving soil health, crop production and environment under changing climate: a review. Front Soil Sci 4:1376159

Patra R, Saha D, Neupane A, Jagadamma S (2024) Deep-rooted winter wheat cover crop promotes subsoil organic carbon storage through improved microbial functional traits. Appl Soil Ecol 199:105413

Paul C, Bartkowski B, Dönmez C, Don A, Mayer S, Steffens M, Weigl S, Wiesmeier M, Wolf A, Helming K (2023) Carbon farming: are soil carbon certificates a suitable tool for climate change mitigation? J Environ Manag 330:117142

Pretty J, Ball A, Xiaoyun L, Ravindranath N (2002) The role of sustainable agriculture and renewable–resource management in reducing greenhouse–gas emissions and increasing sinks in China and India. Philos Trans R Soc Lond Ser A: Math Phys Eng Sci 360:1741–1761

Priori S, Pellegrini S, Vignozzi N, Costantini EA (2020) Soil physical-hydrological degradation in the root-zone of tree crops: problems and solutions. Agronomy 11:68

Rana M, Khan SM, Ali S, Khalid A, Ahmad Z (2024) Carbon credit, trading, green economy, and clean development mechanisms. In: Agroforestry for carbon and ecosystem management. Elsevier, pp 147–159

Rosales RC, Bellino P, Elnahass M, Heubaum H, Lim P, Lemaistre P, Siman K, Sjögersten S (2021) Voluntary carbon markets in ASEAN: challenges and opportunities for scaling up. UK Government/The British High Commission Singapore, London and Singapore

Saeed OA (2023) Effects of heat stress on ruminant physiological changes in dry arid regions: a review. Large Anim Rev 29:271–278

Salzman J, Weisbach D (2024) The additionality double standard. Harv Environ Law Rev 48

Saniga N, Kumar S, Tejaswini R, Siddiqua A, Behera HS (2024) Harnessing Earth's green guardians: exploring carbon sequestration in agricultural practices. Int J Environ Clim Change 14(6):86–104

Santander D, Clariget J, Banchero G, Alecrim F, Simon Zinno C, Mariotta J, Gere J, Ciganda VS (2023) Beef steers and enteric methane: reducing emissions by managing forage diet fiber content. Animals 13:1177

Sarma HH, Borah SK, Das O, Sonowal S, Talukdar N, Das BC (2024) Nano biochar: properties, preparation, key features and its impact on soil and crop ecosystem. J Adv Biol Biotechnol 27:66–78

Sikiru A, Michael AO, John MO, Egena SSA, Oleforuh-Okoleh VU, Ambali MI, Muhammad IR (2024) Methane emissions in cattle production: biology, measurement and mitigation strategies in smallholder farmer systems. Environ Dev Sustain:1–24

Singh S, Luthra N, Mandal S, Kushwaha DK, Pathak SO, Datta D, Sharma R, Pramanick B (2023a) Distinct behavior of biochar modulating biogeochemistry of salt-affected and acidic soil: a review. J Soil Sci Plant Nutr 23(3):2981–2997. https://doi.org/10.1007/s42729-023-01370-9

Singh S, Chaturvedi S, Nayak P et al (2023b) Carbon offset potential of biochar based straw management under rice – wheat system along Indo-Gangetic Plains of India. Sci Total Environ 897:165176. https://doi.org/10.1016/j.scitotenv.2023.165176

Singh SV, Mandal S, Meena RS, Chaturvedi S, Govindaraju K (2024) Biochar production for green economy: agricultural and environmental perspectives. Academic Press, Elsevier, p 500

Singh O, Shahi UP, Dutta D, Rajput VD, Singh A (2024a) Strategic tillage for improved soil health and nutrient dynamics. In: Strategic tillage and soil management-new perspectives. BoD–Books on Demand

Singh S, Kiran BR, Mohan SV (2024b) Carbon farming: a circular framework to augment CO 2 sinks and to combat climate change. Environ Sci: Adv 3:522–542

Srivastav AL, Patel N, Rani L, Kumar P, Dutt I, Maddodi B, Chaudhary VK (2024) Sustainable options for fertilizer management in agriculture to prevent water contamination: a review. Environ Dev Sustain 26:8303–8327

Sultan H, Li Y, Ahmed W, Shah A, Faizan M, Ahmad A, Abbas HMM, Nie L, Khan MN (2024) Biochar and nano biochar: enhancing salt resilience in plants and soil while mitigating greenhouse gas emissions: a comprehensive review. J Environ Manag 355:120448

Suryanarayana A, Rao V (2024) Embracing blockchain technologies for sustainable finance through carbon credits and renewable energy trading. Educ Adm: Theory Pract 30:1151–1155

Thamarai P, Deivayanai V, Saravanan A, Vickram A, Yaashikaa P (2024) Carbon mitigation in agriculture: pioneering technologies for a sustainable food system. Trends Food Sci Technol:104477
Trouwloon D, Streck C, Chagas T, Martinus G (2023) Understanding the use of carbon credits by companies: a review of the defining elements of corporate climate claims. Global Chall 7:2200158
Tumushabe JT, Turyasingura B, Chavula P (2023) The sustainability of carbon markets for climate-smart agriculture among smallholder farmers in Uganda. Asian J Res Agric For 9:337–345
Upadhyay V, Choudhary KK, Agrawal SB (2024) Use of biochar as a sustainable agronomic tool, its limitations and impact on environment: a review. Discov Agric 2:20
van Kooten GC, Zanello R (2023) Carbon offsets and agriculture: options, obstacles, and opinions. Can J Agric Econ/Revue canadienne d'agroeconomie 71:375–391
Van Tam N, Toan NQ, Ngoc PH (2024) Key strategies for achieving net-zero carbon buildings and promoting carbon credits in construction markets: a case of an emerging economy. Energy Sustain Dev 81:101488
Vanlauwe B, Brown P, Bruulsema T, Dobermann A, Hungria M, Majumdar K, McLaughlin M, Wager-Riddle C, Zhang X, Zhao F (2024) Fertilizer and soil health for enhanced productivity and sustainability in sub-Saharan Africa. SPRPN, Paris
Vasan V, Sridharan NV, Feroskhan M, Vaithiyanathan S, Subramanian B, Tsai P-C, Lin Y-C, Lay C-H, Wang C-T, Ponnusamy VK (2024) Biogas production and its utilization in internal combustion engines-a review. Process Saf Environ Prot 186:518
Verschuuren J, Fleurke F, Leach MC (2024) Integrating agricultural emissions into the European Union emissions trading system: legal design considerations. Sustainability 16:5091
Veste M, Sheppard JP, Abdulai I, Ayisi KK, Borrass L, Chirwa PW, Funk R, Kapinga K, Morhart C, Mwale SE (2024) The need for sustainable agricultural land-use systems: benefits from integrated agroforestry systems. In: Sustainability of Southern African ecosystems under global change: science for management and policy interventions. Springer, pp 587–623
Vinodhini S, Manibharathi S, Pavithra G, Sakthivel S (2023) Agroforestry: integrating trees into agricultural systems. Recent Approaches Agric 2:246–272
Vogel E, Dalheimer B, Beber CL, de Mori C, Palhares JCP, Novo ALM (2023) Environmental efficiency and methane abatement costs of dairy farms from Minas Gerais, Brazil. Food Policy 119:102520
Werner CIM (2024) Modeling the global potential and limitations of biomass pyrolysis as a negative emission technology using a dynamic vegetation model. Humboldt Universitaet zu Berlin
Witzgall K, Steiner F, Hesse B, Riveras-Muñoz N, Rodríguez V, Teixeira P, Li M, Oses R, Seguel O, Seitz S (2024) Living and decaying roots as regulators of soil aggregation and organic matter formation—from the rhizosphere to the detritusphere. Soil Biol Biochem:109503
Xia X, Chishti MZ, Dogan E (2024) Transition towards the sustainable development: unraveling the effects of mineral markets, Belt & Road Initiative, and the Paris Agreement on green economic growth. Resour Policy 91:104896
Yang H, He S, Feng Q, Xia S, Zhou Q, Wu Z, Zhang Y (2024a) Navigating the depths of seafood authentication: technologies, regulations, and future prospects. Meas Food:100165
Yang X, Xiong J, Du T, Ju X, Gan Y, Li S, Xia L, Shen Y, Pacenka S, Steenhuis TS (2024b) Diversifying crop rotation increases food production, reduces net greenhouse gas emissions and improves soil health. Nat Commun 15:198
Yao T, Bucka F, Kögel-Knabner I, Knief C (2024) Organic matter type and soil texture shape prokaryotic communities during early-stage soil structure formation. J Plant Nutr Soil Sci 187:89
Zakkour P, Kuijper M, Dixon P, Haszeldine RS, Towns M, Allen M (2024) Carbon storage units and carbon storage obligations: a review of policy approaches. Int J Greenhouse Gas Control 133:104087
Zhang W, Ji C, Liu Y, Hao Y, Song Y, Cao Y, Qi H (2024) Dynamic interactions of carbon trading, green certificate trading, and electricity markets: insights from system dynamics modeling. PLoS One 19:e0304478

Chapter 19
Biochar Integration Sustaining Carbon-Nutrient Cycling and Soil Restoration

Rashmi Sharma, Suprava Nath, S. K. Asraful, Nallagatla Vinod Kumar, Adima Bharadwaj, Rashida Hameed, Adeel Abbas, Shiv Vendra Singh, and Yogeshwar Singh

Abstract Farmers worldwide seek ways to nourish the soil and crops in an environmentally friendly manner. This chapter looks at the outstanding soil enhancer, biochar, as a practical solution in four important areas. It works like a time-release vitamin for the soil by first being a natural fertilizer: a slow release of a feast that feeds crops and prevents the repetition of frequent applications. It can be described as the natural delivery system whereby nutrients are delivered to the plant at the most opportune time. The second one acts as a soil rejuvenator that revives life in tired fields by better soil structure and providing healthy homes for friendly microbes. Benefits are also observable through the harvest stage, such as a healthier

R. Sharma
Department of Agronomy, College of Agriculture, G.B. Pant University of Agriculture and Technology, Pantnagar, Uttarakhand, India

S. Nath
Department of Agronomy and Agroforestry, M. S. Swaminathan School of Agriculture, Centurion University of Technology and Management, Paralakhemundi, Odisha, India

S. K. Asraful
ICAR-Indian Agricultural Research Institute, Pusa Campus, New Delhi, India

N. V. Kumar
Department of Agronomy, College of Agriculture, Professor Jayashankar Telangana State Agricultural University, Rajendranagar, Hyderabad, India

A. Bharadwaj · S. V. Singh · Y. Singh
Department of Agronomy, College of Agriculture, Rani Lakshmi Bai Central Agricultural University, Jhansi, Uttar Pradesh, India

R. Hameed (✉)
Key Laboratory of Marine Chemistry Theory and Technology, College of Chemistry and Chemical Engineering, Ocean University of China, Qingdao, China

A. Abbas
School of Environment and Safety Engineering, Institute of Environment and Ecology, Jiangsu University, Zhenjiang, China

B. Pramanick et al. (eds.), *Climate-Smart Agricultural Technologies*, Sustainability Sciences in Asia and Africa, https://doi.org/10.1007/978-981-96-7699-6_19

crop look and higher yield, which stands out as an appreciated benefit in working with biochar, even in adverse weather situations. Biochar also assists in helping farmers receive more value added from their investments in fertilizers, by conserving nitrogen and phosphorus inputs in ways which save money and are easier on the environment. Using illustrative examples and research results, this chapter demonstrates how biochar can help develop productive farming systems today and pay forward for future generations.

Keywords Biochar · Soil health · Nitrogen-use efficiency · Slow-release fertilizer · Crop productivity

19.1 Introduction

In the twenty-first century, the agricultural sector faces a myriad of challenges, including soil degradation, declining productivity, and increasing environmental concerns. To address these issues, innovative solutions are essential, and biochar technology has emerged as a promising approach. This carbon-rich material, produced through the controlled thermochemical conversion of biomass, offers unique benefits for sustainable agriculture. Biochar represents a convergence of ancient agricultural wisdom and modern scientific innovation, with its historical use dating back to the Amazon's Terra Preta soils. Contemporary biochar production employs precise thermochemical processes, primarily through pyrolysis at temperatures between 300 and 700 °C. This process yields a material with remarkable structural and chemical properties that contribute to its agricultural efficacy (Chaturvedi et al. 2021). Advanced characterization techniques have revealed biochar's intricate pore networks, with surface areas typically spanning 200–400 m^2/g, occasionally reaching beyond 1000 m^2/g (Singh et al. 2022). This structural complexity, combined with variable cation exchange capacities of 5–50 cmol/kg, creates a unique material capable of transforming soil properties at multiple scales. Biochar's effectiveness as a soil amendment is well documented. It enhances soil fertility by improving nutrient retention and availability, which is particularly beneficial in degraded or nutrient-poor soils (Labanya et al. 2024). Studies have shown that biochar application can increase crop yields by 10–30% in various agricultural systems. Additionally, biochar's ability to improve soil physical properties, such as bulk density, water-holding capacity, and soil aggregation, further contributes to enhanced plant growth and resilience (Singh et al. 2019; Singh et al. 2023a).

Recent technological advances have expanded biochar's applications beyond traditional soil amendment. The emergence of engineered biochars, specifically designed for particular agricultural challenges, represents a significant advancement in this field. These materials demonstrate enhanced functional properties, including surface areas exceeding 2000 m^2/g, offering new possibilities for addressing specific agricultural challenges. The development of biochar-based slow-release fertilizers has opened new avenues for nutrient management, potentially reducing the environmental impact of conventional fertilizers by minimizing nutrient leaching and volatilization (Singh C, et al. 2020; Singh SV, et al. 2020; Roy et al. 2022). Biochar's environmental benefits extend beyond soil improvement. Research indicates significant potential for greenhouse gas mitigation, with studies documenting reductions in nitrous oxide emissions ranging from 10% to 90% and methane emissions by 25% to 50% in various agricultural systems. Furthermore, biochar's stable carbon structure, persisting in soil for centuries to millennia, offers a practical approach to carbon sequestration in agricultural landscapes. With a carbon content typically ranging from 50% to 90%, biochar can contribute to long-term carbon storage, making it a valuable tool for climate change mitigation. However, the path to widespread biochar adoption faces several challenges. Production costs, varying between $100 and 500 per ton, remain a significant barrier. Additionally, the heterogeneity in biochar properties, influenced by both feedstock selection and production conditions, necessitates careful matching with specific agricultural applications (Kumar et al. 2021). These challenges underscore the importance of continued research and development in optimizing biochar production and application strategies.

This chapter examines the multifaceted applications of biochar in sustainable agriculture, providing a critical analysis of current evidence and future possibilities. By exploring the intersection of biochar technology with agricultural sustainability, we aim to provide insights into how this versatile material can contribute to the future of farming while addressing pressing environmental challenges. The following sections detail specific applications, mechanisms, and outcomes of biochar use in agricultural systems, supported by quantitative evidence and practical considerations for implementation.

19.2 Biochar Production and Properties

19.2.1 Production Techniques of Biochar

Thermal decomposition is a standard method for biochar production, utilizing thermochemical processes such as pyrolysis, hydrothermal carbonization (HTC), gasification, and torrefaction (Amalina et al. 2022; Tang et al. 2019; Wang and Wang 2019). These techniques apply high heat to break down organic materials (Hassan et al. 2020), with the pyrolysis method significantly affecting biochar yield.

Combustion involves heating biomass in the presence of oxygen, resulting in the exothermic oxidation of materials at temperatures between 700 and 1400 °C, producing CO_2 and H_2O as hot flue gas. The solid by-products include char, which contains organic carbon, and ash, consisting of inorganic oxides and carbonates

(Mamaní et al. 2019). However, combustion is not optimal for biochar production as it converts most carbon to CO_2.

In contrast, gasification occurs at temperatures above 800 °C with limited oxygen, generating gaseous by-products such as carbon monoxide (CO), hydrogen (H_2), and moisture (H_2O). While it avoids total combustion, gasification is also inefficient for biochar production, as it converts a significant portion of carbon to CO and CO_2 (Zaker et al. 2019; Umenweke et al. 2022).

19.2.1.1 Pyrolysis

Pyrolysis is a non-oxidative thermal breakdown process that produces three main product fractions: biochar (solid residue), bio-oil (condensable liquid), and syngas (non-condensable gas) (Rangabhashiyam and Balasubramanian 2019; Senthil and Lee 2020). In the absence of oxygen, combustion cannot occur, allowing pyrolysis to take place at temperatures typically between 300 and 700 °C. It is the preferred method for generating biochar and bio-oil, with lower pyrolysis temperatures and longer residence times leading to higher biochar yields. Extended reaction times promote polymerization, increasing biochar production, while moderate temperatures and short residence times favor liquid yields (Yaashikaa et al. 2019).

Pyrolysis employs various conditions, such as residence times from less than 1 second to several hours, heating rates from under 11 °C/s to over 1000 °C/s, and temperatures above 300 °C (Santoso et al. 2020). Different pyrolysis processes yield varying proportions of by-products, making careful selection crucial. The biomass conversion during pyrolysis involves three pathways: char production, depolymerization, and fragmentation (Daful and Chandraratne 2018; Rangabhashiyam and Balasubramanian 2019; Yaashikaa et al. 2020).

Char generation occurs through molecular rearrangements that enhance thermal stability, often forming aromatic structures (Karimi et al. 2018). Depolymerization dissolves polymer bonds, producing volatile compounds that condense into liquids at lower temperatures. Fragmentation breaks polymer and monomer links, generating gases and condensable organic liquids (Shiung et al. 2019).

Gases and liquids produced during pyrolysis can undergo side reactions, such as cracking, which reduces molecular weight, or recombination, which forms larger, less volatile molecules and secondary char (Dhyani and Bhaskar 2018; Yu et al. 2019). Fast pyrolysis yields more bio-oils due to rapid heating and shorter residence times, while slow pyrolysis favors biochar production (Abdel-Shafy and Mansour 2018; Akdeniz 2019).

The moisture content of the feedstock should ideally be below 30% for effective pyrolysis, although up to 10% is acceptable. Higher moisture content requires additional energy and dilutes the calorific value of produced gases and liquids. Hydrothermal carbonization (HTC) techniques can process wet biomass with over 70% moisture. The most common pyrolysis methods are slow and fast pyrolysis.

Slow pyrolysis operates at low temperatures (300–550 °C) with slow heating rates (0.1–0.8 °C/s) and extended contact times (5–30 minutes to 25–35 hours)

(El-naggar et al. 2019). This method maximizes biochar yields by encouraging secondary reactions through prolonged vapor residence. It produces primary and secondary char (Kumar et al. 2021), with yields influenced by material characteristics and pyrolysis parameters like temperature and reactor type (Sonu et al. 2020). Mineral-rich biomass generally results in lower biochar yields. Slow pyrolysis can process particles sized between 5 and 50 mm (Sakhiya et al. 2020).

Intermediate pyrolysis occurs between 450 and 550 °C, with faster reaction times than slow pyrolysis (10–30 seconds) but less charcoal production (Ge et al. 2021; Sakhiya et al. 2020). It prevents high molecular weight tar formation, producing a mix of biochar, bio-oil, and gases. This method can accommodate a wider variety of biomass sizes and compositions, including materials with nearly 40% moisture (Mbarki et al. 2019).

Fast pyrolysis involves high temperatures (425–600 °C), rapid heating rates (10–1000 °C/s), and short residence times (0.5–2 seconds) (Tomczyk et al. 2020). It minimizes secondary reactions to maximize bio-oil output (Mutsengerere et al. 2019), with by-product distribution influenced by biomass composition and heating conditions. Optimal bio-oil production requires finely ground feedstock (less than 1 mm) to ensure adequate heat transfer (Ravindran et al. 2018). Fast pyrolysis typically yields less biochar, as only primary carbon is produced, and is endothermic due to the lack of secondary reactions. Moisture content should be below 10% to enhance heating efficiency (Lee et al. 2020).

Flash pyrolysis aims to optimize bio-oil production at high temperatures (800–1000 °C) with very rapid heating rates (>1000 °C/s) and brief contact times (<0.5 seconds) (Gaurav et al. 2020). It produces similar outputs to fast pyrolysis and requires very fine biomass feedstock (<0.2 mm).

Microwave-assisted pyrolysis is a newer method designed for efficient energy recovery from waste (Ge et al. 2021). Since the 1990s, it has gained attention for its rapid, uniform heating capabilities (Wang T, et al. 2018; Wang Y, et al. 2018). This method allows for quicker chemical reactions and can handle wet biomass without drying, making it environmentally sustainable (Venkatesh et al. 2022). Microwave pyrolysis heats biomass internally, leading to higher temperatures at the center than the surface of the feedstock (Zaker et al. 2019). The resulting biochar has a higher surface area (BET of 450–800 m^2/g) and more functional groups than traditional biochar (Naji and Tye 2022). However, it requires electricity, which can be more costly than heat generated by conventional pyrolysis (Ao et al. 2018; Ge et al. 2021).

19.2.1.2 Torrefaction

Torrefaction is a medium-temperature pyrolysis process that gradually heats biomass to 200–300 °C (<11 °C/s) in the absence of air at atmospheric pressure (Manyà et al. 2020). This process removes excess water and volatiles while partially degrading biopolymers like cellulose, hemicelluloses, and lignin, resulting in a greater solid yield than liquid or gas. Although biochar is defined as having a molar oxygen-to-carbon (O/C) ratio below 0.4, torrefied biomass typically exceeds this ratio, disqualifying it as biochar (Daful and Chandraratne 2018). Torrefied biomass has properties that are intermediate between raw biomass and char, making it useful as a pre-treatment method to reduce moisture and increase biomass density, thereby lowering transportation costs and enhancing heating value (Yadav et al. 2022; Zoroufchi et al. 2020). Additionally, torrefaction improves hydrophobicity, grindability, and biodegradability, allowing for longer storage without degradation. It typically achieves a mass yield of 70–80% and an energy yield of 80–90% (Enaime et al. 2020).

19.2.1.3 Gasification

Gasification is a thermochemical process that converts carbonaceous materials into biochar and syngas using an oxidizing agent like air, oxygen, or steam at temperatures above 750 °C, often under ambient or high pressure. When air is the oxidizing agent, the resulting gas comprises approximately 85% syngas, which includes H_2, CO, CO_2, CH_4, N_2, and C_2H_2. Steam gasification produces higher hydrogen yields with greater heating value (Nidheesh et al. 2021).

The gasification process consists of four steps: drying, pyrolysis, partial oxidation, and reduction (Umenweke et al. 2022). It can be classified into three contact modes: fixed bed, fluidized bed, and entrained flow. This method primarily generates syngas with low biochar yields, rendering gasification insufficient as a viable biochar production technology (Yaashikaa et al. 2020). Similarly, combustion is not suitable for biochar production, as it results in minimal yields and significant greenhouse gas emissions (Nidheesh et al. 2021).

Gasification operating conditions vary to optimize energy extraction from different feedstocks, influencing its operational parameters (Zuber et al. 2019). Pyrolysis remains the most efficient, sustainable, and preferred technique for producing biochar.

19.2.2 Properties of Biochar

The effectiveness of biochar as a soil amendment is influenced by its physicochemical properties, such as surface area, bulk density, pore structure, particle size, water-holding capacity (WHC), cation exchange capacity (CEC), pH, and functional group composition, which vary by feedstock (Rawat et al. 2019; Tomczyk et al. 2020). Common feedstocks include forest and crop residues, manures, algae, and municipal sewage (Kavitha et al. 2018). Biochar from animal manures and crop residues generally contains more nutrients and has a higher pH than woody biomass (Rawat et al. 2019; Kavitha et al. 2018). These varying properties can significantly impact microbial activity, water retention, nutrient retention, and the sorption potential of organic compounds (Bonanomi et al. 2017; Igalavithana et al. 2017).

19.2.2.1 Surface Area, Porosity, Particle Size, and Pore Distribution

The surface area, porosity, and pore size distribution of biochar are critical for nutrient retention, adsorption capacity, and soil microbial activity (Shakya and Agarwal 2020). A larger surface area improves cation exchange capacity (CEC) and water-holding capacity, making it suitable for agricultural use (Weber and Quicker 2018; Seleiman et al. 2021). Pyrolysis releases volatile compounds that create a porous structure, enhancing water and nutrient retention (Rawat et al. 2019; Shakya and Agarwal 2020). The surface area varies by feedstock; for example, sewage sludge biochar can reach 100 m^2/g, while maize straw biochar often surpasses that of cottonwood or aspen (Zhao et al. 2018). Pore characteristics also depend on feedstock and pyrolysis temperature, with macropores and mesopores aiding water retention, while micropores influence chemical properties (Lu and Zong 2018). Additionally, the size and shape of biochar particles affect soil interactions, promoting aggregation and altering water movement (Blanco-Canqui, 2017; Yao et al. 2017).

19.2.2.2 Hydrophobicity

Hydrophobicity in biochar arises from the condensation of aliphatic compounds (tars) on its surface during pyrolysis, affecting both water retention and microbial interactions. Biochars produced at lower temperatures exhibit higher hydrophobicity (Kameyama et al. 2019), while those made at higher temperatures typically have greater surface area and pore volume, resulting in increased capillary force and hydrophilicity. Using highly hydrophobic biochar can hinder water infiltration, render hydrophilic soils water-repellent, and elevate the risk of soil erosion (Mao et al. 2019).

19.2.2.3 Reactivity of Biochar (pH)

The reactivity of biochar influences soil cation exchange capacity (CEC), nutrient mineralization, mineral precipitation, and nutrient mobility and availability (Novotny et al. 2015; Dai et al. 2017). Generally, biochars are alkaline, and alkalinity increases as pyrolysis temperature rises (Shakya and Agarwal 2020). High temperatures remove functional groups like carboxyl, formyl, and hydroxyl, enhancing the concentration of inorganic elements and basic oxides, which raises biochar pH (Singh C, et al. 2020; Singh SV, et al. 2020). pH also varies by feedstock; for instance, biochar from lignin-rich materials, such as nut shells, has a pH of around 8, while biochar from manure and algae can reach pH levels of 9.5 (Aller 2016). When alkaline biochar (pH 7–9) is added to acidic soils ($pH < 7$), it elevates soil pH and reduces the mobility of cationic metals like copper, zinc, cadmium, and mercury by decreasing competition between metal ions and H^+ ions for cation exchange sites (Novotny et al. 2015; Aller 2016).

19.2.2.4 Cation Exchange Capacity (CEC)

Biochars with high cation exchange capacity can retain more nutrients, reduce nutrient leaching, and enhance the uptake of ammonium (NH^{4+}), potassium (K^+), calcium (Ca^{2+}), and magnesium (Mg^{2+}) by releasing H^+ ions to balance the charge in soil (Seleiman et al. 2021; Palansooriya et al. 2019a, b). CEC generally decreases with increasing pyrolysis temperature due to the loss of negatively charged functional groups, particularly at low temperatures (300–450 °C). Conversely, biochars with high CEC effectively adsorb NH^{4+} and minimize nitrogen leaching (Seleiman et al. 2021). Variability in CEC is also influenced by feedstock type and the specific functional groups formed during pyrolysis, which affect biochar's surface properties (Suliman et al. 2016). For instance, biochar from pig manure has a lower CEC (32.7 cmol kg^{-1}) compared to that from chicken manure (81.4 cmol kg^{-1}) when pyrolyzed at 500 °C (Cely et al. 2015).

19.2.2.5 Surface Functional Groups

During the pyrolysis of biochar, chemical bonds break, leading to the formation of new functional groups, such as nitro ($-NO_2$), methyl ($-CH_3$), hydroxyl ($-OH$), and carboxyl ($C{=}O$) (Zaman et al. 2017). These functional groups can act as electron acceptors or donors, making biochar either acidic, simple, hydrophilic, or hydrophobic. Pyrolysis temperature significantly affects these functional groups; as temperature increases, acidic groups like formyl, carboxyl, and hydroxyl are eliminated, while basic cations become more concentrated, increasing biochar's alkalinity (Weber and Quicker 2018; Naeem et al. 2017). The presence of negative functional groups in biochar enhances soil pH by adsorbing metal ions through proton transfer (deprotonation).

19.3 Biochar as Nutrient Source

19.3.1 Biochar as Slow-Release Fertilizers

Increased global awareness of environmental problems raises increasing concern about the overuse of chemical fertilizers. This will lead to improper application, reducing the fertilizer's effectiveness and triggering severe ecological damage. Nitrogen and phosphorus are critical nutrients for proper growth in crops, while the quantities of nitrogen fertilizers alone used annually are phenomenal. Alarmingly, only 20–30% of this nitrogen is efficiently used by crops, while the rest leads to ammonia volatilization, nitrate leaching, and nitrous oxide emissions, which significantly pollute our surface waters, groundwater, and atmosphere. The sources of phosphorus are equally important, but at an alarming rate, its sources are dwindling.

As the supply of nonrenewable phosphorus rock is limited, the challenge we face is tremendous. Slow-release fertilizers, in recent times, are becoming the next sustainable solution for these urgent concerns. These fertilizers are specially designed to ensure that nutrients are released in a way that reduces environmental loss. Biochar, a new soil amendment, has been shown to improve nutrient supply (Roy et al. 2022), enhance soil structure (Singh et al. 2023a), decrease greenhouse gas emissions (Singh et al. 2023b), and immobilize heavy metals and contaminants (Pandey et al. 2024; Pandey et al. 2025; Singh et al. 2025). The nutrient content of pristine biochar is low, and its effectiveness is significantly enhanced when used in conjunction with fertilizers (Cen et al. 2021; Zhou et al. 2021). Efforts are underway to improve its effectiveness through impregnation, encapsulation, and granulation methods, thereby opening up avenues for a greener agricultural future (Kim et al. 2017; Barbosa et al. 2022; Das and Ghosh 2021). There are a variety of proposed synthesis methods for these effective SRFs: impregnation, encapsulation, granulation, and co-pyrolysis.

Synthesis Strategies for Biochar-Based SR

(a) *In-situ pyrolysis*: Biomass feedstocks rich in nutrients can be directly transformed into biochar-based fertilizers. For example, banana peduncles (Karim et al. 2019), mushroom compost (Lou et al. 2017), and seaweeds (Roberts et al. 2015) have been used to produce potassium-rich biochar. Additionally, some studies have explored phosphorus-rich biochar derived from sources like poultry litter (Cantrell and Martin 2012), human excreta (Liu et al. 2014), sewage sludge, bone waste, and bio-fermentation waste. High phosphorus content in biochar indicates its potential as a fertilizer, and thermally converting bacterial waste into phosphorus-enriched biochar presents a viable strategy for recycling biomass from the fermentation industry (Kim et al. 2018).

(b) *Co-pyrolysis*: Co-pyrolysis of biomass with phosphorus sources can be a great prospective way to produce slow-release fertilizers in the form of biochar. This technique allows phosphorus to be effectively immobilized into the biochar either by being adsorbed or chemically bonded. SRFs prepared using biochar in co-pyrolysis have many advantages, which feature a simple synthesis, high levels of phosphorus loading, and excellent slow-release properties (An et al. 2021). For example, Zhao et al. (2014) investigated the co-pyrolysis of wheat straw with some sources of phosphorus, H_3PO_4, phosphate rock, and triple superphosphate; they realized less carbon lost, lower mineralization of microbial matter, and stabilized biochar at different stages of processing. In addition, they carried out co-pyrolysis of the biomass types of pine sawdust or switchgrass with different fertilizer types in which triple superphosphate or bone meal was included in producing a set of SRFs that were produced from biochars (Zhao et al. 2016). Yet, when the biomass is mixed with an acidic phosphate source, the pH value in the corresponding biochar is lower, thus limiting its utilization in soil. Alkaline materials such as MgO added before pyrolysis can balance pH and increase the nutrient content of phosphorus and magnesium.

(c) *Impregnation*: Due to biochar's low nutrient content, impregnating pristine biochar in nutrient solutions effectively enhances its nutrient levels (Figueiredo et al. 2024). Its porous structure and large surface area create storage sites for nutrients, extending their availability for plants. This method is straightforward and cost-effective. Biochar impregnated with ammonium salts, phosphates, and nitrates has been shown to create slow-release fertilizers, reducing nutrient loss and boosting crop growth.
(d) *Encapsulation*: Encapsulation is the most popular method for manufacturing SRF, which is mainly the coating of fertilizers with inorganic or organic substrates to minimize losses and maximize utilizations. This process is crucial as it balances the agronomical, environmental, and economic requirements of nutrient releases. Formulation of biochar-coated urea with biodegradable polymers and waterborne copolymers has shown many advantages through various research studies (Gonzalez et al. 2015; Chen et al. 2018a). The efficiency of biochar-coated fertilizers is also dependent on the properties of biochar, the materials used for coating, and encapsulation methods. Large surface area helps in regulating the release of nutrients through strong adsorption (Sha et al. 2019). However, the alkaline groups and inorganic salts can improve the loss of nitrogen due to ammonia volatilization (Fidel et al. 2017). Another polymer in use includes polyolefin and wax. They are also efficient though at a higher cost and nonbiodegradable and may cause damage to the structure and microbial activity of soil. Instead, naturally biodegradable polymers such as waste palm oil, starch, and chitosan are gaining importance.
(e) *Granulation*: One of the most used processes in the manufacture of fertilizer is granulation or palletization. Granulation helps avoid the issues resulting from the brittle nature and low density of biochar, making it challenging for storage and transport. It significantly decreases losses and costs in handling and thus enables it to be utilized at a large scale in agriculture (Ye et al. 2019; Chen et al. 2018b). In addition to nutrients, granulation decreases the rate of nutrient release through enhanced resistance to diffusion. Slow-release fertilizers (SRFs) based on biochar can be prepared by carbonizing nutrient-rich biomass or by pelletizing fertilizers along with biochar. For pelletizing, there is a uniform mixing of biochar with fertilizers and then granulation using disc granulation or mechanical extrusion that often contains binders (Cantrell and Martin 2012).

19.3.2 Co-composted Biochar

Severe soil fertility depletion and declining agricultural productivity in tropical regions are primarily due to reduced soil organic matter (SOM) and nutrient imbalances (Pender 2012; Sanchez 2002). Maintaining adequate SOM levels and ensuring effective nutrient cycling are essential for successful soil management and agricultural productivity (Bationo et al. 2007; Vanlauwe et al. 2010). Effective practices involve using organic and inorganic fertilizers tailored to local conditions to enhance nutrient efficiency and crop yields (Vanlauwe et al. 2010). Although mulches, composts, and manures can improve soil fertility, organic matter tends to mineralize quickly in tropical climates (Tiessen et al. 1994), leading to most applied organic compounds being released as CO_2 rather than stabilizing in the soil (Bol et al. 2000; Fearnside 2000). An alternative approach is using more stable carbon compounds, such as carbonized materials or their extracts.

Biochar and biochar-compost mixtures from alternative organic sources are proposed to enhance soil fertility, restore degraded lands, and reduce greenhouse gas emissions from agriculture. Incorporating biochar into compost has been shown to reduce nitrogen loss during sludge composting significantly. Studies indicate that biochar addition improves nitrogen retention in the final compost (Steiner et al. 2010), stabilizes heavy metals (Hua et al. 2009), accelerates volume reduction through increased carbon mineralization, and alters microbial community structures (Jindo et al. 2012). The resulting co-composted product shows promise for enhancing soil fertility and crop yields (Agegnehu et al. 2016b; Dias et al. 2010; Glaser et al. 2015; Lashari et al. 2013; Schulz and Glaser 2012). Additionally, exposing fresh biochar to composting conditions promotes the oxidation of its surfaces (Prost et al. 2013; Wiedner et al. 2015), fostering beneficial plant-soil interactions. Recent research also indicates that co-composting biochar improves nitrogen uptake and reduces potentially phytotoxic PAH compounds (Borchard et al. 2014).

Biochar can serve as an effective bulking agent for composting nitrogen-rich materials, helping to prevent emissions from organic waste and crop residues by inhibiting natural decomposition in soil. For example, Hua et al. (2009) found that adding 9% biochar reduced total nitrogen loss during composting by 64.1% compared to no biochar while also decreasing the mobility of copper and zinc by 44.4% and 19.3%, respectively. Similarly, Dias et al. (2010) reported that biochar application to poultry manure minimized nitrogen losses in mature composts. However, sawdust was noted to be more effective for preserving organic matter and nitrogen. Steiner et al. (2010) indicated that biochar can absorb ammonia and water-soluble ammonium, further reducing nitrogen losses during manure composting. Ammonia emissions decreased by up to 64% when poultry litter was mixed with 20% biochar, resulting in total nitrogen losses dropping by up to 52%. Thus, co-composted biochar offers a valuable product from organic waste, enhancing soil fertility, carbon sequestration, greenhouse gas reduction, and food security.

Biochar and biochar-compost mixtures enhance crop yields through various mechanisms, including direct nutrient supply, improved soil pH, and increased nutrient-use efficiency. They enhance nutrient uptake by increasing soil cation exchange capacity (CEC) and improving water retention in sandy soils or drainage in clay soils (Agegnehu et al. 2015). Immediate benefits arise from the higher availability of potassium, phosphorus, and zinc in biochar, with calcium and copper contributing less (Steiner et al. 2007). Long-term advantages include better stabilization of organic matter, slower nutrient release, and improved cation retention due to enhanced CEC. Biochar applications have been linked to increased soil organic carbon (SOC) and nitrogen concentrations, as well as greater availability of major cations and phosphorus, CEC, and pH (Steiner et al. 2007; Agegnehu et al. 2016a;

Peng et al. 2011). This increased nutrient availability results from direct nutrient contributions from biochar and enhanced nutrient retention, along with potential shifts in soil microbial dynamics (Lehmann et al. 2011).

19.4 Biochar Improving Soil Health

Biochar's role extends beyond carbon sequestration to include long-term improvements in soil physicochemical properties, fertility, and health, as well as supporting diverse microbial communities, making it a unique soil amendment. Research on biochar as a soil amendment has produced mixed results due to the complex interactions between biochar, soil, and crops (Lychuk et al. 2015). Bulk density (BD), a key indicator of soil compactness, is inversely related to water-holding capacity. Meta-analyses by Omondi et al. (2016), Razzaghi et al. (2020), and Singh et al. (2022) revealed that biochar application reduced soil bulk density by averages of 8%, 9%, and 29%, respectively. Biochar incorporation changes the arrangement of soil particles, creating extra soil porosity and increasing the total pore volume, which results in a decrease in soil bulk density (Lim et al. 2016). Research consistently demonstrates that incorporating biochar into the soil increases its water retention ability, which is particularly advantageous in drought-prone areas or regions with limited irrigation. Large specific surface area, high porosity, and the presence of hydrophilic domains of biochar modify the shape and size of soil pores and improve overall soil porosity, soil aggregation, and structure (Kameyama et al. 2012). The pyrolysis temperature of biochar also influences soil aggregation; low-temperature biochar promotes aggregation by enhancing microbial activity and organic matter accumulation, while high-temperature biochar contributes to stability through the formation of stable pore structures (Albalasmeh et al. 2024). Therefore, biochar has the potential to improve the hydro-physical properties of soil, offering promising opportunities for enhancing soil structure, water retention, and overall agricultural productivity. Additionally, it can be particularly advantageous during extreme drought and heavy rainfall cycles, as it helps reduce soil eluviation during wet periods and alleviates soil pore water deficiency during dry periods.

Biochar has come out as one of the transitional soil amendments providing enormous benefits regarding the availability of nutrients in soils and biological health. Its use, based on the pyrolysis process of organic matter, has valuable properties for improvement in soil fertility and promotion of sustainable agriculture. Biochar exerts influence on the availability of soil nutrients through a multiplicity of mechanisms. Primarily, these involve increasing cation exchange capacity (CEC) and nutrient retention. Alkaline nature contributes to the neutralization of acidic soils and optimal nutrient availability for plant uptake. Manure-based biochar has higher primary nutrient concentrations (N, P, K); in contrast, carbon and structural benefits are better seen in wood-based biochar. Chemical gains are, however, supplemented with significant increases in soil biological health through the creation of amenable microbial habitats. The pore structure of biochar, through its structure, provides an ideal habitat for microbial colonization resulting in increased microbial biomass and diversity. These microbes are involved in nutrient cycling, decomposition of organic matter, and disease suppression (Singh et al. 2023a, b). Such modification in soil pH and enhanced retention of moisture favors higher microbial activity and increased enzymatic activities and mineralization of nutrients.

Apart from chemical attributes, biochar promotes soil biological health in terms of encouraging microbial habitats in the soil. Biochar being porous is good for microbial inoculation, giving a boost to the microbial population and diversity; it acts as a critical player in nutrification or the cycling processes, decomposition processes of organic materials, and prevention of diseases as well (Singh et al. 2023a, b). Further microbial proliferation is supported by modification of soil pH and improved moisture retention, which also enhance enzymatic activities and nutrient mineralization. Application rates are taken into account in practical applications, usually ranging from 10 to 40 tonnes per hectare, with specific soil requirements. Integration of biochar into existing agricultural practices, such as composting or fertilizer application, enhances its effectiveness. Long-term impacts include residence times longer than hundreds of years, which would further suggest advantages in soil quality and carbon sequestration over the long term. In summary, this double increase in nutrient availability and biological health from biochar represents a giant leap in the management of sustainable soils. Optimization of the production and application techniques will be required while economically viable for the agricultural practitioners as put forth by Ghosh and Maiti (2023) (Table 19.1).

19.5 Carbon Sequestration Potential and Stability

Biochar serves as a powerful tool for soil carbon sequestration, primarily due to its recalcitrant nature and long-term persistence in soil systems. Its effectiveness in carbon sequestration operates through multiple mechanisms. First, biochar's structural properties enhance soil carbon storage through improved soil aggregation. Its porous nature creates favorable habitats for beneficial microbes, leading to better soil structure and protection of organic carbon from decomposition (Weng et al. 2022). This results in increased carboxyl carbon, aromatic carbon, ketones, and esters within aggregates, reflecting the enhancement of both stable and labile carbon fractions (Hu et al. 2021). Second, biochar's recalcitrant structure provides a protective barrier against microbial attack, effectively reducing carbon mineralization rates and improving carbon stabilization (Fang et al. 2014). This protection mechanism is further enhanced by biochar's influence on soil enzymatic activity, where it reduces β-glucosidase and cellobiohydrolase activity while increasing RubisCO enzyme activity, indicating greater carbon retention (Fang et al. 2014). Additionally, biochar's surface properties play a crucial role in carbon sequestration. Its large surface area and highly porous structure, developed during pyrolysis, increase cation exchange capacity and enhance organic substance adsorption. This

Table 19.1 Biochar and soil physical properties

Soil type	Biochar feedstock used	Application rate of biochar	Crop	Findings	References
Ultisols in Jambi Province (Indonesia)	Coconut shell biochar produced through low-temperature pyrolysis at (300–400 °C) (plus Leucaena compost)	10 kg ha^{-1}; 5, 10 t ha^{-1}	Soybean	↑ Soil organic matter (SOM), total porosity (TP), and aeration pore (AP) ↓ Bulk density (BD) and low drainage pore (LDP)	Endriani and Listyarini (2023)
Entisols of Indonesia	Rice husk and cocopeat biochar produced through low-temperature pyrolysis at (400–600 °C) (plus cow manure)	0, 10 t ha^{-1}	Peanut	↑ Soil pH, organic C, total N, C:N ratio, available-P, and exchangeable-K ↑ Weight of peanut seeds and yield of peanuts	Yunilasari (2020)
Ultisols dryland in Jambi Province (Indonesia)	Coconut husk biochar (simple pyrolysis)	0, 3, 6, 9, 12, 15 t ha^{-1}	Corn	↑ Physical (TPS, SWC) and chemical properties of ultisols (pH, EC, CEC, OC, P, K)	Ichwan et al. (2020)
University of Kurdistan, Iran (semiarid climate)	Cattle manure biochar produced through the process of pyrolysis at a temperature of 600 ± 30 °C for 4 hours under oxygen-limited conditions	0, 15 t ha^{-1}	*Nigella sativa* L.	↑ Oil yield by 23% ↑ Linoleic acid by 2% in conventional tillage, 3% in minimum tillage, and 5% in no-tillage compared to those without biochar	Kiani et al. (2024)

(continued)

Table 19.1 (continued)

Soil type	Biochar feedstock used	Application rate of biochar	Crop	Findings	References
Fine, mesic, Typic Chromoxerert of Southern Italy	Biochar derives from a process of pyro-gasification of wood obtained from forest cutting (pine, oak, holm oak, chestnut, fir) with high pyrolysis temperature above 700–800 °C (plus compost and mineral fertilizers)	10, 30 Mg ha^{-1}	Durum wheat	↑ Soil total organic carbon (TOC) content ↓ Water extractable organic carbon (WEOC)	Leogrande et al. (2024)
Sandy soil with a very low content of organic carbon (Western Slovakia)	Willow feedstock (pyrolysis at 300 and 520 °C)	20, 40 Mg ha^{-1}	–	↑ Available water content (AWC) by 76–168%, contact angle (CA) by 252–489%, porosity (P) by 6–11% ↓ Saturated hydraulic conductivity (*Ksat*) by 37–90%	Vitková et al. (2024)
Reclaimed post-coal mining soil in polybags (Jambi Province, Indonesia)	Corncob biochar (plus Tithonia compost and fertilizers)	0, 5, 10 t ha^{-1}	Soybean	↓ Soil bulk density ↑ Total pore space, aggregate stability, and soil water content capacity ↑ Soybean growth and yield	Wiskandar and Ajidirman (2024)

(continued)

Table 19.1 (continued)

Soil type	Biochar feedstock used	Application rate of biochar	Crop	Findings	References
Red soil pot experiment (China)	Homemade rice straw biochar produced through the process of pyrolysis at a temperature of 500 °C (plus 12% rice straw)	5%	Vegetable cultivation	↑ SOC content by 11.47–24.65% ↓ MWD by 44.97%, 34.32%, and 56.81%; GMD by 52.10%, 42.86%, and 62.19%; and $R_{0.25}$ by 19.13%, 11.63%, and 24.06%, respectively, in 5% biochar, 12% straw, and 5% biochar +12% straw treatment	Liao et al. (2024)
Sandy loam soil incubation study (Jordan)	Olive pomace (Jift) feedstock (pyrolysis temperatures 300, 400, 500, and 600 °C) and incubated for 30, 60, and 90 days	0, 15, 30, 60 t ha^{-1}	–	↑ Soil aggregate stability by 1.8%, 5.6%, and 8.8% ↓ Infiltration rate ↑ Water retention	Albalasmeh et al. (2024)
Hanford sandy loam (HSL) and yolo silt loam (YSiL) soils of USA	Almond shell pyrolysis at 500 and 800 °C; coconut shell pyrolysis at 650 °C; softwood pyrolysis at 500, 650, and 800 °C; and an additional softwood biochar produced at 500 °C	0%, 2%	–	↓ Quantity of ammonium in the leachate by 22–78% ↓ Saturated hydraulic conductivity (*Ksat*) decreased by 64–80%	Gelardi et al. (2021)
Silty clay loam (Pakistan)	Acacia tree pruning pyrolysis at 450°C (plus fertilizers)	0, 5, 10 t ha^{-1}	Legumes	↑ Biomass production, grain yield, SOC, and fertility soil (N, P, and K)	Arif et al. (2021)

(continued)

Table 19.1 (continued)

Soil type	Biochar feedstock used	Application rate of biochar	Crop	Findings	References
Red soil (China)	Peanut shell biochar pyrolysis at 500 °C for 2 hours (plus peanut shell)	1000, 1500, 2000 kg hm^{-1}	Sweet potato	↑ Microbial biomass carbon content ↑ Phospholipid fatty acid (PLFA) contents of total soil microorganisms and different microbial groups	Zhou et al. (2023)
Sandy clay loam pot experiment (Bundelkhand Region, India)	*Parthenium hysterophorus* L. and *Lantana camara* L. pyrolysis at ~400 °C for 2 hours (plus fertilizers)	2.5, 5, 10 t ha^{-1}	Oat	↑ SOC by 5.17–7.99 g kg^{-1} and CEC by 3.36–37.14%	Choudhary et al. (2022)

characteristic reduces the susceptibility of labile carbon molecules to microbial decomposition, effectively accumulating soil organic carbon (Ghosh and Maiti 2023). Through these mechanisms, biochar emerges as a promising option for recarbonizing agricultural soils, offering a sustainable approach to enhancing soil carbon sequestration while improving overall soil health.

19.6 Crop Health and Productivity

Biochar, which is the pyrolysis product of organic matter and has carbon content, is shown to significantly improve crop productivity. Moreover, field experiments and meta-analysis studies have proved that many crops such as rice, maize, wheat, and legumes benefit from positive effects after the addition of biochar. Field trials in tropical regions have recorded yield gains after applying biochar. For instance, rice husk biochar in 4-year studies in the Philippines and Thailand has reported improvements in yield up to 16–35% for rice and maize with increased potassium (K) and phosphorus (P) availability (Haefele et al. 2011). In Ghana, biochar from sorghum and rice husk increased maize yields and positively affected soil pH, organic carbon, and nutrient concentration (Haefele et al. 2011). Maize and bean yield significantly rose in acidic soils after the application of eucalyptus biochar in Madagascar (Yamato et al. 2006).

In the Mediterranean, biochar addition increased durum wheat yields by as much as 30% (Vaccari et al. 2011), whereas in Nepal, it was found that combining biochar with cow urine enhanced squash yield (Schmidt and Noack 2000). Some long-term experiments have also revealed delayed but sizeable benefits. A 4-year Colombian trial reported rises in maize yields of 20%, 30%, and 140% in years two, three, and four, which demonstrates the residual effect of biochar for long-term sustainability in soil production (Major et al. 2010). The patterns were repeated also in Nepal wherein biochar amended maize and mustard yields in other years (Pandit et al. 2018). A meta-analysis of over 100 studies mentioned that the addition of biochar to the soil increased average yield by 13%, with the strongest impacts noted in acidic soils with a pH of 5 (Liu et al. 2013). Another meta-analysis concerning 114 studies concluded a 20% increase in crop yield, thus indicating the ability of biochar to enhance agricultural productivity (Chen et al. 2023). Apart from improving yield, biochar increases the properties of soil nutrient retention, water-holding capacity, and carbon content. This is indeed most useful for degraded and arid soils where nutrient leaching due to excess water can significantly limit crop growth. However, it depends on the properties of biochar, such as feedstock type and pyrolysis conditions, and the characteristics of the soil.

19.7 Enhancing Nutrient-Use Efficiency

Biochar has been proven to enhance nutrient-use efficiency in agricultural production systems through various mechanisms. For instance, biochar with N fertilizers has been shown to increase fertilizer-use efficiencies and reduce environmental losses of N through several processes in the soil (Liu et al. 2017a, b; Huang et al. 2018). These properties, with high surface area, porosity, CEC, and having both acidic and basic functional groups, make specific physicochemical properties of biochar contribute positively to the prevention of nutrient loss from soils in general, which is more obvious in the impact on nitrogen dynamics (Grutzmacher et al. 2018). Application of biochar has been shown by meta-analyses to lower the soil N_2O emission reductions by 38–49% as reported by Borchard et al. (2019), whereas it alters N transformations due to changes in CEC, porosity, aeration, pH, and microbial activity in soils (Edwards et al. 2018). Biochar application at rates of 0.4–1.9% (w/w) along with N fertilizer reduced N_2O emissions by up to 71% compared to sole N application by Abbruzzini et al. (2019). The optimal application rate of biochar in wheat production systems ranges between 5 and 20 Mg ha^{-1}, which can enhance NUE by 5.2–37.9% and grain yield by 2.9–19.4%. It is also reported that higher than 30 Mg ha^{-1} application of biochar adversely affects both NUE and grain yield (Sun et al. 2019). This showed great efficiency for phosphorus because the nutrient availability increased with significant improvement in crop performance resulting from biochar application. The outcomes of this study revealed that the available content of phosphorus (P) increased by 30% and wheat productivity in which an extremely significant improvement is depicted for grain yield at 27% and shoot biomass at 16% compared to the untreated soil (Abbruzzini et al. 2019). It relates to the multidirectional effects of how biochar might reduce losses and increase available nutrient supply due to improved conditions and physical as well as chemical soil properties that are conducive to the uptake of plant nutrients from these substrates.

19.8 Conclusion

Such integration of biochar into agricultural systems comes with many great potential benefits concerning some of the major issues surrounding modern farming. Advanced characterization clearly shows that there is a crucial role for it as a soil amendment, containing intricate pore networks and variable cation exchange capacities, hence beneficially improving its properties. Biochar significantly increases NUE by 5.2–37.9% and grain yield by 2.9–19.4% at the optimal application rate of 5–20 Mg ha^{-1}. Otherwise, rates above 30 Mg ha^{-1} have been observed to have adverse effects, and the right application rates are therefore crucial. Soil-biochar interaction causes drastic alterations in nitrogen transformations within the soil, mainly due to changes in CEC, porosity, aeration, pH, and microbial activity. The nutrient retention and availability for plant uptake, improving crop productivity while reducing environmental degradation, have the simultaneous advantages of increasing agricultural productivity while reducing greenhouse gas emissions; hence, it is an essential input in climate-smart agriculture. The technology of biochar offers a way to practically improve the sustainability and resilience of farm systems. Potential future studies involve optimizing application rates of different types of soils, cropping systems with consideration for both long-term implications and economic viability, and a guarantee that there will be complete integration into prevailing agricultural practice.

References

Abbruzzini TF, Davies CA, Toledo FH, Cerri CEP (2019) Dynamic biochar effects on nitrogen use efficiency, crop yield and soil nitrous oxide emissions during a tropical wheat-growing season. J Environ Manag 252:109638

Abdel-Shafy HI, Mansour MSM (2018) Solid waste issue: sources, composition, disposal, recycling, and valorization. Egypt J Pet 27(4):1275–1290. https://doi.org/10.1016/j.ejpe.2018.07.003

Agegnehu G, Bass AM, Nelson PN, Muirhead B, Wright G, Bird MI (2015) Biochar and biochar-compost as soil amendments: effects on peanut yield soil properties and greenhouse gas emissions in tropical North Queensland, Australia. Agric Ecosyst Environ 213:72–85

Agegnehu G, Bass AM, Nelson PN, Bird MI (2016a) Benefits of biochar, compost and biochar–compost for soil quality, maize yield and greenhouse gas emissions in a tropical agricultural soil. Sci Total Environ 543:295–306

Agegnehu G, Nelson PN, Bird MI (2016b) Crop yield, plant nutrient uptake and soil physicochemical properties under organic soil amendments and nitrogen fertilization on Nitisols. Soil Tillage Res 160:1–13

Akdeniz N (2019) A systematic review of biochar use in animal waste composting. Waste Manag 88:291–300. https://doi.org/10.1016/j.wasman.2019.03.054

Albalasmeh AA, Quzaih MZ, Gharaibeh MA, Rusan M, Mohawesh OE, Rababah SR et al (2024) Significance of pyrolytic temperature, application rate and incubation period of biochar in improving hydro-physical properties of calcareous sandy loam soil. Sci Rep 14(1):7012

Aller MF (2016) Biochar properties: transport, fate, and impact. Crit Rev Environ Sci Technol 46:1183–1296

Amalina F, Razak ASA, Krishnan S, Zularisam AW, Nasrullah M (2022) Water hyacinth (Eichhornia crassipes) for organic contaminants removal in water – a review. J Hazard Mater Adv 7:100092. https://doi.org/10.1016/J.HAZADV.2022.100092

An X, Wu Z, Liu X, Shi W, Tian F, Yu B (2021) A new class of biochar-based slow-release phosphorus fertilizers with high water retention based on integrated co-pyrolysis and co-polymerization. Chemosphere 285:131481

Ao W, Fu J, Mao X, Kang Q, Ran C, Liu Y, Zhang H (2018) Microwave assisted preparation of activated carbon from biomass: a review. Renew Sustain Energ Rev 92:958–979. https://doi.org/10.1016/j.rser.2018.04.051

Arif M, Ikramullah, Jan T, Riaz M, Akhtar K, Ali S et al (2021) Biochar and leguminous cover crops as an alternative to summer fallowing for soil organic carbon and nutrient management in the wheat-maize-wheat cropping system under semiarid climate. J Soils Sediments 21:1395–1407

Barbosa CF, Correa DA, Carneiro JSDS, Melo LCA (2022) Biochar phosphate fertilizer loaded with urea preserves available nitrogen longer than conventional urea. Sustainability 14(2):686

Bationo A, Kihara J, Vanlauwe B, Waswa B, Kimetu J (2007) Soil organic carbon dynamics, functions and management in West African agro-ecosystems. Agric Syst 94(1):13–25

Blanco-Canqui H (2017) Biochar and soil physical properties. Soil Sci Soc Am J 81:687–711

Bol R, Amelung W, Friedrich C, Ostle N (2000) Tracing dung-derived carbon in temperate grassland using 13C natural abundance measurements. Soil Biol Biochem 32(10):1337–1343

Bonanomi G, Ippolito F, Cesarano G, Nanni B, Lombardi N, Rita A, Saracino A, Scala F (2017) Biochar as plant growth promoter: better off alone or mixed with organic amendments? Front Plant Sci 8:1570

Borchard N, Siemens J, Ladd B, Möller A, Amelung W (2014) Application of biochars to sandy and silty soil failed to increase maize yield under common agricultural practice. Soil Tillage Res 144:184–194

Borchard N, Schirrmann M, Cayuela ML, Kammann C, Wrage-Mönnig N, Estavillo JM, Fuertes-Mendizábal T, Sigua G, Spokas K, Novak J (2019) Biochar, soil and land-use interactions that reduce nitrate leaching and N_2O emissions: a meta-analysis. Sci Total Environ 651:2354–2364

Cantrell KB, Martin JH (2012) Poultry litter and switchgrass blending and pelletizing characteristics for biochar production. American Society of Agricultural and Biological Engineers, Dallas

Cely P, Gascó G, Paz-Ferreiro J, Méndez A (2015) Agronomic properties of biochars from different manure wastes. J Anal Appl Pyrolysis 111:173–182

Cen Z, Wei L, Muthukumarappan K, Sobhan A, McDaniel R (2021) Assessment of a biochar-based controlled release nitrogen fertilizer coated with polylactic acid. J Soil Sci Plant Nutr:1e13

Chaturvedi S, Singh SV, Dhyani VC, Govindaraju K, Mandal S (2021) Characterization, bioenergy value and thermal stability of biochar derived from diverse agriculture and forestry lingo cellulosic waste. Biomass Convers Biorefinery 13:1–14. https://doi.org/10.1007/s13399-020-01239-2

Chen S, Yang M, Ba C, Yu S, Jiang Y, Zou H, Zhang Y (2018a) Preparation and characterization of slow-release fertilizer encapsulated by biochar-based waterborne copolymers. Sci Total Environ 615:431e437

Chen L, Chen Q, Rao P, Yan L, Shakib A, Shen G (2018b) Formulating and optimizing a novel biochar-based fertilizer for simultaneous slow-release of nitrogen and immobilization of cadmium. Sustainability 10(8):2740

Chen Z, Liu J, Sun H, Xing J, Zhang Z, Jiang J (2023) Effects of biochar applied in either rice or wheat seasons on the production and quality of wheat and nutrient status in paddy profiles. Plants 12(24):4131

Choudhary P, Prasad M, Choudhary M, Kumar A, Mahawer SK, Srinivasan R, Kumar S (2022) Effect of invasive weed biochar mixed with inorganic fertilizer on soil physical and chemical properties in semi-arid region of India. Int J Plant Soil Sci 34(14):92–98

Daful AG, Chandraratne MR (2018) Biochar production from biomass waste-derived material. In: Encyclopedia of renewable and sustainable materials. https://doi.org/10.1016/B978-0-12-803581-8.11249-4

Dai L, Fan L, Liu Y, Ruan R, Wang Y, Zhou Y, Zhao Y, Yu Z (2017) Production of bio-oil and biochar from soapstock via microwave-assisted co-catalytic fast pyrolysis. Bioresour Technol 225:1–8

Das SK, Ghosh GK (2021) Development and evaluation of biochar-based secondary and micronutrient enriched slow release nano-fertilizer for reduced nutrient losses. Biomass Convers Biorefin:1e12

Dhyani V, Bhaskar T (2018) A comprehensive review on the pyrolysis of lignocellulosic biomass. Renew Energy 129:695–716. https://doi.org/10.1016/j.renene.2017.04.035

Dias BO, Silva CA, Higashikawa FS, Roig A, Sánchez-Monedero MA (2010) Use of biochar as bulking agent for the composting of poultry manure: effect on organic matter degradation and humification. Bioresour Technol 101(4):1239–1246

Edwards JD, Pittelkow CM, Kent AD, Yang WH (2018) Dynamic biochar effects on soil nitrous oxide emissions and underlying microbial processes during the maize growing season. Soil Biol Biochem 122:81–90

El-naggar A, Soo S, Rinklebe J, Farooq M, Song H (2019) Biochar application to low fertility soils: a review of current status, and future prospects. Geoderma 337:536–554. https://doi.org/10.1016/j.geoderma.2018.09.034

Enaime G, Baçaoui A, Yaacoubi A, Lübken M (2020) Biochar for wastewater treatment-conversion technologies and applications. Appl Sci (Switzerland) 10:10. https://doi.org/10.3390/app10103492

Endriani E, Listyarini D (2023) Amelioration of dry land suboptimal using biochar and compost to improve soil physical properties and soybean yield. In: BIO web of conferences, vol 80. EDP Sciences, p 03001

Fang Y, Singh B, Singh BP, Krull E (2014) Biochar carbon stability in four contrasting soils. Eur J Soil Sci 65:60–71

Fearnside PM (2000) Global warming and tropical land-use change: greenhouse gas emissions from biomass burning, decomposition and soils in forest conversion, shifting cultivation and secondary vegetation. Clim Chang 46(1):115–158

Fidel RB, Laird DA, Thompson ML, Lawrinenko M (2017) Characterization and quantification of biochar alkalinity. Chemosphere 167:367e373

Figueiredo CC, Mel, LCA, Silva CA, Fachini J, de Carneiro JSS, de Morais EG, Ndoung OCN, Singh SV, and Nandipamu TMK (2024) Nutrient enriched and co-composted biochar: system productivity and environmental sustainability In Biochar Production for Green Economy; Academic Press, Elsevier 311-331 https://doi.org/10.1016/B978-0-443-15506-2.00010-9

Gaurav GK, Mehmood T, Cheng L, Klemeš JJ, Shrivastava DK (2020) Water hyacinth as a biomass: a review. J Clean Prod:122214. https://doi.org/10.1016/j.jclepro.2020.122214

Ge S, Nai P, Yek Y, Wang Y, Xia C, Adibah W, Shiung S (2021) Progress in microwave pyrolysis conversion of agricultural waste to value-added biofuels: a batch to continuous approach. Renew Sustain Energ Rev 135:110148. https://doi.org/10.1016/j.rser.2020.110148

Gelardi DL, Ainuddin I, Rippner DA, Abou Najm M, Parikh SJ (2021) Biochar alters hydraulic conductivity and inhibits nutrient leaching in two agricultural soils. Soil Discuss 2021:1–23

Ghosh D, Maiti SK (2023) Invasive weed-based biochar facilitated the restoration of coal mine degraded land by modulating the enzyme activity and carbon sequestration. Restor Ecol 31:e13744

Glaser B, Wiedner K, Seelig S, Schmidt HP, Gerber H (2015) Biochar organic fertilizers from natural resources as substitute for mineral fertilizers. Agron Sustain Dev 35:667–678

Gonzalez ME, Cea M, Medina J, Gonzalez A, Diez MC, Cartes P et al (2015) Evaluation of biodegradable polymers as encapsulating agents for the development of a urea controlled-release fertilizer using biochar as support material. Sci Total Environ 505:446e453

Grutzmacher P, Puga AP, Bibar MPS, Coscione AR, Packer AP, de Andrade CA (2018) Carbon stability and mitigation of fertilizer induced N_2O emissions in soil amended with biochar. Sci Total Environ 625:1459–1466

Haefele SM, Konboon Y, Wongboon W, Amarante S, Maarifat AA, Pfeiffer EM, Knoblauch CJFCR (2011) Effects and fate of biochar from rice residues in rice-based systems. Field Crop Res 121(3):430–440

Hassan M, Liu Y, Naidu R, Parikh SJ, Du J, Qi F, Willett IR (2020) Science of the Total Environment Influences of feedstock sources and pyrolysis temperature on the properties of biochar and functionality as adsorbents: a meta-analysis. Sci Total Environ 744:140714. https://doi.org/10.1016/j.scitotenv.2020.140714

Hu F, Xu C, Ma R, Tu K, Yang J, Zhao S, Yang M, Zhang F (2021) Biochar application driven change in soil internal forces improves aggregate stability: based on a two-year field study. Geoderma 403:115276

Hua L, Wu W, Liu Y, McBride MB, Chen Y (2009) Reduction of nitrogen loss and Cu and Zn mobility during sludge composting with bamboo charcoal amendment. Environ Sci Pollut Res 16:1–9

Huang M, Fan L, Chen J, Jiang L, Zou Y (2018) Continuous applications of biochar to rice: effects on nitrogen uptake and utilization. Sci Rep 8(1):11461

Ichwan B, Mapegau M, Marlina M (2020) Coconut husk biochar application on increasing growth and yield of maize plant, and improvement fertility of Ultisol dry land. Sys Rev Pharm 11(2):187–192

Igalavithana AD, Mandal S, Niazi NK, Vithanage M, Parikh SJ, Mukome FND, Rizwan M, Oleszczuk P, Al-Wabel M, Bolan N (2017) Advances and future directions of biochar characterization methods and applications. Crit Rev Environ Sci Technol 47:2275–2330

Jindo K, Sánchez-Monedero MA, Hernández T, García C, Furukawa T, Matsumoto K et al (2012) Biochar influences the microbial community structure during manure composting with agricultural wastes. Sci Total Environ 416:476–481

Kameyama K, Miyamoto T, Shiono T, Shinogi Y (2012) Influence of sugarcane bagasse-derived biochar application on nitrate leaching in calcaric dark red soil. J Environ Qual 41(4):1131–1137

Kameyama K, Miyamoto T, Iwata Y (2019) The preliminary study of water-retention related properties of biochar produced from various feedstock at different pyrolysis temperatures. Materials 12:1732

Karim AA, Kumar M, Mohapatra S, Singh SK, Panda CR (2019) Co-plasma pro-cessing of banana peduncle with phosphogypsum waste for production of lesser toxic potassium sulfur rich biochar. J Mater Cycles Waste Manag 21(1):107e115

Karimi S, Soofiani NM, Mahboubi A, Taherzadeh MJ (2018) Use of organic wastes and industrial by-products to produce filamentous fungi with potential as aqua-feed ingredients. Sustainability (Switzerland) 9:10. https://doi.org/10.3390/su10093296

Kavitha B, Reddy PVL, Kim B, Lee SS, Pandey SK, Kim KH (2018) Benefits and limitations of biochar amendment in agricultural soils: a review. J Environ Manag 227:146–154

Kiani H, Khalesro S, Mokhatssi-Bidgoli A, Sharifi Z (2024) Biochar and conservation tillage affect the agronomic performance and fatty acid composition of Nigella sativa L. under both irrigated and dryland conditions. Sci Rep 14(1):2648

Kim J, Yoo G, Kim D, Ding W, Kang H (2017) Combined application of biochar and slow release fertilizer reduces methane emission but enhances rice yield by different mechanisms. Appl Soil Ecol 117:57e62

Kim J, Vijayaraghavan K, Reddy D, Yun Y (2018) A phosphorus-enriched biochar fertilizer from bio-fermentation waste: a potential alternative source for phosphorus fertilizers. J Clean Prod 196:163e171

Kumar S, Kumar G, Avasthe R (2021) Environmental technology & innovation applications of biomass derived biochar in modern science and technology. Environ Technol Innov 21:101306. https://doi.org/10.1016/j.eti.2020.101306

Labanya R, Sahu P, Mandal S, Singh SV, and Meena RS (2024) Crop waste conversion into biochar: an overview In Biochar Production for Green Economy; Academic Press, Elsevier 1- 23 https://doi.org/10.1016/B978-0-443-15506-2.00027-4

Lashari MS, Liu Y, Li L, Pan W, Fu J, Pan G, Zheng J, Zheng J, Zhang X, Yu X (2013) Effects of amendment of biochar-manure compost in conjunction with pyroligneous solution on soil quality and wheat yield of a salt-stressed cropland from Central China Great Plain. Field Crop Res 144:113–118

Lee DJ, Lu JS, Chang JS (2020) Pyrolysis synergy of municipal solid waste (MSW): a review. Bioresour Technol 318(June):123912. https://doi.org/10.1016/j.biortech.2020.123912

Lehmann J, Rillig MC, Thies J, Masiello CA, Hockaday WC, Crowley D (2011) Biochar effects on soil biota–a review. Soil Biol Biochem 43(9):1812–1836

Leogrande R, Vitti C, Castellini M, Garofalo P, Samarelli I, Lacolla G et al (2024) Residual effect of compost and biochar amendment on soil chemical, biological, and physical properties and durum wheat response. Agronomy 14(4):749

Liao Y, Awan MI, Aamer M, Liu J, Liu J, Hu B et al (2024) Evaluating short-term effects of rice straw management on carbon fractions, composition and stability of soil aggregates in an acidic red soil with a vegetable planting history. Heliyon 10(1)

Lim TJ, Spokas KA, Feyereisen G, Novak JM (2016) Predicting the impact of biochar additions on soil hydraulic properties. Chemosphere 142:136–144

Liu X, Zhang A, Ji C, Joseph S, Bian R, Li L, Pan G, Paz-Ferreiro J (2013) Biochar's effect on crop productivity and the dependence on experimental conditions—a meta-analysis of literature data. Plant Soil 373:583–594

Liu X, Li Z, Zhang Y, Feng R, Mahmood IB (2014) Characterization of human manure-derived biochar and energy-balance analysis of slow pyrolysis process. Waste Manag 34(9):1619e1626

Liu Z, Dugan B, Masiello CA, Gonnermann HM (2017a) Biochar particle size, shape, and porosity act together to influence soil water properties. PLoS One 12(6):e0179079

Liu X, Yang Y, Gao B, Li Y, Wan Y (2017b) Environmentally friendly slow-release urea fertilizers based on waste frying oil for sustained nutrient release. ACS Sustain Chem Eng 5(7):6036e6045

Lou Z, Sun Y, Bian S, Baig SA, Hu B, Xu X (2017) Nutrient conservation during spent mushroom compost application using spent mushroom substrate derived biochar. Chemosphere 169:23e31

Lu S, Zong Y (2018) Pore structure and environmental serves of biochars derived from different feedstocks and pyrolysis conditions. Environ Sci Pollut Res 25:30401–30409

Lychuk TE, Izaurralde RC, Hill RL, McGill WB, Williams JR (2015) Biochar as a global change adaptation: predicting biochar impacts on crop productivity and soil quality for a tropical soil with the Environmental Policy Integrated Climate (EPIC) model. Mitig Adapt Strateg Glob Chang 20(8):1437–1458

Major J, Rondon M, Molina D, Riha SJ, Lehmann J (2010) Maize yield and nutrition during 4 years after biochar application to a Colombian savanna oxisol. Plant Soil 333:117–128

Mamaní A, Ramírez N, Deiana C, Giménez M (2019) Journal of environmental chemical engineering highly microporous sorbents from lignocellulosic biomass: different activation routes and their application to dyes adsorption. J Environ Chem Eng 7(5):103148. https://doi.org/10.1016/j.jece.2019.103148

Manyà JJ, García-Morcate D, González B (2020) Adsorption performance of physically activated biochars for postcombustion CO2 capture from dry and humid flue gas. Appl Sci (Switzerland) 10(1):1–17. https://doi.org/10.3390/app10010376

Mao J, Zhang K, Chen B (2019) Linking hydrophobicity of biochar to the water repellency and water holding capacity of biocharamended soil. Environ Pollut 253:779–789

Mbarki F, Selmi T, Kesraoui A, Se M, Gadonneix P, Celzard A, Fierro V (2019) Hydrothermal pre-treatment, an efficient tool to improve activated carbon performances. Ind Crop Prod 140. https://doi.org/10.1016/j.indcrop.2019.111717

Mutsengerere S, Chihobo CH, Musademba D, Nhapi I (2019) A review of operating parameters affecting bio-oil yield in microwave pyrolysis of lignocellulosic biomass. Renew Sustain Energ Rev 104:328–336. https://doi.org/10.1016/j.rser.2019.01.030

Naeem M, Ansari AA, Gill SS (2017) Essential plant nutrients: uptake, use efficiency, and management. Essent Plant Nutr Uptake Use Effic Manag:1–569

Naji SZ, Tye CT (2022) A review of the synthesis of activated carbon for biodiesel production: precursor, preparation, and modification. Energy Convers Manag 13:100152. https://doi.org/10.1016/J.ECMX.2021.100152

Nidheesh PV, Gopinath A, Ranjith N, Praveen Akre A, Sreedharan V, Suresh Ku-mar M (2021) Potential role of biochar in advanced oxidation processes: a sustainable approach. Chem Eng J 405:126582. https://doi.org/10.1016/j.cej.2020.126582

Novotny EH, de Freitas Maia CM, de Melo Carvalho MT, Madari BE (2015) Biochar: pyrogenic carbon for agricultural use-a critical review. Rev Bras Ciênc Solo 39:321–344

Omondi MO, Xia X, Nahayo A, Liu X, Korai PK, Pan G (2016) Quantification of biochar effects on soil hydrological properties using meta-analysis of literature data. Geoderma 274:28–34

Palansooriya KN, Ok YS, Awad YM, Lee SS, Sung JK, Koutsospyros A, Moon DH (2019a) Impacts of biochar application on upland agriculture: a review. J Environ Manag 234:52–64

Palansooriya KN, Wong JTF, Hashimoto Y, Huang L, Rinklebe J, Chang SX, Bolan N, Wang H, Ok YS (2019b) Response of microbial communities to biochar-amended soils: a critical review. Biochar 1:3–22

Pandey D, Savio N, Naudiyal N, Srivastava RK, Nayak P, Cabañas B, Moreno A, Singh SV (2024) Bioremediation of organic pollutants soil and water through biochar for a healthy ecosystem. In: Biochar production for green economy. Academic Press, Elsevier, pp 479–505. https://doi.org/10.1016/B978-0-443-15506-2.00007-9

Pandey D, Singh SV, Savio N, Bhutto JK, Srivastava RK, Yadav KK, Sharma R, Nandipamu TMK, Sarkar B (2025) Biochar application in constructed wetlands for wastewater treatment: a critical review. J Water Process Eng 69:106713. https://doi.org/10.1016/j.jwpe.2024.106713

Pandit NR, Mulder J, Hale SE, Zimmerman AR, Pandit BH, Cornelissen G (2018) Multi-year double cropping biochar field trials in Nepal: finding the optimal biochar dose through agronomic trials and cost-benefit analysis. Sci Total Environ 637:1333–1341

Pender J (2012) The world food crisis, land degradation, and sustainable land management

Peng XYLL, Ye LL, Wang CH, Zhou H, Sun B (2011) Temperature-and duration-dependent rice straw-derived biochar: characteristics and its effects on soil properties of an Ultisol in southern China. Soil Tillage Res 112(2):159–166

Prost K, Borchard N, Siemens J, Kautz T, Séquaris JM, Möller A, Amelung W (2013) Biochar affected by composting with farmyard manure. J Environ Qual 42(1):164–172

Rangabhashiyam S, Balasubramanian P (2019) The potential of lignocellulosic biomass precursors for biochar production: performance, mechanism and wastewater application – a review. Ind Crops Prod 128:405–423. https://doi.org/10.1016/j.indcrop.2018.11.041

Ravindran R, Hassan SS, Williams GA, Jaiswal AK (2018) A review on bioconversion of agro-industrial wastes to industrially important enzymes. Bioengineering 5(4):1–20. https://doi.org/10.3390/bioengineering5040093

Rawat J, Saxena J, Sanwal P (2019) Biochar: a sustainable approach for improving plant growth and soil properties. In: Biochar—an imperative amendment for soil and the environment. IntechOpen, London

Razzaghi F, Obour PB, Arthur E (2020) Does biochar improve soil water retention? A systematic review and meta-analysis. Geoderma 361:114055

Roberts DA, Paul NA, Dworjanyn SA, Bird MI, de Nys R (2015) Biochar from commercially cultivated seaweed for soil amelioration. Sci Rep 5(1):1e6

Roy A, Chaturvedi S, Singh SV, Govindaraju K, Dhyani VC, Pyne S (2022) Preparation and evaluation of two enriched biochar-based fertilizers for nutrient release kinetics and agronomic effectiveness in direct-seeded rice. Biomass Convers Biorefinery 14:1–12. https://doi.org/10.1007/s13399-022-02488-z

Sakhiya AK, Anand A, Kaushal P (2020) Production, activation, and applications of biochar in recent times. Biochar 2. https://doi.org/10.1007/s42773-020-00047-1

Sanchez PA (2002) Soil fertility and hunger in Africa. Science 295(5562):2019–2020

Santoso E, Ediati R, Kusumawati Y, Bahruji H, Sulistiono DO, Prase-tyoko D (2020) Review on recent advances of carbon based adsorbent for methylene blue removal from wastewater. Mater Today Chem 16:100233. https://doi.org/10.1016/j.mtchem.2019.100233

Schmidt MW, Noack AG (2000) Black carbon in soils and sediments: analysis, distribution, implications, and current challenges. Glob Biogeochem Cycles 14(3):777–793

Schulz H, Glaser B (2012) Effects of biochar compared to organic and inorganic fertilizers on soil quality and plant growth in a greenhouse experiment. J Plant Nutr Soil Sci 175(3):410–422

Seleiman MF, Almutairi KF, Alotaibi M, Shami A, Alhammad BA, Battaglia ML (2021) Nano fertilization as an emerging fertilization technique: why modern agriculture can benefit from its use? Plants 10:2

Senthil C, Lee CW (2020) Biomass-derived biochar materials as sustainable energy sources for electrochemical energy storage devices. Renew Sustain Energy Rev 137:110464. https://doi.org/10.1016/j.rser.2020.110464

Sha Z, Li Q, Lv T, Misselbrook T, Liu X (2019) Response of ammonia volatilization to biochar addition: a meta-analysis. Sci Total Environ 655:1387e1396

Shakya A, Agarwal T (2020) Potential of biochar for the remediation of heavy metal contaminated soil. In: Biochar applications in agriculture and environment management. Springer, Berlin/Heidelberg, pp 77–98

Shiung S, Adibah W, Mahari W, Sik Y, Peng W, Tung C, Tsang DCW (2019) Microwave vacuum pyrolysis of waste plastic and used cooking oil for simultaneous waste reduction and sustainable energy conversion: recovery of cleaner liquid fuel and techno-economic analysis. Renew Sustain Energ Rev 115:109359. https://doi.org/10.1016/j.rser.2019.109359

Singh SV, Chaturvedi S, Datta D (2019) Biochar: an eco-friendly residue management approach. Indian Farm 69(08):27–29

Singh C, Tiwari S, Singh JS (2020) Biochar: a sustainable tool in soil pollutant bioremediation. In: Bioremediation of industrial waste for environmental safety. Springer, Berlin/Heidelberg, pp 475–494

Singh SV, Chaturvedi S, Dhyani VC, Govindaraju K (2020) Pyrolysis temperature influences the characteristics of rice straw and husk biochar and sorption/desorption behavior of their biourea composite. Bioresour Technol 314:123674. https://doi.org/10.1016/j.biortech.2020.123674

Singh H, Northup BK, Rice CW, Prasad PV (2022) Biochar applications influence soil physical and chemical properties, microbial diversity, and crop productivity: a meta-analysis. Biochar 4(1):8

Singh S, Luthra N, Mandal S, Kushwaha DK, Pathak SO, Datta D, Sharma R, Pramanick B (2023a) Distinct behavior of biochar modulating biogeochemistry of salt-affected and acidic soil: a review. J Soil Sci Plant Nutr 23(3):2981–2997. https://doi.org/10.1007/s42729-023-01370-9

Singh SV, Chaturvedi S, Nayak P, Dhyani VC, Nadipamu TMK, Singh DK, Gudapaty P, Mathyam P, Srinivasarao K, Gavindaraju K (2023b) Carbon offset potential of biochar based straw management under rice – wheat system along Indo-Gangetic Plains of India. Sci Total Environ 897:165176. https://doi.org/10.1016/j.scitotenv.2023.165176

Singh SV et al (2025) Biochar's multifaceted role in bioremediation of emerging contaminants and heavy metals in complex rhizospheric ecosystem. Int Biodeterior & Biodegrad 198:106005. https://doi.org/10.1016/j.ibiod.2025.106005

Sonu K, Sogani M, Syed Z, Dongre A, Sharma G (2020) Enhanced decolorization and treatment of textile dye wastewater through adsorption on acid modified corncob derived biochar. ChemistrySelect 5(39):12287–12297. https://doi.org/10.1002/slct.202003156

Steiner C, Teixeira WG, Lehmann J, Nehls T, de Macêdo JLV, Blum WE, Zech W (2007) Long term effects of manure, charcoal and mineral fertilization on crop production and fertility on a highly weathered Central Amazonian upland soil. Plant Soil 291:275–290

Steiner C, Das KC, Melear N, Lakly D (2010) Reducing nitrogen loss during poultry litter composting using biochar. J Environ Qual 39(4):1236–1242

Suliman W, Harsh JB, Abu-Lail NI, Fortuna A-M, Dallmeyer I, Garcia-Perez M (2016) Influence of feedstock source and pyrolysis temperature on biochar bulk and surface properties. Biomass Bioenergy 84:37–48

Sun H, Zhang H, Shi W, Zhou M, Ma X (2019) Effect of biochar on nitrogen use efficiency, grain yield and amino acid content of wheat cultivated on saline soil. Plant Soil Environ 65(2)

Tang S, Shao N, Zheng C, Yan F, Zhang Z (2019) Amino-functionalized sewage sludge-derived biochar as sustainable efficient adsorbent for Cu(II) removal. Waste Manag (Oxford) 90:17–28. https://doi.org/10.1016/j.wasman.2019.04.042

Tiessen H, Cuevas E, Chacon P (1994) The role of soil organic matter in sustaining soil fertility. Nature 371(6500):783–785

Tomczyk A, Sokołowska Z, Boguta P (2020) Biochar physicochemical properties: pyrolysis temperature and feedstock kind effects. Rev Environ Sci Biotechnol 19(1):191–215. https://doi.org/10.1007/s11157-020-09523-3

Umenweke GC, Christianah I, Epelle EI, Okolie JA (2022) Bioresource technology reports machine learning methods for modeling conventional and hydrothermal gasification of waste biomass: a review. Biores Technol Rep 17:100976. https://doi.org/10.1016/j.biteb.2022.100976

Vaccari FP, Baronti S, Lugato E, Genesio L, Castaldi S, Fornasier F, Miglietta F (2011) Biochar as a strategy to sequester carbon and increase yield in durum wheat. Eur J Agron 34(4):231–238

Vanlauwe B, Bationo A, Chianu J, Giller KE, Merckx R, Mokwunye U et al (2010) Integrated soil fertility management: operational definition and consequences for implementation and dissemination. Outlook Agric 39(1):17–24

Venkatesh G, Gopinath KA, Reddy KS, Reddy BS, Prabhakar M, Srinivasarao C, Singh VK (2022) Characterization of biochar derived from crop residues for soil amendment, carbon sequestration and energy use. Sustainability 14(4):2295. https://doi.org/10.3390/su14042295

Vitková J, Šurda P, Lichner Ľ, Výleta R (2024) Influence of biochar application rate, particle size, and pyrolysis temperature on hydrophysical parameters of sandy soil. Appl Sci 14(8):3472

Wang J, Wang S (2019) Preparation, modification and environmental application of biochar: a review. J Clean Prod 227:1002–1022. https://doi.org/10.1016/j.jclepro.2019.04.282

Wang T, Stewart CE, Sun C, Wang Y, Zheng J (2018a) Effects of biochar addition on evaporation in the five typical Loess Plateau soils. Catena 162:29–39

Wang Y, Zeng Z, Tian X, Dai L, Jiang L, Zhang S, Ruan R (2018b) Production of bio-oil from agricultural waste by using a continuous fast microwave pyrolysis system. Bioresour Technol. https://doi.org/10.1016/j.biortech.2018.08.067

Weber K, Quicker P (2018) Properties of biochar. Fuel 217:240–261

Weng Z, Van Zwieten L, Tavakkoli E, Rose MT, Singh BP, Joseph S, Macdonald LM, Kimber S, Morris S, Rose TJ (2022) Microspectroscopic visualization of how biochar lifts the soil organic carbon ceiling. Nat Commun 13:5177

Wiedner K, Fischer D, Walther S, Criscuoli I, Favilli F, Nelle O, Glaser B (2015) Acceleration of biochar surface oxidation during composting? J Agric Food Chem 63(15):3830–3837

Wiskandar W, Ajidirman A (2024) Effect of biochar and Tithonia compost on physical properties of post-coal mining soil. J Degraded Min Lands Manag 11(3):5829–5838

Yaashikaa PR, Kumar PS, Varjani SJ, Saravanan A (2019) Advances in production and application of biochar from lignocellulosic feedstocks for re-mediation of environmental pollutants. Bioresour Technol 292:122030. https://doi.org/10.1016/j.biortech.2019.122030

Yaashikaa PR, Kumar PS, Varjani S, Saravanan A (2020) A critical review on the biochar production techniques, characterization, stability and applications for circular bioeconomy. Biotechnol Rep 28:e00570. https://doi.org/10.1016/j.btre.2020.e00570

Yadav M, Joshi C, Paritosh K, Thakur J, Pareek N, Kumar S, Vivekanand V (2022) Reprint of organic waste conversion through anaerobic digestion: a critical insight into the metabolic pathways and microbial interactions. Metab Eng 71:62–76. https://doi.org/10.1016/j.ymben.2022.02.001

Yamato M, Okimori Y, Wibowo IF, Anshori S, Ogawa M (2006) Effects of the application of charred bark of Acacia mangium on the yield of maize, cowpea and peanut, and soil chemical properties in South Sumatra, Indonesia. Soil Sci Plant Nutr 52(4):489–495

Yao Q, Liu J, Yu Z, Li Y, Jin J, Liu X, Wang G (2017) Three years of biochar amendment alters soil physiochemical properties and fungal community composition in a black soil of Northeast China. Soil Biol Biochem 110:56–67

Ye Z, Zhang L, Huang Q, Tan Z (2019) Development of a carbon-based slow release fertilizer treated by bio-oil coating and study on its feedback effect on farmland application. J Clean Prod 239:118085

Yu X, Han Z, Fang S, Chang C, Han X (2019) Optimized preparation of high value-added activated carbon and its adsorption properties for methylene blue. Int J Chem React Eng:1–16. https://doi.org/10.1515/ijcre-2018-0267

Yunilasari M (2020) Effects of biochar and cow manure on soil chemical properties and peanut (Arachis hypogaea L.) yields in entisol. In: IOP conference series: earth and environmental science, vol 425, No. 1. IOP Publishing, p 012014

Zaker A, Chen Z, Wang X, Zhang Q (2019) Microwave-assisted pyrolysis of sewage sludge: a review. Fuel Process Technol 187:84–104. https://doi.org/10.1016/j.fuproc.2018.12.011

Zaman CZ, Pal K, Yehye WA, Sagadevan S, Shah ST, Adebisi GA, Marliana E, Rafique RF, Johan RB (2017) Pyrolysis: a sustainable way to generate energy from waste. In: Pyrolysis. BoD–Books on Demand, Norderstedt, p 1

Zhao L, Cao X, Zheng W, Kan Y (2014) Phosphorus-assisted biomass thermal conversion: reducing carbon loss and improving biochar stability. PLoS One 9(12):e115373

Zhao L, Cao X, Zheng W, Scott JW, Sharma BK, Chen X (2016) Copyrolysis of biomass with phosphate fertilizers to improve biochar carbon retention, slow nutrient release, and stabilize heavy metals in soil. ACS Sustain Chem Eng 4(3):1630e1636

Zhao B, O'Connor D, Zhang J, Peng T, Shen Z, Tsang DCW, Hou D (2018) Effect of pyrolysis temperature, heating rate, and residence time on rapeseed stem derived biochar. J Clean Prod 174:977–987

Zhou J, Qu T, Li Y, Van Zwieten L, Wang H, Chen J et al (2021) Biochar-based fertilizer decreased while chemical fertilizer increased soil N2O emissions in a subtropical Moso bamboo plantation. Catena 202:105257

Zhou J, Hong W, Feng J, Song L, Li X, Xu S, Zhou S (2023) Effects of applying peanut shell and its biochar on the microbial activity and community structure of dryland red soil. Heliyon 9(2)

Zoroufchi K, Motalebi A, Soltan J, Neil K (2020) Science of the Total Environment Treatment of aqueous arsenic – a review of biochar modification methods. Sci Total Environ 739:139750. https://doi.org/10.1016/j.scitotenv.2020.139750

Zuber MA, Yahya WJ, Ithnin AM, Sugeng DA, Kadir HA, Ahmad MA (2019) A brief review of palm oil liquid waste conversion into biofuel. Environ Rev:1–10. https://doi.org/10.1139/er-2018-0124

Chapter 20
Application of AI Tools for Efficient Resource Management

Medida Sunil Kumar, K. A. Mrudula, M. Sri Surya, Dilip Nandwani, Biswajit Pramanick, B. S. S. Siddartha Naik, and B. Asha Jyothi

Abstract Modern technologies have dramatically changed resource management in agriculture. This chapter is going to explore how the innovation of some of these technologies deals with crucial problems such as water deficiency, land deterioration, and unwise use of available resources. Modern systems can combine sensors, data analytics, and automation in such a way that farmers make smart decisions in regard to irrigation, fertilizer, and pest control. These technologies will help monitor in real time and provide predictive insights on how to reduce waste, increase productivity, and minimize the impact on the environment. This chapter discusses the practical applications of these technologies in optimizing irrigation systems and improving soil health, underlining the importance of sustainability in agriculture. However, the challenges of high costs, limited access to technology, and the need for training of farmers are significant. Despite these problems, the outlook of agriculture seems hopeful because these are believed to reach more areas and be adaptable toward other farming systems. If farmers take up these innovations, their

---

M. S. Kumar · B. A. Jyothi
Geo-spatial Technology Center, Regional Agricultural Research Station,
Guntur, Andhra Pradesh, India

K. A. Mrudula
Saline Water Scheme, Acharya NG Ranga Agricultural University,
Bapatla, Andhra Pradesh, India

M. S. Surya · B. S. S. Siddartha Naik
Department of Agronomy, Agriculture College, Acharya NG Ranga Agricultural University,
Bapatla, Andhra Pradesh, India

D. Nandwani (✉)
Department of Agricultural Science and Engineering, College of Agriculture, Tennessee State University, Nashville, TN, USA
e-mail: dnandwan@tnstate.edu

B. Pramanick
Department of Agronomy, Dr Rajendra Prasad Central Agricultural University,
Pusa, Bihar, India

B. Pramanick et al. (eds.), *Climate-Smart Agricultural Technologies*, Sustainability Sciences in Asia and Africa, https://doi.org/10.1007/978-981-96-7699-6_20

systems will eventually be made efficient, food security will be gained, and therefore, an agricultural system sustainable and resilient for later generations will be ensured.

Keywords Artificial intelligence · Precision agriculture · Resource management · IoT in farming · Data-driven farming; Real-time monitoring

20.1 Introduction

The agriculture sector, one of the world's oldest industries, faces increasing pressure to evolve in response to both present and future global demands. With the world population expanding rapidly, ensuring the sector's sustainable development is essential for securing food supply and sustaining life on Earth. A key factor driving the need for this evolution is the challenge of managing agricultural resources efficiently. Effective resource management in agriculture requires the careful use of vital inputs such as water, soil nutrients, seeds, fertilizers, and energy while simultaneously minimizing waste and mitigating environmental damage. Failure to manage these resources properly can lead to severe consequences such as soil degradation, water shortages, improper fertilizer application, and diminished crop yields. These challenges are being exacerbated by climate change, unpredictable weather conditions, and decreasing availability of arable land. As a result, integrating innovative technologies into agriculture has become crucial for enhancing sustainability and ensuring long-term productivity.

Artificial intelligence (AI) has emerged as a powerful tool to tackle the complex challenges associated with resource management in agriculture. First conceptualized at the Dartmouth Conference in 1956 by John McCarthy, AI was defined as the science and engineering of creating machines capable of performing tasks that typically require human intelligence. Over the years, AI has developed to incorporate various technologies that enable machines to process information, learn from data, and respond intelligently to diverse situations. Subfields of AI, including machine learning, deep learning, natural language processing (NLP), swarm intelligence (SI), expert systems, fuzzy logic, and computer vision, have contributed to significant advancements in agricultural technology. These technologies provide farmers with the ability to analyze large datasets, recognize patterns, and generate actionable insights, which are particularly valuable in precision farming, where decisions must be made based on numerous, often rapidly changing, variables. By leveraging AI, farmers are now able to manage resources more effectively, reduce waste, increase efficiency, and ultimately boost crop productivity.

The incorporation of AI into agriculture has involved a range of technologies designed to improve farming practices and streamline operations. Remote sensing technologies, such as drone-mounted cameras and satellite imaging systems, collect real-time data on crop conditions, soil health, and environmental variables. Global Positioning System (GPS) technologies provide precise location data, which is essential for precision agriculture, allowing farmers to target specific areas for treatment or intervention. Machine learning and deep learning algorithms use the data collected from these systems to identify patterns, forecast outcomes, and support smarter decision-making. Geographic information systems (GISs) enable the integration and visualization of data, giving farmers the ability to interpret the information in relation to historical trends and other data layers. Decision support systems (DSSs), driven by AI insights, offer practical recommendations for farmers, while automated robotics bring together these technologies to execute tasks with high accuracy. Additionally, connectivity solutions such as the Internet, WiFi, and Bluetooth allow for seamless communication between these systems, ensuring efficient data exchange and integration across platforms.

Compared to traditional methods, AI-driven agricultural technologies offer a more precise, data-oriented, and scalable approach. In contrast to conventional farming practices, which often rely on manual observation, generalized approaches, and human intuition, AI automates the processes of data collection, analysis, and decision-making. This automation allows for quicker, more accurate decision-making, leading to optimized resource use, reduced environmental impact, and increased agricultural productivity. AI-driven solutions can also enable proactive management, allowing farmers to address potential issues before they escalate into significant problems. This contrasts with traditional methods, which often rely on reactive measures taken after a problem has already occurred.

In reviewing the current applications of AI in agricultural resource management, it is clear that AI has the potential to significantly enhance efficiency and sustainability within the sector. AI technologies have already shown their value in areas such as water management, soil health optimization, pest and disease control, and crop yield forecasting. However, challenges remain in terms of widespread AI adoption in agriculture, particularly concerning the costs of implementation, technological accessibility for small-scale farmers, and the need for ongoing education and training. Despite these obstacles, future innovations in AI are expected to further transform agricultural resource management, offering even greater opportunities for farmers to improve efficiency, minimize waste, and promote sustainability across the industry. By exploring these areas, we gain a deeper understanding of how AI continues to shape the future of agriculture and contributes to more sustainable and efficient farming practices.

20.2 Different Tools of Artificial Intelligence in Agriculture

Artificial intelligence (AI) has become an essential part of modern agriculture, transforming farming practices and optimizing the use of natural resources to boost both productivity and sustainability. The following discussion delves into the most prominent AI technologies and tools currently utilized across the agricultural sector.

20.2.1 Machine Learning

Machine learning (ML), a critical subset of AI, allows computers to learn from data and make informed decisions without requiring explicit programming. In agriculture, ML has transformed how farmers manage their crops and resources by utilizing different approaches such as supervised, unsupervised, and reinforcement learning. By analyzing extensive datasets like weather forecasts, historical crop performance, and soil conditions, ML algorithms offer highly accurate predictions that help farmers manage their farms proactively.

Supervised learning models, for instance, can predict crop yields and recommend the best planting times based on historical data. In contrast, unsupervised learning helps identify patterns in soil and crop conditions, which can reveal trends or anomalies without predefined categories. Reinforcement learning improves decision-making by learning from trial and error, which is particularly useful for automating tasks such as irrigation control and pest management. These approaches enable farmers to act on real-time data and improve the precision of farming operations.

Moreover, ML plays a pivotal role in early disease and pest detection. By analyzing real-time data from sensors, cameras, and drones, ML algorithms can detect subtle changes in plant health, allowing farmers to take preventive measures before the damage becomes severe. This early intervention minimizes crop loss and maximizes yield. ML also enhances smart farming by optimizing resources such as water, fertilizers, and labor. Whether through supervised, unsupervised, or reinforcement learning, these models provide tailored recommendations, ensuring resource efficiency, reducing waste, and lowering environmental impact. This data-driven methodology empowers farmers to improve sustainability and productivity while minimizing costs and risks associated with traditional farming techniques.

20.2.2 Computer Vision

Computer vision (CV) is another transformative AI technology that enhances agricultural processes by allowing computers to interpret visual data from images. This capability has revolutionized various farming tasks, from crop health monitoring to pest detection and harvesting automation. CV methods such as object detection, image segmentation, and pattern recognition enable machines to process real-time images captured by drones, satellites, and ground-based cameras to make informed decisions. For example, CV-based object detection algorithms identify specific features in plants, such as leaf discoloration or wilting, which are early signs of disease or nutrient deficiency. This allows farmers to address issues with precision before they escalate. Image segmentation further refines this process by dividing images into components like soil, leaves, and fruit, which enhances the accuracy of analysis.

In terms of pest and disease detection, CV systems powered by machine learning models like convolutional neural networks (CNNs) can recognize symptoms of infestation or disease based on image patterns. These systems can distinguish between healthy and unhealthy plants and detect subtle visual changes that indicate the onset of problems. This enables timely intervention, helping farmers to prevent widespread crop damage.

Furthermore, CV plays a crucial role in automated harvesting, where robots equipped with CV systems use object detection algorithms to evaluate the ripeness of fruits and vegetables. These systems assess parameters such as shape, size, and color, allowing robots to harvest produce at the optimal time, thereby maximizing both quality and yield. By improving accuracy and reducing human labor, CV contributes to more efficient and sustainable agricultural practices.

20.2.3 Robotics and Autonomous Systems

Robotics and autonomous systems, when combined with AI, have revolutionized farming by automating labor-intensive tasks, thus lowering operational costs and enhancing overall efficiency. AI-powered autonomous tractors, for instance, can perform tasks like plowing, planting, and fertilizing with incredible precision. Equipped with advanced sensors and GPSs, these tractors can navigate fields autonomously, improving the accuracy of farming operations while reducing the need for human intervention.

Harvesting robots represent another significant innovation. These robots, integrated with AI and computer vision, can detect ripe fruits and vegetables, ensuring timely and efficient harvesting. By operating continuously, these machines increase productivity during peak harvest times. Weeding robots are also becoming prevalent, using ML algorithms and cameras to distinguish between crops and weeds. These robots apply herbicides only where necessary, significantly reducing chemical usage and promoting eco-friendly farming practices.

Overall, robotics and autonomous systems not only increase the efficiency of farm operations but also contribute to sustainability by optimizing the use of resources such as water, fertilizers, and pesticides. These advancements minimize human error, improve crop management, and reduce environmental impact.

20.2.4 Internet of Things (IoT) in Agriculture

The Internet of things (IoT) has further revolutionized agriculture by facilitating real-time data collection and communication between interconnected devices. AI-powered IoT applications, such as smart irrigation systems, use data from soil moisture sensors and weather forecasts to optimize water usage, thus minimizing waste and enhancing crop yield. These systems can adjust irrigation schedules based on real-time data, ensuring that crops receive the right amount of water at the right time.

Soil monitoring sensors, developed by companies like Teralytic and Arable, measure key variables such as soil moisture, temperature, and nutrient content. AI analyzes this data to provide precise recommendations for irrigation and fertilization, improving resource efficiency. Similarly, AI-powered drones from companies like DJI and PrecisionHawk collect high-resolution imagery to monitor crop health, detect pests, and provide farmers with actionable insights in real time.

By integrating AI and IoT technologies, farmers can make data-driven decisions that cnhancc productivity and sustainability. These innovations enable more efficient use of water, fertilizers, and other resources, contributing to more sustainable farming practices.

20.2.5 AI-Driven Precision Agriculture

Precision agriculture, powered by AI, focuses on optimizing resource use while maximizing crop productivity and minimizing environmental impact. A key component of precision agriculture is variable rate technology (VRT), which uses AI to apply resources like water, fertilizers, and seeds in precise quantities based on real-time data from the field. This minimizes waste, ensures that each section of the field gets the treatment it needs, and results in cost savings and higher yields.

AI also plays an essential role in field mapping and zoning. By analyzing satellite imagery and sensor data, AI creates detailed field maps that highlight variations in soil quality, moisture levels, and crop health. These maps allow farmers to make informed decisions about how to allocate resources across different zones within a field, improving the efficiency of farm management.

Yield monitoring systems, another aspect of AI-driven precision agriculture, use data on soil conditions, weather patterns, and historical crop performance to predict future yields. This helps farmers plan their harvests and market strategies, improving overall farm profitability. Through these advanced AI tools, precision agriculture promotes sustainability, reduces resource waste, and enhances crop outcomes.

These AI technologies are reshaping modern agriculture, making farming more efficient, sustainable, and productive. By adopting these innovations, farmers can overcome many of the challenges posed by traditional farming methods, ensuring a more secure and sustainable future for the agricultural industry.

20.3 Application of AI Tools for Effective Resource Management in Agriculture

20.3.1 AI for Soil Resource Monitoring

Effective soil resource monitoring is a vital component of ensuring agricultural productivity and sustainability. AI-powered technologies, such as sensors and remote sensing techniques, play a significant role in assessing soil health by detecting critical factors like nutrient levels, pH balance, and overall soil conditions. These tools analyze real-time data from satellite imagery and ground-based sensors to identify areas where soil quality is deteriorating or where key resources are becoming depleted. Once areas of concern are identified, automated systems, including robotic arms or applicators, can deliver precise treatments like fertilizers, lime, or compost, ensuring efficient and targeted resource management.

An essential element of soil monitoring is soil texture identification, which impacts water retention, crop selection, and overall agricultural productivity. Traditionally, soil texture has been classified using manual methods, such as the hydrometer test, to measure the proportions of sand, silt, and clay in a sample. However, these methods are time-consuming and labor-intensive. Recent advancements in AI, particularly machine learning and digital image processing, have revolutionized soil texture classification. Techniques like multi-class support vector machines (SVMs) now analyze soil images to classify textures based on features such as color histograms and wavelet transforms. This not only enhances accuracy but also speeds up the classification process, providing farmers and land managers with quicker, more accessible solutions (Barman and Choudhury 2020) (Fig. 20.1).

Soil moisture mapping is another critical aspect of agricultural monitoring, especially for understanding hydrological and climatic variables. Traditional soil moisture retrieval techniques are often impractical on a large scale due to the spatial variability of soil and labor-intensive data collection methods. However, advancements in satellite remote sensing, particularly the use of synthetic aperture radar (SAR) systems like Sentinel-1, have facilitated large-scale soil moisture mapping. AI-based algorithms, including artificial neural networks (ANNs), process SAR data to extract near real-time soil moisture content. These models can overcome challenges such as surface roughness and vegetation interference, providing timely, accurate moisture maps essential for agricultural decision-making, water management, and environmental conservation (Paloscia et al. 2013).

Soil fertility mapping is crucial for optimizing crop production and promoting sustainable land use. Traditional soil fertility assessments rely on extensive field sampling and laboratory testing, processes that are both costly and time-consuming, especially across large areas. The integration of remote sensing and geospatial technologies, however, has transformed how soil fertility is assessed. AI models now combine data from multispectral and hyperspectral sensors, such as those on Landsat and Sentinel-2, with GIS technologies to generate detailed nutrient maps. These maps reveal spatial variability in essential soil nutrients like nitrogen,

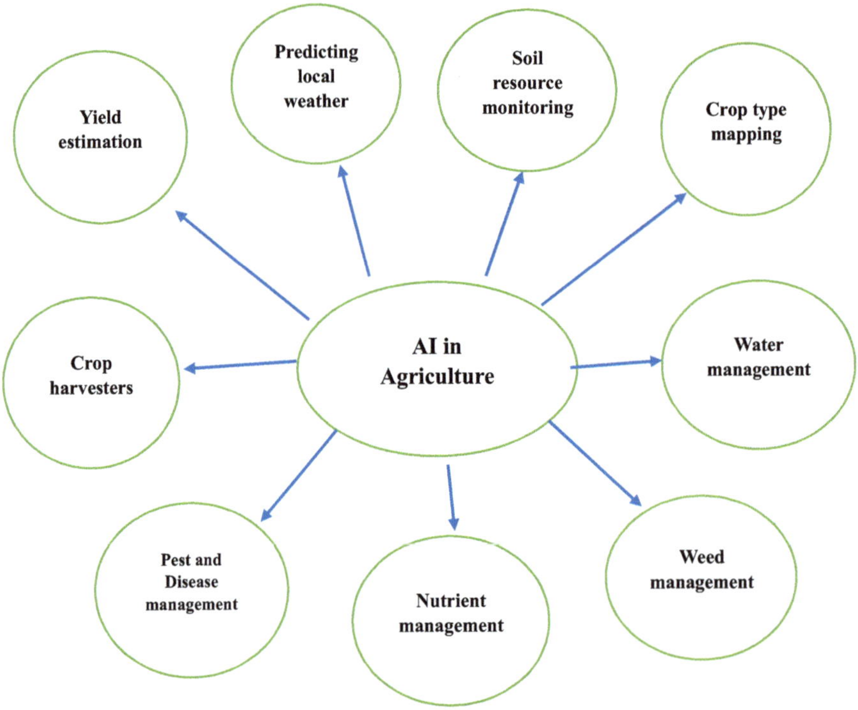

Fig. 20.1 Precision nutrient management tools optimizing resource management in agriculture

phosphorus, and potassium (NPK), allowing farmers to apply fertilizers with precision. AI-powered techniques, particularly machine learning algorithms, enhance the accuracy of these maps, facilitating better resource management and supporting sustainable agricultural practices (Mishra et al. 2014).

Traditional digital soil maps often suffer from issues such as outdated data and poor spatial resolution, limiting their usefulness in accurate soil loss predictions. Recent innovations have addressed these challenges through the development of the Soil Data Cube, which integrates high-resolution Sentinel-2 imagery, open-access databases, ground-truth soil data, and advanced AI architectures. This approach has significantly improved the Revised Universal Soil Loss Equation (RUSLE) model by incorporating fine-grain geospatial layers, enhancing both the reliability and resolution of soil loss assessments (Samarinas et al. 2024).

20.3.2 AI for Crop Type Mapping

Artificial intelligence plays a pivotal role in agricultural mapping by significantly improving efficiency, reducing costs, and enhancing decision-making processes. AI-driven crop type mapping is a vital tool in agricultural management, allowing for more precise resource allocation and the identification of potential farming challenges.

Machine learning algorithms are increasingly applied to satellite imagery and multispectral data from platforms like LiDAR, Sentinel-2, and Landsat to perform crop type mapping. Models such as random forest (RF), support vector machines (SVMs), and decision trees (DTs) are trained on labeled datasets to classify various land cover types, including agricultural lands, forests, water bodies, and urban areas. This automation greatly enhances the speed and accuracy of land use and land cover (LULC) mapping, improving resource management and agricultural planning (Kafi et al. 2014).

As pointed out by Prins and Van Niekerk (2021), traditional manual digitization methods for crop mapping are labor-intensive and prone to errors. AI-based techniques, particularly those using remote sensing, provide a more efficient and accurate alternative. A study conducted at Stellenbosch University utilized LiDAR, Sentinel-2, and aerial imagery combined with machine learning models to classify five different crop types. By applying the XGBoost model to LiDAR data, researchers achieved an accuracy rate of 87.8%. When integrating data from Sentinel-2 and aerial imagery, random forest models improved the classification accuracy to 94.4%. These results highlight the power of combining diverse datasets with AI algorithms to optimize crop type mapping (Prins and Van Niekerk 2021).

Further illustrating AI's potential in crop type mapping, a study by Tariq et al. (2023) combined Sentinel-2 and Landsat-8 imagery with machine learning models to classify crop types and cropping patterns in the Gujranwala District of Pakistan. Using decision tree classifier (DTC) and random forest (RF) algorithms, the researchers successfully differentiated between crops, such as tobacco, wheat, barley, and gram, and identified specific cropping patterns like wheat-tobacco and wheat-maize. The study employed time series data linked to phenological parameters, enabling accurate crop mapping based on NDVI-derived indices. This approach demonstrates how AI can enhance agricultural monitoring, especially in regions vulnerable to climate change.

20.3.3 Application of AI in Water Management

AI-based irrigation systems are revolutionizing water management by optimizing water usage in agriculture. These systems use real-time data collected from soil moisture sensors, weather conditions, and crop requirements to adjust irrigation automatically, ensuring efficient water distribution. Sensors placed throughout the fields continuously monitor parameters such as soil moisture, temperature, humidity, and evapotranspiration rates, allowing AI systems to make informed irrigation decisions (Krishnan et al. 2022). An example of an AI-powered irrigation system is the remote sensor–based automated system developed by Savitha and UmaMaheswari (2018). Utilizing the Arduino technology, their system optimized water usage by delivering the correct amount of water at the right time. This system increased production by up to 40%, demonstrating how AI can enhance agricultural productivity while reducing water waste. Similarly, Shi et al. (2019) conducted research that involved feeding multispectral remote sensing images into a decision-making system powered by unmanned aerial vehicles (UAVs). This advanced system used vegetative indices linked to evapotranspiration models for making irrigation decisions through fuzzy logic. The technique was successful in managing variable rate irrigation in center-pivot systems, further showcasing AI's potential in improving water management in agriculture. The different irrigation automation using various AI technologies are summarized in Table 20.1.

In conclusion, AI's integration into soil resource monitoring, crop type mapping, and water management significantly enhances agricultural productivity and sustainability. By leveraging advanced technologies like machine learning, remote sensing, and real-time data processing, AI helps optimize resource use, improve

Table 20.1 Irrigation automation using various artificial intelligence technologies

S.no	Algorithms	Method of evapotranspiration/ desired calculation	Other technologies	Advantages/results
1	PLSR and other regression algorithms	Evapotranspiration model	Sensors for data collection, IoT hardware implementation	Increased efficiency and economic feasibility
2	Artificial neural network–based control system	Evapotranspiration model	Sensors for measurement of soil, temperature, wind speed, etc.	Automation
3	Fuzzy logic	FAO penman–Monteith method	–	Optimization
4	ANN (multilayer neural model), Levenberg Marquardt, backpropagation	Penman–Monteith method	–	Evaporation decreased due to schedule and savings observed in water and electrical energy
5	Fuzzy logic	–	WSN, Zigbee	Experimental results verification. Can be applied to home gardens and grass
6	ANN feedforward, backpropagation	–	–	Optimization of water resources in a smart farm
7	Fuzzy logic controller	Penman–Monteith method	Wireless sensors	Drip irrigation prevents wastage of water and evaporation
8	Machine learning algorithm	–	Sensors, Zigbee, Arduino microcontroller	Prediction and tackles drought situations

Source: Talaviya et al. (2020)

decision-making, and promote sustainable farming practices. These innovations are reshaping the future of agriculture, providing solutions to some of the most pressing challenges in the industry.

20.3.4 Application of AI in Weed Management Using Robotic Arms

Traditional weed management methods have heavily relied on herbicides, which pose significant concerns regarding environmental sustainability and food safety. The prolonged use of herbicides can result in residues that persist in soil and water, negatively affecting non-target plants, aquatic ecosystems, and even raw food products, potentially compromising human health. Given these challenges, there is an increasing need for more sustainable and precise weed management solutions. Artificial intelligence (AI) is emerging as a crucial tool in addressing these concerns, providing innovative methods to enhance the efficiency, precision, and environmental sustainability of weed control practices (Vasileiou et al. 2023). Robotic arms, integrated with AI, have become a groundbreaking solution in precision weed management, significantly reducing the reliance on herbicides. These AI-driven systems can accurately detect and target weeds, applying herbicides only where needed, thereby minimizing the quantity of chemicals used and reducing their environmental footprint. This targeted approach ensures that crops are protected while significantly lowering the potential for soil and water contamination. By analyzing vast datasets, including weather conditions and historical weed patterns, AI-powered robots optimize the precision of weed control strategies.

In high-value row crops and organic farming, where precise weed management is crucial, robotic swarms have proven to be highly effective. These autonomous robots are capable of targeting specific weeds without harming crops, offering an eco-friendly alternative to traditional herbicide applications. Early prototypes of these spraying robots have demonstrated impressive performance, with accuracy rates of over 80% in image classification and weed detection and up to 96% in weed control efficiency, all while minimizing crop damage (Ruigrok et al. 2020). Several commercial robots, such as the Robotti from Denmark, Bilberry from France, and ecorobotix from Switzerland, have been successfully deployed, showcasing the practical benefits of AI in weed management (Table 20.2).

Additionally, AI-equipped robots are increasingly being used for physical weed control. Instead of relying on chemicals, these robots are designed with advanced tools such as hoeing blades, high-voltage systems, lasers, and flaming technologies to physically remove or destroy weeds. This approach further reduces the environmental impact of weed management. For example, companies like Naïo Technologies in France and Farming Revolution in Germany have developed sensor-based hoeing robots, such as Naio OZ and DINO. These robots use real-time kinematic global navigation satellite systems (RTK-GNSSs) to guide their movements with

Table 20.2 Different weeding robots proposed in the past few years

Name	Country	Scope	Year	Methods	Sensor	Source
Ecorobotix	Switzerland	–	2015	Electricity	RGB camera	Zhang et al. (2022)
Laser Weeder	America	–	2021	Laser	Camera	
BoniRob	Germany	Sugar beet	2015	Mechanical	RGB, NIR cameras, and ultrasonic sensor	
AVO	Switzerland	Bean, cotton, rapeseeds	2020	Chemical	RGB camera	
Ladybird	Australia	Lettuce, cauliflower, and broccoli	2015	Chemical	Hyperspectral and thermal cameras	
AgBotII	Australia	Thistle, feather top Rhodes grass, and wild oats	2015	Mechanical	RGB camera	
TerraSentia	America	Maize, soybean	2018	–	RGB camera	
Weed robot 1	Japan	Paddy field	2018	Mechanical	–	
Weed robot 2	Japan	Paddy field	2018	Mechanical	–	
DINO	France	Vegetables (lettuce, carrot, etc.)	2017	Mechanical	Cameras and LiDAR for high-precision navigation and weed detection	https://www.naio-technologies.com
Oz	France	Vegetables	2014	Mechanical	Cameras and LiDAR for weed detection	
Ecorobotix's Avo	Switzerland	Cereal crops, beet, potato, and vegetables	2020	Chemical	Computer vision and machine learning, combined with multispectral cameras for precision herbicide application	https://www.ecorobotix.com
Farming revolution's Robocrop	Germany	Lettuce, onions, and other row crops	2019	Mechanical	Vision-based cameras, machine learning algorithms, weed detection and removal	https://www.farming-revolution.com
Agxeed's AgBot	Netherlands	Row crops, cereals, vegetables	2021	Mechanical	Vision-based cameras, LiDAR, for navigation and weed detection	https://www.agxeed.com
Agrointelli's Robotti	Denmark	Row crops, including cereals and vegetables	2018	Mechanical	Cameras and LiDAR for weeding	https://www.agrointelli.com

Name	Country	Scope	Year	Methods	Sensor	Source
Bosch's Bonirob	Germany	Row crops like lettuce, spinach, and carrots	2015	Mechanical	Machine learning–based visual recognition systems and camera	Mathes et al. (2017)
WeedBot	Sweden	Vegetables and cereal crops	2019	Hybrid (chemical/mechanical)	High-resolution cameras, AI-based algorithms	https://www.weedbot.com
Garford farm machinery's Robocrop InRow Weeder	UK	Lettuce, brassicas, onions, leeks, and other tightly planted crops	2016	Mechanical	Vision-guided cameras	https://www.garford.com
University of Sydney's RIPPA	Australia	Row crops, orchards, and vineyards	2017	Mechanical	Multispectral and hyperspectral cameras, LiDAR	Underwood and Walsh (2018)
Saga robotics' Thorvald	Norway	Vineyards, soft fruit crops, and field crops	2020	Mechanical	Cameras, GPS, LiDAR	https://www.sagarobotics.com

precision, enabling them to manage weeds around crops without causing harm to the plants (Nørremark et al. 2012). In summary, the integration of AI with robotic arms in weed management represents a significant step toward more sustainable, precise, and efficient agricultural practices. By reducing the use of herbicides and employing eco-friendly methods, these technologies offer a viable solution to the environmental and health concerns posed by traditional weed control methods.

20.3.5 Application of AI in Nutrient Management

Efficient nutrient management is key to enhancing fertilizer use, cutting production costs, and promoting sustainable agriculture while mitigating climate change (Raza et al. 2023a, b). Artificial intelligence (AI) is transforming how nutrient deficiencies are predicted, significantly improving the accuracy and speed at which crops' nutrient needs are identified. Traditional methods like soil testing, while effective, are often slow and expensive. In contrast, AI leverages machine learning models to analyze real-time data from soil sensors, weather conditions, and plant growth stages. This enables precise nutrient management, ensuring that crops receive the exact nutrients they need at the right time, thereby improving yields and fostering sustainable farming practices.

AI models integrate data from various sources, including soil sensors, remote sensing technologies, and geographic information systems (GISs), to generate nutrient maps that provide detailed information on the spatial variability of soil nutrients, such as nitrogen, phosphorus, and potassium (NPK). These nutrient maps allow for site-specific fertilizer application, reducing over-fertilization and preventing nutrient leaching (Mishra et al. 2014). By using historical soil data and real-time sensor readings, machine learning models can predict nutrient deficiencies. These models, trained on large datasets that include soil properties, weather patterns, crop types, and historical yields, provide a comprehensive and cost-effective solution for nutrient management, offering precise decision-making tools for farmers (Ennaji et al. 2023a, b).

AI-driven nutrient management plays a critical role in addressing challenges related to food security and sustainability. By optimizing fertilizer use, minimizing waste, and enhancing both crop productivity and quality, AI ensures a balanced nutrient supply while promoting soil fertility and sustainable agricultural practices.

20.3.6 Application of AI in Pest and Disease Monitoring and Management

The integration of artificial intelligence (AI) in pest and disease management is revolutionizing agriculture, offering advanced tools for early detection and efficient control. AI technologies such as image recognition, geographic information systems (GISs), drones, and predictive analytics are creating innovative methods for identifying pests and diseases early, enabling more environmentally friendly and effective solutions. This approach reduces the dependency on harmful pesticides and enhances overall crop health and productivity (Dong et al. 2019a, b).

AI-enabled drones have become crucial for large-scale monitoring of pest and disease outbreaks. Equipped with high-resolution cameras and multispectral sensors, these drones capture detailed crop imagery, which is then analyzed using AI algorithms to detect early signs of stress caused by pests or diseases (Chen et al. 2021a, b). For example, a study by Yuan et al. (2016) focused on monitoring powdery mildew in winter wheat using SPOT-6 satellite imagery. By analyzing various spectral features sensitive to the disease, such as the green and red bands and the normalized difference vegetation index (NDVI), the researchers were able to map disease outbreaks with 78% accuracy. This demonstrates the significant potential of high-resolution satellite imagery and AI in managing crop diseases on a large scale.

These advanced AI tools and remote sensing technologies are enhancing pest and disease surveillance, enabling early intervention and improving agricultural productivity while reducing the reliance on chemical pesticides.

20.3.7 Application of AI in Crop Harvesting

In recent years, the agricultural sector has faced challenges due to declining labor availability and increasing labor costs (Fountas et al. 2023). In response, AI-powered automated machinery has become indispensable, particularly for tasks like harvesting. By automating the harvesting process, AI reduces dependence on manual labor, boosts productivity, and allows farmers to manage their operations more efficiently, even with a smaller workforce. This shift helps address labor shortages while maintaining production levels, thus contributing to more sustainable farming practices (Sane and Sane 2021).

An example of this is the work by Rao (2013), which presents an innovative solution for automating cotton harvesting using machine vision, image processing, and microcontrollers. This technology enables robotic arms to autonomously pick cotton, maximizing production efficiency and potentially increasing yields per hectare in a single day. The use of AI and robotics in crop harvesting is transforming agricultural operations, addressing labor shortages, and enhancing overall farm productivity and sustainability.

20.3.8 Application of AI in Yield Estimation

AI and remote sensing technologies are reshaping crop yield estimation, a crucial area of precision agriculture. Traditional yield estimation methods, which rely on manual sampling and extrapolation, are often labor-intensive and lack scalability. These methods can be both time-consuming and prone to inaccuracies. The advent of AI has revolutionized this process, offering automated solutions that significantly enhance the efficiency and accuracy of yield predictions (Chlingaryan et al. 2018a, b).

Key biophysical parameters such as the normalized difference vegetation index (NDVI) and leaf area index (LAI) are integral to estimating crop yield through remote sensing. These indices reflect the health and growth of crops by measuring canopy greenness. By analyzing temporal remote sensing data, AI models can monitor crop development and make accurate pre-harvest yield predictions across large areas. This capability is particularly valuable in modern agriculture, where scalable solutions are needed to address the demands of large-scale farming (Rahman et al. 2012a, b). AI-driven yield estimation tools enable farmers to plan harvests more effectively and make informed decisions, helping to improve overall farm management and productivity through the integration of remote sensing.

20.3.9 Application of AI in Predicting Local Weather

Artificial intelligence (AI) has emerged as a powerful tool for enhancing the accuracy and efficiency of local weather predictions. Traditional weather forecasting relies heavily on mathematical models that simulate atmospheric conditions using historical weather data, satellite imagery, and other meteorological inputs. While these models have been successful to a certain extent, they often struggle with accurately predicting localized weather patterns due to the complex nature of atmospheric systems. AI, with its advanced machine learning algorithms and deep learning techniques, offers a significant improvement (Raza et al. 2023a, b) in this area by processing vast amounts of data and identifying patterns that might be missed by conventional models.

20.3.10 Enhancing Prediction Accuracy with Machine Learning

Machine learning (ML), a subset of AI, has proven particularly effective in refining local weather forecasts. By analyzing large datasets from various sources such as radar images, satellite data, ground-based weather stations, and historical records, ML models can uncover hidden patterns and correlations that contribute to more accurate predictions. These models can handle nonlinear relationships between variables, which is crucial in predicting complex weather phenomena like thunderstorms, tornadoes, or localized precipitation (Wang et al. 2018a, b). For instance, ML algorithms can process real-time data inputs and provide accurate short-term forecasts for specific locations, something that traditional models often struggle to achieve. Moreover, AI can learn from past errors, continuously improving its predictive capabilities by updating models based on new data. This ability to self-improve makes AI an invaluable tool (Chen et al. 2021a, b) in predicting localized weather conditions such as sudden temperature drops, heavy rainfall, or extreme wind events that might occur in a specific area over a short period.

20.3.11 Deep Learning for Real-Time Weather Predictions

Deep learning, another branch of AI, takes weather forecasting to an even higher level of accuracy by employing neural networks that mimic the human brain's functioning. These networks can analyze vast and complex datasets, making them particularly effective for predicting highly localized weather conditions in real time. Deep learning models can process satellite images, analyze cloud formations, and predict weather changes with minimal human intervention, significantly improving the speed and accuracy of weather forecasts (Dong et al. 2019a, b). Fountas et al. (2023) demonstrated that convolutional neural networks (CNNs) can analyze satellite and radar images, identifying cloud patterns and formations that might indicate impending storms or rainfall. Recurrent neural networks (RNNs), known for processing time series data, are particularly useful for tracking the evolution of weather systems over time, enabling accurate predictions of local weather changes within hours or even minutes.

20.3.12 AI-Driven Hyperlocal Forecasting

AI's ability to analyze multiple data streams simultaneously makes it an excellent tool for hyperlocal weather forecasting. Hyperlocal forecasts are highly specific, offering predictions for areas as small as a few square kilometers, such as neighborhoods or individual large farms (Rahman et al. 2012a, b). AI-based hyperlocal forecasts are useful to farmers to determine the best times for planting, irrigating, or harvesting crops, ensuring they avoid adverse weather conditions like frost or excessive rainfall.

20.3.13 Integrating AI with Internet of Things (IoT) for Localized Weather Monitoring

The integration of AI with Internet of things (IoT) devices has further enhanced the ability to predict local weather. IoT devices such as weather sensors, drones, and other connected technologies gather vast amounts of real-time environmental data, which AI systems can analyze to deliver highly accurate local forecasts. This data can include temperature, humidity, wind speed, and pressure readings from sensors distributed across a specific area (Rahman et al., 2012). By processing this data through machine learning algorithms, AI can provide highly localized and real-time weather predictions. For instance, drones equipped with AI can fly over agricultural fields, collecting atmospheric data and analyzing it to predict rainfall, temperature fluctuations, or storm movements with high precision (Chlingaryan et al. 2018a, b).

20.3.14 AI in Extreme Weather Prediction

Another critical application of AI in local weather forecasting is in the prediction of extreme weather events. Predicting hurricanes, tornadoes, floods, and other severe weather conditions requires processing vast amounts of data in real time. AI can analyze historical patterns and real-time data from weather stations, satellites, and ocean buoys to predict the likelihood and trajectory of these events with improved accuracy (Ennaji et al. 2023a, b). AI models have already been used to predict hurricanes and typhoons by analyzing variables like sea surface temperatures, wind speeds, and atmospheric pressure. Wang et al. (2018a, b) reported that these models can provide earlier and more accurate warnings, giving people in affected areas more time to prepare and respond, ultimately saving lives and reducing damage.

20.3.15 Improving Climate Models with AI

While AI is often used for short-term weather forecasting, its applications also extend to improving climate models. Long-term climate predictions rely on simulations of how various factors, such as greenhouse gas emissions, ocean currents, and solar radiation, will affect the Earth's climate over decades or centuries. AI helps refine these models by analyzing historical climate data, enabling more accurate predictions of long-term changes in temperature, precipitation patterns, and the frequency of extreme weather events at local levels (Mishra et al. 2014). Incorporating AI into climate models also helps to account for localized variations in climate that are often overlooked by traditional models.

20.3.16 Limitations of AI Adoption in Agriculture

High Initial Cost One of the most significant barriers to AI adoption in agriculture is the high initial investment required for implementing AI-powered systems. In countries like India, where the majority of farmers are small or marginal, the cost of setting up AI infrastructure—such as sensors, high-performance computing systems, and specialized software—can be prohibitive. These technologies, along with their maintenance, require substantial financial outlay, which is often beyond the capacity of small-scale farmers. As a result, AI adoption remains limited to larger agricultural enterprises or those with access to significant capital.

Limited Access to Financing Farmers often face challenges in securing affordable financing or loans to cover the capital investment needed for AI technologies. Unlike inputs like seeds, fertilizers, or pesticides, financing for AI systems may not be readily available or easily accessible. Additionally, many rural areas lack the infrastructure and support systems to access government subsidies or financial aid that could help offset the initial costs of AI implementation. This financial barrier limits the widespread adoption of AI technologies, particularly among smallholder farmers.

Data Availability and Quality Issues The effectiveness of AI in agriculture largely depends on the availability and quality of data. In rural areas, particularly in developing countries, there is often insufficient infrastructure for collecting data on crucial factors like soil health, weather conditions, and crop performance. This data scarcity hinders the ability of AI models to deliver accurate insights and recommendations. Even when data is available, it is often fragmented and scattered across various platforms, including government databases, research institutions, and private companies. Integrating these diverse datasets into a unified AI model presents a significant challenge, affecting the reliability of AI-driven solutions.

Infrastructure Challenges Many AI applications in agriculture rely on cloud computing or require Internet access to collect and process real-time data. However, many rural regions lack reliable broadband or 4G/5G connectivity, making it difficult to deploy AI systems that depend on constant data transmission and real-time analysis. Additionally, AI-powered systems require consistent and reliable power supplies, especially for high-performance tasks like training machine learning models or operating drones. In areas with frequent power outages or inconsistent electricity, maintaining these systems becomes a significant hurdle, further limiting the adoption of AI technologies in rural agriculture.

Lack of Technical Skills and Readiness Among Farmers The successful implementation of AI in agriculture requires a certain level of technical expertise, which many farmers may not possess. Particularly among older farmers or those less familiar with digital tools, there may be hesitation or difficulty in adopting AI-powered systems for tasks such as crop monitoring, disease detection, or yield prediction. Traditional farming methods are often deeply rooted in agricultural communities, and persuading farmers to switch to AI-driven approaches can be challenging. Farmers may also harbor concerns about the reliability of automated systems and may be reluctant to trust AI over their own experience and intuition.

20.4 Future Prospects of AI Tools in Efficient Resource Monitoring

The future of AI in agriculture holds great potential for revolutionizing resource monitoring and management. AI tools are expected to improve efficiency and accuracy through advanced predictive analytics and automation. These tools will be able to analyze historical data to forecast future resource needs, recognize patterns in resource usage, and optimize allocation strategies, thereby minimizing waste and enhancing productivity. Additionally, AI's ability to automate routine tasks will enable real-time monitoring and quicker responses to emerging issues, leading to more adaptive and efficient operations. As AI technology continues to evolve, its integration into agricultural resource management systems will play a crucial role in meeting the growing demands of global food production while promoting sustainability.

20.5 Conclusion

AI technologies are transforming agricultural practices, offering innovative solutions for improving resource management, land use, and sustainability. By integrating AI with tools like IoT, computer vision, machine learning, and robotics, precision agriculture is becoming increasingly effective in addressing key challenges such as water scarcity, soil degradation, and pest control. As AI continues to evolve, it will play an even more vital role in helping farmers make data-driven decisions, boost productivity, and reduce environmental impact. The ongoing research and development of AI applications in agriculture will shape the future of farming, driving more efficient and sustainable practices that are essential for feeding an expanding global population.

References

Barman U, Choudhury RD (2020) Soil texture classification using multi-class support vector machine. Inf Process Agric 7:318–332

Chen CJ, Huang YY, Li YS, Chen YC, Chang CY, Huang YM (2021a) Identification of fruit tree pests with deep learning on embedded drone to achieve accurate pesticide spraying. IEEE Access 9:21986–21997

Chen J, Huang R, Liu Y (2021b) Deep learning in weather forecasting. IEEE Trans Neural Netw Learn Syst 32:4069–4079

Chlingaryan A, Sukkarieh S, Whelan B (2018a) AI models for extreme weather prediction. Weather Clim Sci 14:245–266

Chlingaryan A, Sukkarieh S, Whelan B (2018b) Machine learning approaches for crop yield prediction and nitrogen status estimation in precision agriculture: a review. Comput Electron Agric 151:61–69

Dong L, Wang P, Xu Z (2019a) Predicting localized weather patterns using AI. Environ Data Sci 12:18–30
Dong Y, Xu F, Liu L, Du X, Ye H, Huang W, Zhu Y (2019b) Monitoring and forecasting for disease and pest in crop based on WebGIS system. In: 2019 8th Int. Conf Agro-Geoinformatics (Agro-Geoinformatics), 1–5
Ennaji M, Bennani S, Alaoui ME (2023a) AI in predicting severe weather events. Clim Dyn 60:897–920
Ennaji O, Vergütz L, El Allali A (2023b) Machine learning in nutrient management: a review. Artif. Intell. Agric. 9:1–11
Fountas S, Sorensen CG, Tsiropoulos Z, Mylonas N (2023) AI for hyperlocal weather forecasting. Precis Agric Rev 14:112–130
Kafi KM, Shafri HZM, Shariff ABM (2014) An analysis of LULC change detection using remotely sensed data; a case study of Bauchi City. In IOP Conf. Ser.: Earth Environ Sci, 20, 012056
Krishnan SR, Nallakaruppan MK, Chengoden R, Koppu S, Iyapparaja M, Sadhasivam J, Sethuraman S (2022) Smart water resource management using artificial intelligence—a review. Sustain For 14:13384
Mathes A, Mauser J, Biber P (2017) Autonomous field robot Bonirob for agricultural tasks. Comput Electron Agric 142:234–245
Mishra A, Pattnaik T, Das D, Das M (2014) Soil fertility maps preparation using GPS and GIS in Dhenkanal District, Odisha, India. Int J Plant Soil Sci 3:986–994
Nørremark M, Griepentrog HW, Nielsen J, Søgaard HT (2012) Evaluation of an autonomous GPS-based system for intra-row weed control by assessing the tilled area. Precis Agric 13:149–162
Paloscia S, Pettinato S, Santi E, Notarnicola C, Pasolli L, Reppucci A (2013) Soil moisture mapping using Sentinel-1 images: algorithm and preliminary validation. Remote Sens Environ 134:234–248
Prins AJ, Van Niekerk A (2021) Crop type mapping using LiDAR, Sentinel-2 and aerial imagery with machine learning algorithms. Geo-spat Inf Sci 24:215–227
Rahman A, Khan K, Krakauer NY, Roytman L, Kogan F (2012a) Use of remote sensing data for estimation of Aman rice yield. Int J Agric For 2:101–107
Rahman MA, Chowdhury MA, Islam T (2012b) AI and IoT integration for real-time weather monitoring. Int J Remote Sens 33:1603–1619
Rao USN (2013) Design of automatic cotton picking robot with Machine vision using image processing algorithms. In 2013 Int Conf Control, Autom Robot Embed Syst (CARE), 1–5
Raza I, Zubair M, Zaib M, Khalil MH, Haidar A, Sikandar A, Abbas MQ, Javed A, Liaqat MM, Ain AT, Nafees M (2023a) Precision nutrient application techniques to improve soil fertility and crop yield: a review with future prospect. Int Res J Educ Technol, 5(8), 109 - 123.
Raza M, Ahmad F, Khan T, Ali S (2023b) AI in weather prediction. J Meteorol Res 45:205–225
Ruigrok T, van Henten E, Booij J, Van Boheemen K, Kootstra G (2020) Application-specific evaluation of a weed-detection algorithm for plant-specific spraying. Sensors 20:7262
Samarinas N, Tsakiridis NL, Kalopesa E, Zalidis GC (2024) Soil loss estimation by water erosion in agricultural areas introducing artificial intelligence geospatial layers into the RUSLE model. Land 13:174
Sane TU, Sane TU (2021) Artificial intelligence and deep learning applications in crop harvesting robots—a survey. In 2021 Int Conf Electr Commun Comput Eng (ICECCE), 1–6
Savitha M, UmaMaheswari OP (2018) Smart crop field irrigation in IOT architecture using sensors. Int J Adv Res Comput Sci 9:1
Shi X, Han W, Zhao T, Tang J (2019) Decision support system for variable rate irrigation based on UAV multispectral remote sensing. Sensors 19:2880
Talaviya T, Shah D, Patel N, Yagnik H, Shah M (2020) Implementation of artificial intelligence in agriculture for optimisation of irrigation and application of pesticides and herbicides. Artif Intell Agric 4:58–73

Tariq A, Yan J, Gagnon AS, Riaz Khan M, Mumtaz F (2023) Mapping of cropland, cropping patterns and crop types by combining optical remote sensing images with decision tree classifier and random forest. Geo-spat Inf Sci 26:302–320

Underwood JP, Walsh MJ (2018) RIPPA: robot for intelligent perception and precision application in agriculture. Robot Auton Syst 120:115–128

Vasileiou M, Kyriakos LS, Kleisiari C, Kleftodimos G, Vlontzos G, Belhouchette H, Pardalos PM (2023) Transforming weed management in sustainable agriculture with artificial intelligence: a systematic literature review towards weed identification and deep learning. Crop Prot 176:106522

Wang AX, Tran C, Desai N, Lobell D, Ermon S (2018a) Deep transfer learning for crop yield prediction with remote sensing data. In Proc 1st ACM SIGCAS Conf Comput Sustain Soc, 1–5

Wang H, Zhang L, Zhou X (2018b) Machine learning applications in weather forecasting. Adv Meteorol 2018:Article ID 9481327

Yuan L, Pu R, Zhang J, Wang J, Yang H (2016) Using high spatial resolution satellite imagery for mapping powdery mildew at a regional scale. Precis Agric 17:332–348

Zhang W, Miao Z, Li N, He C, Sun T (2022) Review of current robotic approaches for precision weed management. Curr Robot Rep 3:139–151

Correction to: Climate-Smart Agricultural Technologies – Approaches for Field Crops Production Systems

Biswajit Pramanick, Shiv Vendra Singh, Sagar Maitra, Silvia Celletti, and Akbar Hossain

Correction to:
Chapters 8 and 12 in: B. Pramanick et al. (eds.), *Climate-Smart Agricultural Technologies*, Sustainability Sciences in Asia and Africa, https://doi.org/10.1007/978-981-96-7699-6

The original version of the book was inadvertently published with errors. The following corrections have been made after the original publication.

Chapter 8 :

Description on P161. Published with incorrect author sequence. The complete author sequence have been corrected "Debabrata Nath, Malay K. Bhowmick, Sunil Kumar, Panneerselvam Peramaiyan, Sudhanshu Singh, Biswajit Pramanick, Ranjan Laik and Vandana Kumari" in the chapter.

The corresponding author is Vandana Kumari.

Chapter 12:

Description on P249, Line 2: percentage value of "80%" has now been replaced with "88%".

The updated versions of these chapters can be found at
https://doi.org/10.1007/978-981-96-7699-6_8
https://doi.org/10.1007/978-981-96-7699-6_12

B. Pramanick et al. (eds.), *Climate-Smart Agricultural Technologies*, Sustainability Sciences in Asia and Africa, https://doi.org/10.1007/978-981-96-7699-6_21

MIX
Papier aus verantwortungsvollen Quellen
Paper from responsible sources
FSC® C105338

If you have any concerns about our products,
you can contact us on
ProductSafety@springernature.com

In case Publisher is established outside the EU,
the EU authorized representative is:
Springer Nature Customer Service Center GmbH
Europaplatz 3, 69115 Heidelberg, Germany

Printed by Libri Plureos GmbH
in Hamburg, Germany